STELLAR NUCLEOSYNTHESIS

ASTROPHYSICS AND SPACE SCIENCE LIBRARY

A SERIES OF BOOKS ON THE RECENT DEVELOPMENTS
OF SPACE SCIENCE AND OF GENERAL GEOPHYSICS AND ASTROPHYSICS
PUBLISHED IN CONNECTION WITH THE JOURNAL
SPACE SCIENCE REVIEWS

VOLUME 109
PROCEEDINGS

STELLAR NUCLEOSYNTHESIS

PROCEEDINGS OF THE THIRD WORKSHOP OF THE
ADVANCED SCHOOL OF ASTRONOMY OF THE ETTORE MAJORANA CENTRE
FOR SCIENTIFIC CULTURE, ERICE, ITALY, MAY 11–21, 1983

Edited by

CESARE CHIOSI

Institute of Astronomy, University of Padova, Italy

and

ALVIO RENZINI

Department of Astronomy, University of Bologna, Italy

D. REIDEL PUBLISHING COMPANY

A MEMBER OF THE KLUWER ACADEMIC PUBLISHERS GROUP

DORDRECHT / BOSTON / LANCASTER

Library of Congress Cataloging in Publication Data

Main entry under title:

Stellar nucleosynthesis.

    (Astrophysics and space science library ; v. 109)
    Includes index.
    1. Nucleosynthesis–Congresses. 2. Stars–Congresses.
I. Chiosi, C. (Cesare) II. Renzini, Alvio. III. Ettore Majorana
International Centre for Scientific Culture. Advanced School of Astronomy.
IV. Series.
QB450.S74   1984   523.01'9723   84–1962
ISBN-13: 978-94-009-6350-4   e-ISBN-13: 978-94-009-6348-1
DOI: 10.1007/978-94-009-6348-1

Published by D. Reidel Publishing Company,
P.O. Box 17, 3300 AA Dordrecht, Holland.

Sold and distributed in the U.S.A. and Canada
by Kluwer Academic Publishers,
190 Old Derby Street, Hingham, MA 02043, U.S.A.

In all other countries, sold and distributed
by Kluwer Academic Publishers Group,
P.O. Box 322, 3300 AH Dordrecht, Holland.

TABLE OF CONTENTS

SESSION I: PLANETARY NEBULAE, SUPERNOVAE, SUPERNOVA REMNANTS
            AND COSMIC RAYS

SESSION II: TOPICS ON NUCLEAR REACTIONS

SESSION III: QUASI-STATIC STELLAR EVOLUTION AND RELATED
              NUCLEOSYNTHESIS

Preface

The synthesis of elements in stellar interiors, their ejection into the
interstellar medium and the consequent chemical enrichment of galaxies
are among the most exciting issues of Modern Astrophysics.

Progress in this field has been steady over the past 30 years, passing
from the early identification of the basic processes of Nuclear Astrophysics to the present stage when specific stellar environments for the
production of the various elements and isotopes are being identified,
and detailed quantitative predictions are being attempted. More specifically, current work in this field is aimed at answering two strictly related questions: i) what is the amount of the various elements (isotopes) which are synthetized and ejected by stars of any given initial mass
and chemical composition, and ii) which are all the possible nucleosynthetic environments for any given element (isotope), and which is their
relative importance.

The comparison with the whole pattern of abundances in the solar system,
nearby stars, HII regions, interstellar medium and galaxies indicates
that theoretical models are encouragingly succesful, even if many facets
of the problem are still far from being completely understood. Moreover,
the recent advances in understanding the history of star formation in
our own galaxy and in nearby galaxies, combined with the current results
of stellar nucleosynthesis, are allowing a more detailed approach to the
problem of the chemical evolution of galaxies.

The aim of this meeting was therefore to bring together both observers
and theoreticians working on stellar nucleosynthesis and related fields,
so as to discuss the most recent results, review facts and reasonably
well established conclusions, and help the formulation of sharper questions for what remains to be understood.

The workshop was held in Erice (Sicily), at the "Ettore Majorana Centre
for Scientific Culture", from May 11 through 21, 1983, and was organized

about a sequence of review lectures, each followed by contributed talks
and discussions.

The opening session was devoted to reviewing the most recent results
concerning the determination of chemical abundances in particularly interesting objects providing direct evidence for stellar nucleosynthesis
(like planetary nebulae, supernovae and supernova remnants).

In Session 2 recent results on some relevant nuclear reaction rates have
been presented.

Session 3 grouped the contributions concerning the quasi-static evolution
of normal stars, its relevance for the nucleosynthesis of the various
elements and isotopes, and the comparison of the evolutionary models with
the observations.

Session 4 addressed the important questions of identifying the precursors
of both Type I and Type II supernovae, and of determining the detailed
composition of the ejecta by explosive nucleosynthesis associated with
these events.

Finally, Session 5 was devoted to a presentation of recent calculations
of the evolution of hypothetical pre-galactic very massive objects (including their potential role in pre-galactic nucleosynthesis), and to
reviewing the progress that has been made in understanding the chemical
evolution of galaxies.

The discussion during and after talks was spirited and contributed significantly to the succes of the workshop. We regrett for not having been
able of reporting it "in toto" in the proceedings. Nevertheless, we hope
that what has been recorded may still give to the reader at least part
of the excitement we have experienced during the ten days of the workshop.

We would like to express here our best appreciation to Drs. M. L. Quarta,
L. Sanz Fernandez de Cordoba and M. Tosi for having untiringly recorded
and collected the contributions to the discussion.

The European Physical Society, the Italian National Council of Research
(CNR) and the "Ettore Majorana Centre for Scientific Culture" are gratefully acknowledge for having provided generous finantial assistance to
several participants, thus allowing them to attend this meeting.

Finally, we would like to thank the local staff of the "Ettore Majorana

Centre for Scientific Culture", and in particular Ms. Pinola Savalli and
Dr. Alberto Gabriele for their efficient organization of the logistic
aspects of the meeting.

Last but not least, we like to express our gratitude to Prof. A. Zichi-
chi, Director of the " Ettore Majorana Centre", for the warm hospitality.
While this preface was being written, the announcement came of the Nobel
Price awarded to Drs. W. A. Fowler and S. Chandrasekhar for their inva-
luable contribution given to Astrophysics. Aware of expressing the senti-
ments of all participants, we like to dedicate these Proceedings to Drs.
W. A. Fowler and S. Chandrasekhar as our modest sign of recognition and
friendship. Without their pioneering work, a meeting on "Stellar Nucleosynthesis" could in fact have been hardly imagined at all !

          Cesare Chiosi   Alvio Renzini

List of Participants

R. BEDOGNI, Astronomy Department, Bologna, Italy
G. BERTELLI, Institute of Astronomy, Padova, Italy
J.R. BOND, Institute of Astronomy, Cambridge University, England
D. BRANCH, Dept. of Physics and Astronomy, University of Oklahoma, USA
R. BRAUN, Sterrewacht, Leiden, Holland

A. BRESSAN, SISSA, Trieste, Italy
M. CASSE', Commissariat à l'Energie Atomique, Gif-sur-Yvette Cedex, France
C. CHIOSI, Institute of Astronomy, Padova, Italy
A. CIANI, Institute of Astronomy, Padova, Italy
J.J. COWAN, Dept. of Physics and Astronomy, University of Oklahoma, USA

I.J. DANZIGER, ESO, Garching bei München, F. R. Germany
A. DIAZ, Royal Greenwich Observatory, Herstmonceux, England
C. DOOM, Astrofisisch Instituut, Vrije Universiteit, Brussel, Belgium
M.G. EDMUNDS, University College, Cardiff, England
C. FORIERI, SISSA, Trieste, Italy

C. FU-ZHEN, SISSA, Trieste, Italy
J.S. GALLAGHER III, Dept. of Astronomy, University of Illinois, USA
L. GREGGIO, Astronomy Department, Bologna, Italy
R.B.C. HENRY, Dept. of Physics, University of Delaware, USA
I. IBEN Jr., Dept. of Astronomy, University of Illinois, USA

M. JOHNSTON, Space Telescope Science Institute, Baltimore, USA
U.G. JØRGENSEN, University Observatory, Copenhagen, Denmark
H. KIRBIYIK, SISSA, Trieste, Italy
Z.W. LI, Dept. of Astronomy, University of Texas, Austin, USA
J.H. LUTZ, University College, London, England

A. MAEDER, Geneva Observatory, Switzerland
G. MATHEWS, L. Livermore Laboratory, University of California, USA
F. MATTEUCCI, Istituto Astrofisica Spaziale C.N.R., Frascati, Italy
R. MONTI, TESRE C.N.R., Bologna, Italy
K. NOMOTO, Dept. of Earth Science and Astronomy, Tokyo, Japan

W.W. OBER, Max Planck Institut für Astrophysik, Garching, F. R. Germany
M. PERSIC, SISSA, Trieste, Italy
A. PREITE-MARTINEZ, Istituto Astrofisica Spaziale C.N.R., Frascati, Italy
L. PULONE, Istituto Astrofisica Spaziale C.N.R., Frascati, Italy
M.L. QUARTA, Istituto Astrofisica Spaziale C.N.R., Frascati, Italy

M. RAYET, Phys. Nucl. Théorique, Université Libre de Bruxelles, Belgium
A. RENZINI, Astronomy Department, Bologna, Italy

P. SALVADOR DELBOURGO, Institut d'Astrophysique, Paris, France
L. SANZ FERNANDEZ DE CORDOBA, Inst. Nac. de Tec. Aerosp., Madrid, Spain
F.-K. THIELEMANN, Max Planck Inst. für Astrophysik, Garching, F.R. Germany
S. TORRES-PEIMBERT, Inst. Astronomia, Univ. Nac. Aut. Mexico, Mexico
A. TORNAMBE', Istituto Astrofisica Spaziale C.N.R., Frascati, Italy
M. TOSI, Astronomy Department, Bologna, Italy
H.-P. TRAUTVETTER, Inst. Für Kernphysik, Munster, R. R. Germany
S.E. WOOSLEY, Dept. of Astronomy, University of California Santa Cruz, USA

A. YAHIL, Astronomy Program, State Univ. of New York, Stony Brook, USA

NUCLEOSYNTHESIS AND PLANETARY NEBULAE

SILVIA TORRES-PEIMBERT
Instituto de Astronomía, Universidad Nacional Autónoma de
México

Abstract. The chemical composition of planetary nebulae of different
types are compared with those predicted from stellar evolution models
for intermediate mass stars with $1 \leqslant M/M_\odot \leqslant 8$. The agreement is in
general good, nevertheless there are a few differences, some of the more
conspicuous are: a) Type I PN show an anticorrelation between O and N
abundances with N+O $\simeq$ constant probably indicating that some O has been
converted into N; b) the C abundances of Type II PN and Type III PN
probably indicate that the 3rd dredge-up is more efficient than predicted;
c) Type IV PN present C excesses that are not predicted.

1. INTRODUCTION

The study of PN envelopes allows us to derive information about
nucleosynthesis and stellar evolution processes that the progenitors have
undergone during their lifetimes. Peimbert (1978) has divided PN into
four different types based on their chemical composition, galactic distribution and morphology. In general it is believed that these types
correspond to progenitors of different masses that experience the same
transient phenomenon. The general characteristics that can be found from
each type are presented here and are compared to present ideas of stellar
evolution. Other reviews on this subject are those by Peimbert (1981),
Aller (1983a) and Kaler (1983a).

In a general description by Renzini and Voli (1981, RV81 see also;
Iben and Truran 1978; Becker and Iben 1979, 1980) the evolution of the
surface abundances of He, C, N and O is reviewed for intermediate mass
stars, $1 \leqslant M/M_\odot \leqslant 8$, from the main sequence until the ejection of the PN
envelope, or until C ignition in the core. They take into account two
processes affecting the surface composition: i) convective dredge-ups and
ii) nuclear burning in the deepest layers of the convective envelope.
Iben and Truran (1978) distinguish three dredge-up phases: i) the first
phase corresponds to the inward penetration of the convective envelope
when the stars reach the red giant branch for the first time during the
H-shell burning phase. This dredge-up increases He and N in the surface;

3

the increase of He is modest ($\sim$ 0.01 to 0.02 by mass) and is only
present for M< 3 $M_\odot$, while the N/O increase is of about a factor of 2
(depending on the initial CNO values) for stars in the 1-8 $M_\odot$ range.
Hereinafter the stellar masses quoted correspond to those in the main
sequence and not to those just prior to the PN ejection which are considerably smaller due to stellar winds. ii) The second phase occurs in
stars initially more massive than $\sim$ 3 $M_\odot$ when, after the ignition of the
He-burning shell, the convective envelope penetrates into the He core
dredging-up He and N. In this case the He and N enrichments are substantial and He can increase up to 0.05 by mass and N/O can increase by
factors as large as 30 (this value is sensitive to the adopted hot bottom
efficiency). iii) The third phase occurs during the asymptotic giant
branch and consists of several individual mixing episodes, that is,
after each He shell flash the base of the convective envelope may penetrate through the H-He discontinuity reaching the intershell region
where partial He burning has taken place. Significant amounts of He and
C are convected to the surface following each He flash. This process is
expected to take place in objects with M> 2 $M_\odot$; the He contribution by
mass is of 0.02 to 0.05 depending on the progenitor's mass; while the
C/O enrichment can be as large as a factor of 10 (also depending on the
progenitor's mass). The situation of whether or not 3rd dredge-up occurs
in stars with M< 2 $M_\odot$ is not yet clear (Iben and Renzini 1982a, 1982b,
1983).

2. TYPE I PN

    PN with He/H$\geqslant$0.14 and log N/O$\geqslant$0.0 were originally defined as Type I
(Peimbert 1978; Peimbert and Serrano 1980). More recently the definition
was extended to include PN with with He/H> 0.125 and log N/O> -0.30
(Peimbert and Torres-Peimbert 1983,hereinafter PTP83).

    In PTP83 there is a list of 29 Type I PN selected from their
chemical abundances. In general most Type I PN show a very pronounced
filamentary structure, very strong forbidden lines ranging from [O I],
[N I], [S II] up to [Ne V] and have been defined in the literature as
binebulous, bipolar, biaxial or hourglass. About half of the objects
classified by Greig (1971) as B nebulae (binebulous or filamentary) with
good abundance determinations have turned out to be He and N rich. Greig
(1972) from kinematical properties found that B nebulae are of population
I while A (annular) and C (centric) are of population II.

    A crude estimate of the mass of the progenitors of Type I PN has
been made for some objects. These are NGC 3132, NGC 2346, and NGC 2818
(Peimbert and Serrano 1980; Méndez and Niemela 1981; Calvet and Peimbert
1983). The average mass for these objects is 2.4 $M_\odot$. Since NGC 3132 and
NGC 2346 are mild examples of He-N enrichment, the lower mass limit for
which this process is expected to take place is around 2.4 $M_\odot$. Furthermore, Kaler (1983a,b) in an extensive study has located the nuclei of
PN on the log L-log Te plane, those objects whose envelope shows N/O$\geqslant$ 1.0
are in a lower L_* and higher T_* region than the region defined by those

objects with lower N/O values. The comparison with theoretical tracks (Paczynski 1971; Iben and Renzini 1983; Schönberner and Weidemann 1981) indicate that in general the nuclei of Type I PN are more massive than other PN nuclei.

The predictions for stars of intermediate mass briefly summarized in §I argue that the He and N rich PN are those that have undergone substantial 2nd dredge-up and they correspond to progenitors with M > 3 M$_\odot$. The above mentioned additional evidence is also in agreement with this idea.

The N/O and He/H abundances from PTP83 and the predictions from RV81 are presented in Figure 1. In this figure there is a very satis-factory agreement between the observed values and the predicted ones for models that allow hot burning at the base of the convective envelope (α = 1 and 2 of RV81). The agreement corresponds to progenitors in the 3.3 - 8 M$_\odot$ range.

From the data in PTP83 it is found that for all objects log C/O > - 0.15 (excepting NGC 6302 that has log C/O = -0.35, see discussion in PTP83). In Figure 2 a comparison of C and He is presented between the observations in PTP83 and the predictions in RV81. Here the observations

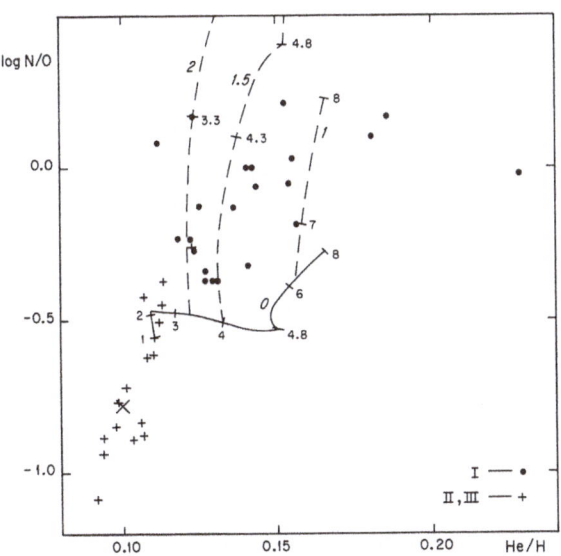

Fig. 1. Observed abundances of PN. Filled circles are Type I objects from PTP83, + symbols are Type II PN from Peimbert and Serrano (1980). Predicted values from RV81: solid lines correspond to models with no nuclear burning at the base of the convective envelope; dashed lines correspond to models for different values of the mixing length α = 1,1.5, 2; numbers along the tick marks are the masses of the main sequence par-ent stars. The cross corresponds to initial composition of stellar models.

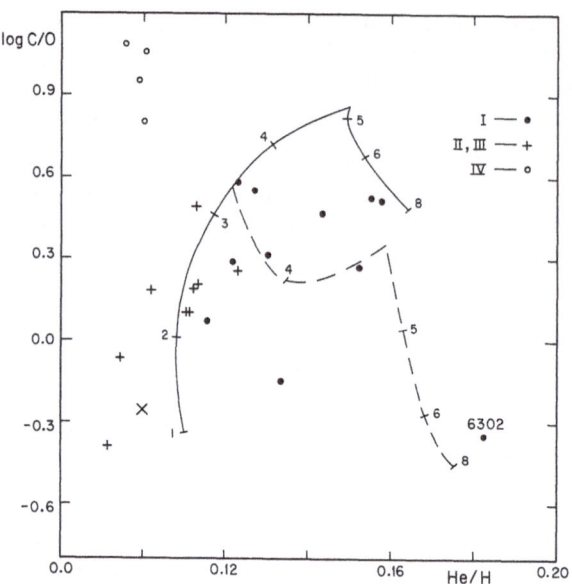

Fig. 2. Observed abundances of PN compared to predictions by RV81. Type
I PN from PTP83, Type II PN from Peimbert and Serrano (1980). Same
symbols as Figure 1.

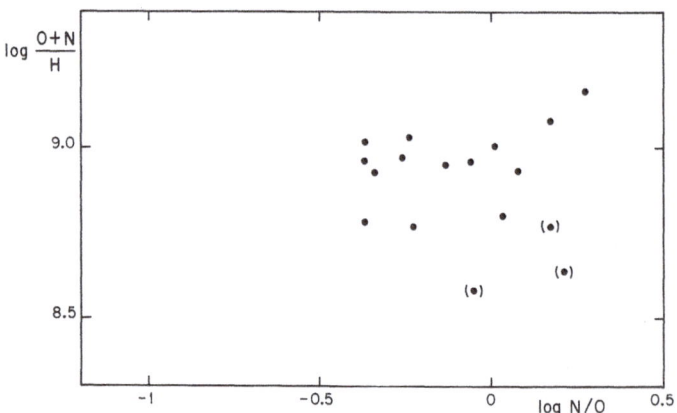

Fig. 3. Observed abundances of Type I PN. Shaded area is location of Type
II PN, filled circles are data by PTP83, parenthesis are uncertain values.

agree in general with the predictions for progenitors of M > 3 $M_\odot$. In
particular the observed C/O and He/H values for NGC 6302 agree with the
expected values for a progenitor of ∿ 8 $M_\odot$. NGC 6853 appears to have
less C or more He than predicted.

   The observed O/H and N/O values are presented in Figure 3. This
figure shows large scatter, however it seems that N + O is constant in
Type I PN and that N and O are anticorrelated. This would be difficult
to explain since it would require O to be partially transformed into N
and dredged-up to the surface. More observations are required to confirm
this suggestion. A similar trend is observed in the PN of the Magellanic
Clouds (Aller 1983a).

   The predicted value of $M \simeq 8 M_\odot$ for NGC 6302 supports the results
of Koester and Reimers (1981) and Reimers and Koester (1982) who found
that white dwarfs do occur up to progenitor masses of $7 M_\odot$. These
results indicate that the dividing line between PN progenitors and SN
progenitors is considerably more massive than what was thought just a
few years ago. With the exception of NGC 6302 all the other PN of Type I
seem to have progenitors in the $3 - 5 M_\odot$ range, other extreme Type I PN
should be found to determine with more precision the limiting mass
between PN and SN.

3. PN OF TYPES II AND III

   Type II PN are intermediate population objects with a distance to
the galactic plane, z, smaller than 1 kpc and a peculiar radial velocity,
v_{pr}, smaller than 60 km s^{-1}; while Type III PN are population II objects
with $|z| > 1$ kpc or $|V_{pr}| > 60$ km s^{-1} that do not belong to the halo
population (Peimbert 1978). In many cases, it is difficult to differen-
tiate whether a PN is of Type II or III, since this distinction is very
sensitive to distance determinations. For the purpose of this discussion we will group Type II and Type III PN together.

   These objects have a mean distance above the galactic plane of
$\langle z \rangle \sim 190$ pc (adopting the distance scale by Cudworth 1974), which
corresponds to $\langle M \rangle \sim 1.4 M_\odot$. As presented in §1, theory predicts that
only the first dredge-up and a modest amount of the third dredge-up,
have taken place in these objects; consequently that N and C have been
moderately enriched and that He/H has increased by about 0.01 by number.

   From the comparison between theory and observations of N/H and
He/H in Figure 1 it can be seen that some Type II and III PN are grouped
together on the predicted zone for $M \leqslant 3 M_\odot$, while the rest extend
towards lower N and He values. This extension seems to be due to different initial C, N and He abundances at the time of the formation of
the progenitor stars produced by galactic abundance gradients (e.g.,
Peimbert 1979).

   The comparison of C/H and He/H in Figure 2 shows good agreement,
Types II and III PN fall in the general location of the $M \leqslant 3 M_\odot$
progenitors; nevertheless their predicted average mass is somewhat
higher than observed, possibly indicating a higher efficiency for the
third dredge-up process.

It is predicted that the surface chemical composition for O and heavier elements has not been affected by stellar evolution, and therefore the composition of the galactic interstellar medium at the time of formation of the parent stars can be traced from the PN results. In particular radial galactocentric gradients for O, A and S can be derived from PN. Gradients from PN have been reported in the literature by several authors (D'Odorico *et al.* 1976; Aller 1976; Torres-Peimbert and Peimbert 1977; Barker 1978; Peimbert and Serrano 1980; Aller 1983b). The reported O/H gradients are in agreement with those derived from H II regions (e.g., Peimbert 1979; Talent and Dufour 1979; Shaver *et al.* 1983). Alternatively the He/H and N/H values are systematically higher than those of H II regions at similar galactocentric distances. This has been attributed to the resulting changes in surface composition undergone by the parent star.

4. TYPE IV PLANETARY NEBULAE

The abundances of halo PN can be more easily interpreted because their initial chemical composition has not been substantially affected by galactic chemical evolution. On the one   hand, there are the chemical elements that have not been altered by the progenitor nuclear evolution and that help establish the conditions of the interstellar medium at the time of formation of the precursor stars, namely: S, Ar and possibly O; and on the other hand, there are those elements whose abundances have been modified on the surface of the star during its lifetime: He, N and C.

At present, there are four known PN of type IV: K 648 in the globular cluster M15 discovered by Pease (1928), H4-1 discovered by Haro (1951), BB-1 found by Boeshaar and Bond (1977), and DD-1 discovered by Dolidzde and Dzhishlejshvili (1966). Their general characteristics are given in Table I. The masses of the progenitor star of these objects are of $\simeq 0.8\ M_{\odot}$.

A compilation of their chemical abundances is presented in Table II. For comparison, abundances of the sun and Orion Nebula are also presented in this table. It can be seen that S, Ar, and O are underabundant relative to population I objects, as is to be expected in population II objects. However, the abundance of O and Ne relative to Ar and S seem to be larger than in the solar neighborhood. For extreme population II objects He is expected to correspond to the pregalactic abundance of He/H = 0.07 and since it is higher, it is consistent with the prediction that there has been He enrichment in the PN. N/O is higher than in Orion which is to be expected from the 1st dredge-up phase in which both He and N are increased. It is also seen that C has been substantially enriched, moreover the total abundance present of C/H is as large as solar.

That is, in these objects H, C and N have been enriched. This can be understood in terms of the 1st and 3rd dredge- ups to have occurred.

The theoretical situation for the low mass limits for the 3rd dredge-up
to take place is not well established. Renzini and Voli (1981) do
not predict any 3rd dredge-up episode to proceed for such low mass
stars. Iben and Renzini (1983) are uncertain about this limit. However
Iben and Renzini (1982a,b) have succeeded in bringing C to the surface
of a star of initial mass 0.7 $M_\odot$ of Z = 0.001; this result is very
sensitive to the opacities used in the computations.

On the observational side, there has been in the literature an
uncertainty about C determinations from the $\lambda 4267$ recombination line;
this line has generally yielded higher CIII abundances than those from
the $\lambda 1908$ CIII] line. This has led to a systematically higher C abundance
derived from optical studies than from ultraviolet work. In a comparative
study of bright planetary nebulae, French (1983) has determined this
systematic difference to be 0.18 dex; also Aller (1983a) finds a
statistical difference of 0.16 dex, while Aller and Czyzak (1983) report
a difference of 0.13 dex.

The excess C obtained from H4-1, BB-1 and DD-1 is based on
observations of $\lambda 4267$ line, the C values presented for K648 were obtained
from IUE data by Adams et al. (1983) showing that the C/O excess is not
based only in the recombination observations but also in the collision-
ally excited UV lines.

In Table III we present the fractional abundance by mass for some
elements, directly derived from Table II. From it we can derive the He/C
enrichment from a hypothetical 3rd dredge-up phase. For this derivation
we will assume: 1) that all C present is due to 3rd dredge-up $X_{12} = \Delta X_{12}^3$; 2) that the observed helium is due to the initial helium at the
time of formation of the progenitor that has been enriched through the
1st and 3rd dredge-up process: $Y_{obs} = Y_i + \Delta Y^1 + \Delta Y^3$; 3) that the
initial helium corresponds to the pregalactic abundance, $Y_i = 0.23$,
4) that the first dredge-up produces uniform enrichment for low mass
objects $\Delta Y^1 = 0.02$. From these assumptions we can derive the ratio of
He to C in the 3rd dredge-up episode, $f \equiv \Delta Y^3/\Delta X_{12}^3$. We list this value
in the last column of Table III. The mean value $\langle f \rangle = 4$ which is

TABLE I

General Characteristics of Halo PN

	ℓ	b	d (kpc)	z (kpc)	V_{rad} (km s^{-1})	Reference
K648	65	$-27°$	10.3	4.6	-114	1
H4-1	49	$+88°$	~ 20	~ 20	-141	2
BB-1	108	$-76°$	~ 25	~ 24	$+196$	3
DD-1	61	$+41°$	12-14	8-9	-304	4

References: 1. Harris and Racine 1979; 2. Miller 1969:
3. Boeshaar and Bond 1977; 4. Barker and Cudworth 1983.

TABLE II

Chemical Composition of Halo PN

	He/H	log O/H	log C/O	log N/O	log Ne/O	log S/H	log A/H	Reference
K648	0.100	7.67	+1.06	-1.2	-1.0	5.15	4.26	1,2,3,10
H4-1	0.098	8.36	+0.95	-0.61	-1.66	5.20	4.69	2,3,10
BB-1	0.095	7.90	+1.19	+0.44	+0.10	5.70	4.59	3,4,10
DD-1	0.100	8.03	+0.80	+0.27	-0.71	6.46	5.58	5
⟨mean⟩	0.099	7.99	+1.00	-0.28	-0.82	5.63	4.78	...
Orion	0.100	8.75	-0.23	-0.99	-0.85	7.41	6.70	6
Sun	...	8.92	-0.25	-0.93	-0.80	7.23	6.7	7,8,9

References: 1. Adams et al. 1983; 2. Torres-Peimbert and Peimbert 1979;
3. Barker 1980; 4. Torres-Peimbert et al. 1981; 5. Barker and Cudworth
1983; Peimbert and Torres-Peimbert 1977; 7. Lambert 1978; 8. Lambert
and Luck 1978; 9. Bertsch et al. 1972; 10. Barker 1983.

TABLE III

Fractional Mass Abundance in Halo PN

	X	Y	Z (X 10^3)	X_{12} (X 10^4)	X_{16} (X 10^4)	f $\Delta Y^3 / \Delta X_{12}^3$
K648	0.711	0.284	5	46	5	7.4
H4-1	0.704	0.276	21	173	26	1.5
BB-1	0.714	0.271	15	106	9	2.0
DD-1	0.718	0.283	9	57	16	4.5

derived for the halo PN is in agreement with the values derived for
different types of stars for the interstellar composition during helium
flashes (population I stars of 1.45 $M_\odot$ initial mass by Schönberner 1979
where   f ≃ 3-6).

K648, H4-1 and BB-1 show a similar pattern in their A and S
behavior in that they have an underabundance of 1.7 to 2.3 dex relative
to solar. The underabundance of O is 0.6 to 1.3 dex while that of Ne
is of 0.1 to 1.6 dex. DD-1 shows a uniform underabundance of A, S, O
and Ne of ∿ 0.8 dex. On the other hand in all 4 halo PN the ratio C/O
is 10 while that of Ne/O is essentially solar (although with large
scatter). Two possibilities have been advanced in the literature: a)
that the enrichment of O and Ne in the interstellar medium proceeded
faster than that of Ar, S and Fe, and b) that the O and Ne excess
relative to Ar, S and Fe are a product of the progenitor of the PN
themselves (Peimbert 1973; Hawley and Miller 1978; Torres-Peimbert and
Peimbert 1979; Barker 1980; Barker and Cudworth 1983). The constancy in

the C/O ratio supports the second possibility.

5. SUMMARY

In general, the changes in surface composition of the progenitor stars of PN appear well understood. Many of the composition differences among PN can be explained in terms of different main sequence masses of the progenitor stars.

For the high mass PN it has been found that there is an anti-correlation between N and O abundance, suggesting that part of the N excess has been produced at the expense of O. Also the composition of NGC 6302 is consistent with that of a progenitor of $\simeq 8$ $M_\odot$. More objects in this mass range should be looked for in an effort to determine the high mass limit for forming PN. From the intermediate mass PN it has been found that the C abundance is relatively high compared to pre-dicted values, perhaps the 3rd dredge-up is somewhat more efficient than predicted.

From the very small sample of halo PN we find that in all four there is evidence of very efficient 3rd dredge-up episode which has yielded log C/O $\sim$ 1. All objects also show O/Ar and O/S higher than in the solar neighborhood, I propose that this high O coupled to the relatively uniform C/O value is an indication that in these objects O has also been dredged-up to the surface in the ratio C/O = 10. The helium abundance is consistent with a 3rd dredge-up that has incorporated material of $\Delta Y / \Delta X_{12} \sim 4$.

Fruitful discussions with M. Peimbert are gratefully acknowledged. This is Contribution No. 110 of Instituto de Astronomía, UNAM.

References

Adams, S., Seaton, M.J., Howarth, I.D., Aurrière, M., and Walsch, J.R.: 1983, Monthly Notices Roy. Astron. Soc., in press.
Aller, L.H.: 1976, Pub. Astron. Soc. Pacific 88, pp. 574-584.
Aller, L.H.: 1983a in D.R. Flower (ed.), "IAU Symposium No. 103, Planetary Nebulae", Dordrecht: Reidel, pp. 1-13.
Aller, L.H.: 1983b, preprint.
Aller, L.H. and Czyzak, S.J.: 1982, Astrophys. J. Suppl. 51, pp. 211-248.
Barker, T.: 1978, Astrophys. J. 220, pp. 193-209.
Barker, T.: 1980, Astrophys. J. 237, pp. 482-485.
Barker, T.: 1983, Astrophys. J., submitted.
Barker, T. and Cudworth, K.M.: 1983, preprint.
Becker, S.A. and Iben, I. Jr.: 1979, Astrophys. J. 232, pp. 831-853.
Becker, S.A. and Iben, I. Jr.: 1980, Astrophys. J. 237, pp. 111-129.
Bertsch, D.L., Fichtel, C.E., and Reames, D.V.: 1972, Astrophys. J. 171, pp. 169-177.

Boeshaar, G.O. and Bond, H.E.: 1977, Astrophys. J. 213, pp. 421-426.
Calvet, N. and Peimbert, M.: 1983, Rev. Mexicana Astron. Astrof. 5, pp. 319-328.
Cudworth, K.M.: 1974, Astron. J. 79, pp. 1384-1395.
D'Odorico, S.D., Peimbert, M. and Sabbadin, F.: 1976, Astron. Astrophys. 47, pp. 341-344.
Dolidzde, M.V. and Dzhimshelejshnili, G.N.: 1966, Astron. Tsirk. No. 385, pp. 7-8.
French, H.: 1983, preprint.
Greig, W.E.: 1971, Astron. Astrophys. 10, pp. 161-174.
Greig, W.E.: 1972, Astron.Astrophys. 18, pp. 70-78.
Hawley, S.A. and Miller, J.S.: 1978, Astrophys. J. 220, pp. 609-613.
Harris, W.E. and Racine, R.: 1979, Ann. Rev. Astron. Astrophys 17, pp. 241-274.
Haro, G.: 1951, Publ. Astron. Soc. Pacific 63, pp. 144-145.
Iben, I. Jr. and Renzini, A.: 1982a, Astrophys J. (Letters) 259, pp. L79-L83.
Iben, I. Jr. and Renzini, A.: 1982b, Astrophys. J. (Letters) 263, pp. L23-L27.
Iben, I. Jr. and Renzini, A.: 1983, Illinois Astrophys. preprint IAP82-2.
Iben, I. Jr. and Truran, J.N.: 1978, Astrophys. J. 220, pp. 980-995.
Kaler, J.B.: 1983a, in D.R. Flower (ed.) "IAU Symposium No. 103, Planetary Nebulae", Dordrecht: Reidel, pp. 245-257.
Kaler, J.B.: 1983b, Illinois Astrophys. preprint IAP83-1.
Koester, D. and Reimers, D.: 1981, Astron. Astrophys 99, pp. L8-L11.
Lambert, D.L.: 1978, Monthly Notices Roy. Astron. Soc. 182, pp. 249-272.
Lambert, D.L. and Luck, R.E. 1978, Monthly Notices Roy. Astron. Soc. 183, pp. 79-100.
Meñdez, R.H. and Niemela, V.S.: 1981, Astrophys. J. 250, pp. 240-247.
Miller, J.S.: 1969, Astrophys. J. 157, pp. 1215-1223.
Paczynski, B.: 1971, Acta Astron. 21, pp. 417-435.
Pease, F.G.: 1928, Publ. Astron. Soc. Pacific 40, p. 342.
Peimbert, M.: 1973, Mem. Soc. Roy. Sci. Liege 6e serie, 5, pp. 391-412.
Peimbert, M.: 1978, in Y. Terzian (ed.), "IAU Symposium No. 76, Planetary Nebulae", Dordrecht: Reidel, pp. 215-224.
Peimbert, M.: 1979 in W.B. Burton (Ed.) "IAU Symposium No. 84, The Large Scale Characteristics of the Galaxy", Dordrecht: Reidel, pp. 307-315.
Peimbert, M.: 1981, in I. Iben Jr. and A. Renzini (eds.) "Physical Processes in Red Giants", Dordrecht: Reidel, pp. 409-420.
Peimbert, M. and Serrano, A.: 1980, Rev. Mexicana Astron. Astrof. 5, pp. 9-18.
Peimbert, M. and Torres-Peimbert, S.: 1977, Monthly Notices Roy. Astron. Soc. 179, pp. 217-234.
Peimbert, M. and Torres-Peimbert, S.: 1983 in R.D. Flower (ed.) "IAU Symposium No. 103, Planetary Nebulae", Dordrecht: Reidel, pp. 233-242.
Reimers, D. and Koester, D. 1982: Astron. Astrophys. 116, pp. 341-347.
Renzini, A. and Voli, M.: 1981, Astron. Astrophys. 94, pp. 175-193.
Schönberner, D.: 1979, Astron. Astrophys. 79, pp. 108-114.
Schönberner, D. and Weidemann, V.: 1981, in I. Iben Jr. and A. Renzini (eds.) "Physical Processes in Red Giants", Dordrecht: Reidel, pp. 463-468.

Shaver, P.A., McGee, R.X., Newton, L.M., Danks, A.C., and Pottasch, S.R.:
1983, Monthly Notices Roy. Astron. Soc.,in press.
Talent, D.L. and Dufour, R.J.: 1979, Astrophys. J. 233, pp. 888-905.
Torres-Peimbert, S. and Peimbert, M.: 1977, Rev. Mexicana Astron. Astrof.
2, pp. 181-207.
Torres-Peimbert, S. and Peimbert, M.: 1979, Rev. Mexicana Astron. Astrof.
4, pp. 341-350.
Torres-Peimbert, S., Rayo, J.F., and Peimbert, M.: 1981, Rev. Mexicana
Astron. Astrof. 6, pp. 315-319.

EDMUNDS: What is the chance that the rare Type IV PN are result of
binary evolution?
TORRES-PEIMBERT: With the available information we cannot rule out this
possibility.
IBEN: Those stars that, in theory, may produce large nitrogen abundances
($M_{main\ seq} \gtrsim 4\ M_\odot$, say, and $M \lesssim 8\ M_\odot$) are expected to produce only a
small fraction of all planetary nebulae, perhaps only 10 - 15%. Further,
they will fade more rapidly, due to the larger mass of the central
star, than will typical PNe in which $M_{central} \sim 0.6\ M_\odot$. Are the
statistics of your Type I, II, ..., IV PNe consistent with this?
MAEDER: Are the observed numbers of planetary nebulae of Type I and Type
II consistent with the estimated mass ranges and the predicted lifetimes for a standard mass spectrum?
TORRES-PEIMBERT: The fraction of PN of Type I in the solar neighborhood
is of the order of 10 to 20%. The mass range for Type I PN is uncertain
but the minimum mass is 2.4 $M_\odot$ (from NGC 2818, that belongs to the
galactic cluster of the same name, NGC 3132 and NGC 2346) and the
maximum mass is $\sim 8\ M_\odot$ from NGC 6302, or $\sim 6\ M_O$, if less extreme cases
are considered. The recombination time in the nebula, which is of the
order of $10^5/N_e$ years, has to be taken also into account; this means
that even if the star has evolved to very low temperatures the gas
could still be partially ionized. To test a standard initial mass
function for stars responsible for the formation of PN of Types II-IV
it is necessary to know: a) the star formation rate as a function of
time, and b) if all stars less massive than 2.4 $M_\odot$ produce a PN (which
is unlikely, see Alloin, Cruz-González and Peimbert 1976, Ap. J. 205,
74). From these considerations it follows that the observed values is
not in contradiction with a standard mass spectrum. It is clear that
there are many factors that enter into this comparison and that all of
them are uncertain.
PERSIC: Is there any difference in the results for the abundances between
the various spectral ranges (visual, ultraviolet, and so forth)?
TORRES-PEIMBERT: The main differences are two: a) in PN of high degree of
ionization the N abundance determinations based on UV data are about a
factor of two higher than in the visual which is due to a better ionization correction factor scheme that can be used with UV data. b) The
C abundances from UV data are ~ 1.5 times smaller than the optical ones

for well observed objects, this could be due to temperature variations
along the line of sight produced by chemical and density inhomogeneities.

PREITE-MARTINEZ: Do you think there might be a correlation between the
size of the vertical error bars in the observational points defining
the abundance gradients in the Galaxy and the slope of the gradients?

TORRES-PEIMBERT: No, but the error bars depend on the element under
consideration. For O and N the errors are larger for smaller galactocentric distances because in general the temperatures are smaller
and more difficult to measure than for larger galactocentric values.
Alternatively, for He the errors are the same because the He/H ratios
are almost temperature independent.

RENZINI: Icko and I have recently found dredge-up in low mass and low
core mass models. So the efficiency of the third dredge-up in such
models was underestimated both in the papers of Iben and Truran, and
Voli and myself. This should improve the agreement between theory and
observations for what you call the Type II Planetary Nebulae. However,
the current theory still predicts that very low mass stars ($M_i \simeq 0.85$
$M_\odot$) should lose their envelope before experiencing the third dredge-up,
which is at variance with the high carbon abundance you find in these
few halo planetary nebulae.

BRANCH: 1) There is evidence that in the local solar neighborhood, dupli-
city is less common among halo population stars than it is among disk
population stars. If the peculiarities of the halo planetaries are
to be attributed to binary evolution, is there a reason why binaries
are more likely than single stars to make planetaries? 2) Do you think
that the differences in your argon to sulphur ratios for the halo
planetaries are real?

TORRES-PEIMBERT: The number of PN observed in the halo is about two to
three orders of magnitude smaller than predicted under the assumption
that all stars become PN. This is not the case for stars in the 1.2 to
6 $M_\odot$ range where a substantial fraction becomes PN. In that sense mass
transfer might increase the mass of the secondary and the likelihood
for it of becoming a PN.   2) The uncertainty in the Ar/S ratios is
of a factor of 2-3, therefore, from the data available I do not think
that a variation in the ratio can be claimed.

EDMUNDS: How good are the ionization corrections for sulphur for the
Type IV observation analysis?

TORRES-PEIMBERT: The results are due to Barker and he has observed [S II]
and [S III] lines; I think that the results are good within a factor of
two.

Cn 1-1 (= HD 330036):   PLANETARY NEBULA OR SYMBIOTIC STAR?

J. H. Lutz
Washington State University and University College London

ABSTRACT

Observations are reported for Cn 1-1 (= HD 330036). This object is one
of many that have been classified both as a symbiotic star and a
planetary nebula. The purpose of this study is to investigate the
physical properties and evolutionary status of Cn 1-1. It is concluded
that Cn 1-1 is a dense planetary nebula in a binary system.

I.   INTRODUCTION

        Generally planetary nebulae are not difficult to distinguish
from symbiotic stars. Classic planetary nebulae have extended nebular
shells and hot central stars (T > 30,000 $^{\circ}$K), while typical symbiotic
stars are point sources and have M-type absorption spectra (T < 3000 $^{\circ}$K).
However, since some of the observed characteristics of these two groups
of objects overlap (see Table 1), there is considerable ambiguity in
the identification of some objects. For example, there are stellar
planetary nebulae and also planetary nebulae with late-type binary
central stars (Lutz 1977, 1978). In addition, there are hotter
symbiotic stars which have F-, G- or K-type absorption spectra. The
ultraviolet and infrared spectra of some planetary nebulae and symbiotic
stars appear to be very similar. Objects that are considered to be
protoplanetary nebulae by some investigators are classified as symbiotic
stars by others (e.g. V1016 Cyg, HM Sge). According to the most recent
definition proposed for symbiotic stars (Boyarchuck 1982), variability
is not a necessary criterion for classifying a star as symbiotic.
Consequently, symbiotic stars have become an even more heterogeneous
group than ever before.

The degree of overlap between objects which are classified as both
symbiotic stars and planetary nebulae has become quite large.
Approximately 1/3 of the objects in the compilations of symbiotic stars
by Allen (1979, 1982) are listed also in the Catalogue of Galactic
Planetary Nebulae (Perek and Kohoutek 1967). Lutz (1983b) has reviewed
the criteria for classifying objects as either planetary nebulae or

C. Chiosi and A. Renzini (eds.), Stellar Nucleosynthesis, 15–18.
© 1984 by D. Reidel Publishing Company.

symbiotic stars and has proposed criteria for distinguishing between low-density symbiotic stars and high-density planetary nebulae.

Table 1: Characteristics of Planetary
Nebulae and Symbiotic Stars

Characteristic	Planetary Nebulae	Symbiotic Stars
Optical Emission Spectrum	Low to high ionization	Medium to extremely high ionization
Stellar Spectrum	Hot star $T > 30,000\ ^{\circ}K$	Cool Star $T < 3000\ ^{\circ}K$
General Infrared Spectrum	Various; some D-type; no S-type	S- and D- infrared types
Angular Size	Most are > 1"	Stellar
Variability	Rare	Common

II.   OBSERVATIONS

Cn 1-1 (= HD 330036=BD-48° 10371=PK 330+4°1) is an object that is found in both planetary nebula and symbiotic star catalogues. Extensive optical and ultraviolet observations of Cn 1-1 were obtained in order to determine the physical parameters and evolutionary state (Lutz 1983a). Optical spectra were obtained between June 1977 and June 1979 to measure radial velocities of the emission and absorption features, relative intensities of the emission lines and the MK spectral class of the absorption spectrum. No variability was found. The radial velocity of both the absorption and the emission features remained constant at approximately - 14 ± 5 km s^{-1}. The absorption spectrum was classified as F5 III-IV.

Low-resolution International Ultraviolet Explorer spectra were obtained in September 1979, June 1980 and April 1981. The IUE fluxes were constant over this period. The emission spectrum is very much like that of a medium excitation planetary nebula, except $\lambda4363$ is strong, which is indicative of relatively high electron density and/or electron temperature for a planetary nebula.

The interstellar extinction and distance can be determined by using the least squares line drawn through the equation of the data points in the diagram of color excess versus distance for the region together with the equation for distance modulus, viz.

$$m_V - M_V = 5 \log r - 5 + 3.2\ E(B-V).$$

The apparent magnitude m_V is 11.03 (Shao and Liller 1972) and the absolute magnitude M_V for an F5 III-IV star is $+1.9$ (Allen 1973). The solution of these two equations gives r = 45 pc and E(B-V) = 0.28.

III.   INTERPRETATION

The electron density and electron temperature as determined from [O III] line ratios are 10^6 cm^{-3} and 1.5×10^4 °K respectively. These values can be used to calculate abundances from the observed line fluxes. The results are C/O ~ 0.8 and N/O ~ 1.1, both well above the solar ratios for these elements.

The size and amount of luminous material can be estimated from the known physical parameters, e.g. distance, electron density and temperature, angular diameter (< 1") and radio flux (Milne and Aller 1982). The conclusion is that the nebula is small (r < 10^{-3} pc) and does not contain much material (M < 10^{-3} M$_\odot$).

If the weak continuum flux at the short ultraviolet wavelengths is attributed to the hot star, then the He II Zanstra temperature is 76,000 °K and the apparent visual magnitude is about 17. At the adopted distance of 450 pc, the absolute magnitude would be approximately $+8$. If Cn 1-1 is a planetary nebula, the hot central star is in a peculiar position (i.e. faint absolute magnitude) on the HR diagram for an object with such a dense nebulosity.

There are two other observations that are useful in resolving whether Cn 1-1 is a planetary nebula or a symbiotic star. First, Cn 1-1 shows no evidence for Si III] $\lambda1892$ emission, whereas this line is usually strong in symbiotic stars. The observed limit of F(C III] $\lambda1909$)/F(Si III] $\lambda1892$) > 1.59 in Cn 1-1 favors its classification as a planetary nebula, since this ratio is observed to be greater than one in planetary nebulae and less than one in symbiotic stars (Feibelman 1983). Second, Cn 1-1 is the only object classified as a symbiotic star which shows strong emission features around 3.3μm (Allen 1981). A number of planetary nebulae show these features (Barlow 1983).

The evidence suggests that Cn 1-1 is a dense planetary nebula involved in a binary system. It is suggested that the dense nebulosity in combination with a faint central star (M=+8) may be due to the evolved star undergoing a "final puff" of mass ejection before evolving to the white dwarf stage.

REFERENCES

Allen, C.W. : 1973, "Astrophysical Quantities", Athlone Press, London.

Allen, D.A. : 1979, in F.M. Bateson, J. Smart and I.H. Urch (eds.), "Changing Trends of Variable Star Research", University of Waikoto, Hamilton, New Zealand, p.125.

Allen, D.A. : 1981, in R.E. Stencel (ed.), "Abstracts of North American
    Symposium on Symbiotic Stars", JILA, Boulder, p.12.

Allen, D.A. : 1982, in M. Friedjung and R. Viotti (eds.), "The Nature
    of Symbiotic Stars", Reidel, Dordrecht, p.27.

Barlow, M.J. : 1983, in D. Flower (ed.), "Planetary Nebulae",
    Reidel, Dordrecht, p.105.

Boyerchuck, A.A. : 1983, in M. Friedjung and R. Viotti (eds.),
    "The Nature of Symbiotic Stars", Reidel, Dordrecht, p.225.

Feibelman, W. : 1983, preprint.

Lutz, J.H. : 1977, Astron. Astrophys., 60, 93.

Lutz, J.H. : 1978, in Y. Terzian (ed.) "Planetary Nebulae: Observations
    and Theory", Reidel, Dordrecht, p.185.

Lutz, J.H. : 1983a, submitted to Astrophys. J.

Lutz, J.H. : 1983b, in preparation.

Milne, D.K., and Aller, L.H. : 1982, Astron. Astrophys. Suppl., 50, 209.

Perek, L., and Kohoutek, L. : 1967, "Catalogue of Galactic Planetary
    Nebulae", Czechoslovakia Academy of Sciences, Prague.

Shao, C. -Y., and Liller, W. : 1972, private communication.

EXTRAGALACTIC SUPERNOVAE

David Branch
Department of Physics and Astronomy
University of Oklahoma

The accelerating pace of research on extragalactic supernovae (SNe) can be illustrated by a very brief history of the topic. The subject began just about a century ago, when S Andromedae, now known also as SN 1885a, appeared near the nucleus of the Andromeda Galaxy. Fifty years later, after only 13 more SNe had been noticed, Zwicky appeared and the discovery rate rose sharply; the total number now exceeds 500. Detailed theoretical discussions of stellar explosions began less than a quarter of a century ago, with Hoyle and Fowler (1960) on the sudden release of thermonuclear energy and Colgate and White (1966) on the release of gravitational potential energy by the collapse of a stellar core. Attempts to make qualitative interpretations of SN observations began to seem worthwhile only a decade ago, when the first spectrophotometric data obtained with modern linear detectors were published (Kirshner, Oke, Penston, and Searle 1973), and the present lively interplay between theoretical predictions and observational constraints has developed just within the last few years (cf. Meyerott and Gillespie 1980, Wheeler 1980, Rees and Stoneham 1982).

In the near future, automated supernova searches promise to make the subject far richer, observationally. For example, the goal of the Berkeley-Monterey search (Kare et al. 1982) is to monitor, to a detection threshold of m_v=18.8, 500 relatively nearby galaxies every night and 6000 others on a three-night cycle. This could yield as many as 100 new SNe per year, resulting in not only a rapidly expanding set of data for statistical studies but also the discovery of pre-maximum, peculiar, and subluminous SNe. Theoretical work is sure to be stimulated by the increasing flow of data.

Advances in our understanding of supernovae will be crucial for nucleosynthesis. Although SNe are believed to be the fundamental nucleosynthesis objects, at least for the elements from magnesium to the iron peak, the amount of each element ejected by each kind of SN is not known. In fact, there is hardly any direct evidence for the ejection of heavy elements by Type II supernovae (SNe II). The evidence that Type I supernovae (SNe I) contribute to nucleosynthesis is much more clear, but

19

C. Chiosi and A. Renzini (eds.), Stellar Nucleosynthesis, 19–33.
© 1984 by D. Reidel Publishing Company.

still quantitatively uncertain. Relatively rare explosions of very
massive stars, possibly exemplified by the (Type V) SN 1961v in NGC 1058,
may be important. The absolute visual magnitudes of SNe I and SNe II
are generally in the range of -17 to -20, but the absolute magnitude of
the 17th-century SN which has produced the Cas A remnant in our Galaxy
was no brighter than -12; the overabundances of heavy elements observed
in Cas A (Kirshner and Chevalier 1978) show that a contribution to
nucleosynthesis is also made by kinds of SNe which, due to their low
optical luminosities, have not yet even been seen in other galaxies.
In this paper, the basic observations of the kinds of SNe which have been
seen in external galaxies are summarized, and current interpretations
of SNe I and SNe II are briefly reviewed. For a comprehensive review
of SNe, their remnants, and related topics, see Trimble (1982, 1983).

1. OBSERVATIONS OF SNe I AND SNe II

          Fundamentally, the classification of a supernova is based on
its spectrum. Almost all of the $\sim$100 SNe whose spectra have been observed
can readily be assigned to one of the two main types. About 2/3 of them
are SNe I. These show characteristic Doppler-broadened ($\sim$10,000 km/sec)
spectral bands which faithfully follow a standard pattern of evolution
in wavelength and intensity. There is no sign of the hydrogen Balmer
lines in the blue part of the spectrum; the appearance at certain phases
of an emission peak near the rest wavelength of Hα in the red may be
coincidental. On the other hand, the defining characteristic of SN II
spectra is the definite presence of the Balmer lines, with P Cygni
profiles (emission at the rest wavelength accompanied by blueshifted
absorption). The bands in SNe II are slightly narrower ($\sim$7000 km/sec)
than in SNe I, and their evolution is a little less standardized. The
differences between SN I and SN II spectra are so obvious that misclas-
sifications are rare. Before 1970, SN spectra ordinarily were observed
photographically, but even a single photographic spectrum, red or blue
and at any phase, is adequate for classification. The fully-calibrated
modern spectra are needed for detailed quantitative studies.

          A reasonably reliable classification can be based also on a
well-observed light curve. For extragalactic supernovae this is not too
helpful, because a SN with a good light curve is likely to have had its
spectrum observed, but it is important for the classification of the
historical SNe in our own Galaxy for which we have light curves but no
spectra. The shape of SN I light curves has been known for a long time
to be remarkably uniform. This is nicely demonstrated by the composite
SN I light curve formed by Barbon, Ciatti, and Rosino (1973) from those
of 38 individual SNe I. The basic features are a "peak" consisting of a
steep rise to maximum light in about 15 days and an initial decline of
about 3 magnitudes in 30 days, followed by a linear "tail" declining at a
rate of only 0.5 magnitudes per 30 days. The tail of one SN I, 1982e
in NGC 5253, has been followed to 700 days after maximum light (Kirshner
and Oke 1975). Whether the family of SN I light curves should be split
into two subtypes ("fast" and "slow"; Barbon et al. 1973, Barbon 1978),

arranged according to a single continuous parameter (Rust 1974,
Pskovskii 1977, Branch 1981, 1982), or regarded as intrinsically identical
(Sandage and Tammann 1982) will remain controversial until the number
of accurate photoelectric light curves increases. The light curves of
SNe II definitely show more variety than those of SNe I, but Barbon,
Ciatti, and Rosino (1979) have introduced some order by defining two
subtypes; about 2/3 are "plateau" SNe II, characterized by a phase of
nearly constant luminosity between 30 and 80 days after maximum light,
and the others are "linear," typically having a decline rate of 1.5
magnitudes per 30 days with little or no plateau. A detailed comparison
of the other properties of plateau and linear SNe II has not yet been
carried out. An important unresolved question is whether SNe II, like
SNe I, eventually settle into a linear tail, as the composite plateau
curve of Barbon et al. (1979) suggests.

On average SNe I are observed to be brighter than SNe II by
about 1 magnitude, but the intrinsic difference may be somewhat less
because SNe II occur in spiral arms and may experience more interstellar
extinction. For both types the observed dispersion about the mean peak
brightness approaches one magnitude, but this is partly due to inaccurate
determinations of the peak brightness and partly to errors in the extinction and the relative distances to the parent galaxies. It is clear
that some of the dispersion for SNe II, but not necessarily for SNe I,
is intrinsic. Some SNe II are as bright as some SNe I, so peak brightness
must not be used as a classification criterion. The average peak absolute magnitude depends on the extragalactic distance scale and can be
expressed as

$$M_B = -18.2 + 5 \log h \qquad \text{(Sne I)}$$

$$M_B = -17.5 + 5 \log h \qquad \text{(Sne II)}$$

(Tammann 1982) where h is the Hubble constant in units of 100 km/sec/Mpc.
The peak integrated luminosity and the total radiated energy depend also
on bolometric corrections. At maximum light the continuous spectra of
both types appear to be like $\sim$20,000 K blackbodies, although the SN I
spectrum is truncated at 4000 Angstroms, with little energy at shorter
wavelengths. The SN II ultraviolet radiation more or less makes up for
the fainter optical luminosity, so that a fair estimate of the peak
integrated luminosity for both types is

$$L = 5.0 \times 10^{42} h^{-2} \text{ ergs/sec.}$$

The total radiated energy, for both types, is on the order of $10^{49} h^{-2}$
ergs. The kinetic energy of one solar mass moving at 10,000 km/sec is
10^{51} ergs; this is a reasonable (distance-independent) estimate for SNe
I, which may be completely disrupting white dwarfs of 1.4 solar masses.
For SNe II, the characteristic velocity is lower but the ejected mass is
probably higher, so the kinetic energy again is on the order of 10^{51}
ergs. These estimates are at least good enough to show that for both
types the total radiated energy is just a few percent of the kinetic energy.

SNe I and SNe II appear in practically mutually exclusive locations. SNe II occur only in spiral arms; SNe I, when in spirals, occur in the disk but with no preference for arms (Maza and van den Bergh 1976), and they also occur in ellipticals and IO galaxies. Not a single SN II has been seen in an elliptical, so an SN which appears in an elliptical may be taken for some purposes as an SN I. But of course an SN which appears on a spiral arm is not necessarily an SN II, since SNe I, being disk objects, sometimes happen to be in the arms. The mass above which stars die in spiral arms is not well known but is consistent with 8 solar masses, the critical mass above which stars undergo non-degenerate carbon ignition in their cores. It is likely that SNe II come from stars above 8 $M_\odot$, and SNe I from accreting white dwarfs in binary systems. This is consistent with the accumulating evidence that single stars initially below 8 $M_\odot$ do not explode at all, owing to extensive mass loss (Weidemann and Koester 1983, Iben 1984).

Three of the historical supernovae in our Galaxy are believed to have been SNe I. The light curves of Tycho's and Kepler's SNe, based on contemporary observations, look like standard light curves of SNe I (Clark and Stephenson 1977, Pskovskii 1978). The SN of 1006 AD also is classified as an SN I, primarily on the basis of the resemblance of its remnant to those of Tycho and Kepler. Since neither pulsars nor thermal X-rays from hot young neutron stars are observed in these remnants (Helfand 1980, Nomoto and Tsuruta 1981), SNe I are suspected to disrupt completely, rather than leave neutron stars. None of the historical SNe are known to have been SNe II. The best candidate is the SN of 1054 AD, which has produced the Crab nebula. If the Crab consists only of the classical nebula, it cannot have been an SN II, because the expansion velocity is too small by a factor of 5, but if it is surrounded by a faint, high-velocity shell (Murdin and Clark 1981) it may have been a normal SN II (Chevalier 1977). The existence of the Crab pulsar, together with the inevitable collapse of the iron cores in highly evolved massive stars, suggests that SNe II do leave neutron stars.

2. OBSERVATIONS OF OTHER SUPERNOVAE

Understandably, most interpretation of supernova observations has been devoted so far to SNe I and SNe II, but a few other kinds of SNe have been seen. One SN showed some of the characteristics of both SNe I and SNe II. Bertola (1964a) observed the light curve of SN 1957a in NGC 2841, classified its shape as that of an SN I, and noted that the peak absolute magnitude appeared to be three magnitudes fainter than normal. But when Zwicky and Karpowicz (1965) observed the spectrum they found Balmer lines and therefore classified it as an SN II. They did suggest that it be regarded as a peculiar SN II, because the Balmer lines were unusually narrow (<2000 km/sec) and were shifted some 1500 km/sec to the red of their rest wavelengths. In retrospect it appears that this SN had very little in common with those we know as SNe II. Not only were the hydrogen emission lines narrow and displaced to the red, but they appear to have lacked the usual blueshifted absorption components. In

fact, the spectrum beneath the Balmer lines appears to resemble that of
an SN I, although it is difficult to be sure because of the noise in the
published spectra. When the photometry of Bertola and of Zwicky and
Karpowicz is combined, the light curve bears no resemblance to those of
SNe II; it is like that of an SN I, but more rapidly declining as well
as fainter. The progenitor of this interesting supernova may have been
an accreting white dwarf which exploded and swept away part of the
hydrogen envelope of its companion. The asymmetric expulsion of the
hydrogen could have produced redshifted emission lines with no associated
blueshifted absorption. The reason other SNe I lack the hydrogen lines
might be that the donor, as well as the exploding star, ordinarily is a
carbon-oxygen white dwarf (Iben and Tutukov 1984). Whatever the true
nature of SN 1957a, it is certain that because it was faint and decayed
so quickly there has been a strong selection effect against the discovery
of objects like it in the searches which have been carried out so far.
Its existence points to the increasing diversity of observed supernovae
we may begin to see when the automated searches are underway.

         In addition to SNe I and II, Zwicky (1965) defined SNe III, IV
and V. It is not clear that the prototypes of SNe III and IV were
spectroscopically different from SNe II (Oke and Searle 1974), although
their light curves may have been unusual. However, the prototype of
SNe V, SN 1961v in NGC 1058, was unique in a number of respects and
deserved some attention. It is the only SN to have been seen before it
exploded. Zwicky (1964) found that the progenitor had been visible on
photographs of NGC 1058 since at least 1937, at $m_{pg} \sim 18$. The bolometric
correction in the pre-outburst phase, the distance, and the extinction
are uncertain, but the pre-explosion luminosity probably was on the order
of 10^{41} ergs/sec (Branch and Greenstein 1971), the Eddington limit for a
star of about 500 $M_\odot$ (Chevalier 1981a). After its discovery at $m_{pg} \sim 14$
in 1961, SN 1961v showed a constant luminosity plateau followed by an
abrupt increase of 2 magnitudes within two months near the end of the
year. No other SN has been seen to brighten significantly after the
initial rise to maximum light. The light curve declined exceptionally
slowly between 1962 and 1970, only 5 magnitudes in 8 years (Bertola and
Arp 1970). This suggests the ejection of a large mass. The spectral
lines were unusually narrow (2000 km/sec); the consequent lack of blending
compared to ordinary SN spectra allowed Bertola (1963) to correctly
identify Fe II lines, the first lines other than hydrogen to be identified
in an SN spectrum. The spectrum can be explained as overlapping P Cygni
lines of H, Fe II, and other ions (Branch and Greenstein 1971) which are
seen in SNe II. SN 1961v is the best candidate we have for an explosion
of a very massive star (cf. Bond 1984, Woosley, Axelrod, and Weaver, 1984).
On the basis of fragmentary light curves only, Zwicky (1965) designated
a few other supernovae as probable SNe V. One recent object, SN 1978a
in NGC 4324 (Elliot et al. 1978) has shown a spectrum similar to that of
SN 1961v. We can hope to see new, well observed SNe like these in the
future. SN 1961v was not particularly faint in absolute terms, so there
has not been a strong selection effect against the discovery of others
like it. Evidently such SNe are relatively rare, but they still may be
important for nucleosynthesis if they eject far more mass than SNE I
and SNE II.

3.   INTERPRETATION OF SNeI

The empirical link between SNe I and nucleosynthesis has
become convincing. Three lines of evidence point to the production of
iron by SNe I. The shape of the light curve is accounted for by the
radioactivity of Ni^{56} and its daughter Co^{56}, which decays to stable Fe^{56}
(Colgate and McKee 1969, Arnett 1982, Chevalier 1981b, Schurmann 1983,
Sutherland and Wheeler 1984, Woosley, Axelrod and Weaver 1984). Optical
spectra during both the early photospheric phases (Branch et al. 1983)
and the later optically thin phases (Axelrod 1980a,b) show very strong
lines of iron, and there is some evidence for cobalt. The ultraviolet
spectrum of a star behind the remnant of SN 1006 contains absorptions
which are attributed to overabundant iron within the remnant (Wu et al.
1983). The absolute amount of iron, however, is uncertain. Accurate
abundances from the early-time spectra will require detailed model
atmospheres, the lower limit on the amount of iron in SN 1006 is only
0.03 $M_\odot$, and the amounts needed for the light curve and late-time spectra
depend on the extragalactic distance scale. If the peak of the light
curve is powered only by radioactivity, an approximate relation between
the amount of synthesized Ni^{56} and the Hubble constant is $M(Ni)=0.2h^{-2}$
$M_\odot$. The early optical spectra and the X-ray spectra of the Tycho remnant
(Becker et al. 1980a,b, Shull 1982) show that overabundant intermediate-
mass elements, from oxygen to calcium, also are ejected, but again the
uncertainties in the abundances are large.

Since an accurate, direct determination of the composition of
SN I ejecta is not in sight, the nucleosynthesis contribution of SNe I
will not be known until the correct theoretical model of the explosion
is identified. Recent discussions of the models and their comparison
with the observations have been given by Wheeler (1982) and Branch
(1983). The favorite class of models is accreting white dwarfs, but
there is a variety of outcomes depending on the initial mass and composition of the white dwarf, the accretion rate, and the composition of
the accreted matter. So far, most attention has been given to the problem
of finding a model whose characteristics match the observations, but
we also will need to understand why that model occurs much more frequently
in nature than the others. A first attempt to estimate the frequencies
of the various accreting white dwarf scenarios has been made by Iben
and Tutukov (1984).

The basic information needed to distinguish among the models -
chemical composition as a function of ejection velocity - can in principle
be derived from the spectra. The most detailed studies are those of
Branch et al. (1983) on the early photospheric phases of SN 1981b in
NGC 4536, and of Axelrod (1980a,b) on the late optically thin phases of
SN 1972e in NGC 5253. These were typical SNe I, and practically identical. The spectra near maximum light are dominated by lines of the
intermediate mass elements, but permitted lines of singly ionized iron
become conspicuous within a week and then dominate the spectrum for at
least 100 days; finally, hundreds of days after maximum light, the ejecta
become optically thin and the spectrum consists mainly of forbidden lines

from several ionization stages of iron. The intermediate-mass elements
occur in the velocity interval 8000-20,000 km/sec, and most of the
nickel, cobalt, and iron moves slower than 8000 km/sec, but there may be
some mixing between the two compositions.

Some of the accreting white dwarf models conflict with the
picture described above. Those based on the ignition of degenerate
helium, whether in helium white dwarfs or in accreted layers on top of
carbon-oxygen white dwarfs, tend to develop detonation waves which burn
practically all of the ejected material to Ni^{56} and eject it at excessive
velocities. A more satisfactory model is a C-O white dwarf which
accretes hydrogen at $\sim 10^{-7}$ $M_{\odot}$/year, converts the hydrogen to helium and
then to carbon and oxygen by shell flashes, approaches the Chandrasekhar
mass, ignites degenerate carbon at its center, and completely disrupts
by means of an outgoing deflagration wave which incinerates the inner
portion of the star to Ni^{56} but burns the outer part only as far as the
intermediate-mass elements (Nomoto 1981, 1984).

This carbon-deflagration model is attractive: it meets the
gross composition-velocity constraints described above, produces a
satisfactory light curve (Chevalier 1981b), and completely disrupts.
There are, however, some outstanding questions:

(1) Can each SN I eject 0.7 $M_{\odot}$ or more of iron? In any model
which completely disrupts, all of the kinetic energy must come from
nuclear burning. To accelerate a significant amount of matter to 20,000
km/sec and to achieve a total kinetic energy of 10^{51} ergs, at least 0.7
$M_{\odot}$ of iron must be synthesized (Sutherland and Wheeler 1984). This is
several times too high for current models of the chemical evolution of
the Galaxy (Twarog and Wheeler 1982) and implies that $h \sim 0.5$, not ~ 1.0 as
determined by some recent studies. Models which leave neutron stars
can derive kinetic energy from the binding energy and may eject less
iron; the problem then, apart from the apparent lack of neutron stars in
SN I remnants, is that if the binary system remains bound, low-mass X-ray
binaries will be overproduced by a large factor (Iben and Tutukov 1984).

(2) Would explosions of carbon-oxygen white dwarfs at the
Chandrasekhar limit produce too much uniformity, even for SNe I? The
usual emphasis on the homogeneity of SNe I is justified, but some differences certainly have been seen, for example the spectroscopic peculiarities of SN 1954a (Branch 1972), 19621 (Bertola 1964b), 19641 (Bertola
et al. 1965), and 1980i (Smith 1981). Perhaps these will be found to
correspond to occasional occurrences of some of the other accreting
white dwarf scenarios.

(3) Is the detailed isotopic composition of the carbon-deflagration ejecta consistent with the composition of our Galaxy (the Sun and
cosmic rays)? According to Woosley et al. (1984), it is not.

(4) If the accreted matter is hydrogen-rich, why don't we
detect hydrogen which has been swept off the atmosphere of the donor star

(Wheeler, Lecar, and McKee 1975, Fryxell and Arnett 1981)? LTE estimates
indicate that hydrogen is deficient by a factor of more than a thousand,
but these refer to hydrogen which is distributed over a large velocity
range in the ejecta; hydrogen confined to a narrower velocity range
might be harder to detect. Nevertheless, the apparent lack of any
detected hydrogen may be an argument in favor of the double C-O white-
dwarf carbon-deflagration model of Iben and Tutukov (1984), as
mentioned above in the discussion of SN 1957a.

        (5) Why is the carbon-deflagration model realized so much
more frequently than the alternatives? The relative frequencies estimated
by Iben and Tutukov are comparable for a wide range of scenarios,
including some based on helium detonations, so the overall homogeneity
of SNe I has not yet been explained.

4. INTERPRETATION OF SNeII

        The role of SNe II in nucleosynthesis is not clear. Observa-
tions of extragalactic SNe II are of radiation emitted by the outer,
hydrogen-rich parts of the ejected shell, not from the inner parts which
contain whatever freshly synthesized heavy elements are ejected.
Theoretically, the evolution of massive stars can be followed to the
point of nuclear fuel exhaustion, but the theory of core collapse does not
unambiguously tell which stars manage to explode. One possible approach
is to accept the theoretical predictions up to the point of fuel exhaustion, and then look to the observations to decide which stars do explode.
The ejected mass can be estimated by modelling the observed light curves,
and the mass of the neutron star can be assumed to be about 1.4 solar
masses, but to determine the initial mass of the star, the mass lost
before it exploded must be included. New observations, primarily in
non-optical wavebands, are showing that the progenitors of SNe II do lose
significant amounts of mass before they explode.

        As discussed above, SNe II probably come from stars initially
more massive than 8 $M_\odot$. Theory predicts that stars having initial
masses between 8 and about 12 $M_\odot$ do explode. The cores of these stars
ignite non-degenerate carbon, develop degenerate oxygen-magnesium-neon
or neon-silicon cores (depending on the precise mass), contract due to
electron captures, undergo degenerate oxygen, neon, or silicon burning
to nuclear statistical equilibrium, and collapse to neutron-star densities
(Nomoto 1982, Woosley, Weaver and Taam 1980). The collapse of these
cores, aided by a very steep density gradient at the core-envelope
boundary, leads to the generation of an outgoing shock wave which is able
to eject the envelope with enough kinetic energy for an SN II (Hillebrandt
1983). The heavy element ejection, however, is almost negligible. The
predicted composition of the ejecta of a star in the 8-9.5 $M_\odot$ range is
consistent with that of the Crab nebula (Nomoto 1982, Hillebrandt 1983).
Thus it seems likely that 8-12 $M_\odot$ stars explode, leave pulsars, and
account for most of the SNe II that we see, but contribute little to
nucleosynthesis. For nucleosynthesis, then, the question is what happens

to more massive stars? Stars above 12 $M_\odot$ burn a succession of non-
degenerate fuels until they develop iron cores, and they lack steep
density gradients at the outside of the cores. Whether these stars
explode or collapse to black holes is not yet known (Hillebrandt 1983,
Woosley et al. 1984, Yahil 1984).

    The optical radiation emitted by an SN II consists of a thermal
continuum from a photosphere and P Cygni-type spectral lines formed by
scattering in the regions just above the photosphere, so optical observations give us information only on these layers. The continuum color
temperatures drop from about 20,000 K at maximum light to 6000 K a
month later. Analysis of the spectral lines gives velocities at the
photosphere ranging from up to 11,000 km/sec at maximum light to about
5000 km/sec after a month, and indicates that the chemical composition
is not drastically different from the solar composition (Kirshner and
Kwan 1975, Branch et al. 1981). The composition uncertainties are large,
and realistic model atmospheres will be needed to answer more detailed
questions such as the precise metals-to-hydrogen ratio, which may have
been increased by mixing of hydrogen-poor CNO-cycled material into an
envelope depleted by mass loss.

    The plateau phase of SN II light curves can be modelled with-
out specifying the details of the explosion mechanism (Arnett 1980,
Chevalier 1976, Schurmann, Arnett and Falk 1979, Weaver and Woosley 1980).
In this phase the radiation comes from a photosphere which recedes in
the Lagrangian sense as a transparency wave propagates inward due to
recombination of hydrogen in the outer layers. Litvinova and Nadyozhin
(1983) have calculated theoretical light curves for a variety of
explosion energies and radii and masses of the pre-supernova envelopes.
These quantities determine the absolute brightness, duration, and
velocity at the photosphere during the plateau phase. Adopting typical
values of -17.5 for the absolute visual magnitude of the plateau, 70 days
for its duration, and 5000 km/sec for the velocity, they derive characteristic properties of plateau SNe II: an explosion energy of 7×10^{50}
ergs, an initial radius of 540 solar radii (4×10^{13} cm), and an ejected
mass of 5.6 $M_\odot$. Lowering the velocity to 3000 km/sec would lower the
ejected mass to 2.3 $M_\odot$, while increasing the velocity to 7000 km/sec and
changing the absolute magnitude to -16.0 (i.e., changing from h=0.5 to
1.0) would increase the mass to 19 $M_\odot$, so we know only that for plateau
SNe II the ejected mass is on the order of 10 $M_\odot$.

    The fact that the progenitors of at least some SNe II undergo
extensive mass loss and that the initial mass must be significantly
greater than the sum of the ejected mass and the mass of the neutron
star is shown by recent observations at non-optical wavelengths
(Chevalier 1983). The ultraviolet spectra of two SNe II, 1979c in M100
and 1980k in NGC 6946, have been observed with the IUE satellite
(Benvenuti et al. 1982). Fransson et al. (1982) derive a nitrogen-to-
carbon abundance ratio of 7.7 and an oxygen-to-carbon ratio <4 from the
strengths of uv lines, assumed to come from a hot region above the
photosphere which is heated and ionized by radiation from the shock front.

Fransson et al. interpret the very high N/C ratio as an indication that
the pre-supernova lost much of its hydrogen-rich envelope before
CNO-cycled material was mixed from the interior. Independent evidence
for mass loss comes from observations at radio wavelengths. Three
SNe II, 1970g in M101, 1979c in M100, and 1980k in NGC 6946, have been
detected as radio emitters within the first year after maximum light
(Gottesman et al. 1972, Allen et al. 1976, Weiler et al. 1982). The
emission is interpreted as synchrotron radiation from a hot region of
interaction above the photosphere where the ejected shell encounters a
pre-existing circumstellar shell resulting from mass loss by the progenitor   red supergiant (Chevalier 1983). This would require mass loss
rates on the order of 10^{-4} $M_\odot$/year, which, if maintained over a red
supergiant lifetime on the order of 10^5 years, could result in a mass
on the order of 10 $M_\odot$ for the circumstellar shell. SN 1980k was also
detected at X-ray wavelengths (Canizares, Kriss, and Fiegelson 1982);
the flux is consistent with thermal emission from the same hot interaction region which produces the radio emission. Finally, SN 1979c and
1980k have been found to have infrared excesses (Merrill 1980, Telesco
et al. 1981, Dwek et al. 1983). The observations can best be explained
as thermal emission from pre-existing dust grains in circumstellar
shells, rather than from newly formed dust grains in the supernova
ejecta (Bode and Evans 1980, Dwek 1983). On this basis Dwek estimates
that the circumstellar shells of SN 1979c and SN 1980k had total masses
greater than 1 and 0.1 $M_\odot$, respectively.

       Given on the one hand the theoretical uncertainties regarding
which stars explode, and on the other the uncertainties in the empirical
determination of the ejected mass and the pre-supernova mass loss, the
extent to which SNe II contribute to nucleosynthesis remains unknown.
At present, the only direct evidence for heavy element production is
the apparent presence of a linear tail on the light curves of a few
SNe II (Barbon et al. 1979). Weaver and Woosley (1980) found that the
tail observed for SN 19691 in NGC 1058 could be modelled by allowing for
the delayed energy input from a few tenths of a solar mass of Ni^{56}, which
is much more than can be ejected by 8-12 $M_\odot$ stars.

       This research has been supported by NSF grant AST 7808672.

REFERENCES

Allen R.J., Goss, W.M., Ekers, R.D., and de Bruyn, A.G.: 1976, Astron.
          Astrophys. 48, p. 253.
Arnett, W.D.: 1980, Ap.J. 237, p. 541.
Arnett, W.D.: 1982, Ap.J. 253, p. 785.
Axelrod, T.S.: 1980a, Ph.D. Thesis, University of California at Santa Cruz.
Axelrod, T.S., 1980b: in "Proceedings of the Texas Workshop on Type I
          Supernovae," ed. J.C. Wheeler, Austin: University of Texas
          Press, p. 80.
Barbon, R.: 1978, Mem. Soc. Astron. Ital. 49, p. 331.
Barbon, R., Ciatti, F., and Rosino, R.: 1973, Astron. Astrophys. 25, p.241.

Barbon, R., Ciatti, F., and Rosino, R.: 1979, Astron. Astrophys. 72,
        p. 287.
Becker, R.H., Holt, S.S., Smith, B.W., White, N.E., Boldt, E.A.,
        Mushotzky, R.F., and Serlemitsos, P.J.: 1980a, Ap.J. (Lett)
        235, p. L5.
Becker, R.H., Holt, S.S., Smith, B.W., White, N.E., Boldt, E.A.,
        Mushotzky, R.F., and Serlemitsos, P.J.: 1980b, Ap.J. (Lett)
        237, p. L77.
Benvenuti, P., Sanz Fernandez de Cordoba, L., Wamsteker, W., Macchetto,
        F., Palumbo, G.C., and Panagia, N.: 1982: "An Atlas of UV
        Spectra of Supernovae," Noordwijk: ESA Scientific and
        Technical Publications Branch.
Bertola, F.: 1963, Asiago Contr. 142.
Bertola, F.: 1964a, A.J. 69, p. 236.
Bertola, F.: 1964b, Ann. d'Ap. 27, p. 319.
Bertola, F. and Arp, H.A.: 1970, P.A.S.P. 82, p. 894.
Bertola, F., Mammano, A., and Perinotto, M.: 1965, Asiago Contr. 174.
Bode, M.F., and Evans, A.: 1980, M.N.R.A.S. 193, p. 21P.
Bond, J.R.: 1984, this volume.
Branch, D.: 1972, Astr. Astrophys. 16, p. 247.
Branch, D.: 1981, Ap.J. 248, p. 1076.
Branch, D.: 1982, Ap.J. 258, p. 35.
Branch, D.: 1983, Ann. N.Y. Acad. Sci., in press.
Branch, D., and Greenstein, J.L.: 1971, Ap.J. 167, p. 89.
Branch, D., Falk, S.W., McCall, M., Rybski, P.,Uomoto, A.K., and Wills,
        B.J.: 1981, Ap.J. 244, p. 780.
Branch, D., Lacy, C.H., McCall, M.L., Sutherland, P.G., Uomoto, A.,
        Wheeler, J.C., and Wills, B.J.: 1983, Ap.J. 270, p. 123.
Canizares, C.R., Kriss, G.A., and Feigelson, E.D.: 1982, Ap.J.(Lett)
        p. L17.
Chevalier, R.A.: 1976, Ap.J. 207, p. 872.
Chevalier, R.A.: 1977, in "Supernovae," ed. D.N. Schramm, Dordrecht:
        Reidel, p. 53.
Chevalier, R.A.: 1981a, Fund. Cosm. Phys. 7, p. 1.
Chevalier, R.A.: 1981b, Ap.J. 246, p. 267.
Chevalier, R.A. : 1983, Ann. N.Y. Acad. Sci., in press.
Clark, D.H., and Stephenson, F.R.: 1977, "The Historical Supernovae,"
        Oxford: Pergamon Press.
Colgate, S.A. and White, R.H.: 1966, Ap.J. 143, p. 626.
Colgate, S.A., and McKee, C.: 1969, Ap.J. 157, p. 623.
Dwek, E.: 1983, preprint.
Dwek, E., A'Hearn M.F., Becklin, E.E., Brown, R.H., Capps, R.W.,
        Dinerstein, H.L., Gatley, I., Morrison, D., Telesco, C.M.,
        Tokunaga, A.T., Werner, M.W., and Wynn-Williams, C.G.: 1983,
        preprint.
Elliot, K.H., Blades, J.C., Zealey, W.J., and Tritton, S.: 1978,
        Nature 275, p. 198.
Fryxell, B.A., and Arnett, W.D.: 1981, Ap.J. 243, p. 994.
Fransson, C., Benvenuti, P., Gordon, C. Hempe, K., Palumbo, G.G.C.,
        Panagia, N., Reimers, D., and Wamsteker, W.: 1982, preprint.
Gottesman, S.T., Broderick, J.J., Brown, R.L., Balik, B., and Palmer,
        P.: 1972, Ap.J. 174, p. 383.

Helfand, D.J.: 1980, in "Proceedings of the Texas Workshop on Type I
        Supernovae," ed. J.C. Wheeler, Austin: University of Texas
        Press, p. 20.
Hillebrandt, W.: 1983, Ann. N.Y. Acad. Sci., in press.
Hoyle, F., and Fowler, W.A.: 1960, Ap.J. 132, p. 565.
Iben, I., Jr.: 1984, this volume.
Iben, I., Jr., and Tutukov, A.V.: 1984, preprint.
Kare, J.T., Pennypacker, C.R., Muller, R.A., Mast, T.S., Crawford, F.S.,
        and Burns, M.S.: 1982, in "Supernovae: A Survey of Current
        Research," ed. M.J. Rees and R.J. Stoneham, Dordrecht:
        Reidel, p. 325.
Kirshner, R.P., and Chevalier, R.A.: 1978, Ap.J. 219, p. 931.
Kirshner, R.P., and Kwan, J.: 1975, Ap.J. 197, p. 415.
Kirshner, R.P., Oke, J.B., Penston, M.V., and Searle, L.: 1973, Ap.J.
        185, p. 303.
Kirshner, R.P., and Oke, J.B.: 1975, Ap.J. 200, p. 574.
Litvinova, I.Y., and Nadyozhin, D.K.: 1983, Astrophys. Sp. Sci. 89, p. 89.
Maza, J., and van den Bergh, S.: 1976, Ap.J. 204, p. 519.
Merrill, K.M.: 1980, IAU Circ., No. 3444.
Meyerott, R., and Gillespie, G.H.: 1980, eds., "Supernovae Spectra," New
        York: American Institute of Physics.
Murdin, P., and Clark, D.H.: 1981, Nature 294, p. 543.
Nomoto, K.: 1981, in "Fundamental Problems in the Theory of Stellar
        Evolution," ed. D. Sugimoto, D.Q. Lamb, and D.N. Schramm,
        Dordrecht: Reidel, p. 295.
Nomoto, K.: 1982, in "Supernovae: A Survey of Current Research," ed. M.
        J. Rees and R.J. Stoneham, Dordrecht: Reidel, p. 205.
Nomoto, K.: 1984, this volume.
Oke, J.B., and Searle, L.: 1974, Ann. Rev. Astron. Astrophys. 12, p. 315.
Pskovskii, Y.P.: 1977, Sov. Astr. -A.J. 21, p. 675.
Pskovskii, Y.P.: 1978, Sov. Astr. -A.J. 22, p. 420.
Rees, M.J. and Stoneham, R.J.: 1982, eds., "Supernovae: A Survey of
        Current Research," Dordrecht:Reidel.
Rust, B.W.: 1974, Ph.D. Thesis, University of Illinois.
Sandage, A.R., and Tammann, G.A.: 1982, Ap.J. 256, p. 339.
Schurmann, S.R.: 1983, Ap.J., in press.
Schurmann, S.R., Arnett, W.D., and Falk, S.W.: 1979, Ap.J. 230, p.11.
Shull, J.M.: 1982, Ap.J. 262, p. 308.
Smith, H.A.: 1981, A.J. 86, p. 998.
Sutherland, P., and Wheeler, J.C.: 1984, preprint.
Tammann, G.A.: 1982, in "Supernovae: A Survey of Current Research," ed.
        M.J. Rees and R.J. Stoneham, Dordrecht:Reidel, p. 371.
Telesco, C., Becklin, E., Koehler, K., and Gatley, I.: 1981, IAU
        Circ., No. 3613.
Trimble, V.: 1982, Rev. Mod. Phys. 54, p. 1183.
Trimble, V.: 1983, Rev. Mod. Phys. 55, p. 511.
Twarog, B.A., and Wheeler, J.C.: 1982, Ap.J. 261, p. 636.
Weaver, T.A. and Woosley, S.E.: 1980, Ann. N.Y. Acad. Sci. 336, p. 335.
Wheeler, J.C.: 1980, ed., "Proceedings of the Texas Workshop on Type
        I Supernovae," Austin: University of Texas Press.
Wheeler, J.C.: 1982, in "Supernovae, A Survey of Current Research," ed.
        M.F. Rees and R.J. Stoneham, Dordrecht:Reidel, p. 167.

Wheeler, J.C., Lecar, M., and McKee, C.F.: 1975, Ap.J. 200, p. 145.
Wiedemann, V., and Koester, D.: 1983, Astron. Astrophys. 121, p. 77.
Wieler, K.W., Dramek, R.A., van der Hulst, J.M., and Panagia, N.: 1982,
        in "Supernovae: A Survey of Current Research," ed. M.R. Rees
        and R.J. Stoneham, Dordrecht:Reidel, p. 281.
Woosley, S.E., Axelrod, T.S., and Weaver, T.A.: 1984, this volume.
Woosley, S.E., Weaver, T.A., and Taam, R.E.: 1980, in "Proceedings of
        the Texas Workshop on Type I Supernovae," ed. J.C. Wheeler,
        Austin: University of Texas Press, p. 96.
Wu, C., Leventhal, M., Sarazin, C.L., and Gull, T.R.: 1983, preprint.
Yahill, A.: 1984, this volume.
Zwicky, F.: 1964, Ap.J. 139, p. 514.
Zwicky, F., and Karpowicz, M.: 1965, A.J. 70, p. 564.
Zwicky, F.: 1965, in "Stellar Structure," ed. L.H. Aller and D.B.
        McLaughlin, Chicago: University of Chicago Press, p. 367.

DISCUSSION

YAHILL: Are Type II supernovae concentrated to the inner or
to the outer parts of spirals?
BRANCH: Studies by McCarthy, Tammann, and Barbon, Ciatti,
and Rosino have shown that the radial distribution of Type II supernovae
is similar to the distribution of luminous matter.

CASSE: In your calculations of the spectra of supernovae,
what do you assume about the radial dependence of the expansion velocity
and the matter density?
BRANCH: The velocity is assumed to be proportional to the
radius. This should be an adequate approximation for the phases of
interest, when the ejected matter is very nearly in free expansion.
For Type I supernovae during the first month after maximum light, while
the photosphere is expanding, the density is assumed to go as r^{-7}. This
is an approximation to the density law given by hydrodynamical calculations for the outer layers of exploded white dwarfs. During the next
year, when a photosphere apparently still exists but recedes deep into
the matter distribution, a simple power-law would be too unrealistic,
so a density law based on one of Sutherland and Wheeler's calculations
of exploded white dwarfs is used. For Type II's, only the early photospheric phases have been studied, and various power-laws were explored.
An r^{-7} law was used for the spectra I showed here, but the value of
the exponent is not well determined.

RENZINI: How many spectra of Type I supernovae are available
for investigation of the homogeneity?
BRANCH: There are only a few Type I's for which long sequences
of spectra are available (SN 1937c in IC 4182, 1972e in NGC 5253, SN
1981b in NGC 4536); their spectra seem to have been practically identical. The total number with spectra is about 100, and almost all of
them are consistent with the spectra of the well-observed ones. There
is evidence for a real spread of a few thousand km/sec in the velocity

at the photosphere near maximum light, but we need low-noise spectra of
more supernovae to be sure about that. Then there is the small number
of supernovae I mentioned in the talk which definitely had spectral
peculiarities.

RENZINI: For Type I supernovae, how much mass is above the
photosphere at each time, and when is the Ni-Co-Fe core exposed?
BRANCH: At maximum light the mass above the photosphere is on
the order of 0.1 $M_\odot$. A month later, when the tail of the light curve
begins, the photosphere has reached its maximum extent and the mass is
perhaps 0.4 $M_\odot$. The core may be exposed near this time. Of course the
definition of the core may be a little hazy if there is much mixing
across the initial composition interface.

COWAN: What can you say about the abundances from the optical
spectra of Type II supernovae?
BRANCH: So far, only that they are consistent with solar
abundances. But the metals-to-hydrogen ratio could be different from
solar by a factor of several and we wouldn't know it. The conclusions
regarding the CNO abundances come from the ultraviolet spectrum.

RENZINI: To assess the nucleosynthetic contribution of Type II
supernovae it is critically important to understand the nature of the
exponential decline following the plateau. If it is due to Ni-Co
decay, Type II spectra at this phase should be similar to those of Type
I. On the other hand, if the exponential decline is powered by the
interaction between the supernova shell and the circumstellar shell,
then the spectra should be different. I s there any observational
evidence in this repsect?
BRANCH: There are only a few noisy, low-resolution spectra at
this phase, and they haven't been analyzed in any detail. They don't
look like the late-time spectra of Type I's, but even if the light curve
is powered by Ni-Co decay, the outer hydrogen-rich layers might absorb
and thermalize the radiation from the core, and emit the spectrum that
we see.

IBEN: Wolf-Rayet stars are thought to be fairly massive (>40
$M_\odot$ if single, >20 $M_\odot$ if in binaries) which have lost their hydrogen-rich
envelopes. Their frequency of formation is not negligible relative to
the rate of formation of all stars of initial mass >20 $M_\odot$. Since we
believe that the cores of all of these stars evolve comparably, ultimately
collapsing to form neutron stars and ejecting whatever residual envelopes
they still have, there should be quite a few neutron-star producing
events distinguished only by the fact that some eject hydrogen-rich matter
and some don't. The question then is, where among the 500 known supernovae are those which had Wolf-Rayet stars as progenitors?
BRANCH: Wolf-Rayet stars can't be Type I's because they die
in the spiral arms and they can't be Type II's because they don't have
hydrogen, so the answer much be that exploding Wolf-Rayet's haven't yet
been seen. Maybe stars of initial mass >20 $M_\odot$ don't explode at all, but
even if they do, there would be a severe selection effect against

exploding Wolf-Rayets because of their faint optical luminosities.
Type II's achieve a high luminosity by the diffusive release of shock
energy deposited in a supergiant envelope, and Type I's by the release
of energy deposited slowly by radioactivity. The compact envelopes of
Wolf-Rayet stars will cool by expansion before becoming large enough
to release much optical radiation, and since their cores collapse
relatively little Ni^{56} is ejected. Maybe the automated searches will
start to turn up some of these subluminous supernovae.

ABUNDANCES IN SUPERNOVA REMNANTS

I.J. Danziger
European Southern Observatory, Garching bei München, FRG

GENERAL CONSIDERATIONS.

Because of the nature of this workshop I will attempt to summarize what direct observations of supernova remnants tell us about these possible sites of nucleosynthesis and their contribution to metal enrichment in our galaxy. Very little of what I describe here is new but some of it is still in press and therefore relatively unknown. Many of the comments, particularly those on the optical properties, have been made in other contributions by Danziger and Leibowitz (1983) and Danziger (1983).

One convenient way to summarize the information obtained from SNR relevant to element production, is to present a table of SNR together with velocity information for characteristic filaments, the spectro-scopic properties and possibly some estimates of the age particularly when they are not known to originate from historical SN. If one does this, one immediately notices a couple of general characteristics. Almost without exception apparent oxygen overabundances occur only in high velocity filaments of what must be presumed to be young remnants (< 2000 years). Examples of this type include Cas A and G292.0+1.8 in our galaxy, N 132 D and 0540-69.3 in the LMC, 0102-72.3 in the SMC, and the object in the external galaxy NGC 4449. In some but not all of these neon may also be overabundant.

Enhancements of nitrogen manifest themselves in a different way. For example, one sees enhanced nitrogen almost exclusively in low velo-city filaments even though those filaments can belong to obviously young SNR such as Cas A. It has been noted by Dopita et al. (1983) that in the vast majority of cases, where one believes one is seeing an enhanced nitrogen abundance, this enhancement is that expected from the known general galactic abundance gradient and the position of the SNR in the galaxy. Since there have been reports in the literature (Ruiz 1981, 1983; Leibowitz and Danziger 1983) of variations of nitrogen line strengths from filament to filament in such objects as RCW 86, RCW 103 and Kepler, and since [NII]/Hα line ratios are supposed to be insen-

35

C. Chiosi and A. Renzini (eds.), Stellar Nucleosynthesis, 35–42.
© *1984 by D. Reidel Publishing Company.*

sitive to physical conditions other than abundance, the situation may
not be as clearcut as it could be.

In any case Puppis A is one clear case where enhancements of nitrogen are pronounced (20-40 × solar) and must originate in the presupernova star or the remnants ejected by the explosion.

These global properties have led of course to the concept of pre-
supernova mass loss of an evolved star giving nitrogen enhanced material
(processed by the CNO cycle) with low velocities shocked by the ejection
of high velocity material, some of it enriched in oxygen as a result of
helium burning reactions in the deeper interior of an evolved star of
mass > 10-12 $M_\odot$.

Some of the remnants deserve more detailed discussion because of
abundance effects peculiar to them.

PARTICULAR PROBLEMS

The Crab.

There are some conflicting results for the heavier elements N, O,
Ne and S which, until the recent work of Péquignot (1983), were computed
to be 2-3 lower than solar values. Now Péquignot with his own photoionized models, has computed oxygen abundances more typical of a Population I star, which makes it easier to accept that the progenitor star
could have had a mass of 10-15 $M_\odot$. An even more intriguing result is the
report by Henry (this workshop) of an enhanced Ni/Fe ratio by a
factor ~ 5. If this apparently enhanced nickel has been produced as a
result of the explosion, it remains to be shown theoretically that such
ratios of Ni/Fe are feasibly produced under the expected conditions.

The enhanced abundance of helium in the Crab filaments along with
its variability from filament to filament has been known for some time.
Clark et al. (1983) on the basis of a new dynamical mapping of the Crab
at optical wavelengths propose a model consisting among other things of
a thick shell with a consistently higher helium abundance along the
brighter inner surface compared to the outer fainter surface. This has
not yet been modelled, but appears to raise interesting questions concerning our understanding of the propagation of SN shock waves as well
as the structure of the exploding star.

Cas A.

Along with the apparent oxygen overabundance in the high velocity
knots Chevalier and Kirshner (1979) have reported enhanced abundances of
calcium, argon and sulphur. Since these three elements can be produced
by oxygen-burning reactions at high temperatures in the interiors of
stars with masses in the range 15-25 $M_\odot$, this together with the oxygen
overabundance has been taken to be a confirmation of the type of events

leading up to a Type II SN and a resulting SNR as we see with Cas A. Has nature been so kind as to lay out so much direct evidence before us? No one has yet produced an alternative explanation in terms of excitation effects though it should not be overlooked that the same CaII lines observed in Cas A are seen enhanced in much older SNR. Could this be calcium released from interstellar grains evaporated by the heat associated with the passage of the SN blast wave?

Puppis A.

This SNR, the Crab and Cas A are the only ones which show an enhanced abundance of helium. Although the overabundance of helium in the observed filaments is not very great (1.5 × solar), if it is as high as this throughout the total swept up volume enclosed by the shell, it represents an enormous amount of helium. The same could be said of nitrogen. However, since the [NII]/Hα ratio is quite variable among the observed filaments, this may help avoid the conclusion that the total amounts of nitrogen (and helium) are inconceivably large. Certainly the variability suggests that these abundance effects are very local effects (due to mass loss?) rather than extending over large volumes of the ISM. Since both [NI] and [NII] lines are enhanced in Puppis A this strengthens the case for an abundance effect rather than an excitation effect. One cannot help wondering whether the extremely nitrogen-enriched nebula recently reported by Ruiz (1983) is not somehow related to the Puppis A phenomenon.

Puppis A is something of an enigma in other respects also. Milne, Goss and Danziger (1983) have shown that there is little correlation between the detailed radio and visible optical morphology. This is more characteristic of young remnants rather than middle-aged or old remnants.

LMC 132 D.

This oxygen-rich remnant has at various times been compared to Cas A. It does however differ in the following ways. There is no evidence in N 132 D for the low velocity nitrogen-enriched filaments seen in Cas A. The morphological structure of high velocity filaments in N 132 D also seems different in that they extend continuously over much larger distances than in Cas A.

Summarizing the evidence for helium production we can say that enhanced helium abundances appear to have been detected in only 3 objects, the Crab, Cas A and Puppis A. They are such a divergent group of objects that it is difficult to identify a pattern or other common properties.

Ring Structures

This discussion of the oxygen-rich high velocity filaments in SNR raises the more interesting question of expanding ring-structures.

Bodenheimer and Woosley (1983) have recently proposed a model for an explosion in a star of 25 M$_\odot$ which is rotating. The subsequent dynamical evolution of material with considerable angular momentum gives rise to the formation of a ring or annulus expanding outwards in the equatorial plane of the star.

At first sight the oxygen-rich SNR seem to be prime candidates for testing these ideas. Indeed evidence for ring structures (as opposed to shell structures) have been given for N 132 D by Lasker (1980), for Cas A by Markert et al. (1981) and more recently for SMC 0102-72.3 by Tuohy and Dopita (1983) and Danziger (1983).

Another example is G292.0+1.8, which, on the basis of its X-ray morphological structure, has been quoted by Tuohy et al. (1982) as a further example of an expanding ring. However Braun et al. (1983) have mapped the velocity structure of the optical oxygen-rich filaments. They find that it does not follow the morphology of the X-ray emitting gas, and even more importantly, the dynamical structure is much more con-sistent with an expanding shell (rather than a ring) accompanied by a break-out in some specified direction.

In spite of contradictory evidence in some cases this does seem like an interesting area to follow the dynamical evolution of young SNR. Once again observers are bedevilled by the problem that one is seeing a 3-dimensional problem projected in 2 dimensions on the sky.

Iron

Because one is eternally interested in the problem if and where iron is produced in SN, it may seem rewarding to look for evidence in SNR. This has been done by Danziger and Leibowitz (1983) for a selection of SNR for which [FeII] line strengths were published or in press. At first the results promised to be exciting because both the young SNR Kepler and the quasi-stationary flocculi of Cas A have quite strong [FeII] lines. Further investigation however reveals a significant positive correlation between [FeII] line strength and density as deter-mined by the [SII] 6717/6731 ratio. Old SNR with dense filaments fit this correlation as much as young dense SNR. The lesson from this seems to be that one must carefully model such filaments before drawing any conclusions about iron abundances. So far to my knowlegde this has not been done for a reasonable range of densities.

Average Properties

As a means of summarizing what spectroscopy of SNR can tell us about the average properties of the interstellar medium in galactic systems, one can look at average values of line ratios such as [OIII] 5007/Hβ, [NII]/Hα, [SII] 6717,31/Hα. If one does this for all SNR excluding the special cases of oxygen-rich and hydrogen Balmer-line emitting remnants, one sees these ratios decrease going from more massive systems such as our Galaxy and M31, through M33 to the LMC and

SMC. This trend in implied metallicity has of course been known from other astronomical objects for some time. What is important to this meeting is the very small dispersion in the [NII]/Hα ratio in the LMC implying, since this ratio is sensitive mostly to N abundance, and the SNR are scattered throughout the LMC, that there is a uniformly low abundance of N throughout the LMC, i.e. the interstellar material is well mixed.

X-Ray Abundances

Most of the recent abundance results in SNR coming from the Einstein Observatory have been summarized in papers by Canizares et al. (1983) and Holt (1983). While there is still an on-going debate about the merits of equilibrium and non-equilibrium modelling, the following results, while quantitatively subject to some change, provide a qualitative idea of what may be happening in the hot gas. Unfortunately there is not a great deal of overlap of information obtained from the optical and X-ray regions. At least where overlap does occur there are encouraging signs of agreement.

In Tycho and Kepler the Si/Fe ratios are ~ 10 and 5 respectively, while the iron seems to be relatively normal. This is known to be something of a disappointment from the theoretical viewpoint since Type I SN are supposed to produce copious quantities of iron-group elements. It has been suggested that the interior may now be so cool that any possible X-ray iron lines are below the limit of detection. Recently however Wu et al. (1983) claim to have detected broadened FeII lines in the UV spectrum of a star lying near the projected center of the remnant of SN 1006. Since these lines have high velocity, it is claimed that they are associated with the ejecta from the SN.

In Cas A it is suggested that the Si/H ratio is about 1.5 × solar. Since it is also suggested that S/Si ~ 2 times, Ar/Si ~ 4 times, Ca/Si ~ 2 times and Mg/Si ~ 0.1 times solar, there is here the possibility of at least qualitative agreement with results for S, Ar and Ca from the optical range. Any statement concerning oxygen seems somewhat more circumspect.

N 132 D appears to have an O/Fe ratio significantly greater than solar values. This could therefore be consistent with the known over-abundance of oxygen from the optical region.

Puppis A with non-equilibrium modelling gives oxygen and neon over-abundant relative to iron. Although it is not an accurate determination there was some suggestion from the optical region that oxygen was over-abundant. Unfortunately X-ray information for nitrogen is unavailable. If this results from a SN in a 20-25 $M_\odot$ star as has been suggested, it implies that it could be an older version of Cas A.

It seems to me that in this area important work for the future will involve the theoretical workers convincing not only themselves but all

other interested parties that reliable abundances have been determined
from X-ray spectra. This is a "sine qua non" for pursuing nucleosynthetic models of exploding stars that bear some relation to reality.

REFERENCES

Bodenheimer, P.B., Woosley, S.E., 1983. Astrophys. J., 269, 281.
Braun, R., Goss, W.M., Danziger, I.J., 1983. IAU Symposium 101,
      Supernova Remnants and their X-ray Emission, Reidel, Dordrecht.
Canizares, C.R., Winkler, P.F., Markert, T.H., Berg, C., 1983. IAU
      Symposium 101, Supernova Remnants and their X-ray Emission, Reidel,
      Dordrecht.
Chevalier, R., Kirshner, R.P., 1979. Astrophys. J., 233, 154.
Clark, D.H., Murdin, P., Gilmozzi, R., Danziger, I.J., Furr, A.W., 1983.
      Mon. Not. Roy. Astr. Soc., 204, 415.
Danziger, I.J., 1983. IAU Symposium 101, Supernova Remnants and their
      X-ray Emission, Reidel, Dordrecht.
Danziger, I.J., Leibowitz, E. 1983. ESO Workshop on Primordial Helium,
      p. 249.
Dopita, M.A., Binette, L., D'Odorico, S, Benvenuti, P., 1983. Astrophys.
      J., in press.
Holt, S.S., 1983. IAU Symposium 101, Supernova Remnants and their X-ray
      Emission, Reidel, Dordrecht.
Lasker, B.M., 1980. Astrophys. J., 237, 765.
Leibowitz, E., Danziger, I.J., 1983. Mon. Not. Roy. astr. Soc., 204,
      273.
Markert, T.H., Canizares, C.R., Clark, G.W., Winkler, P.F., 1983.
      Astrophys. J., 268, 134.
Milne, D.K., Goss, W.M., Danziger, I.J., 1983. Mon. Not. Roy. astr.
      Soc., 204, 273.
Péquignot, D., 1983. IAU Symposium 101, Supernova Remnants and Their
      X-ray Emission, Reidel, Dordrecht.
Ruiz, M.T., 1981. Astrophys. J., 243, 814.
Ruiz, M.T., 1983. Astrophys. J. Lett., 268, L103.
Tuohy, I.R., Clark, D.H., Burton, W.M., 1982. Astrophys. J. Lett.,
      260, L65.
Tuohy, I.R., Dopita, M.A., 1983. Astrophys. J. Lett., 268, L11.
Wu, C.-C., Leventhal, M., Sarazin, C.L., Gull, T.R., 1983. Astrophys. J.
      Lett., 269, L5.

DISCUSSION

EDMUNDS: How confident are you about the relative effects of different
ionization and excitation conditions in these SN remnants, i.e., shock
vs. synchrotron vs. secondary radiation off the shock? Hard power law
photoionization could possibly enhance [NII], and the supposed nitrogen
variations could be variation of excitation?

DANZIGER: It is one thing to speculate on various excitation mechanisms, another to prove their feasibility. One should remember in all of

this that, in spite of the theory of propagation of radiating shock
waves having advanced considerably in the past 10 years, there are still
some obvious outstanding problems in comparison between observation and
theory, particularly in old remnants where the invocation of abundance
effects seems to be the least desirable way out of the dilemma.

GALLAGHER: I just wanted to reinforce your statements concerning the
difficulties in interpreting shock spectra. GK Per MO1 is a typical
fast classical nova which obviously contained hydrogen and helium in its
spectrum during outburst. However, digital spectrophotometry which
R. Williams (Steward Observatory) and I have obtained of the nebula
surrounding the old nova GK Per show a nearly pure forbidden line
emission spectrum consisting of [OIII], [NeIII], [NII], [SII] and [OI].
Thus at first glance the spectrum would seem to imply that the metallicity in the ejecta is very highly enhanced, but, of course, this would
be inconsistent with the properties of the system during outburst. A
poor probable explanation is that the strong forbidden lines primarily
reflect excitation conditions, although modest overabundance of Ne and O
and somewhat larger enhancements of N are not excluded. High detection
temperatures estimated from [NII] and [OIII] are consistent with shock
excitation, which is likely to be a result of a reverse shock in the
ejecta.

GALLAGHER: (In response to Sylvia Torres-Peimbert's question concerning
whether line ratios might reflect abundances in the ejecta or swept up
interstellar matter and whether the metal abundances could be roughly
estimated.) Let me clarify the situation regarding the nebular spectrum
of GK Per. First, we do not yet have a detailed model so these comments
are preliminary in nature. Since the nova ejecta has an expansion
velocity of $\sim$ 1000 km s^{-1}, any interstellar material which has been
swept up and shocked will be very hot and will not produce optical
emission. It will, however, decelerate the outer part of the ejected
matter which can lead to an internal shock as decelerated gas is over-
taken from behind. This reverse shock can have sufficiently low
velocity as to produce gas which cools strongly via optical forbidden
line emission. Thus in my opinion the line ratios should in fact
reflect physical conditions and abundances within the nova ejecta.
Unfortunately, since no recombination lines are measurable on our
digital spectra (H Balmer emission is weakly present on uncalibrated,
long slit photographic image tube spectra, as is [NeV]), it is difficult
to make accurate estimates of abundances of metals vs. H + He in the
ejecta.

RENZINI: 1) Do you think that it could be possible to interpret some of
the N-rich remnants as WN star remnants which collapsed to black holes
rather than exploded? 2) Concerning the Crab, Ken'ichi Nomoto has
rather convincingly suggested that the Crab precursor was a
10 $M_\odot$ star. I remember that in this context you raised the problem of
the large distance of the Crab from the galactic plane. I would like to
hear Ken'ichi arguing about that.

NOMOTO: Since the Crab pulsar is a high velocity star, I think that the Crab's progenitor used to be one of the component stars of a wide binary system and got a high velocity as a result of the sling shot effect at the supernova explosion. In fact, Ostriker et al. obtained a progenitor's mass of $\sim$ 10 $M_\odot$ based on this scenario. So I am not so much worried about the Population problem.

DANZIGER: Concerning point (1) I do not have a point of view except to ask where does the non-thermal radiation come from, and to wonder about the statistics of WN stars. Concerning point (2) we (Clark et al., 1983) found a mean velocity of the filaments of the Crab = -20 km/s, which places little constraint on its present position w.r.t. the galactic plane.

HENRY: Concerning the outer shell of the Crab Nebula, whose filaments you claim have less helium than those of the thick, inner shell; have you been able to establish the He/H ratio for those filaments?

DANZIGER: No. Not yet quantitatively.

WOOSLEY: With regard to the iron abundance in supernova remnants, I would comment, first, as I believe most of us recognize, that not all supernovae need produce large abundances of iron. Type II supernovae, in particular, may or may not eject iron, depending on details of the core bounce mechanism. Type I supernovae, on the other hand, must produce copious iron. Spectral analysis during the first year by Branch, Kirshner, Meyerott, Axelrod and others show emission comprised almost entirely of iron lines. Furthermore, the production of at least 0.3 $M_\odot$ of ^{56}Ni is required to understand the regular exponential tails in the light curves of these events. Therefore, there must be large amounts of iron in SNI remnants. The question then is why it is not readily visible. Possible explanations involve the condensation of the iron into grains or an ionization structure (non LTE?) that is not adequately reflected in the relatively simple abundance analyses that have been carried out to date.

CASSE: Are you inclined to think that grains are totally destroyed by the shock wave; can you exclude a differential depletion of elements in the shocked region?

DANZIGER: Our knowledge of grain structure is sufficiently limited that one cannot make strong claims one way or the other.

BRAUN: In relation to the suggestion that Wolf-Rayet stars may lead to SNR-type remnants, how much energy is involved in the mass loss of these objects?

RENZINI: WR winds run at several thousand km/s, and their total energy is of the order of $\sim 10^{51}$ erg, quite comparable to that of a SN.

THE NICKEL AND IRON ABUNDANCES IN THE CRAB NEBULA FILAMENTS

Richard B. C. Henry
University of Delaware

ABSTRACT

Recent near-infrared observations of the Crab Nebula filaments indicate that Ni is overabundant by an order of magnitude, while Fe is underabundant by roughly 3 times, both by number relative to their solar levels. This suggests that Fe is depleted by grain formation. A lower limit on the mass of Fe in the filaments is 0.01 $M_\odot$.

1. INTRODUCTION

The Crab Nebula, the remnant of SN1054, is a good target for chemical abundance studies of supernova debris. Since the Crab is both young and is positioned more than one scale height out of the galactic plane, the filaments are presumably undiluted with swept-up interstellar material. Recently, Henry, MacAlpine, and Kirshner (1983; hereafter HMK), in reporting results of a near-infrared study of 14 filaments, have determined the Ni and Fe abundances for an "average" filament and found that, by number, Ni is an order of magnitude overabundant while Fe is underabundant by a factor of two, both with respect to solar levels. Their results also suggest that Ni/Fe in the Crab filaments exceeds the solar value (and that predicted by e-process calculations) by more than 30 times.

My purpose in this paper is to re-examine in greater detail the question of the abundances of Ni and Fe and the Ni/Fe ratio in the Crab filaments. Using HMK's data for 12 positions, I have calculated the abundance of Ni and Fe at each location. To test the abundance determination method, I have performed similar calculations using published data for several Herbig-Haro objects and the Orion Nebula, objects containing primarily unprocessed material.

C. Chiosi and A. Renzini (eds.), Stellar Nucleosynthesis, 43–47.

2.   THE ABUNDANCE DETERMINATION METHOD

The abundances of Ni and Fe at 12 positions in the Crab were deter-mined using the emission lines of [Ni II] $\lambda7378$ and [Fe II] $\lambda8617$. The line strengths were taken from the spectrophotometric data of HMK, which cover the region from 6500-10200 Å. Since HMK did not observe Hβ, the Ni and Fe abundances were determined relative to that of S, using [S II] $\lambda6724$ and [S III] $\lambda9532$, lines within their spectral range. Since the relative S abundance by number was found by Henry and MacAlpine (1982; hereafter HM) to be roughly solar, it serves as a good element against which to normalize Ni and Fe.

The assumptions made for deriving the Ni and Fe abundances are:
1.   [Ni II] $\lambda7378$, [Fe II] $\lambda8617$, [S II] $\lambda6724$, and [S III] $\lambda9532$ originate exclusively from the H^{o} region of the filaments; and
2.   In the H^{o} region, all Ni and Fe exists as Ni^{+} and Fe^{+}, respec-tively; and all S exists either as S^{+} or S^{+2}. Both of these assumptions are consistent with ionization potential considerations and/or photoioni-zation model results for the Crab Nebula.

I present here only the formulae used for deriving the abundance ratio Ni/S, but those for Fe/S are analogous. From the above assump-tions,

$$\frac{N(Ni)}{N(S)} = \left[\frac{N(S^{+})}{N(Ni^{+})} + \frac{N(S^{+2})}{N(Ni^{+})}\right]^{-1}. \tag{1}$$

The inverses of the two fractions on the right in eq. (1) were calcu-lated as follows:

$$\frac{N(Ni^{+})}{N(S^{+})} = \frac{I(\lambda7378)}{I(\lambda6724)} \cdot \frac{\chi\, A_{6716}\, E_{6716} + \chi'\, A_{6731}\, E_{6731}}{Ne\, q(Ni^{+})\, (A_{7378}/A_{T})E_{7378}} \tag{2}$$

and

$$\frac{N(Ni^{+})}{N(S^{+2})} = \frac{I(\lambda7378)}{I(\lambda9532)} \cdot \frac{q(S^{+2})\, (A_{9532}/A_{T}')\, E_{9532}}{q(Ni^{+})\, (A_{7378}/A_{T})\, E_{7378}}. \tag{3}$$

The first fraction on the right in eqs. (2) and (3) is the intensity ratio, while the $\underline{A}$'s, $\underline{E}$'s, and $\underline{q}$'s are the transition rates, transition energies, and collisional excitation coefficients for the relevant emiss-ion lines, respectively. χ and χ' in eq. (2) are the upper level popula-tions for the $\lambda6716$ and $\lambda6731$ transitions, respectively, for the [S II] doublet, calculated for a 5-level atom. All other emission lines were assumed to be the result of populating the upper level by collisions from the ground state only. Finally, A_{T} and A_{T}' are the sums of all transition rates from the upper level of the relevant line, and N_{e} is the electron density. Atomic data for Ni and Fe were taken from Nussbaumer and Storey (1980; 1982), while those for other ions came from standard sources. The T_e and N_e for each position were taken from Fesen and Kirshner (1982), Davidson et al. (1982), or determined from the data of HMK.

3. DISCUSSION

The results for Ni/S, Fe/S, and Ni/Fe, all normalized to their solar values (Allen 1973), are listed by position in Table 1. The most striking result is that at all positions Ni/Fe is at least an order of magnitude greater than the solar value, with an average of 43.1 ± 18.0. (The standard deviation is roughly that expected from the observational errors.) In addition, the relative abundance of Ni appears well above solar at all positions, while that of Fe is below its solar value. The large range in Ni/S and Fe/S probably represents a real variation among the filaments of the Ni and Fe abundances, especially since the two ratios are well correlated, with a coefficient of 0.78.

To see if these unexpected results could be due to the abundance determination methods employed, I applied them to numerous Herbig-Haro objects and the Orion Nebula, objects consisting of unprocessed material. (The above procedure was altered slightly to account for the different physical conditions in these objects with respect to the Crab filaments.) For the H-H objects, strengths for [Ni II] $\lambda 7378$, [Fe II] $\lambda 8617$, and [S II] $\lambda 6716$, along with values for T_e and N_e, were taken from Brugel, Böhm, and Mannery (1981); for Orion, similar data were taken from Grandi (1975). (The former were corrected for the blending of P14 of hydrogen with [Fe II].) Although Brugel et al. tentatively identified the line near $\lambda 7378$ in H-H objects as Fe II $\lambda 7376$, it is undoubtedly [Ni II] $\lambda 7378$, since the latter is observed in numerous other objects (see Nussbaumer and Storey 1982) while the Fe II line is not.

The results for Ni/S and Fe/S in H-H objects and Orion are presented in Table 2. Several points are worth noting. First, in all cases Ni/S is found to be very close to its solar value, suggesting that the method used to determine this ratio is satisfactory. Second, Fe/S seems to be systematically low among these objects, having an average value for the H-H objects roughly equal to that in the Crab, while for Orion the abundance of Fe appears significantly lower. However, the Fe abundances

TABLE 1

Ni and Fe Abundances* in the Crab Nebula Filaments

| | Filament | | | | | | | | | | | |
	FK1	FK5	FK6	FK7	FK8	FK9	FK10	M2	HMK1	HMK2	HMK3	HMK4
Ni/S	2.8	11.1	19.0	5.7	13.9	5.4	8.4	4.9	5.5	3.8	23.4	17.3
Fe/S	.04	.64	.33	.14	.23	.41	.17	.09	---	.07	1.0	.51
Ni/Fe	70.0	17.3	57.6	40.7	60.4	13.2	49.4	54.4	---	54.3	23.4	33.9

*by number normalized to solar values (Allen 1973)

TABLE 2

Ni and Fe Abundances* in Herbig-Haro Objects and the Orion Nebula

				Object				
	H-H1(NW)	H-H2A	H-H2G	H-H2H	H-H3	H-H11	H-H32	Orion
Ni/S	1.4	0.6	1.2	1.6	2.0	1.9	0.7	1.2
Fe/S	.38	.23	.36	.63	---	---	.22	.07
Ni/Fe	3.7	2.6	3.3	2.5	---	---	3.2	17.1

*by number normalized to solar values (Allen 1973)

obtained here compare favorably to those determined by Brugel et al. for H-H objects and Cosmovici et al. (1980) for Orion, both of whom used techniques and atomic data which differed from those employed here. Finally, the ratio of Ni/Fe is above the solar value everywhere, and, for Orion, it is similar to the values seen in the Crab.

The results suggest, therefore, that Ni is indeed overabundant by an order of magnitude relative to its solar value in the Crab filaments, but Fe is underabundant in all of the objects tested. On the other hand, how is the large value of Ni/Fe to be interpreted? While it is possible that neglected physical processes or errors in the atomic data could in part explain the calculated Ni/Fe ratio, it is unlikely that they could account for its order of magnitude enhancement observed both in the Crab and Orion. Furthermore, the high values for Ni/Fe derived here for un-processed gases would seem to indicate that an unusual nucleosynthetic process is not responsible.

Perhaps a more tenable explanation is that Fe forms grains more readily than Ni, an interpretation supported by the greater-than-solar values of Ni/Fe found for the H-H objects and Orion. Thus, the relative abundances of the Fe peak elements in the Crab filaments may actually be very high, as indicated by the Ni abundance, yet much of the Fe is in the solid state and is unobservable directly. In fact a high affinity of Fe for the solid state might also explain the apparent absence of Fe in Type I remnants, despite the fact that models of Type I supernovae predict that large amounts of Fe will be ejected, and emission lines of Fe are seen in spectra of these supernovae shortly after they erupt.

If Ni is really a better tracer for Fe peak elements, what does the large Ni/S ratio in the Crab filaments imply about the mass fraction of Fe? To attempt to establish a rough lower limit to the Fe peak mass fraction based on available data, I consider position FK1, which appears to have the least amounts of Ni and Fe of those filaments observed. (More observations will be necessary, of course, to establish if there are many filaments which have lower Ni and Fe abundances.) I assume a relative number abundance for He of 1.0, consistent with the strength of He I $\lambda5876$ at this position (see Fesen and Kirshner 1982 and HM). The mass fraction of Ni is then $\sim6 \times 10^{-4}$ or roughly 5 times the solar level. Furthermore, if the true Ni/Fe by mass (including the solid Fe) in the filaments is actually equal to the solar ratio of .05, and the lower limit on the total mass of the filaments is 1.2 $M_\odot$ (see HM), then $M_{Fe} > 0.01 M_\odot$. This limit is important for determining the amount of core material ejected during the supernova event which formed the Crab Nebula.

This research was supported by NSF Grant AST 81-15095.

REFERENCES

Allen, C.W.: 1973, Astrophysical Quantities, (Athlone), p. 31.
Brugel, E.W., Böhm, K.H., and Mannery, E.: 1981, Ap. J. Supp., 47, pp. 117-138.
Cosmovici, C.B., Strafella, F., and Dirscherl, R.: 1980, Ap. J., 236, pp. 498-507.
Davidson, K., et al.: 1982, Ap. J., 253, pp. 696-706.
Fesen, R.A., and Kirshner, R.P.: 1982, Ap. J., 258, pp. 1-10.
Grandi, S.A.: 1975, Ap. J., 199, pp. L43-L46.
Henry, R.B.C., and MacAlpine, G.M.: 1982, Ap. J., 258, pp. 11-21.
Henry, R.B.C., MacAlpine, G.M. and Kirshner, R.P.: 1983, Ap. J., in press.
Nussbaumer, H. and Storey, P.J.: 1980, Astron. Astrophys., 89, pp. 308-313.
_____.: 1982, Astron. Astrophys., 110, pp. 295-299.

DISCUSSION

Nomoto: Recently, the collapse of a 9 $M_\odot$ star which I proposed as the Crab's progenitor has been calculated by Hillebrandt, Nomoto, and Wolff (1983). We have found that a shock formed at the core bounce is strong enough to eject a relatively small amount of iron peak elements (with probably a negligible amount of S, Si, etc.). Although some Ni isotopes are expected to be ejected because of neutronization processes, their amounts must be smaller than Fe.

Cassé: The Ni/Fe ratio produced in SN II models depends sensitively on the mass cut (the frontier between the collapsing zone and the ejected zone). Thus, the observed Ni/Fe ratio in the Crab Nebula should allow the determination of the mass cut of the Crab supernova.

A STATISTICAL COMPARISON OF GALACTIC SNRs AND GALACTIC GIANT HII REGIONS

Z. W. Li, J. C. Wheeler, and F. N. Bash
Department of Astronomy, University of Texas, Austin

Abstract

In order to examine the quality of the evidence, we test the hypothesis
that all supernovae in the Galaxy lie in narrow spiral arms. The
distributions in longitude of 138 SNRs and 100 giant HII regions are
presented, and compared with a simple Monte Carlo model in which 140
supernovae are distributed either along arms according to a density
wave model, or uniformly in the plane. The model is intended to give
a statistical measure of the ratio of Type II (spiral arms) to Type I
(disk) supernovae in the Galaxy. We wish to compare the distributions
in galactic longitude of supernova remnants (SNRs) with those predicted
by our model in order to allow a comparison which is independent of
the computed distance to the SNR. The observed SNR distribution seem
to correlate with that of the giant HII regions and hence with the
spiral arms.

The Monte Carlo model confirms that there is no obvious population of
SNRs not confined to spiral arms. This may suggest that the ratio of
Type I to Type II events in the Galaxy is less than unity, or that
Type I supernovae do not make long-lived extended remnants.

I. Introduction

The lore of supernova remnants (SNRs) is an important subject in its
own right. These objects are among the most conspious radio and X-ray
sources in our own and other "normal" galaxies. In addition, the
investigation of SNRs may help to constrain models of the original
supernova events.

Recent catalogues of nonthermal radio sources list about 140 objects as
SNRs. Forty-four have been observed to be strong sources of X-rays with
more energy radiated as X-rays than at radio frequencies; 41 SNRs have
been observed optically.

Many questions arise in the study of the distribution, morphology, and
physical characteristics of the SNR. Is there a connection between the
radio and X-ray emission? What is the relationship between the type
of the original supernovae (Type I or Type II) and the type of SNR
(Crablike or pure shell)? Do the SNRs associate with giant HII regions?
The SNRs play a role as probes to help answer such questions.

This study will compare the distribution of SNRs and the distribution
of giant HII regions, and explore the relationships of SNRs and spiral
arms.

Our intention here is to present some preliminary results of this
investigation without discussing the details of analysis which will
appear elsewhere.

II.   The Choice of Data

A catalogue of 125 SNRs (Milne 1979) has resulted from the work of
Milne and Downes (1971) and Clark and Caswell (1976). Van den Bergh
(1982) presents a catalogue of 135 SNRs in the Galaxy. SNRs which
have been detected at optical wavelengths are noted as are those SNRs
that have been detected in X-rays. We reexamined the SNRs which have
been detected in X-rays with new catalogues of X-ray sources (Amnuel
et al. 1982, Nugent et al. 1983) and added three SNRs. Thus our sample
has a total of 138 SNRs. In order to compare the distribution of giant
HII regions and SNRs, Georgelins' (1976) catalogue and Blitz's (1982)
catalogue of HII regions were used.

III.   The Distribution of SNRs and HII Regions

For the SNRs and a selection of 100 giant HII regions, the distances
(R) from galactic center and the perpendicular distance to galactic
plane Z(pc) are very uncertain. So for the present we restrict
ourselves to a study of the distribution in galactic longitude as
shown in Figure 1.

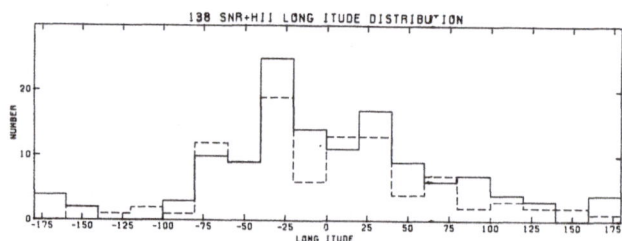

Fig. 1   The longitude distribution of 138 SNRs (——) and
         100 giant HII Regions (---)

Figure 1 suggests that the SNR distribution is similar to that of giant HII regions and hence implies that the SNRs are concentrated in spiral arms like the giant HII regions.

IV. Opportunities to Compare Observed SNRs and Model

Supernovae are rare. There has not, for instance, been a supernova spotted in our Galaxy since 1680. The determination of realistic supernova rates is difficult and the discrimination of different types of events, more so. We utilize the distributions of SNRs and a simple model of the supernova distribution, to put constraints on supernova rate and distribution.

Maza and Van den Bergh (1976) have shown that Type II supernovae (SN II) are correlated with spiral arms but that Type I (SN I) events are not. The popular presumption is that SN I are from an old disk population (Tammann, 1982) although Oemler and Tinsley (1979) have argued for a connection between SN I and active star formation. We adopt a simple model in which one population of supernovae occurs only in spiral arms and the other occurs uniformly in the disk (for now we have neglected the exponential radial distribution of stellar density in the disk). The spiral arms are taken to be a simple two-armed logarithmic spiral as given in Bash (1981). We use a Monte Carlo calculation to distribute a total of 140 supernova events in one of the two populations as illustrated in Figure 2. The model supernovae are distributed throughout the Galaxy, on average. Although distances are not well known the sample of observed SNR is depleted at great distances. No attempt has been made to correct for this factor at this time. The position of the Sun is taken from Bash in order to be consistent with his parameterization of the spiral arms.

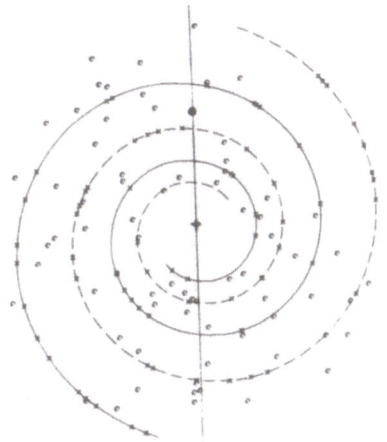

We present the results of three representative calculations in Figures 3-5. In Figure 3 half of the SNR's are in the arms and in Figure 5 all are in the uniform disk. While we have not yet done a formal statistical fit, nor examined the effects of binning systematically we note that Figure 4 appears, to the eye, to be the best fit. This suggests that an adequate model to account for the supernova distribution

Fig. 2   140 SN Distribution on the
         galactic plane 70 SN II (x),
         70 SN I (o)

is one in which all supernovae are tightly confined to spiral arms.

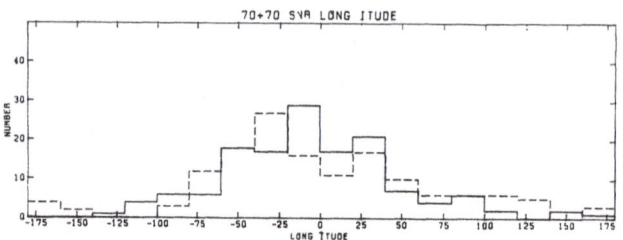

Fig. 3   The longitude distribution of 138 SNRs (---) and
         70 SN II + 70 SN I (——)

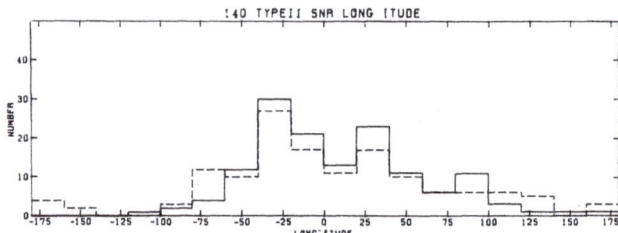

Fig. 4   The longitude distribution of 138 SNRs (---) and
         140 SN II (——)

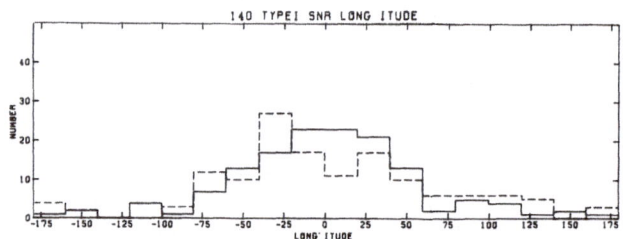

Fig. 5   The longitude distribution of 138 SNRs (---) and
         140 SN I (——)

The real question, of course, is the fraction of disk-distributed
events which could be included without causing an unsatisfactory
deterioration of the fit. Figure 3 suggests that a half and half
distribution may already be deleterious. If the "SN I" events were
distributed according to an exponential disk model there would be
relatively more at small Galactic longitude, and one would expect the
fit to be even worse.

These results are preliminary but suggest that this approach is worth pursuing. The model is susceptible to more careful statistical analysis to set quantitative limits on the allowable fraction of SN I to SN II. The distributions in b, R, and Z of various samples and models can also be compared.

The simple expectation is that the rate of optical supernova events in the Galaxy is equal to the rate of formation of extended SNRs, but this is not necessarily so. The total rate of supernovae in the Galaxy is estimated to be roughly one per 25 years (Tammann 1982). The rate of formation of radio emitting SNR, correcting for the effects of the Z (Caswell and Lerche 1979) distribution, is about one per 80 yr. These may be significantly different. Cas A gives an example of an event which made an extended SNR, but was abnormally dim. Attempting to determine separately the rates of Type I and Type II supernovae is even more difficult. A separate but closely related question is the correlation of compact remnants with optical events and/or extended SNR.

Although the present results do not demand it, they are consistent with a relative paucity of SNR outside of spiral arms. This is hinted at both by the correlation of SNR with the giant HII regions and by the Monte Carlo results.

The implications of this result, if it can be confirmed, are not clear. The rate of explosion of SN I in the Galaxy may be somewhat less than the rate of SN II despite the apparent domination of historical events by SN I. The large uncertainties inherent in determining the SN I/SN II ratio by interpolation between galactic types does not preclude this (Wheeler and Wheeler 1983). If the SN I rate were lower, problems with the over-production of iron in current models of SN I might be alleviated (Twarog and Wheeler 1981). Alternatively SN I may not lead to long-lived extended remnants. This might be the case if SN I selectively exploded in lower density environments so that they expanded and dissolved more quickly.

REFERENCES

Amnuel, P. R. et al., 1982, Astrophys. & Space Science, 82, 3.
Bash, F. N., 1981, Ap. J., 250, 551.
Blitz, L. et al., 1982, Ap. J. (Suppl.), 49, 183.
Caswell, J. L. and Lerche, I. 1979, M.N.R.A.S., 187, 202.
Clark, D. H. and Caswell, J. L. 1976, M.N.R.A.S., 174, 267.
Downes, D. 1971, A. J., 76, 305.
Georgelin, Y. M. and Georgelin, Y. D. (1976) A. & Ap., 49, 57.
Maza, J. and Van den Bergh, 1976, Ap. J., 204, 519.
Milne, D. K., 1979, Aust. J. Phys., 32, 83.
Oemler, A. and Tinsley, B. M. 1979, A. J., 84, 985.
Tammann, G. A., 1982, in Supernovae: A Survey of Current Research,
    ed. M. J. Rees and R. J. Stoneham, p. 371.
Twarog, B. A. and Wheeler, J. C., Ap. J., 261, 636.
Van den Bergh, 1982, IAU No. 101.

Wheeler, J. C. and Wheeler, J. A., 1983, in Science Underground,
   ed. M. M. Nieto, AIP Conference Proceeding No. 96.

DISCUSSION

Braun: The radio distances for the majority of SNRs in your sample
are obtained from the Σ-D relation which is calibrated in our galaxy
with a very small number of reliable determinations. The recent result
of Mills (presented at IAU Symposium No. 101) for SNRs in the Magellanic
Clouds showed very little indication of a correlation between radio
brightness and linear distance.

Li: It is an important question, but Mathewson et al. (1983) still
get the relation of Σ_ν-D for LMC. Our sample is yet available for
statistic analysis.

WOLF - RAYET STARS, COSMIC RAYS AND GAMMA RAYS

Michel Cassé

Service d'Astrophysique, CEN Saclay, France

ABSTRACT

Similarities and differences between the heavy element composition of galactic cosmic ray sources and solar energetic particles, together with cosmic ray isotopic anomalies are discussed in the framework of current theories of the origin and acceleration of the cosmic radiation. It is shown that carbon rich Wolf-Rayet stars could play the role of auxiliary sources of cosmic rays, accelerating to relativistic energies the products of their nucleosynthesis ($^{12}C, ^{16}O, ^{22}Ne, ^{25}Mg, ^{26}Mg, ^{33}S, ^{58}Fe$, s-process elements up to mass 90).

The recent discovery of gamma-ray sources held out the possibility of identifying at least some of the CR sources. It has been suggested that a class of gamma-ray sources is possibly linked with regions of star formation. Stellar winds from WR stars dominate the energetics of OB associations over a period of a few million yr, after which supernovae take over. The connection between gamma-ray sources and OB associations is the most vividely illustrated by the Carina Nebula, for which a self-consistent model of γ ray emission is presented.

1. THE COSMIC RAY SOURCE [1] COMPOSITION: AN OVERVIEW [2]

Knowledge of the elemental and isotopic composition of galactic cosmic rays (CR's) is essential to studies of the origin of these high-energy particles. In the recent years, due to ample technical improvements, ballon and satellite borne spectrometers have provided a wealth of data (for reviews see Mewaldt 1981, 1983, Simpson 1983a,b). In parallel the extrapolation of the composition observed in the earth's vicinity back to the sources has improved in quality thanks to persistent work on high energy nuclear cross sections describing the different modes of breakup of CR nuclei in collisions with interstellar matter (see e.g Silberberg, Tsao and Letaw 1983 and references therein). The refinement is such that for certain elements (all volatile) like He, C, N, O, Ne and Ar the knowledge of CR source abundances is more precise than that of solar system abundances (Table 1).

Since CR particles have lost the memory of their incident direction the only way to identify the still mysterious CR sources is the determination of ordered features in their composition based on a careful analysis of the observational material. The study of systematic trends in elemental and isotopic abundance distributions is of crucial importance in unfolding the nuclear and

C. Chiosi and A. Renzini (eds.), Stellar Nucleosynthesis, 55–75.

atomic effects that should shape the source composition and in constraining the parameters of CR acceleration models (see e.g. Cassé and Goret 1978, Cassé 1981, 1983, Meyer 1983).

ELEM.	LOCAL GALACTIC (LG)	SOLAR CORONA	SOLAR WIND (SW)	SEP (m.u. baseline)	GCRS	ELEM.
H	$2.71.10^6$ (1.10)	$2.55.10^6$ (1.40)	$1.43.10^6$ (1.30)	-	$0.077.10^6$ (1.12) [a]	H
He	$0.26.10^6$ (1.25)	$0.25.10^6$ (3.00)	$0.056.10^6$ (1.30)	$[0.038.10^6$ (1.27)] [b]	$0.0118.10^6$ (1.07)	He
C	1260. (1.26)	600. (3.00)	-	290. (1.30)	420. (1.08)	C
N	225. (1.41)	100. (1.70)	-	81. (1.33)	34. (1.34)	N
O	2250. (1.25)	630. (1.60)	660. (3.00)	650. (1.13)	505. (1.04)	O
F	.093 (1.60)	-	-	-	< 2.5	F
Ne	325. (1.50)	90. (1.60)	100. (1.40)	84. (1.32)	62. (1.14) [c]	Ne
Na	5.5 (1.18)	7. (1.70)	-	8.5 (1.47)	4.6 (1.90)	Na
Mg	105. (1.03)	95. (1.30)	-	123. (1.28)	105. (1.06) [c]	Mg
Al	8.4 (1.05)	7. (1.70)	-	8.9 (1.55)	10.2 (1.45)	Al
Si	≡ 100. (1.03)	≡ 100. (1.30)	≡ 100. (3.00)	≡ 100. (1.37)	≡ 100. (1.06) [c]	Si
P	.94 (1.24)	-	-	-	< 2.5	P
S	43. (1.35)	22. (1.70)	-	20. (1.80)	14.1 (1.20)	S
Cl	.47 (1.60)	-	-	-	< 1.6	Cl
Ar	10.7 (1.50)	5.4 (1.80)	2.6 (1.50)	3.8 (1.70)	3.2 (1.30)	Ar
K	.34 (1.41)	-	-	-	< 1.9	K
Ca	6.2 (1.14)	7.5 (1.60)	-	7.6 (1.55)	6.8 (1.34)	Ca
Sc	.0035 (1.15)	-	-	-	< 0.8	Sc
Ti	.27 (1.16)	-	-	-	< 2.4	Ti
V	.026 (1.21)	-	-	-	< 1.1	V
Cr	1.29 (1.10)	-	-	2.25 (1.90)	< 2.9	Cr
Mn	.77 (1.24)	-	-	-	< 3.7	Mn
Fe	88. (1.07)	100. (1.50)	57. (2.00)	99. (1.47)	91. (1.06)	Fe
Co	.21 (1.15)	-	-	-	0.28 (1.55)	Co
Ni	4.8 (1.13)	5.5 (1.70)	-	4.5 (1.75)	4.7 (1.17)	Ni
Cu	.052 (1.60)	-	-	-	0.060 (1.20)	Cu
Zn	.098 (1.22)	-	-	-	0.058 (1.20)	Zn

Table 1. Elemental abundances in local samples (local galaxy, solar corona, solar wind, solar energetic particles) and in galactic CR sources (Meyer 1983b).
In parenthesis : error factors ("within a factor of...").
The entries are normalized to Si.
a) based on the H/He ratio at a given magnetic rigidity.
b) not really significant, since the m.u.baseline SEP abundance of He cannot be properly defined (Meyer 1983a), see note 3.

a) Elemental CR source composition

Data on the elemental composition of the Galactic Cosmic Ray Sources (GCRS) are summarized in table 1 together with the best estimates of the composition of different samples of matter pertaining to our local environment: local galaxy (LG), solar corona (CORONA), solar wind (SW), and solar energetic particles (SEP).[3] Ratios of abundances SEP/LG, SEP/CORONA, SEP/SW, GCRS/LG and GCRS/SEP are ordered versus the first ionization potential (FIP) of the elements of interest in fig. 1 and 2 from Meyer 1983b. A critical discussion of the data is given in Meyer 1983a,b (see also Cassé 1983). Common features are apparent in the various abundance patterns under study: there is a clear enhancement of low FIP (< 9ev) elements relative to high FIP elements in GCRS,[*] SEP, CORONA and probably SW w.r.t LG abundances. These correlations are remarkably expressive and thought provoking, taken at face value they suggest that a multiple step process is at work in nature boosting particles from thermal to relativistic energies :

(i) The temperature of the primary reservoir of energetic particles (both solar and galactic cosmic rays) is undoubtly moderate ($\simeq 10^4$ K) otherwise FIP will not be the influential parameter. (ii) Galactic CR's are extracted (but presumably not accelerated to relativistic energies) from the coronae of solar-like stars. (iii) Stellar flares, similar to solar flares, inject subrelativistic nuclei into the surrounding medium, preserving on average the composition of the coronal medium. (iv) Occasionally some of the injected particles are boosted to relativistic energy by an external agent of acceleration as e.g. a passing shock wave (section 3c), then they become full members of the CR population.

This view has the merit of simplicity, but looking in more details, beyond the similarities of the GCRS and SEP elemental composition, specific traits come to evidence . Vive la difference ! (Mewaldt 1981)

At first glance abundances in GCRS and SEP seem identical (fig.2) the resemblance is however not perfect,. At a finer level, meaningful differences show up. Leaving aside H and He which behave in a very special way, not yet understood, C is clearly 2 to 3 times as abundant in CR sources as in SEP and O is possibly enhanced in CR sources relative to SEP, by a factor of 1.5 to 1.7 (Meyer 1983 b). These pecularities and the isotopic anomalies in GCRS , that we shall discuss now, are in all likelihood consanguine.

b) Isotopic CR source composition

Thus, the processes governing the <u>ionic</u> abundances in astrophysical plasmas are presumably responsible for most of the pecularities of the <u>elemental</u> CR source composition relative to that of the solar system. However the relative proportions of the <u>isotopes</u> of a given element should be approximately conserved in the course of acceleration since the selective effects seem to involve only the atomic properties of the elements. Spacecraft data on the isotopic composition of solar energetic particles (SEP) confirm this prediction,showing that within measurement uncertainties ($\pm \sim 30\%$) C,N,O and Mg are isotopically consistent with their meteoritic counterpart (Mewaldt et al, 1981, 1983).

Accordingly isotopic ratios are the most genuine clues to the nuclear origin of CR's. Unfortunately these are difficult to obtain and a limited amount of data is available. Salient features of the isotopic CR source composition, as it is presently known (fig.3) , are : i) <u>a conspicuous excess of ^{22}Ne relative to ^{20}Ne</u> w.r.t. solar system (by a factor 3.5 ± 0.6 or 5.8 ± 1.0, depending on wether the solar-flare or solar-wind ^{22}Ne/^{20}Ne ratio is used for normalization, (Mewaldt 1983), ii) <u>a marginal excess of both ^{25}Mg, ^{26}Mg relative to ^{24} Mg and ^{29} Si, ^{30}Si relative to ^{28}Si</u> compared to solar system ratios (by a factor of ~ 1.6) (see e.g. Wiedenbeck and Greiner, Wiedenbeck 1983, Mewaldt 1981, 1983, Simpson 1983a,b and references therein).

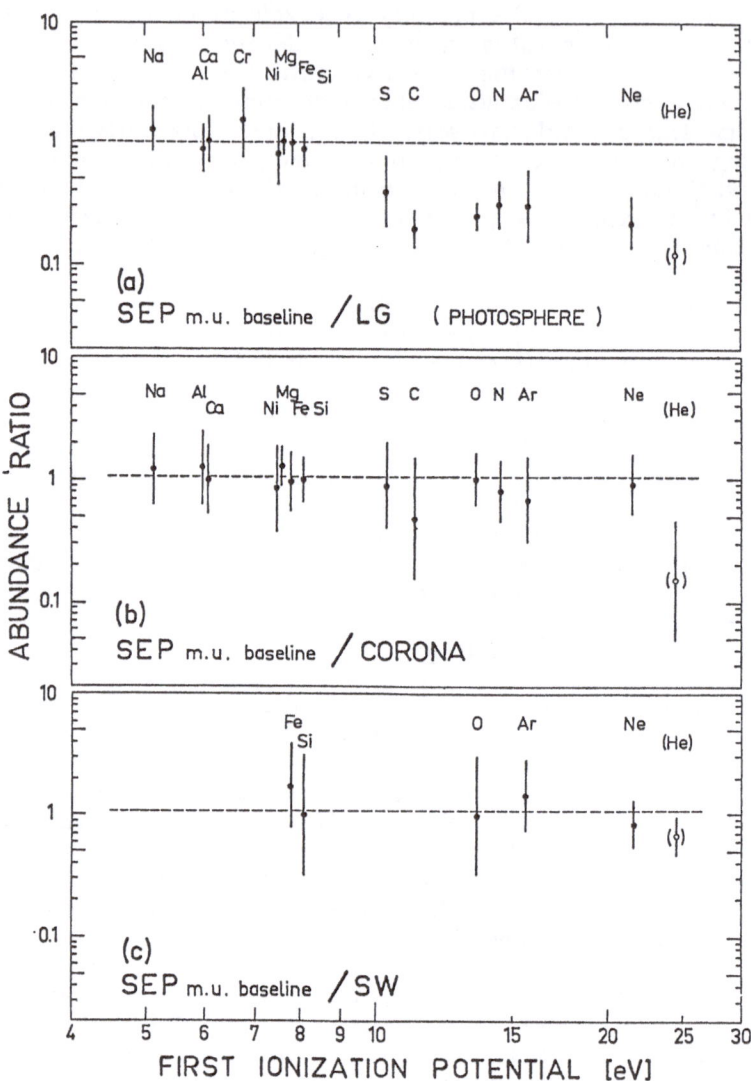

Fig. 1. Ratios of SEP abundances to LG, Coronal and SW abundances versus FIP
The errors have been summed quadratically. The points for He are not really
significant since the SEP mass unbiased baseline abundance cannot be properly
defined (Meyer 1983a).

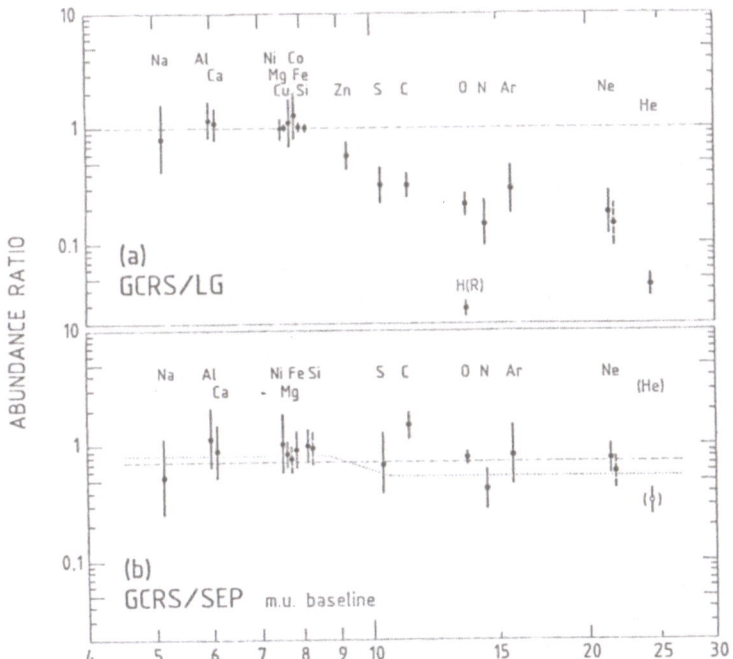

Fig. 2 . Ratios of galactic CR source to LG and SEP abundances versus FIP (Meyer 1983b).

As dashed bars are indicated the abundance ratio of Ne, Mg and Si if specific neutron-rich components are substracted (see section 2). The dashed horizontal line shows that galactic CR source and SEP abundances are, to first order, identical within errors, except for C. He is not significant. The dotted line tentatively suggests a slight oxygen excess in CR sources w.r.t. N and Ne relative to SEP.

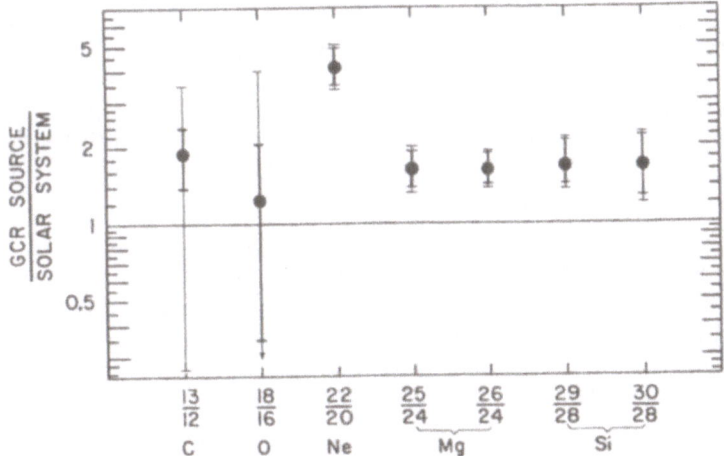

Fig.3. Enhancements of isotope ratios in the cosmic rays sources (from Wiedenbeck and Greiner). Uncertainties are shown with and without propagation uncertainties included.

So far a variety of suggestions have been advanced to explain these observations, all related to stellar explosions (Cassé, Meyer and Reeves 1979, Audouze, Chièze and Vangioni-Flam 1980, Woosley and Weaver 1981). It is only recently that Wolf-Rayet stars, or more precisely WC stars have been called into play.

2. WOLF RAYET's AND COSMIC RAYS

Wolf-Rayet (WR) stars seem to offer, in the same way than supernovae (SN) and supernova remnants (SNR) in the recent past, an interesting connection between astronomy and high energy astrophysics. The interest of the cosmic ray physicist for this peculiar kind of objects is motivated by their unusual power and composition.

a) Power. Wolf-Rayet stars possess stellar winds having high density and high velocity (mass loss rate $\dot{M} \sim 3.10^{-5}$ Mo yr^{-1}, terminal velocity ~ 2500 km s^{-1}, see e.g. Willis 1982, Abbot 1982). Accordingly they are powerful objects which compare favourably with SN. During its lifetime[5] a WR star returns $\sim 5.$ 10^{50} ergs of kinetic energy to the interstellar medium (see section 3c for a comparison of the collective power of SN and WR).

b)Composition. The composition of the material flowing from WC stars is somewhat reminiscent of the CR source isotopic anomalies as we shall see. The WC winds are thought to contain gas which has been processed through core helium-burning (see e.g. Chiosi 1981, 1982, 1984 this conference, de Loore 1981a,b, Maeder 1983 b, Maeder 1984 this conference, Greggio 1984, this volume, Maeder and de Loore 1982). WC stars indeed, believed to be massive stars that have been stripped of their hydrogen-rich envelope to sufficient depth for uncovering the helium core, can be considered as opencast mines of helium burning products of total purity[a]. Moreover this enriched material is automatically extracted by the wind. Furthermore, the wind in question being highly supersonic will inevitably trigger a strong shock wave in the surrounding ISM, which would have the virtue of accelerating the incoming wind particles to relativistic energy via the first order Fermi acceleration mechanism (see section 3c). In this sense Wolf-Rayet stars would be ideal CR machines. Altogether WR stars could be important channels for the acceleration to CR energies of ^{4}He, ^{13}C and ^{14}N in the WN stage and ^{12}C, ^{16}O, ^{22}Ne, ^{25}Mg, ^{26}Mg and s-process elements up to mass 90[7] in the WC (and WO)[8]stage (Cassé and Paul 1982, Maeder 1983b, Prantzos and Arnould 1983, Meyer 1983b, Cassé and Maeder 1983, Prantzos, Arnould and Cassé 1983).

Theoretical estimates of the abundances (by mass) at the surface (in the wind) of the WN and WC stars, Xi (WN) and Xi (WC), together with average abundances of WR winds as a whole, Xi (WR), are shown in table 2 (adapted from Maeder 1983b). The main uncertainty concerns the contribution of the most massive WC stars (initial mass ≥ 50 Mo). The role of convective overshooting and turbulent diffusion need to be clarified to get more refined estimates (see Maeder 1983a).

Note that these results are representative of the WR accelerated component only if the acceleration process conserves the wind abundance distribution. In the following we assume that it is approximately the case[9] . The balance between WN and WC has been estimated by Maeder (1983b) under the assumption that the number of CR's accelerated by both populations scales with the net rate of mass input[10] . The estimate is based on observations made within 3 kpc from the sun, thought to be complete. However we cannot exclude that some of the sources contributing to the observed cosmic radiation are located further away (Woosley and Weaver 1981, see however Ormes 1983). At this point a thorough discussion of the size of the containment volume of CR's and of the effect of the various galactocentric gradients [11]so far observed would be necessary. Unfortunately up to now, no physical propagation model has been worked out in sufficient details to provide the beginning of a solution. Anyway, the cumulated yield of WR stars obtained under these assumptions presents interesting features : altogether WC stars should give rise to excesses of ^{12}C, ^{22}Ne, ^{16}O, ^{25}Mg and ^{26}Mg in the proportions 0.6/ 1 / 0.3 (max) / 0.4 (max) / 0.4 (max) w.r.t solar system (from table 2).

The production of ^{16}O increases with the mass of the WR progenitor at the expense of ^{12}C as does the production of neutron-rich isotopes[12] like ^{25}, ^{26}Mg, ^{36}S and ^{58}Fe (Couch, Schmiedekamp and Arnett 1974, Lamb et al 1977, Prantzos and Arnould 1983). Consequently we can predict that if massive WC stars contribute significantly to the ^{25}Mg and ^{26}Mg CRS excesses they should also enhance significantly the abundance of ^{36}S and ^{58}Fe (Prantzos, Arnould and Cassé 1983).

	$X_i(WN)$	$X_i(WC)$	$X_i(WR)$	$X_i(Sun)^{(d)}$	$\varepsilon_i(WR)$
^{4}He	0.96	0.60	0.74	0.25	3.0
^{12}C	$3.9\ 10^{-4}$	0.34	0.20	$4.3\ 10^{-3}$	46.5
^{14}N	$1.35\ 10^{-4}$	$1.0\ 10^{-7}$	$5.4\ 10^{-3}$	$1.1\ 10^{-3}$	5.0
^{16}O	$3.6\ 10^{-4}$	$3.5\ 10^{-2}$	$2.1\ 10^{-2}$	$9.6\ 10^{-3}$	2.2
		$0.40^{(b)}$	$0.24^{(b)}$		$25^{(b)}$
^{20}Ne	$1.7\ 10^{-3}$	$2.0\ 10^{-3(c)}$	$1.9\ 10^{-3}$	$1.7\ 10^{-3}$	1.1
^{22}Ne	$1.5\ 10^{-4(a)}$	$2.0\ 10^{-2}$	$1.2\ 10^{-2}$	$1.5\ 10^{-4(a)}$	80
^{24}Mg	$6.5\ 10^{-4}$	$6.5\ 10^{-4}$	$6.5\ 10^{-4}$	$6.5\ 10^{-4}$	1.0
^{25}Mg	$8.8\ 10^{-5}$	$3.0\ 10^{-4}$	$2.2\ 10^{-4}$	$8.8\ 10^{-5}$	2.5
		$5.0\ 10^{-3(b)}$	$3.0\ 10^{-3(b)}$		$34^{(b)}$
^{26}Mg	$1.0\ 10^{-4}$	$4.8\ 10^{-4}$	$3.0\ 10^{-4}$	$1.0\ 10^{-4}$	3.3
		$5.0\ 10^{-3(b)}$	$3.0\ 10^{-3(b)}$		$30^{(b)}$

Table 2. Average enhancement factor ε_i(WR) of different species in WR accelerated particles (adapted from Maeder 1983).
The last column displays the predicted excesses in CR originating from WR stars before dilution with CR's of common origin (extracted in all likelihood from the coronae of normal stars, see e.g. Meyer 1983, Cassé 1983).
(a) Based on the ratio ^{22}Ne/^{20}Ne~0.10 measured in solar energetic particles (Mewaldt, Spalding and Stone 1983).
(b) Extreme values corresponding to WC stars with very massive progenitors (M = 120 $M_\odot$ on the ZAMS).
(c) Includes the product of ^{17}O-burning.
(d) From Cameron (1982) in good agreement with Meyer (1983a,b).

$\dfrac{\varepsilon_i}{\varepsilon_j}$ (32M$_\odot$)	Cosmic ray source/solar system
$^{18}O/^{16}O$ 0.4	$\lesssim$ 4
$^{22}Ne/^{20}Ne$ 67.5	3.5 ± 0.6 or $5.8 \pm 1.0^{(1)}$ [1]
$^{25}Mg/^{24}Mg$ 22.7	$1.6^{+0.4}_{-0.3}$
$^{26}Mg/^{24}Mg$ 23.5	1.6 ± 0.25
$^{29}Si/^{28}Si$ 2.4	$1.6^{+0.5}_{-0.3}$
$^{30}Si/^{28}Si$ 4.1	$1.6^{+0.5}_{-0.4}$
$^{58}Fe/^{56}Fe$ 40.6	$\lesssim$ 10

Table 3. Isotopic ratios at the surface of a very massive WC star :
Time integrated excesses w.r.t. solar system for species of interest for CR studies (Prantzos, Arnould and Cassé 1983).
These ratios should not be drastically affected by selective acceleration effects (see however, Mullan 1983). Estimates of the time integrated excesses in particles accelerated by the most massive WC stars, examplified by a 32 M$_\odot$ helium-core model, are compared with cosmic ray source enhancements (Mewaldt 1983). For the sake of clarity no dilution with "normal" cosmic rays has been taken into account. Therefore only relative values are significant.
(1) Depending on wether the solar-flare or solar wind $^{22}Ne/^{20}Ne$ ratio is used for normalization.

While encouraging agreement is obtained with observations in the mass range $12 \leq A \leq 26$ after adequate dilution with "normal" CR's (see Maeder 1983b), Si poses a problem. The nucleosynthetic yield of WC stars does not exhibit $^{29,30}Si/^{28}Si$ enhancements comparable to the $^{25,26}Mg/^{24}Mg$ enhancements , and thus does not mimic the CR sources. This failure of helium-burning models was already apparent in the work of Lamb et al (1977). This could be an intrinsic limitation of WR models. But before concluding in this sense, the data on Si isotopes, based on a unique experiment (Wiedenbeck and Greiner 1981), though remarkably precise, need to be confirmed by independent observations of similar quality. It would be interesting, also, to check a prediction of the WR model: that of a slight enhancement of the $^{58}Fe/^{56}Fe$ ratio at CR sources.

In spite of the $^{29,30}Si$ deficiency[13], these results suggest the existence of a galactic CR source component originating from WR stars. Indeed the expected abundances of ^{12}C and ^{22}Ne at the surface of WR stars is ~ 30 times higher than the corresponding CR source abundances (table 2). Thus no more than a few percent of CR's need to originate from this class of object, which is a moderate requirement.

3. WOLF RAYET's AND GAMMA RAYS

a) The diffuse galactic synchroton and gamma ray emissions

A tight correlation has been found between the integrated non-thermal radio flux of spiral galaxies and their total Hα emission line flux (Kennicut 1983) suggesting, but only suggesting, that the relativistic electrons responsible for the synchroton emission originate principally from young and massive stellar populations and are confined close to their sources. This result might reveal a strong coupling between the current star formation rate in a galaxy and the sources of relativistic electrons. Unfortunately we cannot make a strong case of this observation since the distribution of the synchroton emission does not translate directly into a distribution of potential CR sources[14] . From the distribution of the diffuse non thermal radio emission and that of γ -rays (see below) the distribution of the sources of cosmic ray electrons and protons can be inferred if, and only if, the propagation characteristics of the particles, depending principally on the strength, the configuration and the small scale fluctuations of the interstellar magnetic field are known, which is obviously not the case. Radio observations, at least, have revealed that CR electrons spread throughout galaxies. But there is no reason to suppose that electrons and protons have common sources and the same certitude has not even been reached concerning CR protons due to the ambiguity of the observed spectrum of the diffuse galactic γ radiation (10 MeV-10 GeV). The expected bump around 70 MeV - a clear signature of high energy protons inducing π° production -is absent (fig.4) (see e.g. Bertsch and Kniffen 1983, Sacher and Schönfelder 1983). In addition to the expected π° component from nuclear interactions of CR protons with interstellar nucleons (proton -proton collisions), an additional component must exist - either from interaction of CR electrons (bremsstrahlung, inverse Compton) or from unresolved γ ray sources or a combination of both (see e.g. Cesarsky, Paul and Shukla 1978, Stephens 1981, Fichtel and Kniffen 1982).

There is however some evidence that a class of γ ray sources -i.e. hot spots in the gamma-ray sky - are physically linked to young stellar aggregates (OB associations). This is not too surprising if we adopt the view that CR's are accelerated by shock waves[15] associated with supernova remnants (SNR) and supersonic stellar winds (Montmerle 1979, Cassé and Paul 1980). This point deserves a special discussion.

b) γ -ray sources

Following the pionering work of SAS-2 (Fichtel et al 1975), the European COS B satellite has now isolated 25 hot spots of high energy ($\geq$100 MeV) gamma emission satisfying the operational definition of a γ ray source [16] (Hermsen 1980, Swanenburg et al 1981, Bignami and Hermsen 1983), fig.5.

The identification of these sources with known objects is certain only in the case of 2 pulsars, Crab and Vela, because of the temporal signature, but most of them have not found an undisputed counterpart (see e.g. Montmerle 1979, Bignami and Hermsen 1982, Morfill and Tenorio-Tagle 1983, Wolfendale 1983). The position of a discrete source cannot be appreciated with a precision

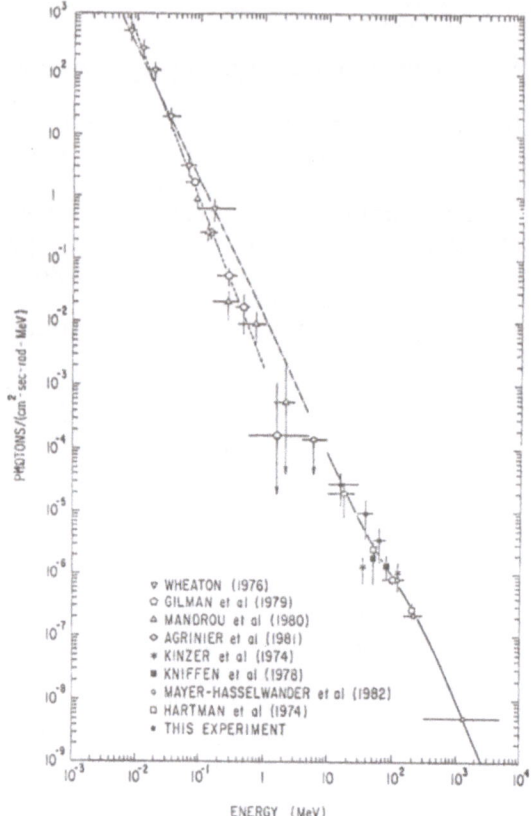

Fig.4 The energy distribution of high energy photons in the galactic plane (Bertsch and Kniffen 1982).

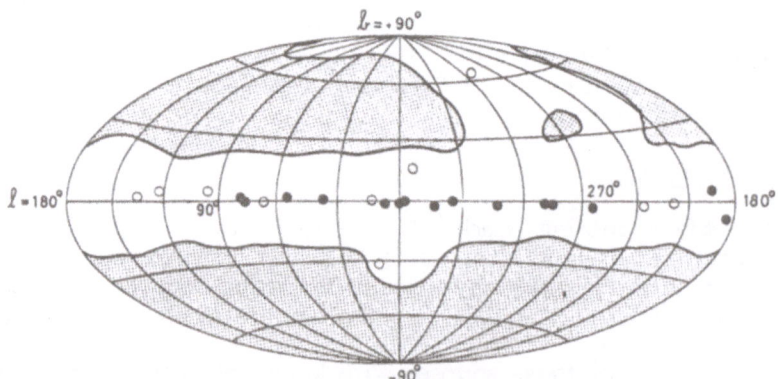

Fig.5. Region of the sky searched for gamma-ray sources (unshaded) and sources detected above 100 MeV by spatial analysis. (Swanenburg et al 1981, Bignami and Hermsen 1983). The filled circles denote sources with measured fluxes $\geq 1.3 \times 10^{-8}$ photon cm^{-2} s^{-1}. Open circles denote sources below this threshold.

better than 1 or 2° depending on the source intensity and spectrum relative to the surrounding background. Then, observationally speaking no difference exists between compact sources (except for pulsars) and relatively extended sources (angular diameter up to $\sim$ 1.5°). Moreover, in the inner galaxy (-30°$\lesssim$ l$\lesssim$ + 30°) the sources are less contrasted, therefore more difficult to isolate, then the sample is not free from selection effects, especially below 100 MeV (see e.g. Mayer-Hasselwander et al 1980). The energy spectrum of a given source will depend on the physical conditions prevailing at the source (matter density, CR proton density and spectrum, CR electron density and spectrum, photon field) through the respective contributions of the different gamma ray production mechanisms ($\pi°$ decay, electron bremsstrahlung, Compton scattering). Unfortunately the lack of precise energy spectra of point-like sources do not allow to determine the main emission mechanisms in each individual case.

Up to now there has been no claim of variability from individual COS B sources. The observed flux from the localized maxima in the γ ray maps range between $\sim 10^{-6}$ and $\sim 10^{-5}$ photons cm^{-2} s^{-1} and their latitude distribution -reminiscent of very young galactic populations - is extremely peaked at b = 0. The lack of source identification and hence the lack of knowledge about their distance prohibits the estimate of individual luminosity and allows only gross estimates derived from collective properties (table 4).

Average |b|	$\sim$ 1.5°
Angular size	0° – 2°
Photon-flux range (100 MeV<E<1 GeV)	$\sim$ (1-5) x 10^{-6} ph cm^{-2} s^{-1}
Distance range	2 – 7 kpc
Scale height <|z|> (300° $\lhd$ 60°)	$\lesssim$ 130 pc
Spectral shape	different for individual objects, average dN/dE $\sim$ E^{-2}
Average photon energy	$\sim$ 250 MeV in the 100 MeV – 1 GeV range
Energy-flux range (100 MeV <E <1 GeV)	$\sim$ (4-20) x 10^{-10} erg cm^{-2} s^{-1}
Luminosity range (100 MeV<E<1 GeV)	$\sim$ (0.4-5) x 10^{36} erg s^{-1}

Table 4. Average properties of the unidentified galactic gamma-ray sources (Swanenburg et al 1981, Bignami and Hermsen 1983)..

Thus it is entirely possible that many of the unidentified gamma objects are of extended nature. Links between a number of COS B sources and OB associations have been advocated for some times (Montmerle 1979, Cassé, Montmerle and Paul 1981, Paul 1981), the Carina complex being perhaps the best candidate.

c) The Carina gamma-ray source[17]

c 1) Empirical evidences. The γ-ray source at $l=288°$ and $b=0$ displays a striking positional coincidence with the Carina nebula, located at ~ 2.7kpc. Identifying the source and the nebula yields a γ ray luminosity $L_\gamma \sim 5.10^{35}$ erg s^{-1} (a rather low luminosity by statistical standards, see table 4). The Carina complex, indeed, is rather exceptional. It contains the richest aggregate of young and massive stars known in the galaxy. This region is noted in particular for the compact star clusters Tr 14, Tr 16 and Cr 228 (Humphreys, 1978), altogether comprising 6 of the 7 03 stars observed in the galaxy and the extraordinary η Car object. It is also remarkable that 3 Wolf-Rayet stars (WN 7) are associated with the complex.

The total mechanical energy input in the form of supersonic stellar winds by OB stars and WR stars respectively amounts to $L_{ob} \sim 10^{38}$ erg s^{-1} and $L_{wr} \sim 2.10^{38}$ erg s^{-1} (Montmerle , Cassé and Paul 1981, 1983 in preparation).[18] Altogether one has $L_{wind}/L_\gamma \sim 10^3$. There is a priori no energy problem, but we need other ingredients to make a good γ ray source! The very existence of the Carina γ-ray source at its observed level implies values of the CR energy density ~ 10 to 20 times that in the solar neighborhood (Montmerle 1981, Issa et al 1981, Wolfendale 1983) i.e. in-situ CR acceleration and strong confinement of the newly accelerated CR's near their birthplace.

c 2) In- situ acceleration. Diffusive shocks in the steady state approximation are believed to accelerate particles to relativistic energies[19] . Accordingly stellar wind terminal shocks around OB and especially WR stars are appealing agents (sites) of particle acceleration (Cassé and Paul 1979, 1980, Dorman 1979) since recent work has shown that mass loss, especially from early type stars is a very common occurence (see Cassinelli 1982, Abbot 1982, Barlow 1982 for recent reviews). Whether this acceleration is continuous or intermittent is still open to discussion (see Völk and Forman 1981 for criticisms and Cesarsky and Montmerle 1983 for possible solutions to the difficulties raised). Assuming that the acceleration is continuous, several estimates have been made on the production of high energy γ rays through the interaction of the newly accelerated particles with the dense region surrounding the accelerating stars.

The confinement problem has been considered by Montmerle and Cesarsky 1981a,b, and Cesarsky and Montmerle 1983. They propose the following scenario:

c 3) Trapping. The shock accelerated particles must be well trapped in the HII region NG 3372 surrounding the hot and massive stars by Alfvén-wave scattering (Wentzel 1974)[20] and are strongly confined close to their birth place. This trapping in turn allows CR to produce γ rays very efficiently, given the high density of the ionized medium ($\sim 10^2$ cm^{-3}). Then they calculate that the proton acceleration efficiency required to explain the observed gamma ray luminosity is $\sim 2\%$, not an unreasonable demand. A sketch of the Montmerle-Cesarsky model is shown in fig.6. Note that the HII regions plays a central role i) ensuring the developement and survival of Alfvén-waves which confine the CR's and ii) offering the target material to produce the γ's. This model is open   to   tests since it clearly predicts that the dominant γ ray emission must come from the

HII region itself and not the underlying molecular cloud. Future γ ray experiments with improved spatial resolution, like the French-Soviet-Italian " GAMMA I" and the French-Soviet "SIGMA" experiments will help to clarify this question. A direct observation of the Carina spectrum at gamma ray energies could clear up any doubt concerning the assumption of a purely proton-induced γ ray emission.

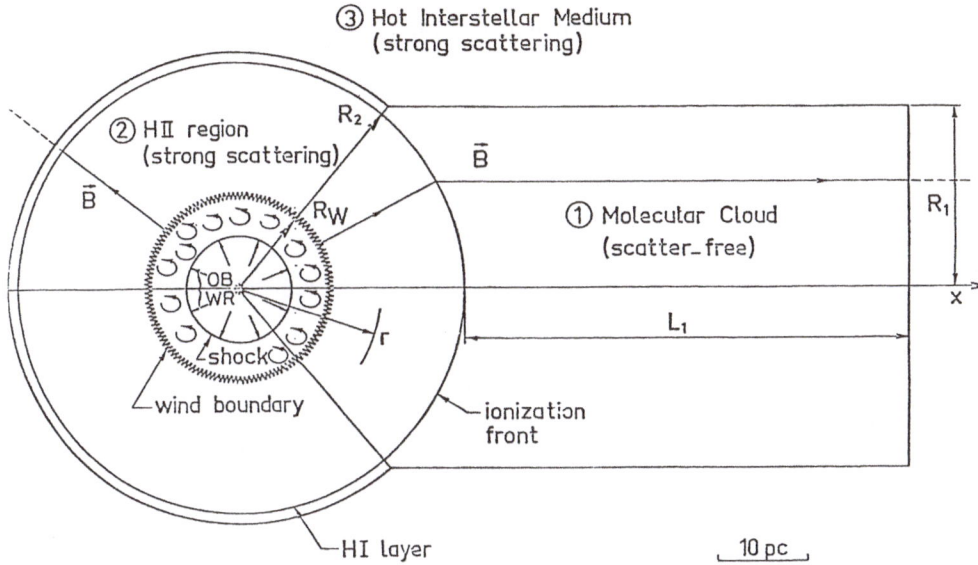

Fig.6. A model for the Carina γ ray source (Montmerle and Cesarsky 1981a,b,, Montmerle 1981).
The picture, drawn to scale, features the 3 major components of the region.
Component 1: the molecular cloud, approximately cylindric
Component 2 : The HII region at the border of the molecular cloud (radius R_2 = 25 pc, mass 1-2 10^5 $M_\odot$. The ionizing agents are the 03 stars dwelling within the nebula, WR's provide comparatively little ionization, but their wind offer a major fraction of the kinetic energy necessary to accelerate CR's. The supersonic winds drive a shock front (radius R_w = 10 pc) well within the ionized region, a prerequisite for diffuse CR acceleration.
Component 3: The nebula is embedded in the hot interstellar medium. CR protons possibly accelerated at the wind boundary are sufficiently trapped in the giant HII region to generate a copious flux of γ rays. The acceleration efficiency needed to explain the γ luminosity (i.e. the conversion factor between the mecanical energy and the energy inparted to CR's) is moderate ($\sim$2%).

d) Other OB associations

In order to check whether the stellar wind hypothesis is compatible with the γ-ray data elsewhere in the Galaxy, we have applied the same procedure to 72 OB associations (Humphreys 1978), using empirical laws for mass-loss when required, Cassé el al. 1981). Table 5 lists the expected γ-ray surface brightness of the 4 brightest OB associations, normalized to that of the Carina complex. The WR stars bring a dominant contribution. The figures obtained are not inconsistent with the γ- ray data. We have assumed that the acceleration and confinement properties are as in the case of the Carina nebula.

Rank	Association	Number of WR stars	Number of Of stars	γ-ray surface brightness
2	Cr 121	0~1	0	0 ~1.3
3	Sco OB1	2	3	1.1
4	Carina	3	1	1.0
5 to 72	-	0	< 3	< to ≪ 1.0

Table 5. Expected γ ray surface brightness of the 4 brightest galactic OB associations relative-to the Carina complex

We concluded that OB associations may stand above the COS B visibility threshold only if they are powered by WR winds or supernovae. Most of the remaining associations must be undectable as individual γ-ray source by the COS B satellite, but they must contribute significantly to the diffuse galactic background.

e) Wolf-Rayet's versus Supernovae

Now the question arises, which are the most powerful sources of kinetic energy in the galaxy, supernovae or Wolf- Rayet stars ?

In the solar neighborhood, the answer is rather clear from direct arguments (star counts, Abbot 1982) or indirect arguments (concerning the initial mass function and the mass of the WR and SN progenitors, Cesarsky and Montmerle 1983). The average energy input from stellar winds within a radius of 3 kpc from the sun is $<P_w> \sim 2.10^{38}$ erg $s^{-1} kpc^{-2}$, half of this power coming from ~ 50 WR stars (Abbot 1982),whereas the SN contribution in the same zone is estimated to $< P_{sn}> \sim 10^{39}$ erg $s^{-1} kpc.^{-2}$

The results derived within a radius of 3 kpc from the sun may not be valid in the inner galaxy. The observed ratio (surface density of WR stars)/(surface density of O stars) increases by about a factor 3 from $\sim$ 8 kpc to ~ 10 kpc (galactocentric distance) (Azzopardi, Maeder and Lequeux 1980, Meylan and Maeder 1983). If this ratio continues to increase inward up to a distance of say 4 kpc from the galactic center, WR would compete more favorably with SN since the radial distribution of SN remnants levels off at ~ 5 kpc (Kodaira 1974, Clark and Caswell 1976).

Although WR winds are of secondary importance to the global input of kinetic energy to the local interstellar medium, they dominate the energy deposition in the vicinity of OB associations, as shown by Abbot 1982 and Cesarsky and Montmerle 1983. The supremacy of WR stars is however ephemeral (~4 millions of years) and comes to its end when the most massive (short lived) of them explodes in the cavity excavated by the winds (Cesarsky and Montmerle 1983, Doom, Dorland and Montmerle 1983, in preparation) . Thus it is expected that only a small but meaningful fraction of the COS B ray sources may be energized by WR winds.

CONCLUSION

Cosmic rays are an inexhaustible matter of reflexion: where are they born? How do they acquire their immense energy? This communication has been concerned with these questions. We have tried to unfold the nuclear and atomic effects governing the makeup of source abundances in the light of analogies and differences between the abundance pattern in galactic CR sources and solar energetic particles. The data suggest that most of the galactic CR's are extracted from the coronae of solar-like stars and are subsequently accelerated to high energy without alteration in their composition. Irrespective of the acceleration mechanism, isotopic ratios are the most genuine clue to the nuclear origin of CR species. In this respect, a detailed comparison between CR source and solar system isotopic proportions sheds light on possible differences in the nucleosynthesis and history of these CR species for which CR source abundances are available and that of the matter around us. The presence of excesses of ^{12}C, ^{16}O, ^{22}Ne, ^{25}Mg, ^{26}Mg at the CR sources suggests the existence of an extra CR component especially enriched in helium-burning products and highly diluted with "normal" CR's. This component might be produced and energized by WC stars, provided the surface material is accelerated in the shock triggered by the wind. This scenario however do not explain the enhancements of ^{29}Si and ^{30}Si at the sources (if real). Therefore a composite model aimed at explaining the isotopic CR anomalies seems difficult to avoid.

The recent discovery of gamma-ray sources migh offer the opportunity to locate at least some of the CR sources, like WR-dominated stellar associations. Wolf-Rayet stars release in the ISM a large amount of kinetic energy. In addition to SN, they should be considered as potential candidates for CR energization through the diffusive shock acceleration mechanism. The estimated rate of energy input by SN within 3 kpc from the sun is a factor of ~5 greater than the average energy input from WR and 0 stars. However the situation is uncertain in the inner galaxy due to the incompleteness of the stellar catalogues. Close to young OB associations WR's dominate as long as no SN explodes i.e. during the first millions of years after star birth, after which SN taker over. The Carina complex seems to be in the WR dominated phase. A self-consistent model of the γ-ray source 2CG 288-0 associated with the Carina Nebula has been worked out by Montmerle and Cesarsky (1981a,b). The efficiency required to sustain the γ-ray emission is found to be moderate (~2%) and hence plausible. This remains, of course, only speculative until more observational evidences for in-situ CR acceleration are obtained.

Acknowledgements

I am grateful to Cesare Chiosi and Alvio Renzini for allowing me to develop this subject at length.
I thank Claudine Belin for careful typing.

(1) By definition the CR source abundances refer to the relative abundances of the various nuclear species prior to propagation through interstellar space. On their tortuous paths, during their travel from the sources to us, CR's undergo nuclear collisions with interstellar matter that change their mean composition.

(2) This presentation is restricted to the sources of the most common cosmic-ray nuclei of galactic origin (H to Ni) observed in the 100 MeV-100 GeV energy range. Cosmic ray electrons and ultra-heavy nuclei, which are other important clues to the origin of the galactic cosmic radiation, are discussed at length by Webber (1983) and Israel (1983).

(3) All satellite observations of solar energetic particles (1 to 50 MeV/n) can be explained in terms of a unique baseline composition distorted by highly variable biases depending only upon the mass/charge ratio of the SEP ions (Meyer 1983a). This ever-present baseline SEP composition (mass unbiased baseline) serves here as a standard.

(4) This trend is confirmed by observations of trans-iron elements (Israel et al 1983 and references therein).

(5) Thought to be of the order of the helium-burning lifetime of a massive star, i.e. $\sim 3.10^{5}$ yr.

(6) In red giants, planetary nebulae or supernova ejecta, the ashes of helium burning are mixed with large quantities of material of discording composition (see e.g. Iben and Renzini 1983, Woosley and Weaver 1982).

(7) Relying on Lamb et al 1977 and Couch, Arnett and Pardo 1974.

(8) WO stars might be the most evolved (stripped off) of the WR stars. In these objects He/C/O is about 20/10/1 and there is no N (Hummer and Barlow 1982).

(9) A preliminary analysis indicates that selection effects at acceleration between the different ions would not be too severe (Cassé and Maeder 1983). But this conclusion needs to be substanciated by better ionization models of WR winds.

(10) Assuming that it scales with the input of kinetic energy makes little difference.

(11) Abundance gradients (Shaver et al 1983), asymmetry of the WR distribution, radial dependence of stellar wind energy (Abbot 1982), variation of the WR/O, WR/SNR and WN/WC number ratios with the galactocentric distance (Maeder, Lequeux and Azzopardi 1980, Meylan and Maeder 1983, Clark and Caswell 1976).

(12) The production of n-rich isotopes takes place at the end of helium-burning when neutrons are liberated by the destruction of ^{22}Ne through the reaction $^{22}Ne(\alpha,n)^{25}Mg$.

(13) A Theoretical deficit is always easier to explain than an undesirable excess since it is often possible to invoke an additional process specific enough to fill the gap.

(14) The same observation of a correlation between synchroton and H α emissions could have been interpreted in term of a strenghtening of the magnetic field - hence of a better CR trapping - in denser regions of the interstellar medium where massive objects form preferentially (see e.g. Paul, Cassé, and Cesarsky 1976).

(15) Up to now a large amount of work has been devoted to demonstrate that shock waves are good candidates for the acceleration of protons and nuclei to relativistic energies, but to our knowledge, little work has been dedicated to electrons. One might speculate that at a given energy electrons are less efficiently accelerated than protons because they "feel" the shock transition less abruptly than proton of the same energy due to their much smaller gyroradius (see note 19).

(16) It is necessary to state clearly that gamma ray telescopes, like COS B, have a very poor resolution power by astronomical standards. The resolution power is limited by the intrinsic angular resolution of the instrument (a few degrees at 100 MeV) and by background problems. A so call gamma-ray source is operationally defined as a localized enhancement on the 2-dimensional (l,b) sky map, compatible with a standard source profile (namely the profile of the stronger point source: the Vela pulsar).

(17) More exotic explanations are presented by Morfill and Tenorio-Tagle (1983)

(18) Based on empirical mass loss rates and terminal velocities of OB and WR stars. The peculiar η Car object, which shed mass at a probably unprecedented rate has not been taken into account in the energy budget. With mass-loss estimates ranging from 10^3 to 0.075 M$\odot$ yr^{-1}, and terminal velocity $\sim$600 km s^{-1}, the kinetic power of this enigmatic object is P$\eta \sim 2.4 \ 10^{38}$ to $1.8 \ 10^{40}$ erg s^{-1}. In the lattest case, η Car would energetically dominate the whole region

(19) Theories of CR acceleration involving repeated scattering of particles across a shock front have been extensively discussed (reviews are given by Axford 1981a, b, Blandford 1979, Toptygin 1980, Völk 1981, and Drury 1983). These theories, when treating the CR's as test particles - i.e. neglecting the CR pressure - yield a power low energy spectrum with a spectral index close to 2. This value is similar to the index of the source spectrum and not to the observed value as often said (= 2.7). The difference between the observed and source spectra is due to preferential leakage out of the galaxy of high energy particles (escape probability α E.$^{0.6}$, see e.g. Koch et al 1983, Ormes and Protheroe 1983). However the presence of a finite CR pressure, if not negligible compared to the ram pressure of the inflowing gas (in the shock rest-frame) complicates considerably the problem (see e.g. Mc Kenzie and Völk 1982, Drury 1983, Ellison and Eichler 1983, Heavens 1983, Webb 1983). The self-consistent determination of the shock structure and of the resulting spectrum is extremely difficult due to strong non-linearities. The non-linear problem has not been solved exactly but progresses have been made using Monte Carlo simulations (Ellison 1981, Ellison and Eichler 1983).

(20) Since Alfvén waves are strongly damped in dense and neutral media, high energy protons are strongly confined in the ionized media (and not in the placental molecular cloud) where they interact, producing γ rays through the pp$\rightarrow \pi^{0} \rightarrow 2\gamma$ mechanism.

REFERENCES

Abbot, D.C., 1982a, Ap.J. 263, 723.
Abbot, D.C., 1982b, in "Wolf-Rayet Stars: Observations, Physics, Evolution".
         eds. C.W.H. de Loore and J.A. Willis (Reidel) p.185.
Arnett, W.D., 1972, Ap.J., 176, 681.
Arnett W.D. and Thielemann, 1984, this volume.
Audouze, J., Chièze, J.P., and Vangioni-Flam, E., 1980, Astron. Astrophys.
         91, 49.
Axford, W.I., 1981a, 17th Int. Cosmic Ray Conf. 12, 155.
Axford, W.I., 1981b, Proc. 10th Texas Symp. on Relativistic Astrophysics.
         Ann. NY. Acad. Sci., 375, 297.
Barlow, M.J., 1982, in "Wolf-Rayet Stars: Observations Physics, Evolution",
         eds. C.W.H. de Loore and J.A. Willis (Reidel) p.149.
Bertsch, D.L., and Kniffen, D.A., 1982, NASA report, TM 83992.
Bignami, G.F., and Hermsen, W., 1983, to appear in Ann.Rev.Astron. Astrophys.
         vol.21.
Blandford, R.D., 1989, in "Particle Acceleration Mechanisms in Astrophysics"
         AIP Conf. Proc. 56, 333.
Cameron, A.G.W., 1982, in " Essays in Nuclear Astrophysics" eds C.A. Barnes,
         D.D.Clayton, and D.N. Schramm (Cambridge U.Press), p.23.
Cassé, M., 1981, 17th Int. Cosmic Ray Conf., 13, 111.
Cassé, M., 1983, in "Composition and Origin of Cosmic Rays" ed. M.M. Shapiro,
         (Reidel), p. 193.
Cassé, M., and Goret, P., 1978, Ap.J., 221, 703.
Cassé, M., and Maeder, A., 1983, 18th Int. Cosmic Ray Conf. Bangalore, India.
Cassé, M., Meyer, J.P., and Reeves, H., 1979, 16th Int.Cosmic Ray Conf.,
         12, 114.
Cassé, M., Montmerle, T., and Paul, J.A., 1981, in "Origin of Cosmic Rays",
         eds. G. Setti, G.Spada, and A.W. Wolfendale (Reidel), p.323.
Cassé, M., and Paul, J.A., 1979, 16th Int. Cosmic Ray Conf. 2, 103.
Cassé, M., and Paul, J.A., 1980, Ap.J. 237, 236.
Cassé, M., and Paul J.A., 1982, Ap. J., 258, 860.
Cassinelli, J.P., 1982, in "Wolf-Rayet Stars: Observations, Physics, Evolution".
         Eds. C.W.H. de Loore and J.A. Willis (Reidel), p.173.
Cesarsky, C.J., Paul, J.A., and Shukla, P.G., 1978, Astrophys. Sp. Sci., 59,73.
Cesarsky, C.J., and Montmerle, T., 1983, Sp. Sci. Rev. 36, 173.
Chiosi, C., 1981, Proc. ESO Workshop "The Most Massive Stars", p.27.
Chiosi, C., 1982, in "Wolf-Rayet Stars Observations, Physics, Evolution",
         eds. C.W.H. de Loore and A.J. Willis (Reidel).
Clark, D.H., and Caswell, J.L., 1976, MNRAS, 174, 267.
Couch, R.G., Schmiedekamp, A.B., and Arnett, W.D., 1974, Ap.J. 190,95.
Dorman, L.I., 1979, 16th Int. Cosmic Ray Conf. 2,49.
Drury, L.O'C., 1983, to appear in Rep.Prog.Phys.
Dufton, P.L., Kane, L., and Mc Keith, e.d., 1981, MNRAS, 194,85.
Eberhardt, P., Junck, M.H.A., Meier, F.O., Niederer, F., 1979, Ap.J. 234,L169.
Eberhardt, P., Junck, M.H.A., Meier, F.O., Niederer, F.R., 1981, Geochim.
         Cosmochim. Acta, 45, 1515.
Ellison, D.C., 1981, Ph.D. Thesis, The Catholic U. of American, Washington,
Ellison, D.C., and Eichler, D., 1983. preprint

Fichtel, C.E., Hartman, R.C., Kniffen, D.A., Thompson, D.J., Bignami, G.F., ögelman, H., özel, M.E., Tümer, T., 1975, Ap.J., 198, 163.
Fichtel, C., and Kniffen, D., 1982, NASA report, TM 83992.
Heavens, A.F., 1983, MNRAS, 204, 699.
Hermsen, W. 1980, Ph.D. Thesis, Univ. of Leiden
Hummer, D. and Barlow, W., 1982, in "Wolf-Rayet Stars: Observations, Physics, Evolution" eds. C.W.H. de Loore and A.J. Willis (Reidel)
Humphreys, R.M., 1978, Ap.J.Suppl. 38, 309
Isaa, M.R., Riley, P.A., Li Ti pei, and Wolfendale, A.W., 1981, 17th Int. Cosmic Ray Conf., 1, 150.
Kennicut, R., 1983, Astron. Astrophys. 120, 219.
Lamb, S.A., Howard, W.M., Truran, J.W. , and Iben, I., 1977, Ap.J.,217,213.
de Loore, C., 1981a, 17th Int. Cosmic Ray Conf., 12, 127.
de Loore, C., 1981b, Ann. Phys. Fr., 1981, 6, 149.
Iben, I. Jr., and Renzini, A., 1983, to appear in Ann. Rev. Astron. Astrophys.
Israel, M.H., 1983, in "Composition and Origin of Cosmic Rays" ed. M.M. Shapiro (Reidel), p.83.
Kane, L., Mc Keith, C.D., and Dufton, P.L., 1982, Ap.J. 252, 461.
Koch-Miramond, L., Englemann, J.J., Goret, P., Masse, P., Soutoul, A., Perron, C., Lund, N., and Rasmussen, I.L., 1983, 18th Int. Cosmic Ray Conf. Bangalore, India.
Maeder, A., 1983a, Astron. Astrophys. 120, 113.
Maeder, A., 1983b, Astron. Astrophys. 120, 130.
Maeder, A., and de Loore, C., 1982, Proc. 3rd European IUE Conf., p.21.
Maeder, A., Lequeux, J., and Azzopardi, M., Astron. Astrophys., 207, 209.
Mayer-Hasselwander, M.A., Bennett, K., Bignami, G.F., Buccheri, R., d'Amico N., Hermsen, W., Kambach, G., Lebrun, F., Lichti, G.G., Masnou, J.L., Paul, J.A., Pinkau, K., Scarsi, L., Swanenburg, B.N., Wills, R.D., 1980, Proc. 9th Texas. Symp. on Relativistic Astrophysics Ann. NY. Acad.Sci., 336, 211.
Mc Kenzie, J.F., and Völk, H.J., 1982, Astron. Astrophys., 116, 191.
Mewaldt, R.A., 1981, 17th Int. Cosmic Ray Conf., 13, 49.
Mewaldt, R.A., 1983, to be published in Rev.Geophys.Sp.Sci.
Mewaldt, R.A., Spalding, J.D., and Stone, E.C., 1983, 18th Int. Cosmic Ray Conf., Bangalore, India.
Mewaldt, R.A., Spalding, J.D., Stone, E.C., and Vogt, R.E., 1981, 17th Int. Cosmic Ray Conf., 3, 131.
Meyer, J.P., 1983a, submitted to Ap.J.
Meyer, J.P., 1983b, submitted to Ap.J.
Meylan, G., and Maeder, A., 1983, Astron. Astrophys., 124, 84.
Montmerle, T., 1979, Ap.J. 231, 95.
Montmerle, T., 1981, Phil. Trans. R. Soc. Lond. A 301, 505.
Montmerle, T., Cassé, M., and Paul, J.A., 1981, in "Effects of Mass Loss on Stellar Evolution" ed. C.Chiosi and R. Stalio (Reidel) p.155.
Montmerle, T., and Cesarsky, C.J. 1981a, 17th Int. Cosmic Ray Conf. 1,173.
Montmerle, T., and Cesarsky, C.J., 1981b, Proc. Int. School and Workshop on Plasma Astrophysics, Varenna, ESA SP-161,319.
Morfill, G.E., and Tenorio-Tagle, G., 1983, Sp.Sci.Rev. 36, 93.
Mullan, D.J., 1983, Ap.J. 268, 385.
Ormes, J., 1983, 18th Int. Cosmic Ray Conf. Bangalore, India
Ormes, J., and Protheroe, R.J., 1983, 18th Int. Cosmic Ray Conf. Bangalore, India.

Paul, J.A., Cassé, M., and Cesarsky, C.J., 1976, Ap.J. 207, 62.

Prantzos, N., and Arnould, M., 1983, in "Wolf-Rayet Stars - Precursors of Supernovae", Observatoire de Paris, Meudon, ed. M.C. Lortet.

Prantzos, N., Arnould, M., and Cassé, M., 1983, 18th Int. Cosmic Ray Conf., Bangalore, India.

Sacher, W., and Schönfelder, V., 1983, Sp.Sci.Rev. 36, 249.

Shaver, P.A., Mc Gee, R.X., Newton, L.M., Danks, A.C., and Pottasch, S.R., 1983, preprint.

Simpson, J.A., 1983, in "Composition and Origin of Cosmic Rays", ed.M.M.Shapiro, (Reidel), p.1.

Simpson, J.A., 1983, to appear in Ann.Rev.Nuc.Part.Phys.

Silberberg, R., Tsao, C.H., and Letaw., J.R., 1983, in "Composition and Origin of Cosmic Rays" ed. M.M.Shapiro (Reidel), p.321.

Stephens, S.A., 1981, Astrophys. Sp.Sci., 79, 419.

Swanenburg, B.N., Bennett, K., Bignami, G.F., Buccheri, R., Caraveo, P., Hermsen, W., Kanbach, G., Lichti, C.G., Masnou, J.L., Mayer-Hasselwander, H.A., Paul, J.A., Sacco, B., Scarsi, L., and Wills, R.D., 1981, Ap.J., 243, L69.

Toptygin, T., 1980, Space Sci.Rev., 27, 157.

Völk, H.J., 1981, 17th Int.Cosmic Ray Conf., 13, 131.

Völk, H.J., and Forman, M., 1982,Ap.J., 253, 188.

Webb, G.M., 1983, Astron. Astrophys. 124, 163.

Webber, W.R., 1983, in "Composition and Origin of Cosmic Rays" ed.M.M.Shapiro, (Reidel) , p.83.

Wiedenbeck, M.E., 1983, in "Composition and Origin of Cosmic Rays", ed. M.M.Shapiro, (Reidel),p.65.

Wiedenbeck, M.E., and Greiner, D.E., 1981, 17th Int.Cosmic Ray Conf., 2,76.

Wentzel, D., 1974, Ann. Rev. Astron. Astrophys. 12, 71.

Willis, A.J., 1982, MNRAS, 98, 897.

Wolfendale, A.W., 1983, Q.J.IR. astr.Soc.24, 122

Woosley, S.E., and Weaver, T.A., 1981, Ap.J., 243, 651.

Woosley, S.E., and Weaver, T.A. , 1982, in "Essays in Nuclear Astrophysics" eds. C.A.Barnes, D.N. Schramm, and D.D.Clayton (Cambridge U.Press), p.377.

DISCUSSION

G. Mathews: Would you comment on another mechanism for a ^{22}Ne overabundance which is enrichment of the ISM by evolving low mass stars which have changed the ISM since the solar system condensed.

M.Cassé: Considering that the production of ^{22}Ne in low mass stars goes together with that of ^{14}N, it seems doubtful that galactic evolutionary effects alone can explain an increase by a factor of ~ 4 of the ^{22}Ne/^{20}Ne ratio in the last 4.6 billion yr in the local ISM since the N abundance has not appreciably varied since the birth of sun as evidenced by the similarities between the abundance of this element in the solar system and in young (B) stars (Kane, Mc Keith and Dufton 1980, Dufton, Kane and Mc Keith 1981).

M.G. Edmunds: Can this model be linked with explaining ^{22}Ne anomalies observed in meteorites?

M.Cassé: The so call Ne-E meteoritic component is (almost) pure ^{22}Ne (see e.g. Eberhardt et al 1979, 1981). In WC stars there is no way to supress the ^{20}Ne initially present and the ^{22}Ne/^{20}Ne ratio, though very high (~ 100) do not match the meteoritic observations (see Maeder 1983b).

S.E.Woosley: 1. Could you comment on the uncertainty in "solar" ^{22}Ne/^{20}Ne

2. Since you have difficulties explaining the $^{29,\,30}$Si and $^{25,\,26}$Mg excesses, why not assume an evolved metallicity (1.6 times solar) to explain these values (which implies also a factor of 1.6 in ^{22}Ne/^{20}Ne) and then explain only the extra ^{22}Ne any way you please?

M.Cassé: 1. Spacecraft measurements of the isotopic composition of solar flare (SF) neon in the energy range 8 to 51 MeV/n give ^{22}Ne/^{20}Ne = (0.109 + 0.026, - 0.019 Mewaldt Spalding and Stone 1983) clearly not consistent with solar wind (SW) neon (^{22}Ne/^{20}Ne= 0.073, Geiss 1972). The SF ratio is marginally consistent with the Ne-C meteoritic component. The difference between the isotopic composition of SF and SW neon is not understood (see however Mullan 1983).

2. This is a pertinent question. I agree, to explain the ^{29}Si and ^{30}Si excesses we might have recourse to the supermetallicity model (Woosley and Weaver 1981), but before concluding to this necessity the Berkeley results need to be confirmed by independent experiments of similar quality. Anyway, I admit that a composite picture of the CR sources aimed at explaining their isotopic anomalies seems difficult to avoid. Therefore it is important, to narrow down the theoretical possibilities, to propose definite predictions that can be tested observationally as e.g. those concerning the ^{54}Fe/^{56}Fe and ^{58}Fe/^{56}Fe source ratios (Woosley and Weaver 1981, Prantzos, Arnould and Cassé 1983). Also, in the light of the results presented by F. Thielemann in this meeting (Arnett and Thielemann 1984) concerning the SNII yields, I am inclined to revise my pessimistic position (Cassé 1981) concerning the direct acceleration of CR's from SN ejecta.

SESSION II: TOPICS ON NUCLEAR REACTIONS

H. P. TRAUTVETTER, J. GÖRRES, K. U. KETTNER and C. ROLFS: The Ne-Na
        Cycle and the ^{12}C+α Reaction

M. RAYET: Nuclear Effective Forces and Their Use in Astrophysical
        Problems

THE NE-NA CYCLE AND THE ^{12}C+α REACTION

H.P. Trautvetter, J. Görres, K.U. Kettner
and C. Rolfs

Institut für Kernphysik, University Münster
West Germany

I. NUCLEAR REACTION RATES

Some problems concerning the measurement of reaction rates involving charged particle reactions are indicated in Fig. 1. For a nondegenerate gas the energy distribution of the particles follows the Maxwell-Boltzman law. The maximum of such a distribution is reached e.g. for T = 15 x 10^6 °K at kT = 1.3 keV. For this reason a nuclear reaction relevant to stellar burning proceeds far below the Coulomb barrier and its energy dependence is hence controlled by the barrier penetrability. In folding those two functions one obtains the so-called Gamow peak (Fig. 1) whose integral represents the non-resonant reaction rate. The effective energy is higher than kT and amounts to e.g. E_{eff} = 5.9 keV for the p+p reaction at T = 15 x 10^6 °K. This is still much too low for any direct yield measurement using present techniques.

For extrapolation procedure one usually converts the cross section function into the astrophysical S-factor function defined by the relation

$$S(E) = \sigma(E) \times E \times \exp(2\pi\eta)$$

where the major energy dependencies due to the barrier penetration ($e^{-2\pi\eta}$) and the de Broglie wavelength ($\lambda^2 \sim 1/E$) are removed. A relatively flat S(E) curve results for a non-resonant reaction, which facilitates extrapolation from experimentally accessible regions to the effective energy of the Gamow peak (dashed line in the bottom graph of Fig. 1). Special care has to be taken not to miss the contribution of any low energy resonances or the high energy tail of a subthreshold resonance, which fall into the Gamow peak region (Fig. 1). Since here again, these resonances can not be studied in a direct way, all properties of the corresponding compound nuclear states have to be investigated via complimentary nuclear reactions reaching these states.

This situation is encountered for the proton induced reactions on the

C. Chiosi and A. Renzini (eds.), Stellar Nucleosynthesis, 79–89.

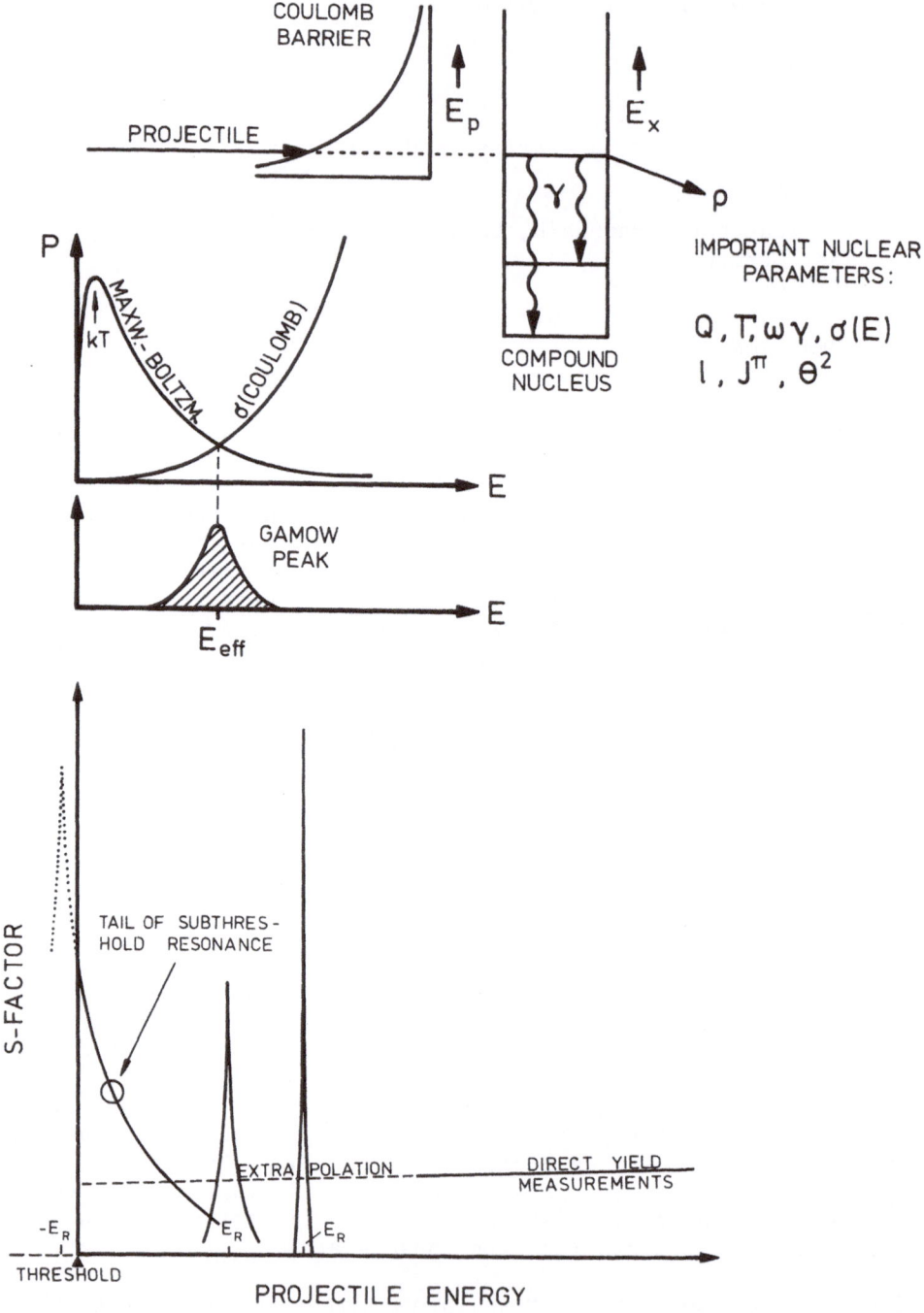

Figure 1. Features of charged particle reaction rates.

Ne-isotopes as well as for the ^{12}C+α capture reaction, which is of importance in the He-burning phase.

Important quantities to measure are the absolute cross section, resonance strengths and widths. Such absolute determinations require a precise knowledge of target stochiometry, target thickness, absolute detection efficiency and absolute beam current integration. Consequently, discrepancies of a factor of two and more have been noted in the literature (Becker, 1982).

A new method for calibration of astrophysical relevant cross sections and resonance strengths was achieved (Becker, 1982) by reversing the otherwise usual reaction X(^{1}H,Z)Y into ^{1}H(X,Z)Y. This is accomplished by using heavy ion beams on ^{1}H- or ^{4}He-windowless gas jet targets. (For details see Becker, 1982).

The ^{1}H(^{19}F,$\alpha\gamma$)^{16}O reaction was chosen as a calibration standard. The cross section of this reaction was determined relative to Rutherford scattering by observing the recoiling ^{1}H-particles together with the reaction-α-particles in the same detector (Fig. 2). There are several advantages using this method:

   i) high purity of the target (impurities < 0.01 %, Fig. 2) and
      projectile, ii) independence of variation in pressure and
      number of projectiles, iii) same geometry for reaction and
      recoil particles, iv) absolute efficiency determination of
      a γ-detector via simultaneous observation of the E_γ = 6.13 MeV
      decay in ^{16}O.

Other reaction cross sections of resonance strengths can then be determined relative to this standard by simply changing the heavy ion beam and/or the target gas.

The method of reversing target and projectile was also used for the ^{12}C+α reaction. In this way the ^{13}C(α,n)-background reaction was avoided and thus the detection sensitivity could be increased by up to a factor of 1000 as compared with previous work (Dyer et al., 1974).

II. THE NE-NA CYCLE

In the presence of any seed material of Na and the Ne-isotopes, hydrogen burning can also proceed via the NeNa cycle:

$$^{20}Ne(p,\gamma)^{21}Na(\beta^+\nu)^{21}Ne(p,\gamma)^{22}Na(\beta^+\nu)^{22}Ne(p,\gamma)^{23}Na(p,\alpha)^{20}Ne.$$

This cycle will not play a significant rôle in the energy production in stars as compared to the CNO-cycles because of the increased height of the Coulomb barrier in the charged particle reactions. However, it might be of importance for elemental abundance considerations.

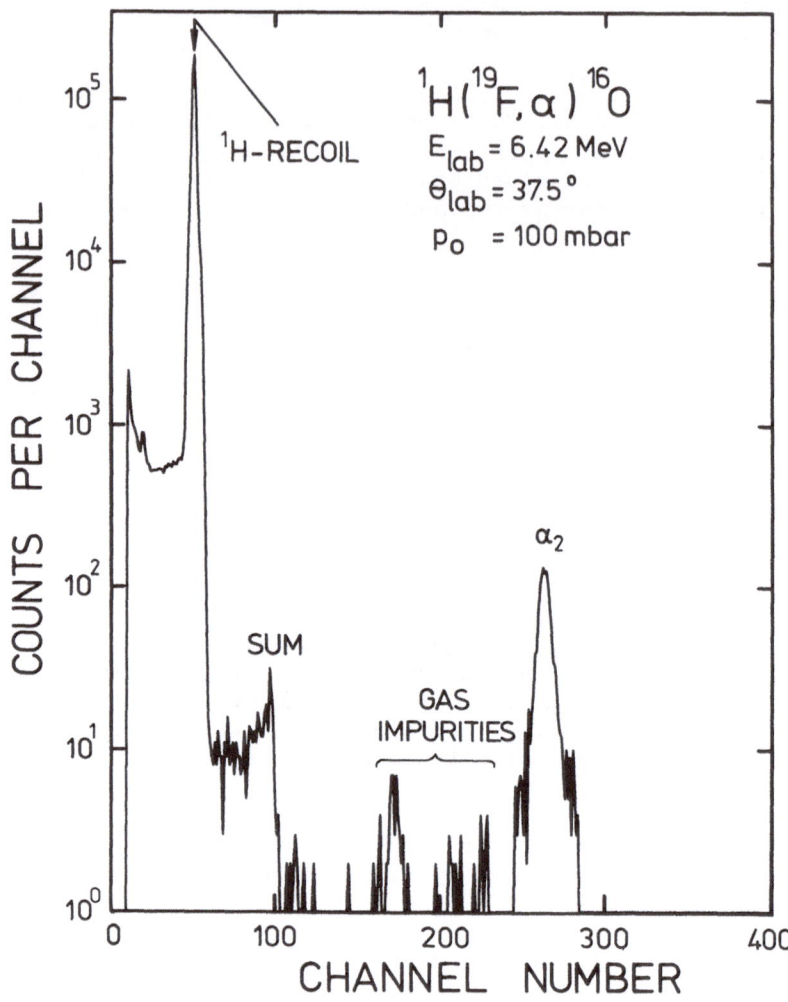

Figure 2. Sample particle spectrum at the $E_{c.m.}$ = 324 keV
resonance in ^{2}He(^{19}F,α)^{16}O, gas impurities are
less than one part in 10^4.

It has been shown(Zyskind etal) that at $T_9 \sim 0.08$ the ^{23}Na(p,α) reaction
rate is about a factor of 130 stronger than the competing ^{23}Na(p,γ)
rate. For T_9 < 0.05 the ratio $\langle\sigma v\rangle$ (p,α)/$\langle\sigma v\rangle$(p,γ) can range between
50 and 20000 depending on the absence or presence of a E_p = 39 keV-
resonance. From these results the cycling is guaranteed for $T_9 < 0.1$.

The other reactions involved in the cycle are all (p,γ)-reactions on
the Ne-isotopes. The ^{20}Ne(p,γ)-reaction has been studied extensively

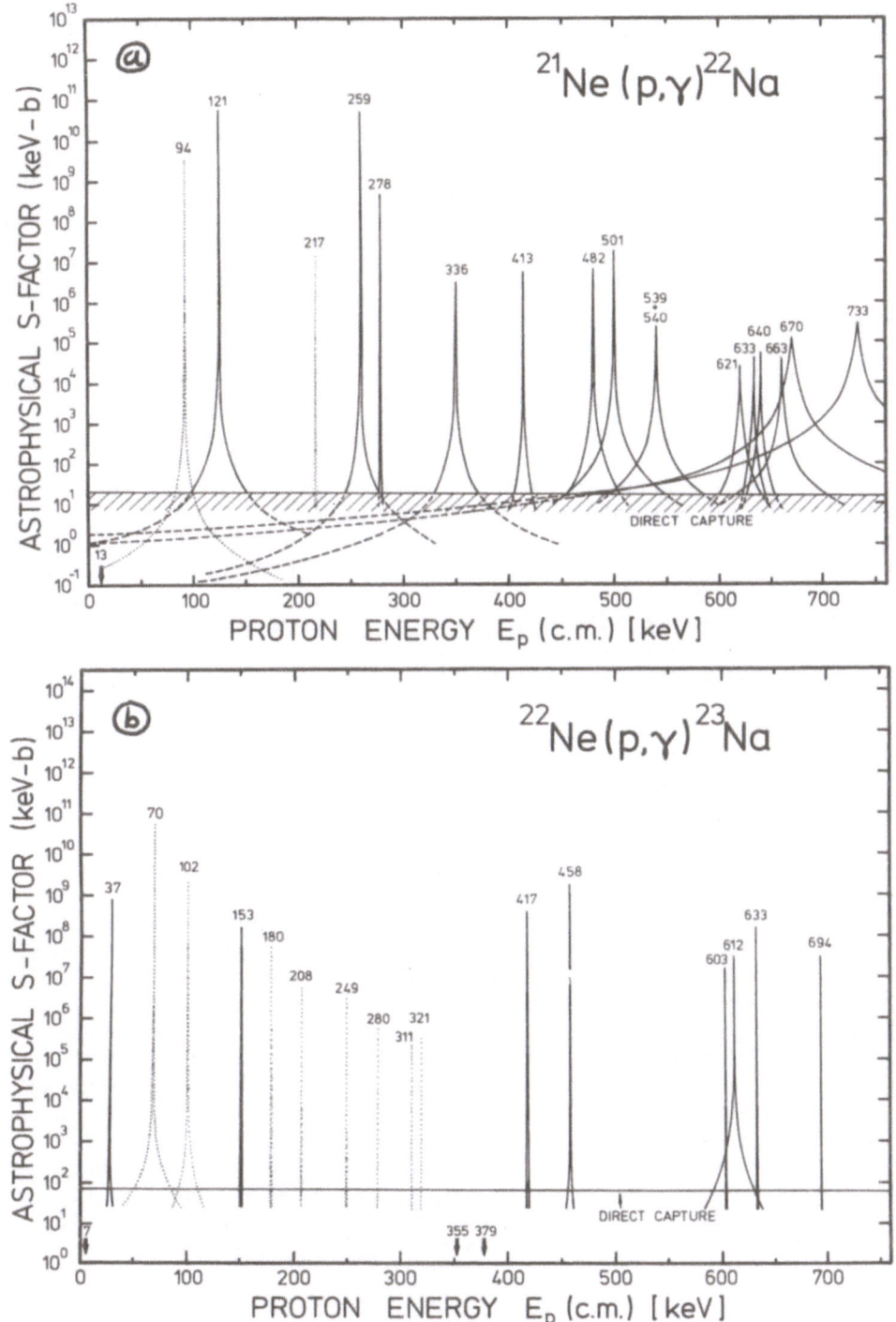

Figure 3. S-factor curve for the ^{21}Ne(p,γ)^{22}Na reaction (a) and the ^{22}Ne(p,γ)^{23}Na reaction (b).

(Rolfs etal.75) down to $E_{\bar{p}}$ = 0.37 MeV. From this work no major uncertainty in the reaction rate is expected.

For the other isotopes, ^{21}Ne and ^{22}Ne, (p,γ)-resonances were observed
down to E_p = 400 keV, where the relevant stellar temperature corresponds
to T9 ~ 0.5. From the compound level scheme in ^{22}Na and ^{23}Na one expects
in principle 6 and 13 further low-lying resonances, respectively. These
low energy regions were studied extensively at the Münster and Stuttgart
accelerators in an attempt to determine the resonance-strengths (or upper limits) and the amplitude of the non-resonant direct capture process.

In Fig. 3 the S-factor curves for the two reactions are shown. For the
^{21}Ne(p,γ)-reaction (top of Fig. 3) four new resonances at $E_{c.m.}$ = 121,
259, 278 and 336 keV have been found (Görres, 1982). For the resonance
expected at $E_{c.m.}$ = 217 keV an upper limit for its strength was determined. This limit indicates that this resonance, if present, plays no
significant rôle in the reaction rate. In the vicinity of the expected
$E_{c.m.}$ = 94 keV resonance no evidence for its existence has been found.
However, with the assumption of a reduced particle width Θ_p^2 = 1 for
this state an even lower upper limit on the resonance strength could be
assigned (Fig. 3) compared to the experimentally determined limit. The
direct capture contribution was calculated using the reduced particle
widths of the final bound states as obtained from stripping data (Görres,
1983). The resulting curve (hatched line in Fig. 3) is not in contradiction with experimental findings at higher energies (Görres, 1983).
The resulting reaction rate from the present experiment (Görres, 1982,
and Görres 1983)   differs by as much as five orders of magnitude at
T9 ~ 0.1 from the rate given by Fowler et al. 1975).

The bottom graph of Fig. 3 shows the S-factor curve of the ^{22}Ne(p,γ)^{23}Na
reaction. None of the expected 10 resonances could be found in a direct
measurement below $E_{c.m.}$ = 321 keV and only upper limits could be extracted (dotted lines in Fig. 3). However, it was possible to determine the
strengths of the $E_{c.m.}$ = 36 keV and 153 keV-resonances from the observation of the direct capture transition into the corresponding compound
states in ^{23}Na at E_x = 8829 keV and 8945 keV respectively, which allows
to deduce reduced particle widths. Together with the likely assumption
(for this low energy domain) of $\Gamma_p \ll \Gamma_\gamma$, it follows $\omega\gamma \sim \omega\Gamma_p$ and one obtains $\omega\gamma(E_p$ = 36 keV) = 2.0×10^{-14} eV and $\omega\gamma(E_p$ = 158 keV) = 6.5×10^{-7} eV.
Similarily, a search for direct capture transition into the 8866 keV
($E_{c.m.}$ = 70 keV) state yielded an upper limit of $\omega\gamma \leq 4.2 \times 10^{-9}$ eV.

Direct capture transitions to bound states in ^{23}Na have also been studied extensively and the resulting S-factor curve is shown in Fig. 3.
The major uncertainty in the reaction rate could be reduced from previously (Görres, 1982) seven orders of magnitude down to a factor of
330 at T9 ~ 0.1.

Using the new reaction rates, abundance ratios for the nuclei involved
in the Ne-Na cycle were calculated. For the condition of equilibrium
and of stellar burning times $\tau_s \gg \tau_\beta$, the ratio of abundances of two

species N_i and N_j is simply given by the inverse ratio of the corresponding reaction rates: N_i/N_j = $<\sigma v>_j/<\sigma v>_i$. The resulting curves are
shown in fig. 4a – 4c. The solid (dashed) line represents the results
obtained for the uncertainty factor f = 0 (f = 1). These results together with the condition $\sum N_i$ = 1 lead to the abundance distributions
as shown in fig. 4d (f = 0) and fig. 4e (f = 1).

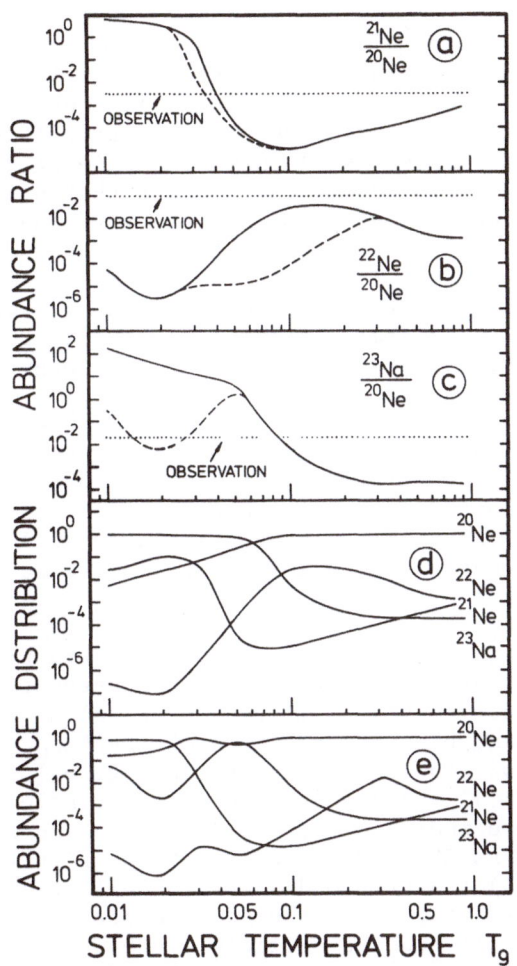

Figure 4. Abundance ratios (a – c) and distributions (d + e)
          from the operation of the NeNa-cycle assuming
          equilibrium condition.

One feature is immediately evident: above $T_9 \sim 0.06$ the major ash from
the operation of the NeNa-cycle is ^{20}Ne whereas below $T_9 \sim 0.06$ ^{23}Na
and/or ^{21}Ne dominates, depending on the uncertainty factor f.

The ^{22}Ne-isotope plays at no temperature a significant rôle, ruling out the possibility that the operation of this cycle could be the origin for the Ne-E component found in meteorites (Black, 1972).   However, Hillebrandt et al., 1982)   used in our reaction rates for calculations on a hot NeNa-cycle ($\tau_\beta > \tau_s$) and found that sufficient ^{22}Na may survive to yield ^{22}Ne via the subsequent decay ^{22}Na($\beta^+\nu$)^{22}Ne.

III. HE-BURNING OF ^{12}C

For nucleo-synthesis considerations and stellar evolution models the ^{12}C(α,γ)^{16}O reaction is of great importance. In contrast to the triple α-process, whose reaction rate is known to about 20 %, the capture rate of α-particles on ^{12}C was less well known in spite of great experimental efforts (Dyer etal.74). This is due to the fact that at relevant stellar energies the rate is mainly determined by the high energy wing of the $E_{c.m.}$ = -45 keV (J^π = 1$^-$) subthreshold resonance as well as by the low energy tail of the $E_{c.m.}$ = 2418 keV (J^π = 1$^-$) resonance. Since both incoming waves have the same orbital angular momentum, interference effects between the two amplitudes are expected. One has to study therefore the excitation curve precisely over a wide range of energy, in particular as low as technically possible.

We used the technique described in section I by reversing the rôle of target and projectile. In this way we could make use of the full ^{12}C-beam of up to 50 μA particle current in conjunction with γ-detectors in close geometry to the target. In addition a coincidence set up was used to search for a cascading E2 component via the E_x = 6.92 MeV, J^π = 2$^+$ state in ^{16}O.

The resulting S-factor curve for the groundstate transition is shown in Fig. 5a. From this figure it is evident that constructive interference of the E1 amplitude for both J^π = 1$^-$ resonances alone can not explaine the data at the low and high energy tails of the $E_{c.m.}$ = 2418 keV J^π = 1$^-$ resonance (dotted line labeled E1 capture in fig. 5a). The contributions of the individual J^π = 1$^-$ resonances alone using single level Breit-Wigner formalism are also indicated by dashed lines. The missing amplitude could be interpreted as E2 capture due to the high energy tail of the E = -245 keV (J^π = 2$^+$) subthreshold state as well as direct E2-capture to the ground state (dotted line labeled E2-capture in fig. 5a). This amplitude implies a full reduced particle width $\Theta_\alpha \sim$ 1 for the E_x = 6.92 MeV (J^π = 2$^+$) state and $\Theta^2 \sim$ 0.25 for the ground state.

From the coincidence experiment also a direct capture transition into the 6.92 MeV, J^π = 2$^+$ state (dotted line in fig. 5b) was found to be necessary to explain the data (dashed line in fig. 5b represents single level Breit-Wigner shapes for the J^π = 1$^-$ and 4$^+$ resonances, solid line is the incoherent sum of all amplitudes). It should be pointed out that from the observation of the direct capture into the 6.92 MeV, J^π = 2$^+$ state a reduced particle width of $\Theta^2_\alpha \sim$ 1 is deduced for this state in excellent agreement with the findings from the groundstate transition (see above).

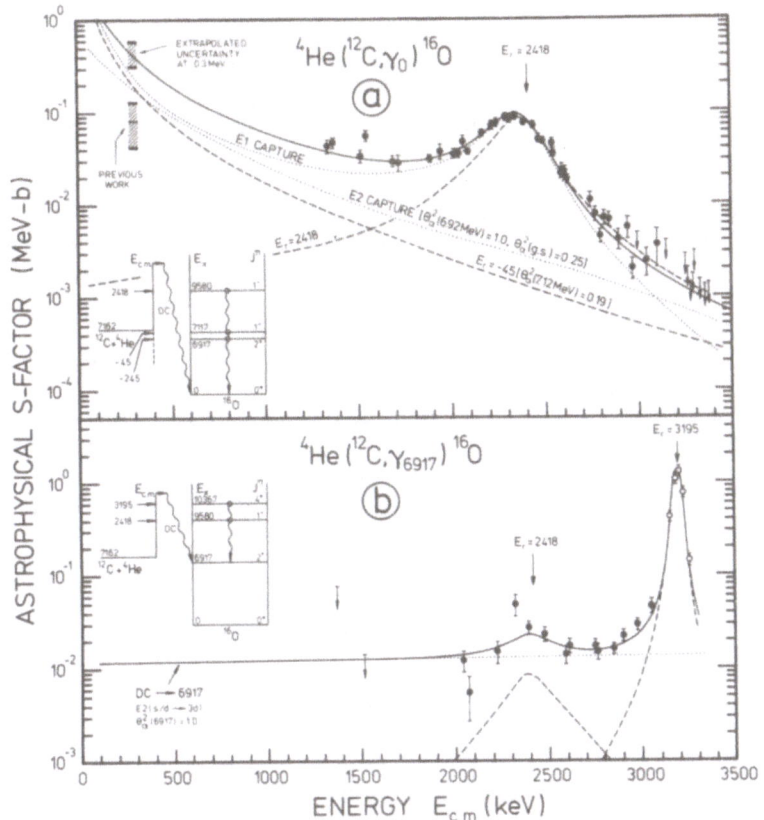

Figure 5. S-factor curve for the ^{12}C+α capture reaction.

In conclusion it is found (Kettner et al.) that the new experiment results in an extrapolation of the S-factor curve (solid line in Fig. 5a) to stellar energies (E_0 = 0.3 MeV) leading to S(0.3 MeV) = $0.42^{+0.16}_{-0.12}$ MeV barn, which has to be compared with previous results (Dyer et al.,74) (Koonin et al.,74) :S(0.3 MeV) = $0.08^{+0.05}_{-0.04}$ MeV barn.

The influence of the new rate is discussed on the basis of first calculations by K.F. Thielemann and W.D. Arnett in these proceedings.

REFERENCES

Becker, H.W., Kieser, W.E., Rolfs, C., Trautvetter, H.P. and Wiescher,
    M., 1982, Z. Phys. A, 305, 319
Black, D.C., 1972, Geochim. Cosmochim. Acta 36, 377
Dyer, P., and Barnes, C.A., 1974, Nucl. Phys., A233, 495
Endt, P.M., and VanderLeun, C., 1978, Nucl. Phys. A310, 1
Fowler, W.A., Caughlan, G.R., and Zimmermann, B.A., 1975, Ann. Rev.
    Astron. and Astrophys. 13, 69
Görres, J., Rolfs, C., Schmalbrock, P., Trautvetter, H.P., and Keinonen,
    J., 1982, Nucl. Phys. A385, 57
Görres, J., Becker, H.W., Buchmann, L., Rolfs, C., Schmalbrock, P.,
    Trautvetter, H.P., Vliecks, A., Hammer, J.W., and Donoghue, T.R.,
    submitted to Nucl. Phys. A
Hillebrandt, W., and Thielemann, F.K., 1982, Astrophys. J. 255, 617
Jaszczak, R.J., Gibbons, H.H., and Macklin, R.L., 1970, Phys. Rev.
    C2, 63 and 2452
Kettner, K.U., Becker, H.W., Buchmann, L., Görres, J., Kräwinkel, H.,
    Rolfs, C., Schmalbrock, P., Trautvetter, H.P., and Vlieks, A.,
    1982, Z. Phys. A308, 73
Koonin, S.E., Tombrello, T.A., Fox, G., 1974, Nucl. Phys. A220, 221
Rolfs, C., Rodney, W.S., Shapiro, M.H., and Winkler, H., 1975,
    Nucl. Phys. A241, 460
Zyskind, J., Rios, M., and Rolfs, C., 1981, Astrophys. J. Lett. 243,
    L53 and 245, L97

DISCUSSION

Mathews:
    I would like to ask if you use a natural hydrogen target for your
^{1}H(HI,γ) measurements? Deuterium, if present, even at an abundance of
10^{-4} could lead to an erroneously high cross section.
Trautvetter:
  i) We have used H_2-gas depleted in deuterium to a degree of 1 ppm.
 ii) The Q-value for the D(^{19}F,nα_1)^{16}O* reaction is Q = -0.24 MeV which
     leads to $E_{\alpha_1} \leq 0.38$ MeV for E_F = 6.45 MeV. This can be clearly
     distinguished from E_{α_1} = 2.3 MeV in the ^{1}H(^{19}F,α_1)^{16}O* reaction.
     Furthermore, because of this low α-energy in the D-induced reaction
     the cross section is greatly reduced for penetrability reasons
     and hence even with natural H-targets the cross section for γ-de-
     tection will not be enhanced.
iii) The Q-value for D(^{14}N,n)^{15}O is Q = 5.07 MeV. Therefore the charac-
     teristic γ-transitions for the E_x = 7.55 MeV state in ^{15}O are not
     influenced by the ^{14}N+D reaction.
Woosley:
    What are the prospects for measuring ^{22}Na(p,γ)^{23}Mg and ^{26}Al(n,p)^{26}Mg?
Trautvetter:
    During the course of preparation of our ^{26}Al-target we also got a lot
of ^{22}Na made. It should be possible to produce a ^{22}Na-target along similar lines ((p,n)-reaction on a cyclotron accelerator). However, in con-

trast to ^{26}Al, the ^{22}Na(p,γ)-measurement will be hampered by the intense 511 keV anihilation radiation.

With our ^{26}Al-targets we are planning to do the ^{26}Al(n,p)^{26}Mg experiment in the near future.

Iben:

Why didn't you measure the ^{4}He(^{12}C,γ)^{16}O reaction using an α-beam instead of a ^{12}C-beam?

Trautvetter:

By using the ^{12}C beam we could increase the sensitivity significantly and also overcome the ^{13}C(α,n) problem. An other ^{12}C(α,γ)^{16}O measurement does exist in the literature (Jaszczak et al., 1970.) This experiment is essentially in agreement with our results.

Iben:

What $\Theta_\alpha^2(7.12)$ value should one use according to the result of your experiment.

Trautvetter:

Our simple extrapolation procedure yields $\Theta_\alpha^2(7.12) = 0.19^{+0.14}_{-0.10}$ as compared with $\Theta_\alpha^2(7.12) = 0.1$ from (Dyer and Barnes, 1974). There are however values in the literature ranging from 0.03 to 0.4 (see Kettner et al. 1982). Our extracted reduced α-particle widths of levels in ^{16}O are in good agreement with results from α-transfer reaction. For a more reliable value, one should wait for improved theoretical analysis of our data. Such work is in progress by Koonin in Cal Tech, as far as I know.

NUCLEAR EFFECTIVE FORCES AND THEIR USE IN ASTROPHYSICAL PROBLEMS

Marc RAYET , Physique Nucléaire Théorique, Université Libre de Bruxelles, CP 229, B-1050 Bruxelles, Belgium.

1.   Introduction

Although more and better nuclear data are most often claimed by nuclear astrophysicists, the need for better nuclear theories is rarely expressed with the same conviction. The success of phenomenological, easy-to-use, formulae based on popular macroscopic theories may explain this self-confident attitude. Many astrophysical processes however deal with physical conditions or quantities which are not at present, and some will never be, accessible to terrestrial experiments and the nuclear physics involved in those processes will depend - sometimes critically - on the extrapolation of expressions established on the basis of the existing experimental data.

An example of exotic nuclear physics is encountered in the collapsing iron cores of massive stars. Being only partly neutronized, these cores are expected to retain, up to nuclear matter density and for temperatures which can exceed 10^{11} K, nucleus like structures of bound nucleons, embedded in a surrounding nucleon gas. Other unusual conditions may also arise in various burning regimes where reactions on unstable and/or excited nuclei play a non-negligible role. Another kind of difficulty is met when nucleosynthetic processes involve nuclei well off the stability region, like very neutron-rich isotopes in the r-process or neutron deficient nuclei in the p-process. From this - non exhaustive - enumeration it is clear that detailed theoretical evaluation of nuclear properties cannot be avoided in various astrophysical scenarios.

We would here support the idea that microscopic nuclear theories have recently become available from which one can reasonnably hope to get a coherent and uniform description of both bulk and surface properties of nuclei, of single particle features (s.p. excitations, level densities, shell effects ...) and collective excitations, etc. In particular, extensive Hartree-Fock calculations of nuclear static properties have been made possible by the use of effective nucleon-nucleon forces of the Skyrme type, and have given the Hartree-Fock model

C. Chiosi and A. Renzini (eds.), Stellar Nucleosynthesis, 91–95.
© *1984 by D. Reidel Publishing Company.*

a real predictive power. Recently the development of dynamical models
based on the time dependent Hartree-Fock (TDHF) equations has extended
the applicability of microscopic theories to nuclear collective motions.
The phenomenological content of these theories is essentially reduced to
the determination of effective nucleon-nucleon force parameters. They
are thus expected to be more reliable than macroscopic theories, like
the droplet model, which have often been proved to lead to questionable
extrapolations.

2.   Parametrization of an effective Skyrme force

To illustrate how a relatively simple and limited phenomenological input can efficiently drive a microscopic theory, we will limit
ourselves to the parametrization of the Skyrme interaction which is
known to be particularly handful and powerful in the treatment of many
nuclear problems. We will then consider two applications of astrophysical interest.

The Skyrme force is a contact (δ function) force containing one
central term, two gradient (non-local) terms giving the nucleons a density dependent effective mass, and a density dependent term $\rho^\gamma \delta$.
Altogether the strength and spin-exchange parameters, plus the exponent
γ , amount to 9 free parameters. A spin-orbit term is also added to the
interaction but is not crucial for the following discussion.

Early parametrization of Skyrme forces [1], with $\gamma = 1$, resulted
in a large nuclear matter incompressibility of K=360 MeV. It was later
recognized that a weaker density dependence (γ =1/6 to 1/3) was needed
to yield a value of K consistent with the more generally accepted value
of 240 MeV[2]. It was also remarked that the existing Skyrme forces did
not correctly fit the energy per particle in pure neutron matter (W_n),
as was obtained in "realistic" calculations (using a bare NN interaction
in correlated nuclear matter), for example by Friedman and Pandharipande
[3]. This deficiency was mended by Ravenhall (quoted in ref. 4) for calculating the equation of state of the neutron rich matter encountered
in collapsing iron cores.

Recently, Rayet et al.[5] performed a new Skyrme parametrization
on the basis of very straightforward constraints : apart from the usual
equilibrium conditions of infinite (symmetric) nuclear matter (energy
per nucleon W = -16 MeV,at equilibrium density ρ_o= .16 fm^{-3},K= 240 MeV),
the force is made to reproduce the zero-temperature results of ref. 3 :
the effective masses for symmetric and nucleon matter at $\rho = \rho_o$, and the
density dependence of W_n. These constraints fix 7 of the 9 parameters,
and the remaining two are determined by a fit to the radius and mass of
^{208}Pb. The resulting force (named RATP) was then shown to give a surprisingly good description of other finite nuclei.

The neutron skin of heavy nuclei is the only terrestrial example of a very neutron rich nuclear matter. The neutron skin thickness,
$r_n - r_p$ is known experimentally in a few cases, the clearest of which is
^{208}Pb, where $r_n - r_p$ = .14 ± .04 fm. It is interesting to note [5] (see
also ref. 6 for additional evidence) that a Skyrme force which fits

FP's results for W_n gives a neutron skin thickness in ^{208}Pb close to the central value .14, while forces which depart from this value give bad predictions for W_n (ρ).

Finally, it must be mentioned that Hartree-Fock calculations favour large values of the effective nucleon mass($m^*/m = 1$ rather than 0.7 used for RATP) which correspond to higher (and more realistic) level densities near the Fermi surface. Different sets of parameters were obtained by Tondeur from a fit to nuclear masses and radii [7] with the condition $m^*=m$. Those forces which fitted the neutron skin thickness in ^{208}Pb were also found to fit FP's neutron matter energy. They therefore essentially differ from RATP through the effective mass, and the choice between one or the other type of parametrization can be decided on the basis of the most appropriate choice of effective mass in a given problem.

3.   Hot and dense matter in the two bulk phases approximation

Lattimer and coll. [4] have proposed a schematic model to describe the properties of a mixture of neutron and proton gases in the temperature and density regimes prevailing in collapsing iron cores, i.e. $\rho \simeq 10^{11}$ to $3 \cdot 10^{14}$ gr cm^{-3} and $T \simeq 1$ to 15 MeV. This model shows that when a hot nucleon gas is compressed isothermically, it becomes unstable and a condensed phase appears to restore stability. This coexistence of two phases of bulk matter(i.e. surface and Coulomb effects are neglected) illustrates in which region of the (T, ρ) plane nuclei can coexist with neutron and proton gases. This region lies inside the curve shown in the figure and contains, according to the current models, the adiabatic path followed by the collapsing core. Coexistence curves have been

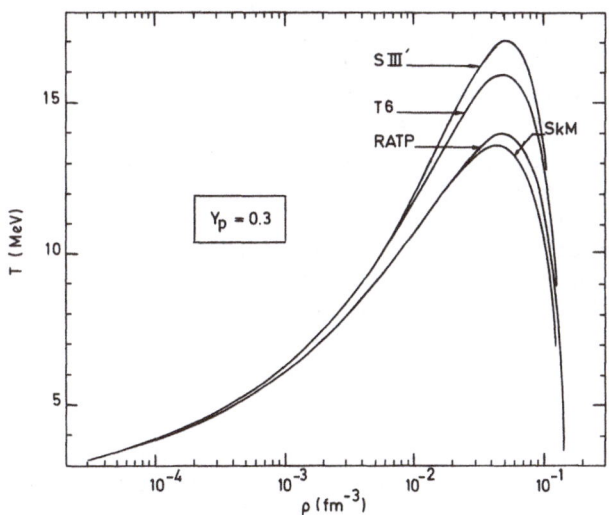

calculated here with different Skyrme forces for an average proton fraction of 0.3. The maximum temperature T_m for coexistence is seen to depend noticeably on the forces. These can easily be related to the above discussion: SIII' is an old force (with large K) modified by Ravenhall to fit the neutron matter energy, T6 is one of Tondeur's force with $m^* = m$ (and fitting neutron matter), while SkM, derived in ref.[2], has a realistic K but is not fitted to neutron matter and has bad fits to static properties of finite nuclei.

It can be seen that two effects compete in determining T_m : a large incompressibility increases T_m (compare SIII' and RATP) and a large effective mass also tends to increase T_m (compare T6 with RATP). As concluded in section 2, the choice of effective mass (small m^*/m for bulk matter of $m^* \simeq m$ for finite nuclei) is still a matter of controversy and indicates the sort of uncertainty which may still be expected

from effective forces. In fact a more realistic treatment of the core
would reveal other force dependent effects.

4.   A microscopic calculation of giant dipole resonances

The giant dipole resonance (GDR) dominates the total photon
transmission function in nuclei and is therefore of prime interest in
reactions such as (n, γ), (γ, xn), or (γ, p) which govern the synthesis of the r- and s-processed and/or p-processed nuclei.

If one wishes to extrapolate the GDR energies and widths to
neutron rich (or deficient) isotopes for which experimental information
is lacking, semi-empirical formulas based on macroscopic models, like
the liquid drop or the droplet model [8], must be taken with some care,
even if they provide a good fit to the overall set of experimentally
measured quantities. In particular, a possible isospin dependence of
the GDR energies, an effect which is not accounted for in those models,
must be seriously investigated.

Starting from the TDHF equations, Krivine et al.[2] have constructed a fluid dynamical Lagrangian density from which the GDR energy,
E_G, is, as usual, expressed as the ratio of a restoring force over a mass
parameter. Both quantities are entirely determined by the effective
force parameter and by the microscopic densities of the equilibrium
nucleus and their derivatives. We have followed this formalism but
contrary to Krivine et al. , have made use of self-consistent Hartree-
Fock densities for the equilibrium nucleus.

For ^{208}Pb (exp. E_G = 13.5 ± 0.1 MeV [9], we obtain GDR energies ranging from 11.9 for one of Tondeur's force (T6) to 14.8 for
RATP. This spread which, as already discussed in ref. 2, is explained
by the different parametrization of the non-local terms of the Skyrme
interaction, confirms the necessity to take into account both dynamical
and static properties of nuclei in the parametrization of Skyrme forces.

For lower mass nuclei, measurements of GDR energies have been
made in a few cases on different isotopes of the same element. For some
elements like Zr (90 to 94) and Mo (92 to 100), the decrease of E_G with
increasing neutron number appears to be much faster than the average
decrease of E_G with the nucleon number A, which is very well represented by the droplet model [8]. If this tendency were shown to persist for
heavier isotopes, the droplet model predictions for very neutron rich
nuclei (encountered in the r-process) would be disastrous. Our microscopic calculation does fairly well reproduce, with all the forces which
have been considered, the observed variation of E_G with the neutron
number for the Zr and Mo isotopes. It also shows however that the negative isospin effect observed when going from ^{92}Mo to ^{100}Mo, i.e. when filling the neutron 2d5/2 and 3s1/2 shells, turns into a positive effect during
the g7/2 shell filling, etc., negative effects being associated with
smaller orbital momenta and conversely. The net effect when reaching
N = 82 is to bring the ^{124}Mo GDR energy near to the droplet model value.
From this preliminary investigation it thus seem that the observed isospin effects in GDR energies might be related to local shell effects
rather than to a cumulative effect due to neutron excess. This statement

however has to be confirmed and cannot be generalized before a thourough study. A microscopic derivation of GDR widths, which present experimentally much stronger variations than E_G is also highly desirable and will be studied later.

We have investigated here two situations encountered in nuclear astrophysics which are beyond the scope of phenomenological models. They demonstrate the utility of microscopic theories and suggest that relatively simple effective forces can be made operative in a wide variety of problems provided they have been parametrized according to carefully and clearly chosen physical criteria.

References

1) Beiner, M., Flocard, H., Nguyen van Giai, Quentin, Ph. 1975, Nucl. Phys. A238, 29

2) Krivine, H., Treiner, J., Bohigas, O. 1979, Nucl. Phys. A336, 155

3) Friedman, B., Pandharipande, V.R. 1981, Nucl. Phys. A361 , 502

4) Lattimer, J.M. 1981, Ann. Rev. Nucl. Phys. Sci. 31, 19 and references therein

5) Rayet, M., Arnould, M., Tondeur, F. 1982, Astron. Astrophys. 116, 183

6) Tondeur, F.1981, in Proc. 4th Int. Conf. on Nuclei far from Stability, Helsingør, CERN 81-09, p. 81

7) Tondeur, F., Brack, M., Håkansson, H.-B. Pearson, J., in preparation and private communication

8) Myers, W.D., Swiatecki, W.J., Kodoma, T., El-Jaick, L.J., Hilf, E.R. 1977, Phys. Rev. C15, 2032 ; Thielemann, F.-K., Arnould, M. 1983, Proc. of the Int. Conf. on Nuclear Data for Science and Technology, Antwerp, Belgium, p. 762

9) Berman, B.L. 1976, Atlas of photoneutron cross sections obtained with nonenergetic photons, Lawrence Livermore Laboratory Report No UCRL-78482 (1976);

   Buenerd, M., Lebrun, D., Martin, P., Perrin, G., de Saintignon, P., Chauvin, J., Duhamel, G. 1982, in Dynamics of Nuclear Fission and Related Collective Phenomena, Proc. 1981, P. David, T.Meyer-Kuckuk, and A. Van der Woude eds. , Springer Vol. 158

SESSION III: QUASI-STATIC STELLAR EVOLUTION AND RELATED NUCLEOSYNTHESIS

A. RENZINI: Some Aspects of the Nucleosynthesis in Intermediate
       Mass Stars

A. MAEDER: Hydrostatic Evolution and Nucleosynthesis in Massive Stars

L. GREGGIO: Nucleosynthesis During the Quasi-Static Evolution of
       Massive Stars

W. D. ARNETT and F.-K. THIELEMANN: Presupernova Yields and Their
       Dependence on the ^{12}C (α,γ) O^{16} Reaction

J. J. COWAN, A. G. W. CAMERON and J. W. TRURAN: r-Process Production
       in Low Mass Stars

G. J. MATHEWS, S. A. BECKER and W. M. BRUNISH: Synthetic H-R Diagrams
       as an Observational Test of Stellar Evolution Theory

G. BERTELLI, A. BRESSAN and C. CHIOSI: Evidence for a Large Main
       Sequence Widening: A Plea for Overshoot and Higher Heavy
       Element Opacity

C. DOOM: On the Structure of the Upper HRD of Humphreys

SOME ASPECTS OF THE NUCLEOSYNTHESIS IN INTERMEDIATE MASS STARS

Alvio Renzini
Dipartimento di Astronomia, CP 596, I-40100 Bologna

ABSTRACT

Intermediate mass stars (IMS) are potentially important contributors to the nucleosynthesis of several elements and isotopes. The relevant nucleosynthetic processes are briefly described, and then attention is focused on CNO isotopes and ^{26}Al. Some IMS may also experience a carbon deflagration, which is believed to release important quantities of iron. The (uncertain) parameters controlling the nucleosynthesis in IMS are then discussed, a comparison with pertinent observations is presented, and future investigations are suggested.

1. INTRODUCTION

By intermediate mass stars (IMS) one intends those stars developing an electron-degenerate C-O core following the exhaustion of helium at the center. This sets an upper limit to the stellar initial mass (M_{up}) in the vicinity of 8 $M_\odot$, the precise value being a function of the initial composition (Becker & Iben 1979). It is worth realizing that M_{up} marks a sharp discontinuity in the evolutionary behaviour of stars. For initial masses (M_i) smaller than M_{up}, stars experience the asymptotic Giant Branch phase (AGB) and eventually die, either as C-O white dwarfs (having failed to ignite carbon), or ignite carbon under degenerate conditions, thus experiencing a carbon deflagration which causes the total disruption of the star (cf. Nomoto, this volume) giving rise to what is called a supernova event of type $I\frac{1}{2}$ (Iben & Renzini 1983). One usually refers to M_w as the critical mass such that for $M_i < M_w$ stars produce a WD remnant, and for $M_w < M_i < M_{up}$ stars produce a SN event of type $I\frac{1}{2}$. Conversely, it is generally accepted that for $M_i > M_{up}$ stars ignite carbon non-degeneratly, thus failing to experience the AGB phase, and eventually undergo a core collapse giving rise to a SN event of type II, and leave a neutron star

C. Chiosi and A. Renzini (eds.), Stellar Nucleosynthesis, 99–114.
© 1984 by D. Reidel Publishing Company.

remnant. Such a striking difference in the evolution implies a seemingly
abrupt change in the quality and quantity of the nucleosynthetic products.
In other words, the chemical yields suffer a sharp discontinuity across
$M_i = M_{up}$, an aspect which has been frequently overlooked in the literature.

Following the above considerations, one can distinguish two major
aspects in the nucleosynthesis in IMS: a) the nucleosynthesis during the
quasi-static evolution, and b) that due to the core explosion triggered
by the degenerate carbon ignition. Section 2 of this review is devoted
to the first aspect, Section 3 to the second, and, finally, Section 4
deals with the problems rised by the comparison of theory with the observations, and suggests future observational and theoretical studies.

2. NUCLEOSYNTHESIS DURING QUASI-STATIC EVOLUTION

During the quasi-static evolution of IMS a variety of thermonuclear
transmutations takes place in the stellar interior. The newly synthetized elements may find their way to the surface (thanks to convective
dredge-up episodes), and eventually to the interstellar medium (thanks to
the stellar wind and envelope ejection). Following Iben & Truran (1978)
one distinguishes three dredge-up phases, each corresponding to an inward
penetration of the envelope convection which engulfs deeper and deeper
layers, where various nuclear transmutations have previously occurred.

2.1 The Three Dredge-ups

The first dredge-up takes place when a star first reaches the red
giant branch (RGB), while burning hydrogen in a shell. Correspondingly,
the surface abundance of the light elements Li, Be, and B is drastically
reduced (Iben 1965), ^{3}He is strongly enhanced (Rood et al 1976), ^{4}He increases by (at most) 10 % (Sweigart & Gross 1978), ^{12}C is depleted and ^{13}C
and ^{14}N are enhanced (Iben 1964, 1965, 1977; Dearborn & Eggleton 1976).
No other important changes are predicted by the canonical theory, and a
critical comparison with the pertinent observations will appear elsewhere
(Iben & Renzini 1984).

The second dredge-up takes place shortly after the exhaustion of He
at the stellar center, when the inner edge of the convective envelope can
penetrate through the H-He discontinuity, reducing the mass of the helium
core, and bringing to the surface relevant quantities of ^{4}He and ^{14}N.
The surface abundance of ^{12}C, ^{13}C, ^{16}O decreases by a nearly identical
factor, but these latter changes are very modest. For a systematic study
of the second dredge-up see Becker & Iben (1979).

The third dredge-up, first encountered by Iben (1975), works during
the subsequent AGB phase, and consists in a number of mixing episodes:

following each helium-shell flash the inner edge of the convective envelope can penetrate through the hydrogen-helium discontinuity, thus reaching into the intershell region where incomplete helium burning has previously taken place during the flash peak. Therefore, following each
flash, significant amounts of ^{4}He and ^{12}C, and some ^{16}O, are brought to
the surface. Moreover, during the pulse peak, practically all the ^{14}N
present in the intershell region is converted to ^{22}Ne, thanks to two consecutive α-captures, and the third dredge-up brings to the stellar surface
part of this fresh ^{22}Ne. The fate of the ^{22}Ne in the intershell region
depends on the mass of the hydrogen-exhausted core (M_H). As investigated
in detail by Iben & Truran (1978), in AGB stars with rather massive cores
($M_H > \sim 0.9 \ M_\odot$) the reaction ^{22}Ne(α,n)^{25}Mg effectively operates, thus re-
leasing one free neutron per every destroyed nucleous of ^{22}Ne. Part of
these neutrons are then captured by ^{22}Ne and ^{25}Mg themselves, thus producing light s-process isotopes (^{23}Na, ^{26}Mg, ^{27}Al, and their neutron-capture
progeny). The remaining neutrons are captured by ^{56}Fe seeds and their
neutron-capture progeny, thus producing the heavy s-process isotopes.
The most attractive characteristics of this process is that the heavy s-
process isotopes are naturally produced *in solar system proportions*, as
emphasized by Iben & Truran. In AGB stars with smaller M_H only a small
fraction (a few percent) of the ^{22}Ne is burned into ^{25}Mg plus a neutron,
but, as shown by Iben & Renzini (1982), another neutron source is likely
to operate in such stars. Indeed, following the pulse peak a semiconvec-
tive zone now appears at the top of the carbon-rich intershell region,
and protons from the temporarily inactive hydrogen shell are mixed inward,
into the carbon-rich intershell. The subsequent reignition of the hydrogen shell combines these protons with ^{12}C, to form a ^{13}C-rich layer which
does not suffer further nuclear processing until it is engulfed by the
convective shell generated by the next flash. The reaction ^{13}C(α,n)^{16}O
then goes to completion, thus producing nearly as many neutrons as protons
were previously introduced into the ^{12}C-rich region. In this case, as emphasized by Iben & Renzini (1982), the abundance pattern of the resulting
neutron-capture isotopes is *not* expected to resemble to the solar-system
distribution. The nucleosynthesis resulting from the thermal pulses in
AGB stars is further reviewed by Iben & Renzini (1983, 1984) and by Iben
(this volume).

2.2 The Envelope Burning Process

Another nucleosynthetic process is potentially active in AGB stars.
Indeed, in bright AGB models the temperature at the base of the convective
envelope can exceed $4 \ 10^7$ K (Sugimoto 1971) while the hydrogen-burning
shell is active, and Iben (1975) found that in his 7 $M_\odot$ model the just
dredged-up ^{12}C was rapidly converted first to ^{13}C and then to ^{14}N. One
now refers to nuclear processing at the base of the convective envelope

as the envelope burning (EB) process.

Renzini & Voli (1981) have extensively investigated the effects of the EB process in AGB stars. In particular, they have shown that, for every choice of the composition and α (= ℓ/H, the mixing length parameter), there is a value of the stellar initial mass M_{EB} below which the conversion of ^{12}C into ^{14}N does not occur at all. For masses in the range between M_{EB} and ~ 1.25 M_{EB} the conversion proceeds at a rate that increases very steeply with M_i, and for even larger masses the conversion occurs at a steady rate. For Z = 0.02, M_{EB} decreases from 6.8 $M_\odot$ to 3.3 $M_\odot$, as α is increased from 1 to 2. M_{EB} also decreases with decreasing metal abundance. For example, for α = 2, M_{EB} decreases from 3.3 to 2.5 $M_\odot$ when Z is decreased from 0.02 to 0.001. The EB process also affects the envelope abundance of other isotopes, like ^3He, ^4He, ^7Li (recently reviwed by Renzini 1983), and ^{26}Al. One important aspect of the EB process is that it operates only in AGB stars of sufficiently large core mass. For instance for α = 2 it works only when M_H is larger than 0.8-0.9 $M_\odot$, which corresponds to AGB stars brighter than $M_{bol} \simeq -6$. In the Sections 2.4-2.7 I shall concentrate on the combined effects that the three dredge-ups and the EB process have on the nucleosynthesis of CNO isotopes and ^{26}Al.

2.3 Effects of Mass Loss on Nucleosynthesis

AGB stars lose mass through what is called a regular red giant wind, and, for M_i < M_w they should eject their remaining envelope thanks to a *superwind* which is thought to be responsible for the origin of planetary nebulae (cf. Renzini 1981a; Iben & Renzini 1983). Both mass loss processes are still poorly known. In particular, the dependence of the wind mass loss rate (MLR) on the stellar parameters (luminosity, mass, radius, surface composition) is uncertain, and we do not precisely know for which combination of these parameters the transition wind-superwind should take place. Therefore, both processes are currently parametrized in order to explore the effects of mass loss on the evolution of AGB stars. The so-called regular wind is parametrized in such a way that the MLR is given by a parameter η times the Reimers' rate (Reimers 1975), and the superwind is assumed to start when the envelope mass decreases below a critical value, function of the stellar luminosity and then of the core mass M_H (see for instance Renzini & Voli 1981).

Mass loss affects the nucleosynthesis during the AGB in two major ways. First, the larger the assumed mass loss the less the core grows before the termination of the AGB phase, the smaller the number of pulses and dredge-up episodes, and then the smaller the total amount of freshly synthetized elements which is dredged-up and shed into the interstellar medium. Second, the larger the assumed mass loss the larger the resulting value of M_w, and then the narrower the range of initial masses (M_w < M_i < M_{up}) of stars giving rise to a carbon-deflagration SN (SN of type $I\frac{1}{2}$). I shall return to this point in Section 3.

2.4 Carbon

As noted by Iben & Truran (1978), IMS produce large amounts of ^{12}C, when the EB process is neglected, to the extent that they can rival with massive stars in determining the total yield of ^{12}C per stellar generation. However, when the EB process is included the amount of ejected ^{12}C rapidly drops with increasing α. For example, from the tabulations in Renzini & Voli (1981) we see that (for η = 1/3) a 5 M_o star ejects ~ 0.09 $M_\odot$ of fresh ^{12}C when the EB process does not operate ($\alpha \lesssim 1.2$, for this particular choice of the mass). This figure is reduced to ~ 0.04 $M_\odot$ when α = 1.5, and to ~ -0.015 $M_\odot$ for α = 2, i.e. stellar ejecta become somewhat depleted in ^{12}C compared to the initial composition. Therefore, the contribution of IMS to the yield of ^{12}C declines with increasing α, and Serrano & Peimbert (1981) conclude that the observed C/O ratio in the solar neighborhood requires a large value of α($\gtrsim 2$), i.e. an efficient EB process. We note however that the same result can be achieved by increasing the MLR parameter η.

As anticipated in Section 2.2, the EB process converts ^{12}C first into ^{13}C and then into ^{14}N. Figure 1 shows the mass $M(^{13}C)$ of fresh ^{13}C that is ejected, as a function of the initial mass and for various values of α. The data are derived from the tabulations in Renzini & Voli (1981) for the case Z = 0.02 and η = 1/3. When the EB process is neglected, only the contribution from the first dredge-up is present. In this case, the typical $^{12}C/^{13}C$ ratio in the ejecta of one generation of IMS should be around 500, ^{12}C obviously coming from the third dredge-up. Since a seemingly large ratio is expected for the total ejecta of massive stars, one is forced to conclude that the low $^{12}C/^{13}C$ ratio in either the solar system (~ 90) or the interstellar medium (~ 60, cf. Wannier 1980) requires an additional source of ^{13}C which should be at least 10 times more productive than the canonical first dredge-up.

Figure 1 shows that when the EB process is allowed to operate, large amounts of ^{13}C are generated. As anticipated by Renzini & Voli, most of this ^{13}C is of *primary* nature, largely resulting from the conversion of ^{12}C produced within the star itself by triple-α reactions. Note also that, for every given initial mass, the amount of produced ^{13}C is not a monotonic function of α. In fact, as α increases above a certain value ^{13}C is efficiently processed into ^{14}N. Correspondingly, the yield of ^{13}C per generation of IMS has a maximum for α around 1.5, the precise value

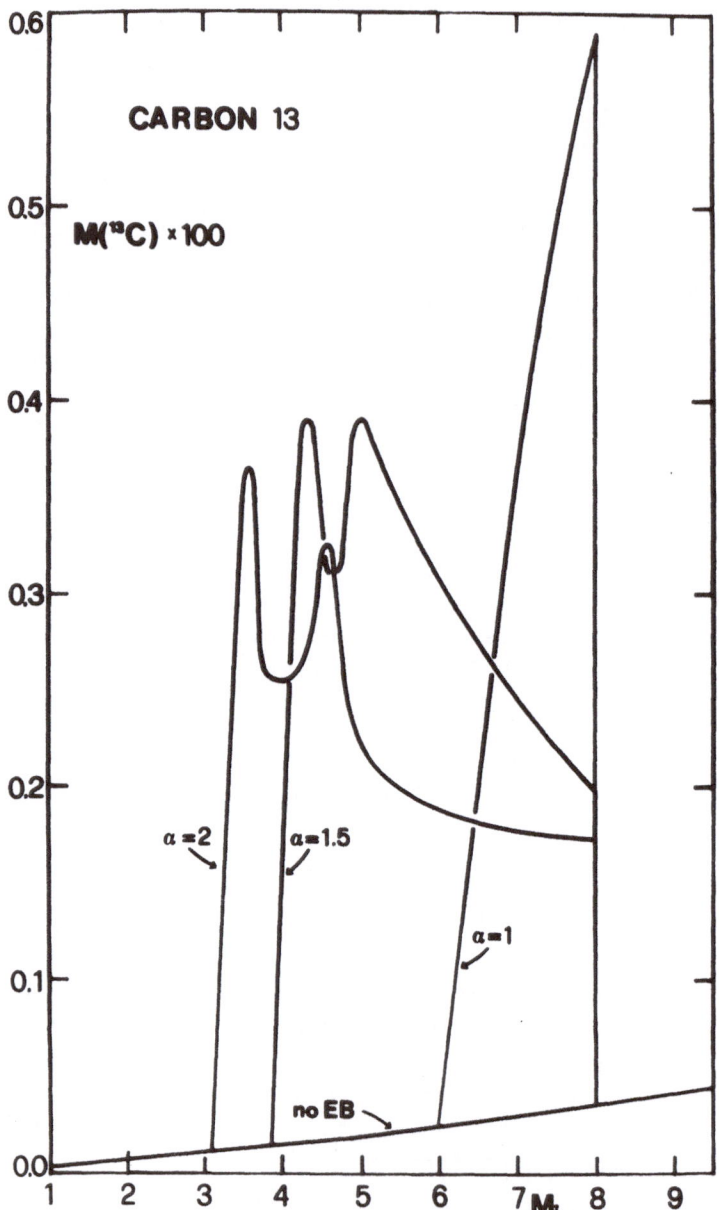

Figure 1: The mass of ejected ^{13}C per star, as a function of the initial mass and for various values of the mixing-length parameter α. The lower line refers to the case when the EB process is neglected. The mass loss parameter η is 1/3 and Z = 0.02.

depending on the adopted mass loss parameters and initial mass function. Finally, note the discontinuity in $M(^{13}C)$ at $M_i = 8\ M_\odot$, which corresponds to the sharp transition from IMS to massive stars ($M_{up} = 8\ M_\odot$, in this case). Similar discontinuities will be seen in figures 2 and 3.

2.5 Nitrogen

The case of ^{14}N is illustrated in Figure 2, for the same choice of the parameters η and Z. When the EB process is ignored, ^{14}N is produced only by the first two dredge-ups, as indicated in the figure. This ^{14}N is of purely *secondary* nature. Then, large amounts of ^{14}N are produced when the EB process is allowed to operate in conjunction with the third dredge-up. In this case the production of ^{14}N increases monotonically with α, reaching values as large as $\sim 0.12\ M_\odot$ per star. The corresponding mass fraction of ^{14}N in the ejecta can reach values as high as 0.03. Note again that most of this ^{14}N is of *primary* nature.

One important aspect of the nucleosynthesis of ^{14}N by the EB process is its dependence on the metal abundance Z. Indeed, as mentioned before, M_{EB} decreases with decreasing Z, and then the mass range of stars experiencing the EB process ($M_{EB} < M_i < M_{up}$) becomes wider. For example, in the case $(\eta,\alpha) = (1/3,1.5)$ and $M_i = 4\ M_\odot$, $M(^{14}N)$ increases from ~ 0.003 to $\sim 0.056\ M_\odot$ when Z is decreased from 0.02 to 0.004 (cf. Renzini & Voli 1981). For $M_i = 5\ M_\odot$ the corresponding variation is from ~ 0.063 to ~ 0.10 $M_\odot$. Moreover, one may expect a decrease in mass loss with decreasing Z, and this effect would further enhance the production of primary nitrogen in metal poor stars. All in all, we should then expect the yield of primary ^{14}N per stellar generation to be a decreasing function of metallicity. On the contrary, the production of secondary ^{14}N increases with metallicity (actually C+O), thanks to the first two dredge-ups and the EB process itself. The trend with Z of the primary vs secondary production of ^{14}N can be derived from the tabulations in Renzini & Voli (1981).

2.6 Oxygen

The surface abundance of ^{16}O is not affected by the first dredge-up, and is only marginally reduced by either the second or the third dredge-up. Also the EB process does not affect very much the surface ^{16}O of AGB stars, the reduction reaching at most 30 % for $\alpha = 2$ (cf. Figure 3). Therefore, IMS are neither important producers or destroyers of ^{16}O.

More interesting is the case of the isotope ^{17}O. Figure 3 shows the maximum temperature at the base of the convective envelope T_b^{max}, as a function of M_i and for $(\eta,\alpha,Z) = (1/3,2,0.02)$. One sees that for $M_i > \sim 4\ M_\odot$ T_b^{max} becomes insensitive to M_i, and remains close to $71\ 10^6$ K.

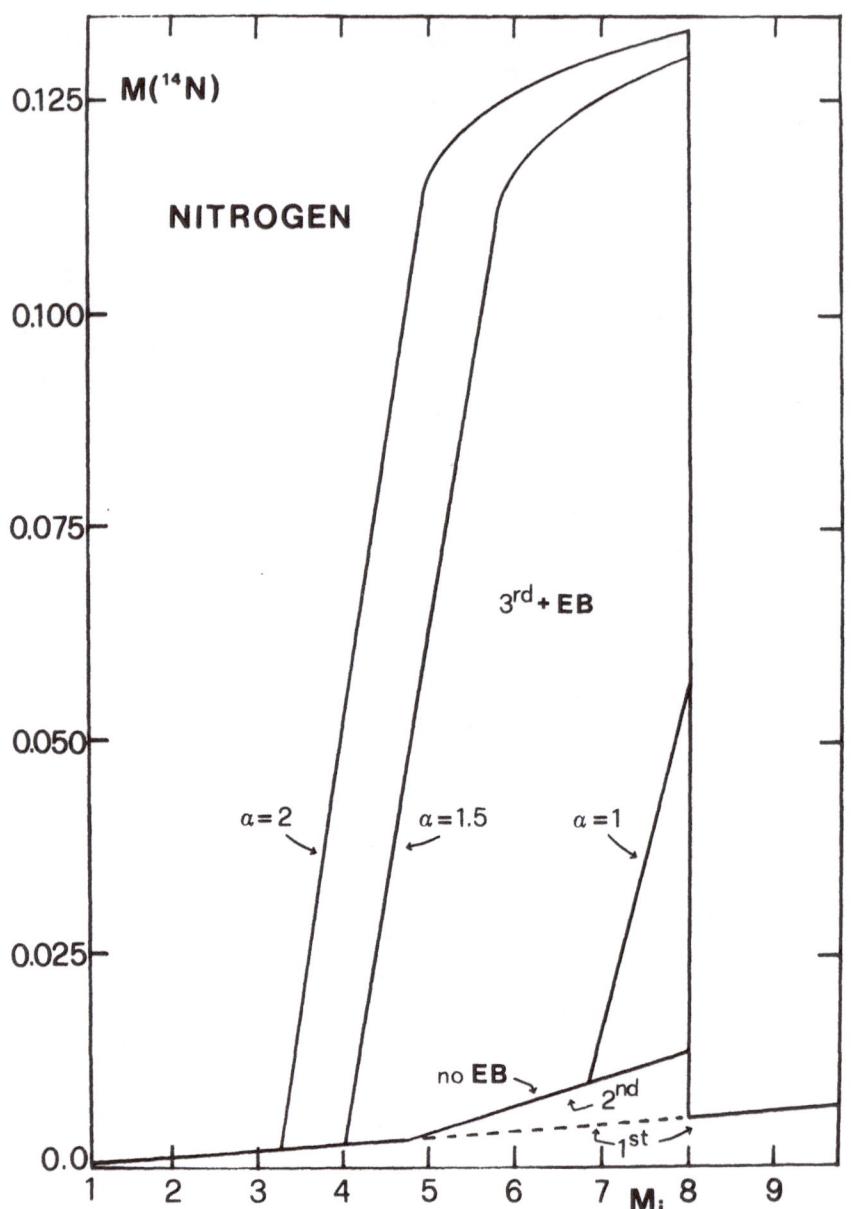

Figure 2: *The same as in Figure 1 but for* [14]N. *The contribution of the fist two dredge-ups is shown.*

From Figure 3 one also sees that for $M_i > 4 \ M_\odot$ [16]O starts being depleted by the EB process, thus producing [17]O. In turn, [17]O is destroyed by (p,α) and (p,γ) reactions. The equilibrium abundance ratio $([17]O/[16]O)_e$ is reported in the insert of Figure 3, for two representative temperatures, and one can conclude that for M_i slightly in excess of $4 \ M_\odot$ [17]O in the

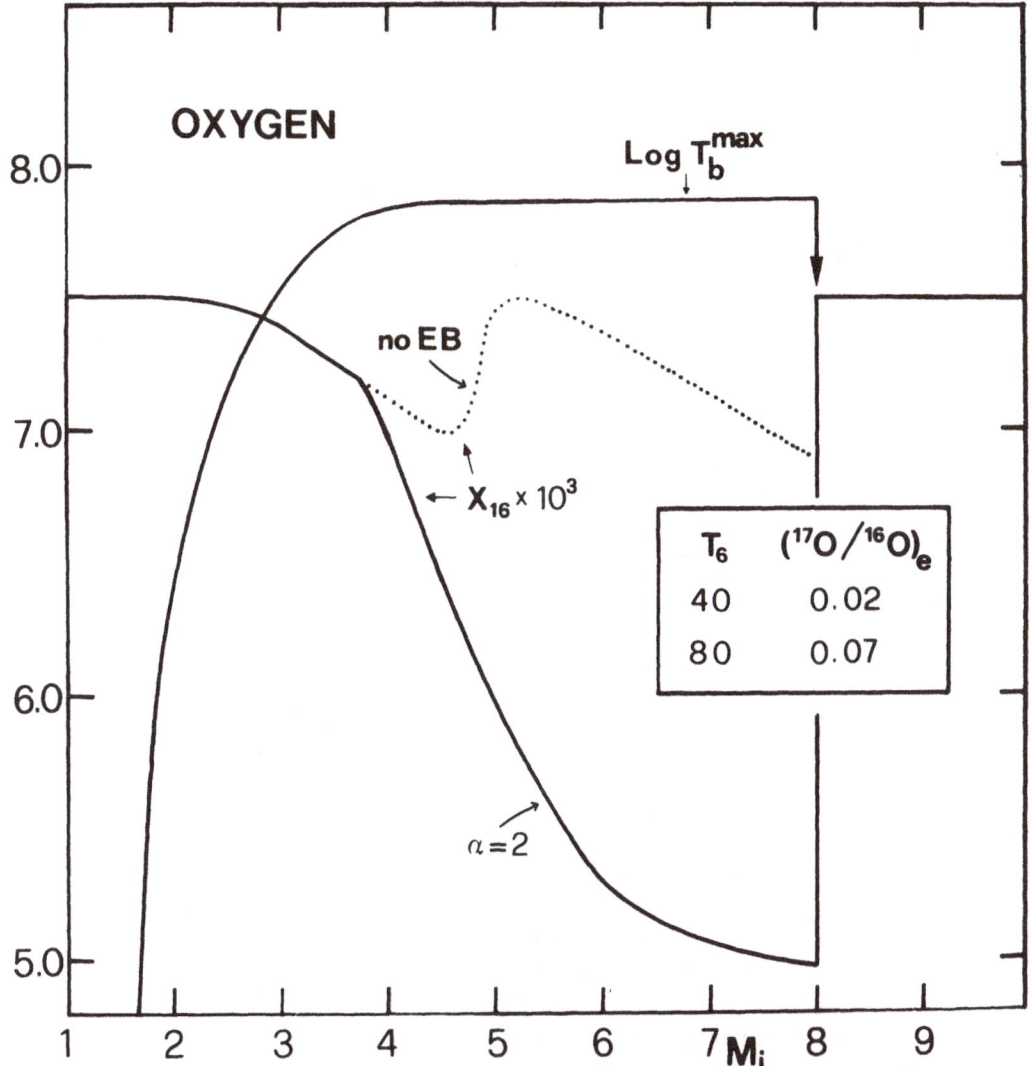

Figure 3: *The final surface abundance of ^{16}O, X_{16}, and the maximum temperature at the base of the convective envelope, T_b^{max}, are shown as a function of the initial mass. The dotted line refers to the case when the EB process is neglected.*

stellar envelope must reach equilibrium between producing and destroying reactions. One further expects that the equilibrium ratio should correspond to the highest temperature in the convective envelope, i.e. to T_b^{max}. Therefore, the ^{17}O abundance should be close to $X_{17} \simeq 0.05\ X_{16} \simeq 3\ 10^{-4}$, an unusually large value for any stellar environment (cf. Greggio, this volume). This is so because in the EB process the CNO hydrogen burning does not approach completion, in spite of the very high temperatures at

the base of the convective envelope. Therefore, IMS with envelope burning are potentially important contributors to the galactic nucleosynthesis of ^{17}O, the yield being obviously a very sensitive function of α.

2.7 Aluminum 26

Nørgaard (1980) noted that the ^{25}Mg brought into the envelope by the
third dredge-up could be efficiently transformed into the unstable isotope
^{26}Al (halflife = 7.2 10^5 yr), thanks to the EB process, provided that T_b
exceeds 70 10^6 K. Indeed, at such temperatures the reaction $^{25}Mg(p,\gamma)^{26}Al$
should be rather efficient, while only at temperatures in excess of about
90 10^6 K is ^{26}Al destroyed by the reaction $^{26}Al(p,\gamma)^{27}Si$. Nørgaard then
suggested that the ^{26}Mg anomalies found in some meteorites could ultimately be ascribed to the EB process in IMS. From Figure 3 one sees that,
indeed, T_b is just close to 70 10^6 K, but, even more interestingly, Champagne et al (1983) have recently redetermined the cross section of the reaction $^{25}Mg(p,\gamma)^{26}Al$, concluding that its rate is up to ten orders of magnitude larger than previously estimated. This ensures that ^{25}Mg is almost
totally converted into ^{26}Al in all stars experiencing the EB process, the
rate of conversion being practically identical to the rate at which ^{25}Mg
is dredged-up from the intershell region.

3. CARBON DEFLAGRATION SUPERNOVAE IN SINGLE STARS

As anticipated in the introduction, stars in the mass range $M_w < M_i$
$< M_{up}$ ignite carbon in a degenerate C-O core, once the mass of this core
approaches the Chandrasekhar limit ($\sim 1.4\ M_\odot$). Owing to the core mass-
luminosity relation for AGB stars, this is expected to happen when an AGB
star reaches the critical luminosity $M_{bol} \simeq -7.3$. The detailed nucleosynthesis in deflagrating cores being discussed by others at this conference (cf. the contributions by Nomoto and Woosley), I will restrict the
discussion to the critical quantity M_w, and to some (qualitative) aspects
of explosive hydrogen burning in these hypothetical stars.

From the models of Renzini & Voli (1981), Iben & Renzini derive the
following expression:

$$M_w \simeq 1.0 + 9.33\ \eta^{0.35} - 3.53\ \eta^{0.27} + 0.8(b - 1.0) \tag{1}$$

relating M_w to the mass loss parameters η and b, the latter parameter
describing the contribution of the "superwind" (cf. the quoted papers for
more details). In practice, for $b = 1$, M_w increases from 4.7 to 8 $M_\odot$
when η is increased from 1/3 to 2. For $\eta > 2$ Equation (1) formally implies
$M_w > M_{up}$, which simply means that mass loss would prevent the core of IMS
from reaching the Chandraskhar limit, and carbon deflagration in single
stars would not occur at all. Unfortunately, empirical MLR's can hardly

provide an interesting constraint to the value of M_W, since they are at least as uncertain as implied by a range in η from 1/3 to 2. On the other hand, for $\eta = 1/3$ the predicted number of SNe of type $I\frac{1}{2}$ would be so high that iron would perhaps be overproduced with respect to the requirements of galactic nucleosynthesis (Iben 1981), while for $\eta \geq 2$ such events would not occur at all. Other ways must then be envisaged in order to assess the contribution of IMS to the nucleosynthesis of iron , if any. I shall return to this point in the final section.

However, it has been frequently noted that M_W may be a function of Z (e.g. Fusi Pecci & Renzini 1976; Iben 1983), M_W being in case smaller at low Z, if the mass loss rate decreases with decreasing Z. Therefore, it is conceivable that SNe of type $I\frac{1}{2}$ played an important role in establishing the abundances of population II stars, and that their frequency later declined with the increasing overall metallicity.

When M_H reaches $1.4\ M_\odot$ in an AGB star the exploding core is still surrounded by an extended and massive H-rich envelope. This, by the way, justifies the nick-name of SN $I\frac{1}{2}$ given to this event, as the core explosion is similar to a SN I event, while the presence of the H envelope will make both the SN spectrum and early light-curve very similar to those of a SN of type II. Has such an event been never observed? To my knowledge, the SN 1979c in M100, with its very high N/C ratio (~ 30 times solar, Panagia 1981), may be a candidate. Such a high N/C ratio can be easily reached in IMS with envelope burning, while can hardly be achieved in massive red supergiants.

In any case, as in SNe of type II, the core explosion will generate a strong shock wave sweeping the envelope, and then some explosive hydrogen burning is expected to take place in the deepest layers of this envelope. The difference with respect to a normal type II supernova is that the envelope composition has now been affected by both the third dredge-up and the EB process. In particular, the envelope has been substantially enriched in ^{14}N and ^{17}O. Correspondingly, explosive H burning in SN $I\frac{1}{2}$ events could lead to a substantial production of the rare isotopes ^{15}N and ^{18}O, to an extent which remains to be investigated by detailed models.

4. DISCUSSION

These results can be summarized as follows: theoretical IMS in the mass range $2.5 \lesssim M_i \lesssim 8\ M_\odot$, core mass in the range $0.9 \lesssim M_H \lesssim 1.4\ M_\odot$, and correspondingly luminosities in the range $-6 \gtrsim M_{bol} \gtrsim -7.3$, potentially produce relevant quantities of ^4He, ^{12}C, ^{13}C, ^{14}N, ^{17}O, ^{22}Ne, ^{25}Mg, ^{26}Al, and of heavy s-process isotopes (in solar proportions). Moreover, those AGB stars reaching $M_{bol} \simeq -7.3$ explode as SN $I\frac{1}{2}$, then producing an amount of ^{56}Fe (and other elements) similar to that of a SN I event, and potentially interesting quantities of ^{15}N and ^{18}O. The production of these isotopes and elements depends on two main (uncertain) parameters: the mass loss parameter η and the convection parameter α. The production of all the above elements decreases with increasing η, while α affects primarily the relative distribution of ^{12}C, ^{13}C, and ^{14}N, ^{16}O and ^{17}O, ^{25}Mg and ^{26}Al, and (indirectly) the abundance of ^{15}N and ^{18}O. In princi- ple, models for the chemical evolution of galaxies should be able to pick- up the "best values" of both α and η. However, before embarking in such a laborious game a preliminary, crucial question should receive a satis- factory answer: do these stars really exist?

Judging from the absence in the Magellanic Clouds of carbon stars significantly brighter than $M_{bol} = -6$, several authors (e.g. Richer 1981; Frogel et al 1981; Cohen et al 1981; Persson et al 1983) argued that IMS may actually fail to populate the upper AGB. Were this the case, all the interesting nucleosynthesis discussed so far simply would not occur in the real world. However, the quoted authors did not apparently realize that for $M_{bol} < \sim-6$ the EB process can prevent the formation of carbon stars, and therefore the absence of bright carbon stars in the Clouds does not necessarily imply that IMS are unable to reach luminosities in the crucial range $-6 > M_{bol} > -7.3$. Indeed, most AGB models in this lu- minosity range are oxygen rich (corresponding to spectral type M or K) when the EB process is taken into account (Renzini & Voli 1981; Iben & Renzini 1983). Moreover, the absence of bright M-type stars in the MC survey of Blanco et al (1980, hereafter BMB) is seemingly inconclusive, since their grism method is able to detect only AGB stars of spectrum M5.5 or later. The "missing" bright AGB stars could then have a spec- trum earlier than M5.5. MC stars with spectral type from M0 to M5, and with the appropriate luminosity, have been detected in the survey of Westerlund et al (1981), but from these very low dispersion spectra it is impossible to decide whether they are AGB stars or massive stars ($10 \stackrel{<}{\sim} M_i \stackrel{<}{\sim} 15$ $M_\odot$) in their core helium-burning phase. Only high disper- sion spectra could tell us which among these stars are AGB stars, as they should exhibit the typical signatures of the third dredge-up and the EB process, namely, enhanced s-process elements, strong CN and ^{13}C isoto- pic features. These high-dispersion spectroscopic observations are of crucial importance for deciding the issue.

Bona fide bright AGB stars have been eventually identified by Wood et al (1981, 1983) among the long period variables (LPV's) in the MC's. These are bright M-type stars, with $-6 > M_{bol} > -7$, pulsational masses up to ~7 $M_\odot$, and strong ZrO bands (Zr is an s-process element). So, at least some bright AGB stars apparently exist, but the number of these LPV's seems still rather small compared to the expectations. The ques- tion is that not all LPV's might have been detected, or not all bright AGB stars are LPV's. Only a *complete* survey of red giants in the MC's,

coupled with infrared observations (to get M_{bol}) and high-dispersion
spectroscopy of the individual objects, will give us the final answer.
Moreover, this type of survey should be extended to several MC fields,
as the recent history of star formation may have been different in different parts of the Clouds (see below).

Frogel & Richer (1983) have recently surveyed about half of the BMB
"bar west" (BW) field in LMC. They have scanned the field at 2.2 (K) and
3.5 μm (L) without finding AGB stars brighter than M_{bol} =-6.4, and conclude that mass loss evaporates AGB stars before they reach this luminosity.
I find their conclusion rather premature, for the following reasons.
The BMB-BW field contains 9 cepheids (Becker 1982): 3 of them have a period shorter than 6 days, 5 have a period between 8 and 14 days, and one
has a period of 18 days. From the period-mass relation given by Becker
et al (1977) one can infer that the short period cepheids have a mass
less than 5 $M_\odot$, those with intermediate periods have a mass around 7 $M_\odot$,
and the long period cepheid is around 9 $M_\odot$. Conversely, the BMB "0"
field in LMC contains 34 cepheids, *none* of which has a period longer than
8 days! Clearly, the BW field has recently experienced a burst of star
formation (some 20 million years ago), while the "0" field did not probably produce stars during the past 10^8 yr or so. (Note that the period-mass relation has a statistical meaning: the intrinsic width of the
instability strip implies that cepheids of given mass must exhibit a period range ΔLog P ≃ 0.3 around the mean period.) Therefore, one can reasonably conjecture that stars produced in the recent burst in the BW field
did not have time to populate the AGB (one 9 $M_\odot$ star is still in the cepheid phase!), and we are left with just 3 short-period cepheids as representative of the older population expected to populate the AGB. Since
the cepheid lifetime is somewhat longer than the time spent during the
AGB phase, one expect perhaps one bright AGB star in the whole BW field.
There is therefore no surprise if Frogel & Richer did not find any in
half the field. To reach firmer conclusions the Frogel & Richer infra-
red survey should be extended to the "0" field, which contains a statisti-
cally significant number of cepheids.

Another method to gather indirect information on the extent of nucleosynthesis in IMS is to look at the relation between the final mass M_f
of white dwarf remnants in a cluster and their initial mass, as inferred
from the mass of the evolving cluster members, then bearing in mind that
important nucleosynthesis takes place only for M_H > ∿0.9 $M_\odot$. This method
is currently used by Reimers & Koester (1982), and from the available
data Weidemann & Koester (1983) conclude that M_W ≃ 8 $M_\odot$, and that the
$M_f(M_i)$ relation runs very flat (with M_f < ∿0.6 $M_\odot$) till at least 3 $M_\odot$.
This result, however, is based on a quite limited number of white dwarfs,
whose individual masses are rather uncertain (by at least 50 %), and the-

refore it cannot be taken as conclusive evidence against the occurrence
of significant nucleosynthesis in IMS. Moreover, were the Weidemann &
Koester $M_f(M_i)$ relation applicable also to MC stars, then rather young
ages (a few 10^8 years) would be implied for the so-called intermediate
age globular clusters, which are rather believed to be a few billion yr
old (cf. Iben & Renzini 1983, and references therein). In any case, this
method deserves further attention, so as to increase the number of WD
cluster members, and to improve the accuracy of their mass determinations.

But what if, after all, bright AGB stars really don't exist or are
extremely rare? From an aesthetic point of view, I don't find very attractive the explanation in terms of mass loss preventing the population
of the upper AGB, at least in the form this idea has been expressed so
far. Indeed, there is strong indirect evidence from the HR diagram of
galactic globular clusters (Renzini (1981b), and from the extension of
the AGB in the intermediate age globular clusters of the MC's (Iben &
Renzini 1983), that η cannot exceed 0.5, at least for $M_i < \sim 2\ M_\odot$. For
this value of the MLR parameter a 2 $M_\odot$ star terminates its AGB evolution
at just M_{bol} = -6. But to prevent stars of 4 to 8 $M_\odot$ from exceeding this
luminosity, their envelope should be removed *before* they enter the thermally pulsing phase of the AGB (cf. figure 7 in Iben & Renzini 1983).
This would imply an unbelievable jump in η when M_i exceeds 2 $M_\odot$. Such a
discontinuous behaviour can hardly be ascribed to something happening
in the atmosphere of these stars, and a much deeper cause should be found.

The fact that the current scheme for the AGB evolution apparently
breaks just around M_i = 2 $M_\odot$ may indeed suggest a possible solution.
Stars with $M_i < \sim 2\ M_\odot$ develop a degenerate helium core during their RGB
evolution. More massive stars do not. Moreover, real IMS are fast ro-
tators on the main sequence. Low mass stars ($M_i < \sim 2\ M_\odot$) will tend to
develop a rapidly spinning core during the *slow* growth of their degenerate helium core, while in IMS ($M_i > \sim 2\ M_\odot$) such a rapidly spinning core
will *suddenly* appear only after the He exhaustion at the center and the
development of a degenerate C-O core. In the two cases the possible appearance of nonaxial symmetric instabilities in such spinning cores
(Kippenhahn et al 1970) might have rather different consequences. For
instance, such instabilities could just spin down the core in the former
case, while in the latter the interaction of the nonaxial symmetric core
with the envelope could be so violent as to lead to the rapid ejection
of the envelope itself. These considerations are admittedly rather speculative, and may eventually turn out even unnecessary if bright AGB stars
arc more abundant in the sky than in existing catalogues.

REFERENCES

Becker, S.A. 1982, Ap. J. 260, 695
Becker, S.A., Iben, I.Jr. 1979, Ap. J. 232, 831
Becker, S.A., Iben, I.Jr., Tuggle, R.S. 1977, Ap. J. 218, 633
Blanco, V.M., McCarthy, M.F., Blanco, B.M. 1980, Ap. J. 242, 938
Cohen, J.G., Frogel, J.A., Persson, S.E., Elias, J.H. 1981, Ap. J. 249,481
Champagne, A.E., Howard, A.J., Parker, P.D. 1983, Ap. J. 269, 686
Dearborn, D.S.P., Eggleton, P.P. 1976, Quart. J.R.A.S. 17, 448
Frogel, J.A., Richer, H.B. 1983 (preprint)
Frogel, J.A., Cohen, J.G., Persson, S.E., Elias, J.H. 1981, Physical Pro-
    cesses in Red Giants, ed. I. Iben Jr., A. Renzini (Reidel: Dordrecht)
    p. 159
Fusi Pecci, F., Renzini, A. 1976, Astron. Astrophys. 46, 447
Iben, I.Jr. 1964, Ap. J. 140, 1631
Iben, I.Jr. 1965, Ap. J. 142, 1447
Iben, I.Jr. 1975, Ap. J. 196, 525
Iben, I.Jr. 1977, Advanced Stages in Stellar Evolution, ed. P. Bouvier,
    A. Maeder (Geneva Observatory), p. 1
Iben, I.Jr. 1981, Effects of Mass Loss on Stellar Evolution, ed. C. Chio-
    si, R. Stalio (Reidel: Dordrecht), p. 373
Iben, I.Jr. 1983, Mem. S.A.It. 54, 321
Iben, I.Jr., Renzini, A. 1982, Ap. J. Lett. 263, L188
Iben, I.Jr., Renzini, A. 1983, Ann. Rev. Astron. Astrophys. 21 (in press)
Iben, I.Jr., Renzini, A. 1984, Phys. Rep. (in press)
Iben, I.Jr., Truran, J.W., 1978, Ap. J. 22°, 980
Kippenhahn, R., Meyer-Hofmeister, E., Thomas, H.-C. 1970, Astron. Astro-
    phys. 5, 155
Nørgaard, H. 1980, Ap. J. 236, 895
Panagia, N. 1981, The Universe at Ultraviolet Wavelengths, ed. R.D.
    Chapman (NASA), p. 521
Persson, S.E., Aaronson, M., Cohen, J.G., Frogel, J.A., Matthews, K.
    1983, Ap. J. 266, 105
Reimers, D. 1975, Mem. Soc. Roy. Sci. Liège, 6e Ser. 8, 369
Reimers, D., Koester, D. 1982, Astron. Astrophys. 116, 341
Renzini, A. 1981a, Physical Prpcesses in Red Giants, ed. I. Iben Jr.,
    A. Renzini (Reidel: Dordrecht), p. 431
Renzini, A. 1981b, Effects of Mass Loss on Stellar Evolution, ed. C. Chio-
    si, R. Stalio (Reidel: Dordrecht), p. 319
Renzini, A. 1983, Primordial Helium, ed. P.A. Shaver, D. Kunth, K. Kjär
    (ESO), p. 109
Renzini, A., Voli, M. 1981, Astron. Astrophys. 94, 175
Richer, H.B. 1981, Physical Processes in Red Giants, ed. I. Iben Jr., A.
    Renzini (Reidel: Dordrecht), p. 153

Rood, R.T., Steigman, G., Tinsley, B.M. 1976, Ap. J. Lett. 207, L57

Serrano, A.P., Peimbert, M. 1981, Rev. Mexican Astron. Astrophys. 6, 41

Sugimoto, D. 1971, Progr. Theor. Phys. 45, 761

Sweigart, A.V., Gross, P.G. 1978, Ap. J. Supp. 36, 405

Wannier, P.G. 1980, Ann. Rev. Astron. Astrophys. 18, 399

Weidemann, V., Koester, D. 1983, Astron. Astrophys. (in press)

Westerlund, B.E., Olander, N., Hedin, B. 1981, Astron. Astrophys. Supp.
    43, 267

Wood, P.R., Bessell, M.S., Fox, M.W. 1981, Proc. A. Soc. Australia, 4,203

Wood, P.R., Bessell, M.S., Fox, M.W. 1983, (preprint)

DISCUSSION

Edmunds: If the primary N production is a significantly decreasing function of the overall metallicity (as indicated by the oxygen abundance)
then [N/O] should decrease with increasing [O/H], which is not the impression we get from the diagram I have shown!

Gallagher: You might have to be careful in interpreting yields directly
in terms of abundances. In this case the N production will be at least
10 times lower than that of O, so it is not clear what the ratios should
look like. Also, aren't N/O ratios variable, or at least rather poorly
known at low metallicities?

Edmunds: Yes, we need to explain the different [N/O] ratios for systems
with similar [O/H]. So the real situation is complicated, perhaps it's
just as well we don't see these stars!

Gallagher: I agree with Mike that most of the data indicate [N/O] =
X[O/H] + const, with $X \gtrsim 1$ but certainly considerably less than 2. However, Howard French has argued that initially $X \simeq 0$, i.e. primary production at low Z's, then switching to dominance by secondary production at
high Z's.

Renzini: I certainly agree with Jay, and this is indeed what models predict, at least in a qualitative sense. At low Z's primary N production
dominates, because the EB process is more efficient, the MLR may be lower,
and there is less CO around to be (secondarily) processed to N. The opposite happens at high Z's: the EB process is less efficient in producing
primary N, while there is a lot of C and O ready for being (secondarily)
processed to N!

HYDROSTATIC EVOLUTION AND NUCLEOSYNTHESIS IN MASSIVE STARS

André Maeder
Geneva Observatory
CH-1290 Sauverny, Switzerland

1. INTRODUCTION

A great renewal in our knowledge of the evolution and nucleosynthesis in massive stars (M $\gtrsim$ 9 M$_\odot$) has occurred in recent years due
to the inclusion by several groups of the effects of mass loss by stellar winds. These works have important consequences for stellar structure,
stability and evolution, appearance of nucleosynthetic products at stellar surfaces, advanced nucleosynthesis, galactic enrichments, final
stages of stellar evolution and galactic distributions of massive stars
(OB stars, red supergiants and Wolf-Rayet stars). Here we shall concentrate on nucleosynthesis and related problems in stellar structure
and evolution of massive stars.

2. AN OVERALL VIEW OF THE EVOLUTION OF MASSIVE STARS WITH MASS LOSS

Let us start by an overview of the evolution of massive stars
with mass loss. An interesting result has emerged from the model computations of massive stars (cf. Chiosi, 1981; Maeder, 1981 a,c): according to initial stellar mass and mass loss rates, the stars may go
through very different evolutionary sequences. Let us firstly recall
that in case of constant mass evolution, the massive stars leave the
main-sequence and end their life, whatever may be the exact nature of
the core collapse, as red supergiants (cf. Lamb, Iben and Howard, 1976).

In case of mass loss (at the currently observed rates) stars with
an initial mass above 50 - 60 M$_\odot$ never reach the red supergiant (RSG)
stage. For these stars, the peeling off by the winds during the main-
sequence and the blue supergiant (BSG) phases is high enough to remove
all the outer stellar layers. Zones which initially were in the core
are revealed at the stellar surface. These stars, especially the
extremely massive ones, remain quasi-homogeneous (cf. Maeder, 1980) and
always stay at the blue side of the HR diagram, firstly as BSG and
Hubble-Sandage variables and then as Wolf-Rayet (WR) stars, a stage

115

C. Chiosi and A. Renzini (eds.), Stellar Nucleosynthesis, 115–136.
© 1984 by D. Reidel Publishing Company.

in which they reach core collapse.

For initial stellar masses between about 50 and 25 M$_\odot$, mass loss at the observed rates on the main-sequence and in the BSG phase is not sufficient to remove all the outer envelope and the star rapidly reaches the RSG stage (cf. Fig. 1). Then the high stellar winds in the red supergiant stage progressively eject the outer envelope.

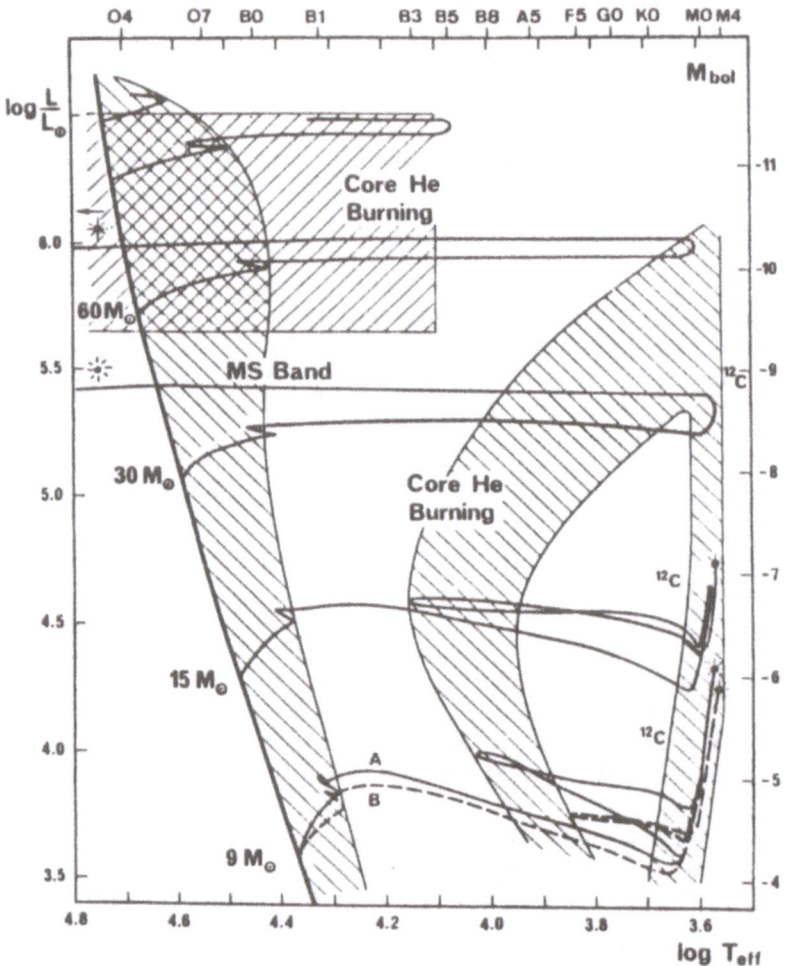

Fig. 1   The HR diagram for evolution with mass loss up to the end of central carbon burning (models B, 30 M$_\odot$ case C, Maeder, 1981c; 1983a, for 120 M$_\odot$)

When the mass fraction of the helium burning core, as a result of the decreasing total stellar mass, becomes larger than some critical value q_c, the star moves back to the blue in the HR diagram and then may become a WR star (cf. Chiosi et al., 1978; Maeder, 1981c, q_c=0.67 for 60 $M_\odot$, 0.77 for 30 $M_\odot$). Thus, there may exist WR stars which are in a post-RSG phase. Clearly high mass loss rates during this RSG phase and during previous phases favour the formation of WR stars, since then the critical q_c is reached earlier. We also note that for most stars with initial mass above 25 $M_\odot$ the final core collapse occurs when the star is in the WR stage.

For initial masses below about 25 $M_\odot$, the mass loss both in the blue and in the red is never large enough to remove the outer layers. After the main-sequence, the star normally becomes a RSG, then it undergoes blue loops during which a Cepheid phase may occur; later the star again becomes a RSG. We can notice that the blue extension of the loops is significantly reduced by mass loss (cf. Maeder, 1981c); in case of high mass loss, the loops may even be suppressed.

As illustrated in Fig. 1 the following evolutionary sequences may be distinguished:

For $M \gtrsim 60\ M_\odot$
O star – Of – BSG and Hubble–Sandage variables – WN – WC – (WO) – SN

For $25\ M_\odot \lesssim M \lesssim 60\ M_\odot$
O star – BSG – yellow and RSG – (BSG) – WN – (WC) – SN

For $M \lesssim 25\ M_\odot$
O star – (BSG) – RSG – yellow supergiant and Cepheid – RSG – SN

The exact values of the mass limits depend on the mass loss rates as well as on the amplitude of the possible mixing processes (overshooting, diffusion etc.). We notice that for initial masses above M $\simeq$ 25 $M_\odot$ the pre-supernova model is a bare core in the WR stage, while below that mass limit the pre-supernova is a red supergiant likely to give rise to a type II explosion since it contains large amounts of hydrogen.

No adequate theory of the stellar winds exists at the present time which predicts mass loss rates $\dot{M}$ for stars in various evolutionary stages. Thus, most authors adjust some parametrized expressions on the observed mass loss rates. Different expressions have obviously to be taken for O-stars (Garmany et al., 1981), for B, A and F supergiants (Barlow and Cohen, 1977), for RSG (Reimers, 1976; Bernat, 1977). This was done in the models of Fig. 1 which correspond to average M-rates. Let us also recall that many grids of models with mass loss have been computed over recent years (cf. Chiosi et al., 1978; de Loore et al., 1978;

Stothers and Chin, 1979; Maeder, 1980, 1981a,c; de Loore, de Greve, 1981
Chiosi, 1981; Brunish and Truran, 1982a,b). The details of the effects
of mass loss on main-sequence evolution have been discussed by most
authors and we only quote these effects here: lower luminosity, but
overluminosity with respect to actual mass, decrease of the core mass,
while the core mass fraction is larger (cf. Fig. 2a and b),

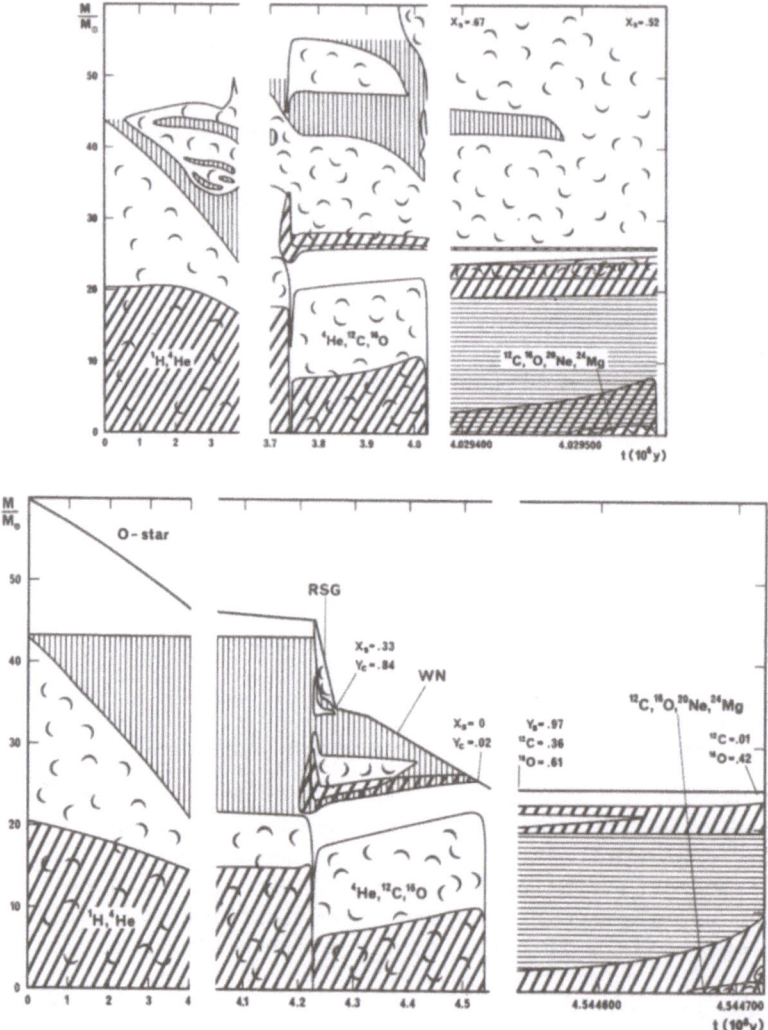

*Fig. 2a, b   Evolution of the internal structure of
stars of 60 M$_\odot$ up to central C-exhaustion; Fig. 3a
with constant mass, Fig. 3b with mass loss. Cloudy
regions are convective zones, heavy diagonals are for
regions where the nuclear energy rates are larger
than 10^3 erg/sec. Fine lines indicate the zone of
variable chemical composition.*

disappearance of semi-convection, increase of lifetimes, main-sequence widening for low and intermediate mass loss rates or main-sequence narrowing for very large mass loss rates (cf. Fig. 1) and also possible changes of surface abundances.

Let us now add a few comments on the helium burning phase, taking a 30 M$_\odot$ star as a typical example. In case of constant mass evolution all the He-phase is spent in the blue. This is due to a large fully convective zone (FCZ; cf. Fig. 2a) which homogenizes the intermediate stellar layers and thus keeps the star to the blue (cf. Stothers and Chin, 1976; Lamb et al., 1976). Thus stars with constant mass spent no time or only a negligible time in the RSG stage. This is the reason why Stothers and Chin, Lamb et al. concluded that there were too many observed RSG with respect to what had been predicted.

Mass loss, even small, reduces the extension of the FCZ and thus favours the redwards motion. This allows the stars to reach the RSG stage early enough during the He-burning phase and to experience lifetimes in the RSG longer by a factor 4 to 5 with respect to the case of constant mass, thus bringing about agreement with the observations (see Fig. 6 in Maeder, 1981c). We do not support the conclusions by Brunish and Truran (1982a,b) who find that stars with initial masses smaller than or equal to 30 M$_\odot$ spend very little time (less than 1%) in the RSG stage. Their result is probably due to their small mass loss rates (we notice that their $\dot{M}$-values on the main-sequence decrease with increasing metal content Z and also decrease during main-sequence evolution). Indeed, the absence of RSG from 20 - 25 M$_\odot$ progenitors predicted by Brunish and Truran is not at all confirmed by the observations of red supergiants in young clusters and associations (cf. Humphreys, 1978; Humphreys and Davidson, 1979). These observations show that the major fraction of red supergiants originates from stars initially less massive than 30 M$_\odot$.

If mass loss in the red supergiant (RSG) stage is small the star remains a RSG until the core collapse. However, as we have seen for $\dot{M}$-values high enough in the RSG stage, the evolution again proceeds to the blue and the star becomes a WR star. The fraction of the helium burning lifetime spent in the WR stage rapidly increases with initial mass and with mass loss rates (cf. Fig. 7 in Maeder, 1981c). Since the total duration t_{He} of the helium burning-phase does not change significantly, a substantial increase of the WR phase implies a corresponding reduction of the RSG phase as $t_{He} \simeq t_{RSG} + t_{WR}$. However, the RSG phase for stars with initial M $\lesssim$ 50 M$_\odot$ still remains larger than in the case of constant mass evolution. The increase of t_{WR} (lifetime in the WR stage) and the corresponding decrease of t_{RSG} (lifetime in the RSG phase) were invoked as being responsible for the observed strong anticorrelation of the frequency of WR stars and red supergiant (RSG) in the Galaxy (cf. Maeder, Lequeux and Azzopardi, 1980; Meylan and Maeder, 1983). Finally,

we also note that the scheme of Fig. 1 allows us to explain
(cf. Humphreys, 1978) why the brightest O-type stars are about 2 mag.
more luminous than the brightest RSG stars.

3. EVOLUTION OF THE STRUCTURE IN THE SLOW PHASES OF NUCLEAR BURNING:
   INCIDENCE OF MASS LOSS ON THE FINAL FATE

The typical lifetimes of the evolution of massive stars with average mass loss rates are given in Table 1. One notices that typically
the He-burning phase represents 8 to 10% of the main sequence lifetimes,
this fraction being much shorter for the C-burning phase. Among the
400 supergiants brighter than M_{bol} = -7.5 and within 2.5 kpc from the
Sun, there should be 1 or 2 such C-burning stars emitting copious floods
of neutrinos.

a) Mass loss and internal structure

The comparison of the internal structure of models of initial mass
60 $M_\odot$ with and without mass loss is made in Figs. 2a and b, in which
the three phases of slow nuclear burning are successively represented
(cf. Maeder, 1981a). Among other points, the most noticeable fact in
these pictures is that the He-burning phases end with almost similar
He + C/O cores in models with and without mass loss: we can see that
the structure of the inner layers is the same in the C-burning phases
of Fig. 2a and b, although more than 25 $M_\odot$ have disappeared in this
last case. In Fig. 2b (with mass loss) the core mass increases much
more during the He-burning phase than during constant mass evolution.
Thus, although the core mass is initially smaller at the beginning of
the He-burning phase in case of mass loss, it finally becomes as large
as in case of constant mass evolution.

The physical reason is that in case of constant mass the H-burning
shell (during central He-burning) is topped by a sizeable convective
zone continuously replenishing the H-shell with hydrogen, so that the
shell cannot migrate outwards. With mass loss, the intermediate convective zone is much smaller or even non-existant, thus the shell does not
undergo any significant replenishment and it progressively migrates outwards, thus increasing the mass both of the whole region interior to
the H-burning shell (frequently called M_α) and of the fully convective
He + C/O core.

In the case of the models shown in Fig. 2a and b a comparison of
the detailed structures at the end of the C-burning phase shows that
the runs of central temperature, density (cf. Fig. 3) and the distribution of chemical elements are the same in the interior of the RSG
stars having suffered no mass loss and in a 25 $M_\odot$ bare core (WR star)
remaining from the evolution with mass loss, both having started from
the same initial mass. Thus, as well as a red giant may be considered

Table 1

LIFETIMES OF NUCLEAR PHASES

$M/M_\odot$	t(H-burn.)	t(He-burn)	t(C-burn.)
120	$2.929 \cdot 10^6$		
60	$4.226 \cdot 10^6$	$3.15 \cdot 10^5$ *7.5%*	$3.51 \cdot 10^3$ *.083%*
30	$6.17 \cdot 10^6$	$5.07 \cdot 10^5$ *8.2%*	$1.60 \cdot 10^4$ *.26 %*
15	$1.1654 \cdot 10^7$	$1.19 \cdot 10^6$ *10.3%*	$1.74 \cdot 10^5$ *1.5 %*
9	$2.3115 \cdot 10^7$	$3.43 \cdot 10^6$ *14.8%*	$8.35 \cdot 10^5$ *3.6 %*

Remark: The numbers in italic are the ratios
to the H-burning lifetimes.

a white dwarf surrounded by an extended envelope, as much can a red
supergiant be considered a WR star surrounded by a very large envelope.
In the case of the white dwarf the He + C/O core is degenerate and
smaller than the Chandrasekhar mass, while the WR star is not degenerate
and has a larger mass.

From what preceeds we may conclude that for a rather wide range of
mass loss rates there are very little effects on physical conditions and
nucleosynthesis in central stellar regions. This is illustrated by the
evolution in the log T_c vs. log ρ_c diagram in Fig. 3 (cf. Maeder and
Lequeux, 1982). As another example, an initial 30 $M_\odot$ star containing
only 10.1 $M_\odot$ at the end of the C-burning stage has exactly the same
central conditions as the model with constant mass. After that stage
the evolutionary time scales are so short that further mass loss is un-
significant: thus core contraction proceeds the same way to core collapse.

This absence of sensitivity of central conditions to mass loss is
the general rule as long as mass loss does not lead to a significant
reduction in the mass of the He + C/O convective core; such a reduction
occurs if the "surface line", as in Fig. 2b, declines so steeply due to
mass loss that it intersects regions which previously were included in
the convective core (such cases have been calculated by Maeder, 1981a
and Maeder, 1983a). This situation corresponds to the case where Wolf-
Rayet stars of type WC are formed which, according to the models, only
seems to occur for stars initially more massive than about 50 $M_\odot$. In
this case, some heavy elements are ejected into the stellar winds, and
the core reduction may influence the subsequent systhesis of further

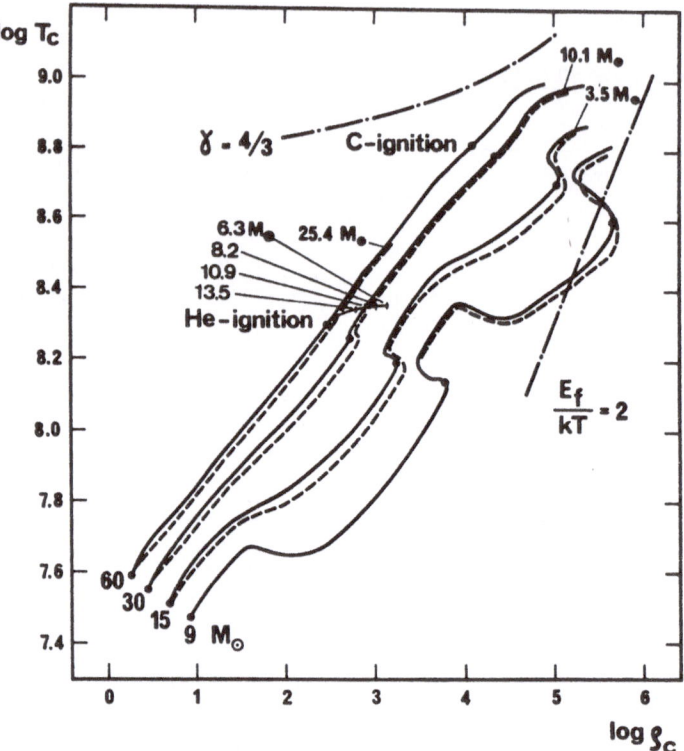

*Fig. 3 Evolution of the central temperatures
and densities for stars of various initial
masses. The continuous lines refer to constant
mass evolution and the broken lines to evolution
with mass loss (case C in Maeder, 1981c). Some
indications on the remaining stellar masses are
given. For 60 M$_\odot$, various cases of mass loss
have been considered.*

heavy elements (cf. § 4).

   Indeed, we must note that, even in the last mentioned case of WC
stars, the departures from the standard tracks in the log T_c vs. log ρ_c
diagram remain rather small for the standard mass loss rates (quoted by
above-mentioned authors). However, there is an appreciable scatter in
the $\dot{M}$ -values at various stages (particularly for OB stars and supergiants) and there can be extreme cases of mass loss. Fig. 3 also shows
such an extreme case where an initially 60 M$_\odot$ star with mass loss rates
higher than the average is followed until only 6.3 M$_\odot$ remain (then log
L/L$_\odot$ = 4.92, Y$_c$ = 0.12; cf. last example in Table 3 by Maeder, 1983a).
We notice in this example that only for remaining masses smaller than
about 11 M$_\odot$ the departure in log T_c at a given ρ_c is larger than 0.1;

for a remaining mass of 15 $M_\odot$ the departure is only about 0.05. If heavy mass loss would go on, the star could become some "dead nuclei" deprived of nuclear reactions. We wonder whether this is the situation of some of the WR stars which are nuclei of planetary nebulae.

In summary we can conclude (cf. Maeder and Lequeux, 1982) that most mass-losing OB stars, which may become WR stars after various other stages, are likely to reach the phase of core collapse. For initial stellar masses smaller than 50 $M_\odot$ there will be practically no change in their central evolution and thus in their final fate and nucleosynthetic production (see also § 4). Above that limit, significant departures in the log T_c vs. log ρ_c diagram only occur if the remaining stellar mass becomes very small with respect to the initial mass (for example, if a 11 $M_\odot$ star results from an initial 60 $M_\odot$ star); this is probably not the general rule in view of the known masses of WR stars. (Let us recall that WR stars have masses which range from 10 to 50 $M_\odot$, with an average of 20 $M_\odot$; cf. Massey, 1981; van der Hucht, 1981).

b) WR stars exploding as supernovae

Estimates of the fraction of supernovae originating from WR progenitors have been performed (cf. Maeder and Lequeux, 1982). The result shows that the average time interval between two consecutive WR stars exploding as supernovae in the Galaxy lies in the range of 2.5 to 5 centuries. This estimate rests on a total number of WR stars in the Galaxy of the order of 1200 (result which is based on the distributions of WR stars by Hidayat et al., 1981, and of the giant H II regions in the Galaxy); it also rests on an average WR lifetime estimated to be of the order of $3-6\cdot10^5$ yr. If the average interval between two supernova explosions is 70 yr (cf. Lerche, 1981), the above result means that out of 3 to 7 supernovae in the Galaxy one could originate from a WR star.

The properties of these supernovae originating from WR precursors are expected to be characterized by several noticeable features: 1) A lack of hydrogen; 2) The presence of two winds, a slower one from the WR star and a fast one from the supernova (cf. Cas A, Peimbert and van den Bergh, 1971); 3) The slow wind must contain elements typical for WR stars (cf. Fig. 4): enhanced ^{14}N, ^{17}O and depleted ^{12}C for WN progenitors, enhanced ^{12}C, ^{16}O, ^{22}Ne and depleted ^{13}C and ^{14}N for WC progenitors; 4) Reduced thermal and optical effects (cf. Chevalier, 1981); the fast material must be enriched by elements of the onion-skin stellar model.

4. NUCLEOSYNTHESIS AND APPEARANCE OF NUCLEAR PRODUCTS IN THE STELLAR WINDS

a) Surface enrichments

Massive stars and especially Wolf-Rayet stars, which are generally identified as bare cores (cf. Conti, 1982; cf. IAU Symposium 99) offer us the most valuable possibility to observe nucleosynthesis in the products of some nuclear reactions revealed at stellar surfaces as a result of mass loss and maybe of some mixing processes. The nucleo-synthetic products, unlike in red giants, are not highly diluted within a large envelope. The WC stars are the only kind of stars in which the products of the 3α reaction and other He-burning reactions prominently manifest themselves at the stellar surfaces. This may provide a very useful comparison basis with nucleosynthetic predictions.

We briefly recall some properties of the chemical composition of Wolf-Rayet stars (cf. Smith and Willis, 1982; Nugis, 1982). WN stars exhibit He- and N-enhancements and C- and O- depletion, which is typi-cal of the products of hydrogen burning through the CNO-cycle. The late WN stars (called WNL: types WN6 - WN8) generally still contain hydrogen while the early WN (called WNE: types WN2 - WN6) contain in general no hydrogen. The WC stars exhibit the products of He-burning, particularly carbon, some extreme WC stars with O/C enhancements exist and are called WO stars (cf. Hummer and Barlow, 1982).

The variations of the abundances of various elements during the evolution of massive stars have been calculated recently by Noels and Gabriel (1981) and Maeder (1983a,b). Fig. 4 from the last reference illustrates the changes of chemical abundances at stellar surface due to the fact that matter originally in the convective core is revealed by the removal of the outer layers by mass loss. At the appearance of nucleosynthetic products, important changes occur. The elements ^{3}He, ^{15}N and ^{18}O disappear. As the CN-cycle rapidly reaches equilibrium, the C/N ratio almost abruptly changes from about 4.1 to 0.03 (in mass). The O/N ratio slowly changes from 9.1 to less than 0.1, while the ON-cycle takes a longer time to reach equilibrium. The abundance of ^{13}C usually keeps a factor of 3.3 below that of ^{12}C. (The departures from this value in Fig. 4 as well as the various "plateaux" in the supergiant phase are due to convective mixing in various parts of the star model). We also notice the very different behaviours of the isotopes ^{16}O, ^{17}O and ^{18}O.

A very large discontinuity marks the appearance of the various products of He-burning at the surface, which is likely to correspond to the beginning of the WC stage. The physical reason for this disconti-nuity is that the convective core in the He-burning phase does not con-tinually decrease (which would leave a smooth transition at the border),

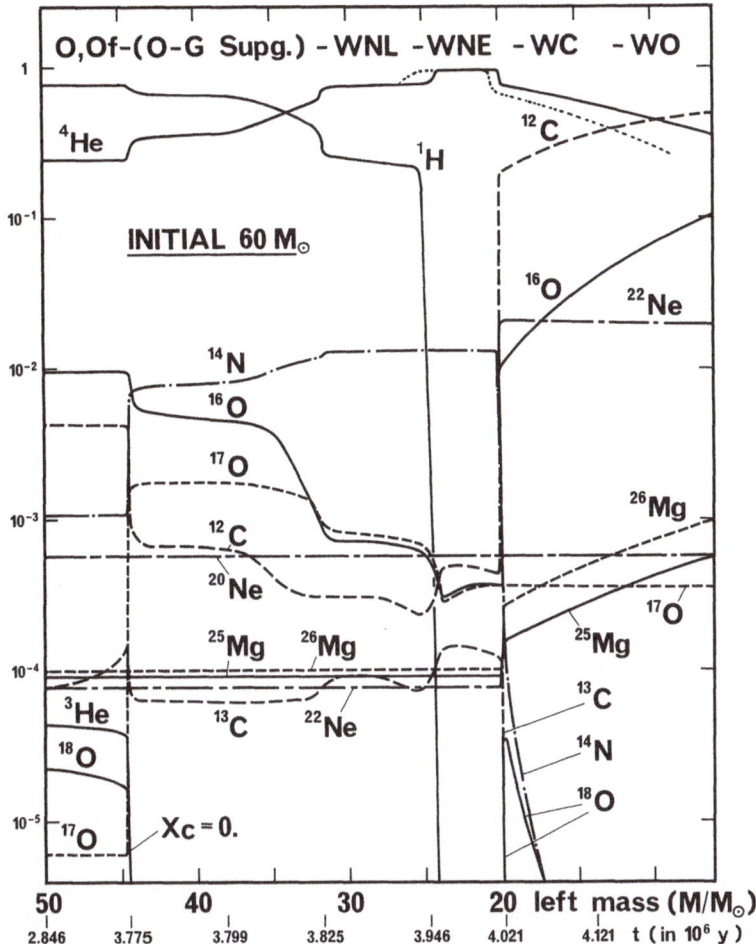

Fig. 4 *Changes of the surface abundances (in mass fraction) in terms of the remaining mass for the model with initial mass of 60 $M_\odot$. The ages are also indicated on the lower axis. At the top of the figure the corresponding evolutionary status is given. The dotted line for ^{4}He in the WN and WC stage corresponds to a lower mass loss rate in the WR stage. $X_c = o$ indicates that the central hydrogen content is zero; it is just a coincidence that the central H-depletion occurs for this model just when CNO-products appear at the surface.*

Table 2   *Galactic chemical enrichments*
by the winds of WR stars

Elements	X_i in WN stage	X_i in WC stage	$\dot{M}_{X_i}$ $(M_\odot yr^{-1} kpc^{-2})$	yield	$\dfrac{yield}{X(ISM)}$
^{1}H	.02	0	---	---	---
^{3}He	1.6 (-11)	7.4 (-14)	-1.1 (- 9)	-1.4 (-7)	-.003
^{4}He	.96	.60	2.4 (- 5)	3.0 (-3)	.012
^{12}C	3.9 (- 4)	.34	9.7 (- 6)	1.2 (-3)	.28
^{13}C	1.2 (- 4)	3.7 (-17)	-1.5 (- 9)	-1.8 (-7)	-.002
^{14}N	1.35 (- 2)	1.3 (- 7)	2.2 (- 7)	2.8 (-5)	.026
^{15}N	6.5 (- 7)	6.5 (- 7)	-7.5 (-11)	-9.4 (-9)	-.003
^{16}O	3.6 (- 4)	3.5 (- 2)*	5.5 (- 7)	6.9 (-5)	.007*
^{17}O	3.6 (- 4)	---	7.1 (- 9):	8.9 (-7):	.14 :
^{18}O	1.5 (- 9)	2.1 (- 7)	-1.1 (- 9)	-1.4 (-7)	-.006
^{20}Ne	5.7 (- 4)	5.7 (- 4)	0	0	0
^{22}Ne	7.7 (- 5)	2.0 (- 2)	5.8 (- 7)	7.3 (-5)	.95
^{24}Mg	6.5 (- 4)	6.5 (- 4)	0	0	0
^{25}Mg	8.8 (- 5)	3.0 (- 4)*	6.1 (- 9)	7.6 (-7)	.009*
^{26}Mg	1.0 (- 4)	4.8 (- 4)*	1.1 (- 8)	1.4 (-6)	.014*

* Abundances as high as $X(^{16}O) = 0.4$, $X(^{25}Mg)$ and $X(^{26}Mg) =$
$5 \cdot 10^{-3}$ are reached in WC models having a large actual
mass. For these elements, the yields are uncertain and
could even be up to an order of magnitude larger.

but this convective core increases in mass during most of its evolution
and a chemical discontinuity is thus built at the border of the core at
its larger extension. We note then a very steep disappearance of ^{13}C
and ^{14}N, a very temporary peak of ^{18}O and above all a vertiginous rise
by more than 2 orders of magnitude of ^{12}C, ^{16}O and ^{22}Ne. The abundances
of ^{25}Mg and ^{26}Mg also rise strongly, particularly in the most massive
WC stars, where s-elements are therefore to be expected.

Comparisons have been made between the theoretical C/He, N/He and
C/N ratios of Fig. 4 and those observed by Smith and Willis (1982) and
by Nugis (1982) for WNL, WNE and WC stars. The general agreement strongly
supports the advanced evolutionary stage of WR stars as left-over cores
resulting from the peeling of massive stars by stellar winds. Fig. 4
suggests the observations of the abundances of further elements in WC
stars, in particular of neon and magnesium.

b) <u>Contribution of the stellar winds to the galactic enrichment
and galactic cosmic rays</u>

The contributions of WR star winds to the enrichment of the interstellar material (ISM) have been estimated (Abbott, 1982; Maeder, 1981b, 1983a). Abbott finds that the net rate of mass input in the ISM by WR stars is $2.0 \cdot 10^{-5}$ $M_\odot yr^{-1} kpc^{-2}$ for WN stars and $2.9 \cdot 10^{-5}$ $M_\odot yr^{-1}$ for WC stars. From the models like those given in Fig. 4 we can estimate the average surface abundances X_i in WR stars and then obtain the net rates $\dot{M}_{x_i}$ of enrichment of the ISM in the various relevant species i, $\dot{M}_{x_i} = [X_i(WR) - X_i(ISM)]\dot{M}_{WR}$. The term in brackets is the excess mass fraction of element i over that of the ISM and $\dot{M}_{WR}$ refers to the rate of mass input per kpc^2 given above for WN and WC stars. The contributions of the winds to the net yields (see Tinsley, 1980, for definitions) may be written

$$y_{x_i} = \frac{\dot{M}_{x_i}}{\psi_1(1-R)}$$

where ψ_1 is the present star formation rate. For $\psi_1(1-R)$ we take $8 \cdot 10^{-3}$ $M_\odot kpc^{-2} yr^{-1}$ (cf. Miller and Scalo, 1979; Tinsky, 1980). The various quantities characterizing the galactic enrichments due to the winds of WR stars are given in Table 2 (cf. Maeder, 1983a); this table also contains the ratios of the yields to the standard abundances in the ISM, which is an indication for the relative importance of the wind to the overall galactic enrichment (let us recall that in a closed model $X_i = y_{x_i} \ln \mu^{-1}$, μ being the mass fraction of the gas in the Galaxy).

Table 2, which only gives rough indications since there are numerous uncertainties involved in such estimates about galactic chemical evolution, suggests that WC stars are great contributors to the galactic enrichments in ^{12}C and ^{22}Ne and have modest relative contributions in ^{16}O, ^{25}Mg, ^{26}Mg and thus s-elements (due to the ^{22}Ne (α, n) ^{25}Mg reaction) WN stars seem to be important for the galactic enrichment in ^{17}O; however we must remember that the rate of the reaction ^{17}O (p, α) ^{14}N is very uncertain (cf. Fowler et al. 1975, 1981).

The case of ^{22}Ne is remarkable and can be brought in relation with the large excess of ^{22}Ne in the galactic cosmic rays, as interestingly suggested by Cassé and Paul (1981). A detailed comparison of the excesses of ^{12}C, $^{22}Ne/^{20}Ne$, $^{25}Mg/^{24}Mg$ and $^{26}Mg/^{24}Mg$ in GCR with the large excesses in WR stars (a factor of 111 for ^{22}Ne in WC stars) has been performed (cf. Maeder, 1983b). This comparison shows that the mentioned excesses in GCR can be very well accounted for by a unique dilution factor: out of 100 particles in the GCR, 2.8 could originate from WC stars and 4.7 from the totality of WR stars. We will not follow up this problem hereupon but call the reader's attention to the interesting contribution on this subject by M. Cassé, contained in this volume.

5. THE SYSTHESIS OF HELIUM AND HEAVY ELEMENTS IN
   MASSIVE STARS

The production of heavy elements by massive stars of constant mass
has been studied by Arnett (1978) who followed the evolution of helium
bare cores. A relation between the mass M_α of the bare core and the
initial stellar mass M was then used to transfer the results of bare
cores to those of standard stars.

The production of helium and heavy elements by massive stars with
mass loss was studied by Dearborn and Blake (1979), Chiosi and Caimmi
(1979), Chiosi (1979) and Chiosi and Matteucci (1982). These authors
used different relations, modified by mass loss, between the initial
stellar mass M and the helium core mass M_α, to operate the transfer of
Arnett's calculations to complete stars. This straightforward procedure
was critically discussed by Maeder (1981b).

Let us now consider the production of helium and heavy elements in
massive stars with mass loss. Fig. 5 shows the various contributions to
the stellar production of new helium and new metals (cf. Maeder, 1981b).
The contributions from winds in various stages of evolution are
distinguished from the contributions by supernova explosions. The models
have been followed up to the end of the C-burning phase. The remnants
are assumed to have 1.4 $M_\odot$ (cf. Tinsley, 1980); however this is very
uncertain and we do not know above which initial mass black holes are
formed and thus remove some fraction of the initial stellar mass from
our visible universe. In this case only the new elements ejected in the
stellar winds would contribute to the galactic enrichment.

For helium we note the increasing importance of the part ejected
in the wind for large initial stellar masses, while the helium ejected
by supernovae becomes negligible for large masses. Below an initial
mass of about 50 $M_\odot$ the overall production of helium is not very much
changed by mass loss at the currently observed rates: there is a compensation (cf. Maeder, 1981b) between the helium ejected early in the
wind and the helium ejected later on in supernovae. Above about 50 $M_\odot$
moderate mass loss reduces the helium production because of the smaller
stellar cores. For very high mass loss rates, however, there is again
an increase in the amount of helium synthetized. This is due to the
fact that in this case a large amount of new helium is ejected and thus
preserved from further destruction. Such behaviour was already qualitatively recognized by Hoyle and Tayler (1964) as being one of the
possible ways to synthetize large amounts of helium in stars.

For heavy elements, Fig. 5 shows that the part ejected in stellar
winds is small in relation to that in supernovae: it only occurs in WC
stars from progenitors likely to be initially larger than about 50 $M_\odot$.
The composition of these heavy elements ejected in the winds is highly
peculiar and has been discussed above (cf. Table 2); it differs from

the composition of the onion-skin models (cf. Arnett, 1978; Weaver and
Woosley, 1978) which will not be considered here. The stellar yields in
metals are reduced by heavy mass loss, this effect being only significant for large initial masses ($M \gtrsim 50\ M_\odot$). In this case, the fact that
much helium is ejected and is not transformed into metals evidently
contributes to a large reduction in the metal yields. This is also the
reason why in Fig. 5 the envelopes of the yield in heavy elements turn
down for large stellar masses.

Fig. 6 illustrates the change of stellar yields in helium and
metals over the whole range of stellar masses. These curves are based
on the results by Renzini and Voli up to 8 $M_\odot$ (their case A), on models
B (Maeder, 1981b) from 15 to 60 $M_\odot$ and on model C^1 for 120 $M_\odot$. The value
for 9 $M_\odot$ is taken from Fig. 12 in the last reference, with a remnant
equal to the actual C/O core.

Small bumps of helium and metal production are to be seen in Fig. 6
at the limit of intermediate mass stars (MS) and massive star models.
This discontinuity is thought to be real. In fact, IMS experience

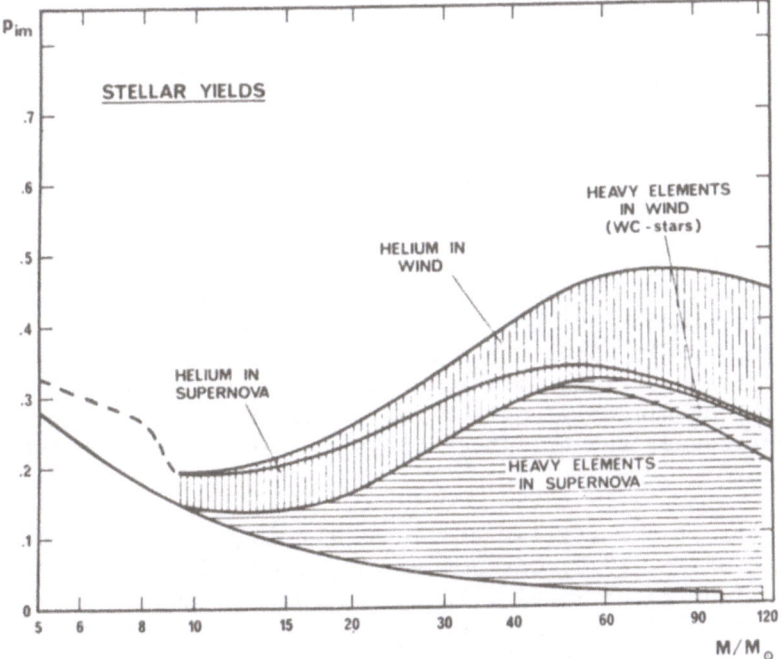

*Fig. 5   Representation of the stellar yields
expressed in mass fractions. The helium and
heavy elements ejected by the stellar wind
are distinguished from those ejected in the
supernova.*

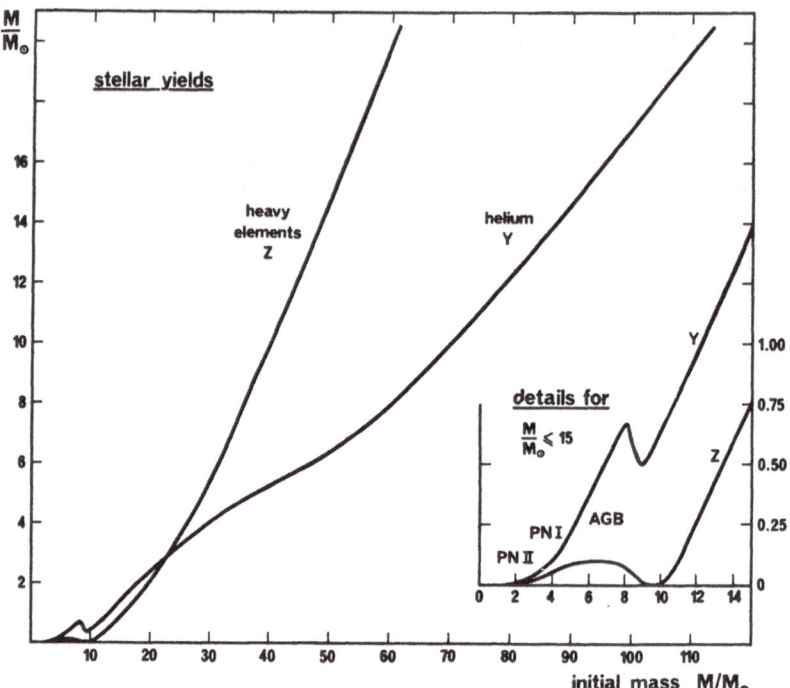

<u>*Fig. 6*</u>   *The amounts of helium and heavy elements*
produced for an average case of mass loss and
with details for low and intermediate mass stars.

helium shell flashes and the associated mixing produces an increased
amount of ^{4}He, which is responsible for the bump in Fig. 6; the flashes
also lead to the production of ^{12}C and ^{14}N as well as s-elements (cf.
Iben and Truran, 1978). Above 8-9 $M_\odot$, stars experience a different evolution without flashes. This is why their yields are initially lower
than that of IMS of slightly smaller initial masses; then, as larger
masses are considered, the helium and metal yields rapidly increase
again.

The products of the stellar yields by the initial mass functions
(cf. Miller and Scalo, 1979; Garmany et al., 1982, above an initial mass
of 25 $M_\odot$) are illustrated in Fig. 7. For helium, we notice the very
large total contribution of small stellar masses: the curve culminates
near 6 $M_\odot$. The small jump near 9 $M_\odot$ existing in the helium curve of
Fig. 6 is essentially smoothed out by the steepness of the IMF. Below
6 $M_\odot$, the helium curve decreases due to the declining stellar yields in
helium; however, near 2.5 $M_\odot$, the growth of the IMF becomes so important
that it produces a second peak. Then the curves fall to near 1 $M_\odot$ because
there the stellar lifetimes become equal to the age of the Galaxy (a
correction for this effect has been brought between 1 and 2 $M_\odot$).

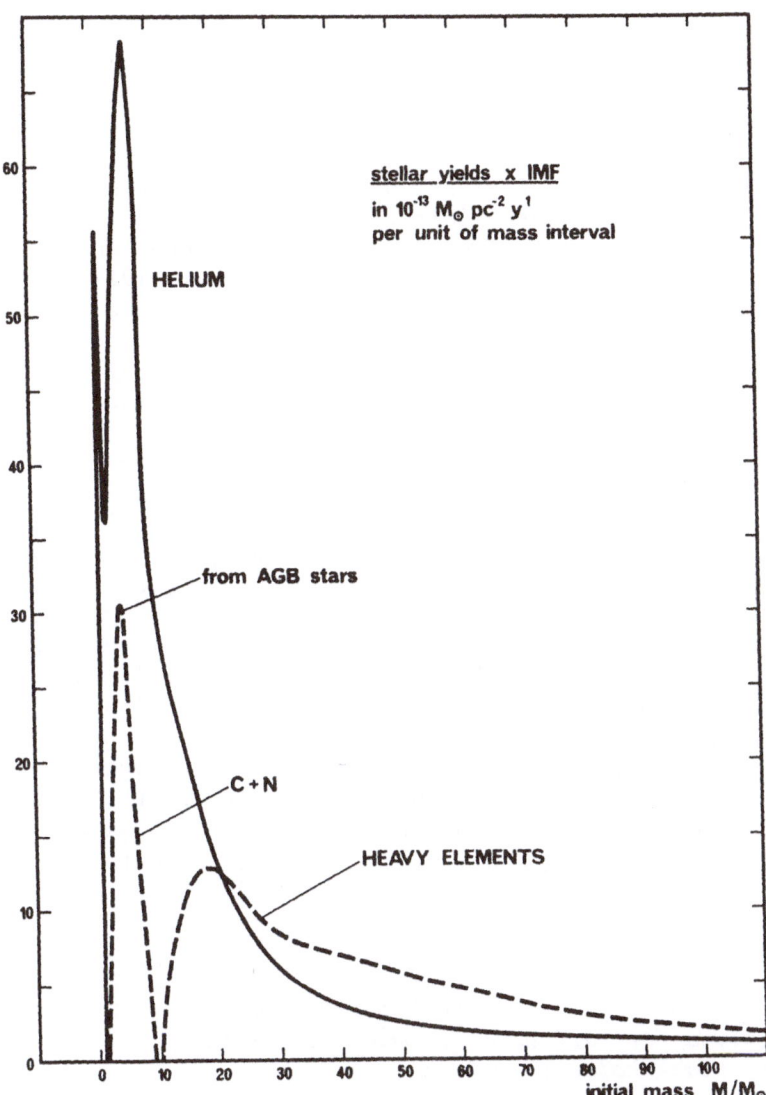

Fig. 7   Product of the stellar yields of Fig. 6 by the IMF. For heavy elements, one recognizes the contribution in ^{12}C and ^{14}N by stars with initial $M < 8$ $M_\odot$ and the second bump due to massive stars.

For the heavy elements, one recognizes in Fig. 7 the two bumps corresponding to those of Fig. 6. For an upper mass limit of 150 $M_\odot$ and a value of 1.4 $M_\odot$ for the remnants, the median and average mass for helium production are 10.8 and 22.6 $M_\odot$ respectively, for metals they are 34.2 and 44.5 $M_\odot$.

The ratio y_Y/y_Z of the net yields of helium and metals, which is obtained from Fig. 7 with a mass limit of 150 $M_\odot$, is then 1.3. Various estimates of this ratio ranging from 0.7 to 3.3 have been made: Arnett's (1978) results lead to 0.7, Chiosi and Caimmi (1979) found 1.8, Chiosi (1979) 3.3, Maeder (1981, without the effects of shell flashes) 1.0, Serrano and Peimbert (1981) 3.3, Chiosi and Matteucci (1982) obtained a value of 2. Recent results by Kunth (1981) on blue compact galaxies and by Shaver et al. (1982) on H II regions appear to exclude a high ratio $\delta Y/\delta Z$ of the relative helium to metal galactic enrichments. Let us also remember that a model of galactic chemical evolution is needed to obtain the ratio $\delta Y/\delta Z$ from the ratio of the net yields y_Y/yz. In the simplest galactic model, one obtains (cf. Maeder, 1981b) the following relation: $\delta Y/\delta Z = (y_Y/yz)(1-Y)$, which shows that $\delta Y/\delta Y$ is 25% smaller than the ratio of the yields.

At present, the mass limit M_{BH} above which core collapse produces black holes is still very uncertain. Above that mass limit we can as a first approximation assume that only stellar winds contribute to the galactic enrichment and that all the matter remaining in the star just before the core collapse is removed from our visible universe and is merged into the black hole. Table 3 shows the ratios of the helium to metal yields in the case of a normal upper limit of initial stellar masses of 150 $M_\odot$ and various mass limits M_{BH} for black holes formation.

Table 3   _Ratios of the helium to metal yields for various lower mass limits M_{BH} of black holes formation._

M_{BH}	y_Y/y_Z
20	3.10
40	2.06
60	1.66
80	1.51
100	1.43
150	1.32

The effects of the winds have been properly accounted for and the
results are different from those obtained from simple cutoffs in the
distributions illustrated in Fig. 7. The matter caught into the black
holes consists essentially of heavy elements, while the matter ejected
in the winds is mainly helium, except for the case of WC stars. Thus,
we understand that the lower M_{BH} is, the higher is the ratio of helium
to metal yields. Obviously, the effects of uncertainties in M_{BH} influence not only the helium and metal yields but also the details of
the galactic enrichment in elements from the onion skin models.

As a general conclusion, let us emphasize that the inclusion of
mass loss has already considerably modified our views about the evolution of massive stars. However, the effects of non-convective mixing
processes remain major uncertainties in stellar evolution and nucleosynthesis and potential changes could stem from it that might guide our
course in the years to come.

REFERENCES

Abbott, D., 1982, Astrophys. J. 263, 723
Abbott, D.C., Bieging, J.B., Churchwell, E., 1981, Astrophys. J.250,645
Arnett, D.W., 1978, Astrophys. J. 219, 1008
Barlow, M.J., Cohen, M., 1977, Astrophys. J. 213, 737
Bernat, A.P., 1977, Astrophys. J. 213, 756
Brunish, W.M., Truran, J.W., 1982a, Astrophys. J. 256, 247
Brunish, W.M., Truran, J.W., 1982b, Astrophys. J. Suppl.Ser. 49, 447
Cassé, M., Paul, J.A., 1982, Astrophys. J. 258, 860
Chevalier, R.A., 1981, Fundamentals of Cosmic Physics 7, 1
Chiosi, C., 1979, Astron. Astrophys. 80, 252
Chiosi, C., 1981, in The Most Massive Stars, ESO Workshop, p. 27
Chiosi, C., Nasi, E., Sreenivasan, S.R., 1978, Astron. Astrophys. 63,103
Chiosi, C., Caimmi, R., 1979, Astron. Astrophys. 80, 234
Chiosi, C., Matteucci, F., 1982, Astron. Astrophys. 105, 140
Conti, P.S., 1982, in WR Stars: Observations, Physics, Evolution,
      IAU Symp. 99, eds. C. de Loore and A. Willis, p. 3
Dearborn, D.S., Blake, J.B., 1979, Astrophys. J. 231, 193
de Loore, C., de Greve, J.P., Vanbeveren, D., 1978, Astron.Astrophys.
      67, 373
de Loore, C., de Greve, J.P., 1981, in The Most Massive Stars,
      ESO Workshop, p. 85
Fowler, W.A., Caughlan, G.R., Zimmermann, B.A., 1975, Ann. Rev.
      Astron. Astrophys. 13, 113
Fowler, W.A., Caughlan, G.R., Zimmermann, B.A., 1981, Corrigenda et
      Addenda to 1975 paper (private communication)
Garmany, C., Olson, G.L., Conti, P.S., 1981, Astrophys. J., 250, 660
Garmany, C.D., Conti, P.S., Chiosi, C., 1982, Astrophys. J. 263, 777
Hidayat, B., Supelli, K., van der Hucht, K.A., 1981, IAU Symp. 99,
      Eds. C. de Loore and A.J. Willis, p. 27

Hummer, D.G., Barlow, M.J., 1981, in Wolf-Rayet Stars, IAU Symp. 99,
     eds. C. de Loore and A. Willis, p. 387
Humphreys, R.M., 1978, Astrophys. J. Suppl. Ser. 38, 309
Humphreys, R.M., Davidson, M.K., 1979, Astrophys. J. 232, 409
Hoyle, F., Tayler, R.J., 1964, Nature 203, 1108
Iben, I., Truran, J.W., 1978, Astrophys. J. 220, 980
Kunth, D., 1981, Thesis, Institut d'Astrophysique, Paris
Lamb, S.A., Iben, I., Howard, W.M., 1976, Astrophys. J. 207, 209
Lerche, I., 1981, Astrophys. Space Sci. 74, 273
Maeder, A., 1980, Astron. Astrophys. 92, 101
Maeder, A., 1981a,b,c, Astron. Astrophys. 99, 97; 101, 385; 102, 401
Maeder, A., 1983a,b, Astron. Astrophys. 120, 113; 120, 130
Maeder, A., Lequeux, J., Azzopardi, M., 1980, Astron. Astrophys. 90, L17
Maeder, A., Lequeux, J., 1982, Astron. Astrophys. 114, 409
Massey, P., 1981, Astrophys. J. 246, 153
Meylan, G., Maeder, A., 1983, Astron. Astrophys. in press
Miller, G.E., Scalo, J.M., 1979, Astrophys. J. Suppl. Ser. 41, 3
Noels, A., Gabriel, M., 1981, Astron. Astrophys. 101, 215
Nugis, T., 1982, in WR Stars: Observation, Physics, Evolution,
     IAU Symp. 99, Eds. C. de Loore and A.J. Willis, p. 127, 131
Peimbert, M., van den Bergh, S., 1971, Astrophys. J. 167, 233
Reimers, D., 1975, in Problems in Stellar Atmospheres and Envelopes,
     Ed. B. Baschek et al., New York, Springer Verlag
Serrano, A., Peimbert, M., 1981, Revista Mexicana de Astron.
     Astrophys. 5, 109
Shaver, P.A., Mc Gee, R.X., Newton, L.M., Danks, A.C., Pottasch, S.R.,
     1982, Preprint ESO No. 210
Smith, L.J., Willis, A.J., 1982, Monthly Notices Roy. Astron. Soc. 201, 451
Stothers, R., Chin, C.W., 1976, Astrophys. J. 204, 472
Stothers, R., Chin, C.W., 1979, Astrophys. J. 233, 267
Tinsley, B., 1980, Fundamentals of Cosmic Physics 5, 287
van der Hucht, K., 1981, in The Most Massive Stars, ESO Workshop, p. 157
Weaver, T.A., Zimmermann, G.B., Woosley, S.E., 1978, Astrophys. J. 225, 1021

DISCUSSION

Bond: In your evolutionary scenario for stars with M > 60 $M_\odot$, you
have them passing through an Of and a Hubble-Sandage variable phase.
Could you give the characteristics of these two types? Also if the Hubble-
Sandage variables are truly variable can this fit into your picture?

Maeder: Of stars are O-type stars in which some emission lines of
NIII and the λ 4686 HeII line are in emission. Finer subdivisions in the
f, (f), ((f)) classification have been introduced by Walborn and by Conti
when the HeII line is in absorption. Of stars usually are the brightest
O-stars and their spectrum shows some similarity with the Wolf-Rayet

spectrum.

Hubble-Sandage variables are also called S Dorados variables or hypergiants. These are the brightest supergiants, they are usually variables in all their properties. They exhibit both secular and short term variations.

The origin of the variability of these extreme supergiants seems to be due to mass ejection. Indeed, Appenzeller et Wolf have shown that the changes of visual brightness are due to changes of the bolometric corrections resulting from Teff variations during shell ejection. Thus, these stars perfectly fit into the proposed scheme of extreme supergiants with very high mass loss rates.

Branch:1)Do you expect a significant dependence of the absolute magnitude of a galaxy's brightest red supergiants on the galaxy's metallicity, resulting from a dependence of mass-loss rate on metallicity? 2) Your estimate of the frequency of explosion of Wolf-Rayet stars is based on the assumption that all such stars explode. Suppose that only stars which develop iron cores ≤ 1.35 M$_\odot$ are able to explode by core bounce. How much smaller would the Wolf-Rayet explosion rate become?

Maeder: The observations by Humphreys suggest that the upper luminosity of red supergiants is independent of metallicity. Recently I proposed (cf. Astron. Astrophys. 120, 113) that the upper limit of supergiants in the HRD is due to instabilities associated to convective processes. For red supergiants, these processes appear to be not too dependent on metallicity.

As to the final fate of WR stars, my point is that most WR stars are likely to reach the phase of core collapse just as their constant mass counterparts do. It would thus be most interesting to learn up to which initial mass this core collapse leads to supernova explosion. If this mass limit is high enough it will not change our estimate of the number of exploding WR stars.

Torres-Peimbert: What are the present ideas about WN and WC objects being of the same mass range? Are these WC stars that are nuclei of planetary nebulae of the same mass or although they have the same spectroscopic characteristics of normal WC they are of different origin?

Maeder: The masses of WR stars span a large range and for now the masses of WC stars, according to Massey, do not seem to be smaller than those of WN stars.

The Wolf-Rayet stars which are nuclei of planetary nebulae have much lower luminosities ($10^{3.5}$ - $10^{4.1}$ L$_\odot$) than the so-called Pop. I WR stars which have luminosities in the range of 10^5 to $10^{6.5}$ L$_\odot$.

Cassé: I am not sure to understand your last point concerning the
gradients since i) it seems that according to Alvio Renzini, the mass
loss rate in the red supergiant stage does not depend sensitively on Z
and ii) the mass loss rate of WR is almost independent of Z. Does it
mean that mass loss in the O and B stages is the only significant
phenomenon?

Maeder: The values of mass loss rates $\dot{M}$ in the O and B stages are
certainly very significant for the course of the evolution. A linear
dependence of $\dot{M}$ on the metal content Z was proposed by Abbott in his
stellar wind theory. For red giants, a milder dependence of $\dot{M}$ on Z was
suggested by Renzini at the Trieste meeting in 1980 ($\dot{M} \sim Z^{\alpha}$, with
$\alpha \simeq 1/4$). However, the evolutionary models clearly show that even small
changes of $\dot{M}$ (well within the observed scatter) may strongly change the
expected number of WR stars (cf. Maeder, 1981c). Thus mass loss rates
do not need to be very dependent on metallicity in order to affect the
number of WR stars.
    Concerning the mass loss rates in the WR stars, there is no indi-
cation at all and even no suspicion that they could be dependent on the
initial stellar metallicity.

Renzini: I have a comment and a question: One should bear in mind
that probably there is both a direct and an indirect effect of Z on
total mass loss, that is $\dot{M}$ may depend on Z for given stellar L, M and Te
(direct effect), and varying Z one may drive the star into a region in
the HR diagram where $\dot{M}$ is larger (indirect effect). I think that is
rather difficult to disentangle the two effects and to assess their
relative importance. My question is the following: from your models one
would expect the relative time spent in the WN and WC phases to be a
strong function of initial mass, and then the WC/WN ratio should depend
on the luminosity of these stars. Is this actually observed?

Maeder: I agree with your distinction between direct and indirect
effects. Let me add, however, that we can expect the indirect effects
for OB stars to be rather limited since the opacity is mainly due to
electron scattering.
    As you mention the models suggest that only the most massive stars
may reach the WC stage. What has been observed is that there is a
higher number ratio WC/WN in the direction towards the galactic center,
where, according to Garmany et al., the IMF indicates the presence of
relatively more very massive stars. Thus, this feature goes in the right
direction. R. Schild and myself have now completed an examination of
Wolf-Rayet stars in clusters. The results show that WC stars are generally found in younger clusters rather than in those containing only WN
stars. Again, this fact is in agreement with the tendency indicated by
the models.

NUCLEOSYNTHESIS DURING THE QUASI-STATIC EVOLUTION OF MASSIVE STARS

Laura Greggio
Dipartimento di Astronomia
CP 596, I-40100 Bologna, Italy

1. INTRODUCTION

The distribution of massive stars in the HR Diagram (HRD) is characterized by the following main features:
a) a group of very luminous O-type stars, next to the ZAMS;
b) an upper limit to the luminosity of blue and yellow supergiants,
   which is decreasing with decreasing effective temperature and which
   ends up with the lack of M supergiants brighter than $M_b \simeq -9.5$;
c) an almost continous distribution of blue supergiants, up to the spectral type A0 (Humphreys 1981).

Since standard evolutionary sequences are not able to account for
the overall appearance of the upper part of the HRD, the effects of a
variety of potentially relevant physical processes have been explored
(Bertelli et al. 1983 and references quoted therein). These processes
include: mass loss, during the H and He burning stages, overshooting
from convective cores and opacity enhancement for temperatures around
10^6 K. Since these phenomena are not yet satisfactorily understood, par
ametrized formulae are often used for taking them into account in model
computations. It turns out that the location of theoretical models in
the HRD, as well as the time spent in different T_{eff}-ranges, are fairly
sensitive to these parameters.

In this context, the comparison between observations and theoretical
predictions for the chemical abundances on the surface of massive stars
may provide hints for the understanding of the evolutionary scenario.
In particular, when nuclear processed material is observed on the surface of massive stars, constraints on the mass loss taking place during
the different evolutionary stages can be derived.

C. Chiosi and A. Renzini (eds.), Stellar Nucleosynthesis, 137–144.

To this end, evolutionary tracks corresponding to initial masses of
20 $M_\odot$ and 60 $M_\odot$ have been computed up to core He-exhaustion, following
a detailed nucleosynthesis and taking into account the physical process
es already mentioned.

The calculations can also provide chemical yields through the stellar winds from massive stars and help in the understanding of their fi
nal fate.

2. INPUT PHYSICS

Changes in the abundances of $^1H, ^3He, ^4He, ^{12}C, ^{13}C, ^{14}N, ^{15}N, ^{16}O, ^{17}O, ^{18}O,$
$^{20}Ne, ^{22}Ne, ^{25}Mg, ^{26}Mg$ have been followed in detail, making use of an explicit scheme of integration. For the H-burning reactions, the three
pp chains and the CNO tri-cycle have been considered; while, for He-
burning, the following reactions have been taken into account:
^4He (α,γ) ^8Be (α,γ) ^{12}C (α,γ) ^{16}O (α,γ) ^{20}Ne

^{14}N (α,γ) ^{18}F $(\beta\nu)$ ^{18}O (α,γ) ^{22}Ne $\begin{cases} (\alpha,\gamma) & ^{26}Mg \\ (\alpha,n) & ^{25}Mg \end{cases}$.

The nuclear reaction rates were taken from Fowler et al. (1975), using
an intermediate factor of 0.1 when cross sections were uncertain.
Overshooting from covective core was taken into account during both H
and He-burning stages, following the procedure described by Bressan et
al. (1981).
Following Chiosi and Olson (1981), a mass loss rate given by
$$\dot{M} = 3.47 \times 10^{16} L^{1.75} \times \{RM/(1-\Gamma)\}^{1.03} \quad M_\odot/yr$$
was adopted during the MS phase, while a constant value of $\dot{M} = 10^{-5} M_\odot/yr$
was applied during the red supergiant stage. After this stage, the models return to the blue side of the HRD with a low surface H content: I
have assumed that this final stage corresponds to the WR stage and rates
of mass loss equal to 5×10^{-6} and 3×10^{-5} $M_\odot/yr$ were adopted for the 20
and 60 $M_\odot$ models, respectively.
The 20 $M_\odot$ track was computed taking into account the effect of an opaci
ity bump, due to the ionization of the elements heavier than He, through
the Bertelli et al. (1983) algorithm, i.e. $\kappa = \kappa_{cs}\{1+4\exp[-10(5.8-LogT)^4]\}$
where κ_{cs} stands for the Cox and Stewart (1970) opacity.

3. RESULTS

The 20 $M_\odot$ model, first moves from its initial ZAMS location crossing
the HRD on a nuclear timescale, until the T_{eff} decreases to $\approx 10^4$ K.
Then, a rapid expansion follows and the star enters the red supergiants
region while still burning Hydrogen in its core. During this stage, the
mass loss decreases the stellar mass to ≈ 12 $M_\odot$, and, when the central H

content is sufficiently low, an overall contraction causes the model to
return to the blue, where the central fuel is eventually exhausted.
During the subsequent shell H-burning phase a second loop occurs, until,
due to the mass removal, the star looses its whole H envelope and returns
to high temperatures. From this stage on the model is considered to repre
sent a WR star.

The 60 $M_\odot$ model was evolved using normal Cox Stewart opacities. There
fore, the whole core H-burning phase is spent on the blue side of the HRD.
H-ignition in a shell causes a redward excursion, which is however stopped and reversed by the large mass loss experienced by the model during
this stage. Eventually, the He core is exposed and the model enters the
WR phase.

Following the representation adopted by Maeder (1983), Figures 1 and
2 show the behaviour of the surface abundances as functions of the remaining stellar mass, for the 20 and 60 $M_\odot$ models, respectively.
When the star enters the red supergiants region, due to the inward penetration of the convective envelope, the surface abundances of the 20 $M_\odot$
model are enriched in CNO processed material. The combined effect of mass
loss and a deeper and deeper penetration of the envelope causes the steady
growth of ^{14}N and ^{13}C surface abundances, as well as the decline of ^{12}C
and ^{18}O. As the inner edge of the convective envelope reaches a region
which has been interested by the core convection during the previous
stages, even the $^{16}O, ^{17}O, ^{1}H$ and ^{4}He abundances are affected. Though the
CN-cycle processed material exhibits equilibrium abundances over the inn
er 70% of the star, when its mass is about 15 $M_\odot$, mixing with the outer
layers prevents the display of equilibrium ratios, and in this stage one
has: $C/N \simeq 1.9$, $^{12}C/^{13}C \simeq 19.2$, where the abundances are given by num
ber. The $^{16}O/^{17}O$ decreases steadily from $\simeq 2.4 \times 10^3$ to $\simeq 6.7 \times 10^2$. When the
whole envelope is removed, partial CNO-burning products appear at the sur
face: a rapid variation of the surface abundances occurs and, since H is
still present, the model may represent a WNL star. Carbon, Nitrogen and Ox
ygen exhibit equilibrium abundances and the predicted number ratios are:
$C/N \simeq 0.02$; $^{12}C/^{13}C \simeq 3.5$; $^{16}O/^{17}O \simeq 10^2$ while O/N steadily decreases
from $\simeq 0.08$ to $\simeq 0.03$. Further removing of the outer layers brings complete
CNO processed material to the surface: the model represents a WNE star,
characterized by a C-enhancement of a factor of $\simeq 1.5$ with respect to the
previous stage. When He-burning products appear at the surface, the abun
dances of the various elements suffer an abrupt variation: the ^{12}C and
^{22}Ne abundance increase respectively by a factor of $\simeq 500$ and 200, while
^{13}C and ^{14}N virtually disappear. This is due to the behaviour of the He-
burning convective core, which developes from a small initial size, reach
es a maximum extension and then retreats. Therefore, the first appearance

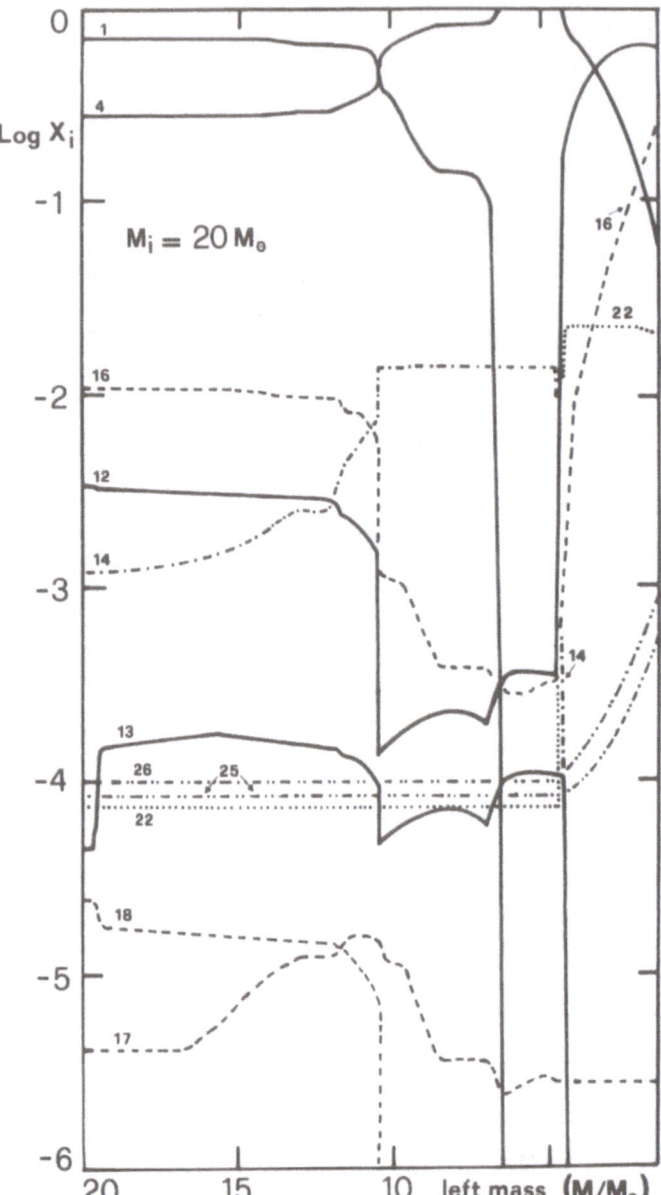

$$\text{Log } X_i$$

$$M_i = 20 \, M_\odot$$

left mass $(M/M_\odot)$

Fig. 1 Evolution of the surface chemical abundances, in mass fraction, as functions of the remaining stellar mass for the 20M⊙ model. The mass number of the various species is indicated.

at the surface of the star of the He-burning products corresponds to the chemical composition of the convective core in a somewhat advanced stage of its evolution. In this stage, the model may represent a WC star, and may eventually turn into a WO star, as the Oxygen surface abundance keeps steadily growing.
Even the surface ^{25}Mg and ^{26}Mg are enhanced during this stage.

As a consequence of overshooting and mass loss, layers which were once in the convective core are exposed at the surface of the 60M⊙ model, during the latest stages of the MS phase. The surface ^{12}C abundance suddenly drops to its equilibrium value, while ^{14}N is enhanced; ^{15}N and ^{18}O virtually disappear and ^{16}O gradually decreases.
An interesting feature is represented by the sudden growth by more than one order of magnitude of the ^{17}O, as it approaches its equilibrium with ^{16}O.
In this stage, the predicted abundance ratios

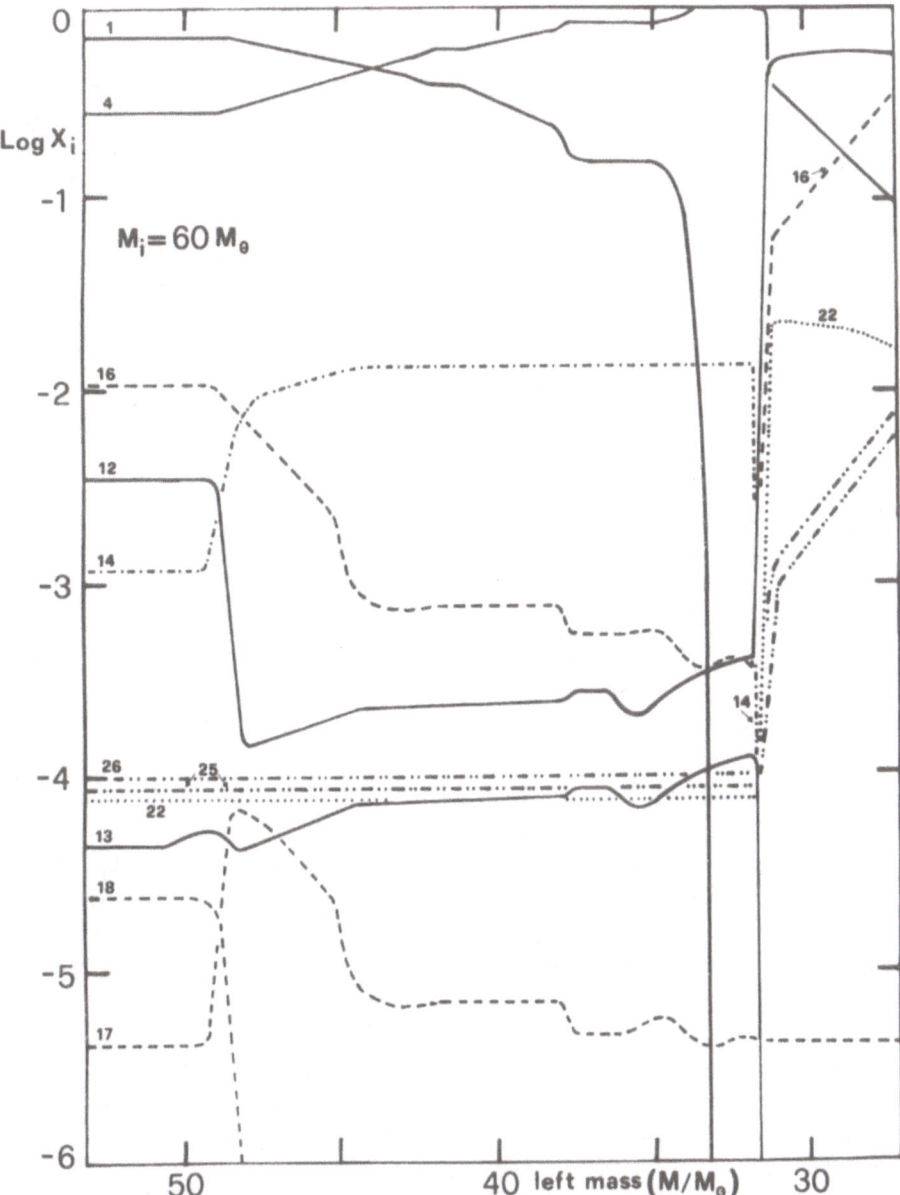

Fig. 2 *The same as in Figure 1, for the 60M$_\odot$ model.*

are C/N≅ 0.02; 0.08 ≤ O/N ≤ 1. The following plateaux, which can be
seen in Figure 2, is due to the external convection, occurring during
the redward motion of the model. When the surface H abundance has dropped
below ≃ 0.4, the model returns to the blue side of the HRD and may represent a WNL star. The subsequent evolution of the surface chemical abundances is very similar to the general behaviour already described for
the 20M$_\odot$ model.

4. CONCLUSIONS

The general agreement between the predicted abundances and the observations for WR stars (Smith and Willis 1982) supports the evolutionary
scheme according to which, due to mass loss, an initially massive star be
comes WNL, WNE and, eventually, a WC star, during the core He-burning
phase. Since the observations give only an upper limit to the N/C ratio
in WC stars, a little room is left for an alternative evolutionary path,
leading to the formation of these objects. Measurements of the Ne and Mg
surface abundances would eventually confirm the interpretation of WC stars
as left-over He-burning cores.

Constraints on the mass lost by massive stars during the MS phase can
be derived by comparing the surface abundances of OB stars and of red
supergiants with the theoretical predictions. In particular, we expect
some OB stars to show a N enhancement on their surfaces, if high mass loss
rates apply during the MS phase (cf. Dearborn and Eggleton 1977).
It is worth noticing, however, that the results depend also on the inclu
sion of the overshooting, which alters appreciably the mass size of the
convective core. Indeed, compared to the 60 $M_\odot$ sequence of Maeder (1983),
which does not include overshooting, the present 60 $M_\odot$ model enters the
WNL, WNE and WC stages with a left mass of 38, 33 and 31 $M_\odot$, respectively,
while the corresponding Maeder's values are 33, 24 and 20 $M_\odot$. On the other
hand, the same predicted WR masses are much closer to the analogous values
given by Maeder for the 85 $M_\odot$ model.
Therefore, it seems that similar final results can be obtained by simply
enhancing the rate of mass loss or the extent of the overshooting and
good constraints on each of these phenomena can hardly be derived by the
mere comparison between the predicted properties of WR stars and the obser
vations.

The results of the present calculations are in general agreement with
those found by Maeder (1983), apart from the high surface abundance of
^{17}O in WN stars. Indeed, the present 60 $M_\odot$ model never exhibits a ^{16}O to
^{17}O ratio smaller than $\simeq 10^2$, while Maeder finds an extended region where
this ratio is near unity. When the CNO cycle operates, ^{17}O is produced
from ^{16}O up to the equilibrium ratio, which depends only on the nuclear
reaction rates. Having used the same sources for the nuclear reaction rates, the discrepancy between the present results and those obtained by
Maeder can hardly be understood, unless Maeder has used rates other than
those formally stated. Alternatively, the difference could be due to the
use of an implicit method for the solution of the network equations, while
the present models have been obtained using an explicit method.
In any case, these models produce much less ^{17}O than those of Maeder.

REFERENCES

Bertelli,G.,Bressan,A.G.,Chiosi,C. 1983, preprint
Bressan,A.G.,Bertelli,G.,Chiosi,C. 1981, Astron. Astrophys. 102, 25
Chiosi,C.,Olson,G.L. 1981, private comunication
Cox,A.N.,Stewart,J.N. 1970, Astrophys. J. Suppl. 19, 243
Dearborn,D.S.P.,Eggleton,P.P., Astrophys. J. 213, 448
Fowler,W.A.,Caughlan,G.R.,Zimmermann,B.A. 1975, Ann. Rev. Astron.
    Astrophys. 13, 113
Humphreys, R.M. 1981, in "The Most Massive Stars", ESO Workshop, eds.
    S. D'Odorico, D. Baade and K. Kjar, p.5
Maeder, A. 1983, Astron. Astrophys. 120, 113
Smith,L.J.,Willis,A.J. 1982, Monthly Notices Roy. Astron. Soc. 201, 451

DISCUSSION

Bond: You chose a mass loss rate for red supergiants of $10^{-5}M_\odot$/yr.
Why? This choice is not very far from the $3\times10^{-5}M_\odot$/yr chosen for WR's.
Have you any comment on the near coincidence?

Greggio: Since mass loss rates for red supergiants are poorly determined, I have chosen the value quoted for αOri by Jura and Morris (1981,
Astrophys. J. 251, 181). On the other hand, current estimates of $\dot{M}$ for
WR stars are all on the order of few$\times10^{-5}M_\odot$/yr: wether or not the two
values are intimately related is hard to decide, since theoretical models for stellar winds are still controversial.

Torres-Peimbert: You mentioned that it would be important to determine Mg and Ne abundance in WC stars, but you did not explain what those
abundance determinations would tell us. Can you give us some details?

Greggio: As already mentioned, we expect that, if WC stars show
on their surface triple-alpha processed material, the N abundance should
be virtually undetectable. Since the observations only give an upper li
mit for the N abundance, a little room is left for different evolutiona
ry paths, leading to the formation of WC stars (cf. Chiosi 1981, in
"The Most Massive Stars", ESO Workshop, eds. S. D'Odorico, D. Baade and
K. Kjar, p.27; Maeder 1982, in "Wolf Rayet Stars: Observations, Physics,
Evolution", IAU Symp. 99, eds. C. de Loore and A. Willis, D. Reidel Pub.
Co., Dordrecht, Holland, p.405). On the other hand, measurements of an
overabundance of Ne and Mg would unquestionably confirm the interpreta-
tion of WC stars as bare He-burning cores.

Cassé: A comment: since the ^{22}Ne (α,n) ^{25}Mg reaction liberates a
significant number of neutrons, s-process elements should appear at the
surface of WC stars. It would be interesting
i) to calculate the yield of light s-process elements (A < 90) produced
   in this kind of process (see e; g. Lamb,Iben and Howard 1976, Astrophys. J. 207, 209);
ii) to know if some of these species are observable in the IR, visible
    or UV spectra of WC stars;
iii) to estimate the effect of WC stellar winds on the evolution of s-
     process elements in the Galaxy;
iv) to consider these s-process elements as seeds for the r-process syn
    thesis that should occur at explosion (see for instance the Helium
driven r-process model of Truran, Coward and Cameron 1983, Astrophys. J.
265, 429).

PRESUPERNOVA YIELDS AND THEIR DEPENDENCE ON THE $^{12}C(\alpha,\gamma)^{16}O$ REACTION

W. David Arnett
Enrico Fermi Institute, University of Chicago
Chicago, Illinois, U.S.A.

F.-K. Thielemann
Max-Planck-Institut für Physik und Astrophysik
Garching b. München, Fed. Rep. Germany

ABSTRACT

Massive stars (M $\gtrsim$ 20 $M_\odot$) produce abundance patterns during the high temperature shell burning phases of hydrostatic evolution before the core collapse which are little changed by explosive processing of the ejected envelope in a type II supernova explosion. Thus a prediction for pre-supernova abundances can be taken as an approximation to the final products of explosive nucleosynthesis. A calculation using a detailed nucleosynthesis network (254 nuclear species), which adds shell burning products at typical burning conditions (temperature, density) weighted in mass according to stellar evolution calculations, reproduces the qualitative behavior of previous yield curves from postprocessing calculations, including the (too) large enhancement of carbon burning products and (too) small enhancement of oxygen burning products. When the $^{12}C(\alpha,\gamma)^{16}O$ rate is increased, as indicated by recent experiments, this leads to a lower ^{12}C and higher ^{16}O abundance after core helium burning. Consequently the subsequent burning phases give less carbon burning products and more oxygen (and silicon) burning products. These changes are shown to lead to a uniform enhancement of nucleosynthesis products ($2 < Z \lesssim 32$ at least) from an "average" type II supernova.

1. A SIMPLE MODEL FOR CALCULATION OF PRE-SUPERNOVA ABUNDANCES

It is a feature of massive stars (M $\gtrsim$ 20 $M_\odot$) that the products of high temperature shell burning are very close to the final abundances after explosive processing during the ejection of the outer layers in a type II supernova explosion (Arnett and Wefel, 1978; Weaver & Woosley, 1980); thus we focus on the prediction of abundances yields from hydrostatic stellar evolution. Detailed analysis of evolutionary calculations suggests some simplifications.

In hydrogen and helium burning, extended convective cores develop.

145

C. Chiosi and A. Renzini (eds.), Stellar Nucleosynthesis, 145–150.
© *1984 by D. Reidel Publishing Company.*

Compared to the time scales of subsequent core burning phases, the outward movement of the hydrogen and helium burning shells is slow. In the
pre-supernova stage the matter inside those shells can be regarded consisting essentially of core (hydrogen and helium) burning products. Starting with carbon burning, cores of the size of a Chandrasekhar mass (M_{Ch})
are formed (which is thought to be the size of the neutron star remnant
after a type II supernova explosion). Thus the ejecta after a SN explosion contain only <u>shell</u> burning products of C, Ne, O, and Si burning.

A simple way to predict the abundance yields is to take material inside the H and He burning shell from typical core burning conditions while
matter inside the convective C, Ne, O, and Si burning shells is considered
to be the product of typical shell burning conditions. Those conditions
(ρ,T) and the positions of the convective burning shells in the mass
coordinate can be taken from evolutionary calculations (e.g. Arnett, 1977;
Weaver et al., 1978). Fig. 1 shows the typical burning conditions for
which the nucleosynthesis calculations of this paper were performed; they
are close to the conditions in the evolutionary calculations of Arnett
(1972, 1977). The abundances after core hydrogen burning were taken from
Arnould and Nørgaard (1978).

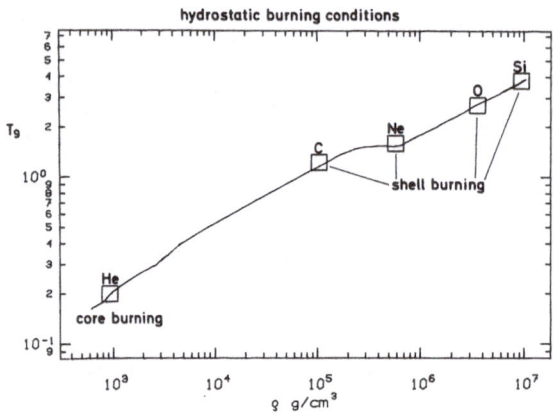

Fig. 1: Open squares denote
the burning conditions for
subsequent burning phases
used in the present calculation.

2. THE NUCLEOSYNTHESIS NETWORK

The aim of the present paper (for more details see Thielemann and
Arnett, 1983) is to discuss nucleosynthesis in detail with simplified
structure assumptions, and to explore nucleosynthesis effects which cannot be treated in evolutionary calculations. The network employed,
contains 254 nuclear species, from neutrons and protons to ^{74}Ge and is
essentially suited for all burning stages.

The nuclear reaction rates are, whenever possible, taken from Fowler
et al. (1975) including the update by Fowler (1981). For heavier nuclei
the statistical model calculations from Woosley et al. (1978) are included and complemented in some cases by results obtained with the statistical
model code SMOKER (Thielemann, 1980). Weak interaction rates applied
here are those from Fuller et al. (1982) for the β-decay and electron

capture of 166 nuclei, complemented by experimental (terrestrial) β-half lives (Seelmann-Eggebert et al., 1981).

The influence of most recent experimental work, especially the new value obtained for the $^{12}C(\alpha,\gamma)^{16}O$ rate by Kettner et. (1982), will be discussed in the following section.

3. NUCLEOSYNTHESIS YIELDS FROM A 8 $M_{\odot}$ HE-CORE

The nucleosynthesis calculations performed here took into account the conditions for the evolution of a 8 $M_{\odot}$ He-core regarding the position in mass coordinates, temperatures and densities as described in detail by Arnett (1972, 1977). Following the recipe given in section 1, and adding up the results of several burning stages leads to an overabundance curve of elements as given in Fig. 2a when the original $^{12}C(\alpha,\gamma)^{16}O$ rate of Fowler et al. (1975) is applied. The original stellar evolutionary calculations by Arnett used a value roughly twice as large. The two solid lines include an area with overproductions from 7-28 (14 with an uncertainty of a factor of 2) which was indicated as an average overproduction for a 25 M_O star ($\approx$9 $M_{\odot}$ He-core) by Woosley and Weaver (1982) who used the Fowler rate. Qualitatively a similar behavior exists for both results in form of an s-shaped curve in Z or A. The carbon burning products including Ne, Na, Mg are enhanced too much while O and Si shell burning products (including P, S, Cl, Ar, K, Ca, Sc, Ti) are too low. The reason that V, Cr and Mn behave differently from Woosley and Weaver is probably due to our choice of the mass cut (1.32 $M_{\odot}$) for the neutron star remnant being different from that of Woosley and Weaver (1982). This reflects a fundamental uncertainty. How might we obtain a constant overabundance which produces relative abundances in solar ratios? An enhanced $^{12}C(\alpha,\gamma)^{16}O$ reaction rate would enhance the $^{16}O/^{12}C$ ratio after core He-burning. This leaves less fuel for shell Carbon burning and more fuel for shell oxygen burning and eventually silicon burning. Another possibility would be that more massive stars produce mainly explosive oxygen burning products (see Woosley, this volume).

However, the recent experimental result by Kettner et al. (1982) suggests an enhancement by a factor of 3-5 for $^{12}C(\alpha,\gamma)^{16}O$ over Fowler et al. (1975) which supports the first possibility. A reevaluation of the Münster Data by Langanke and Koonin (1983) results in roughly a factor of 3. Figs. 2b,c show results when the $^{12}C(\alpha,\gamma)^{16}O$ rate was enhanced by a factor of 3 and 5, respectively. The overabundance curve flattens out remarkably, except for ^{23}Na. This could reflect uncertainties in the reactions $^{12}C(^{12}C,p)^{23}Na$ and (or) $^{23}Na(p,\alpha)^{20}Ne$ or indicate a slightly higher temperature is needed for shell carbon burning. The nucleosynthesis calculations all were performed with the same stellar structure, an approximation which has to be checked by evolutionary calculations. Fig. 3 shows the neutron excess parameter $\eta = (\bar{N}-\bar{Z})/(\bar{N}+\bar{Z})$ (total excess of neutrons over protons, free and in nuclei) and its change during subsequent evolutionary phases. There is a large increase in He burning due to $^{18}F(\beta^+)^{18}O$ (in the reaction sequence $^{14}N(\alpha,\gamma)^{18}F(\beta^+)^{18}O(\alpha,\gamma)^{22}Ne$) followed by a minor decrease due to β^--decay of s-process products. The increase in later

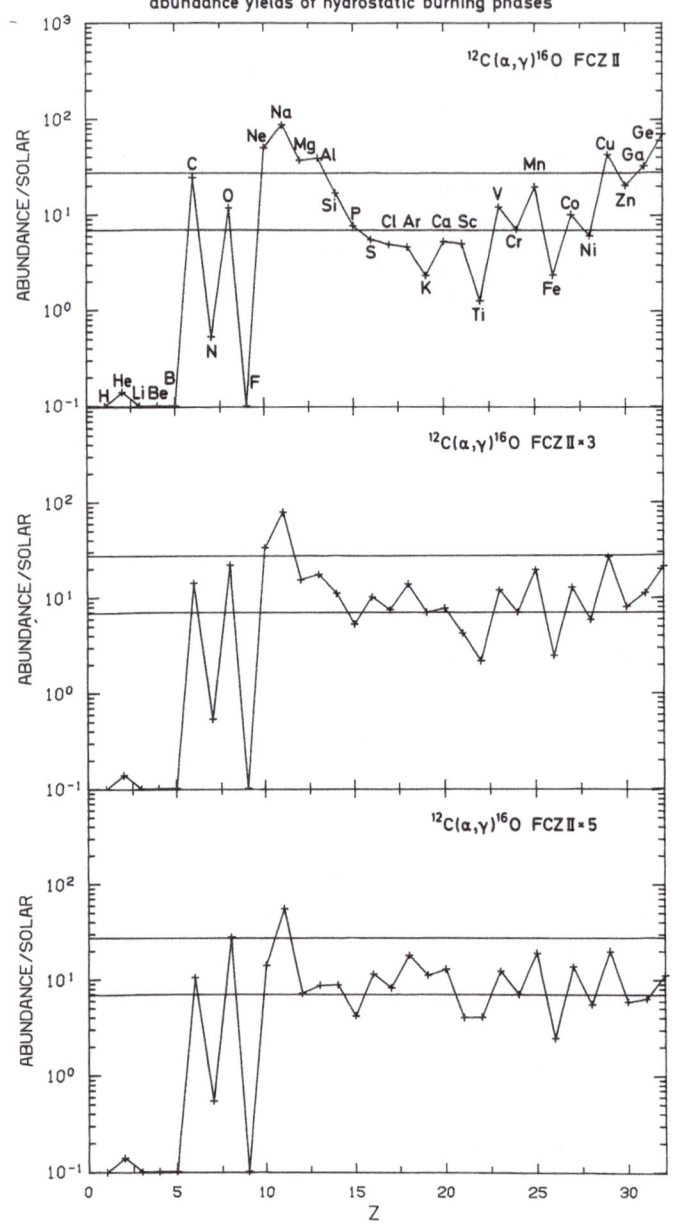

Fig. 2: Overproduction factors with respect to solar for the elements up to Ge, produced during hydrostatic burning phases, before the onset of core collapse. (a),(b) and (c) result from multiplying the $^{12}C(\alpha,\gamma)^{16}O$ rate by Fowler et al. (1975) with a factor of 1,3, and 5. The curve flattens with an enhanced $^{12}C(\alpha,\gamma)^{16}O$ rate resulting in <u>relative</u> abundances close to solar.

The additional consideration of shell He-burning products (this approximation takes only <u>core</u> He-burning products into account, see Sect. 1) would mainly effect the abundances of ^{4}He, ^{12}C, ^{16}O, ^{19}F... and s-process nuclei.

burning stages is given by β^+-decays of proton-rich nuclei produced by fusion reactions of matter with essentially Z=N, while the line of stability bends towards more neutron-rich nuclei. Having less fuel for carbon burning leads to a smaller increase of η. Thus a fast $^{12}C(\alpha,\gamma)^{16}O$ rate results in a small η after carbon burning which meets much better the requirements by Woosley, Arnett, and Clayton (1973) for reproducing solar abundances in oxygen burning.

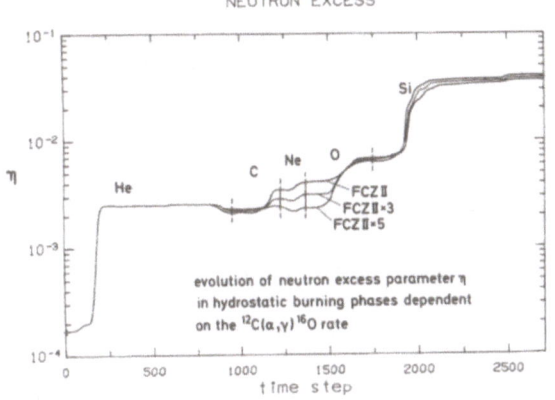

Fig. 3: Evolution of the neutron excess parameter $\eta = (\bar{N}-\bar{Z})/(\bar{N}+\bar{Z})$ during different burning stages. An enhanced $^{12}C(\alpha,\gamma)$ ^{16}O rate results in a lower value of η in oxygen burning.

We conclude that a changed $^{12}C(\alpha,\gamma)^{16}O$ rate results in overabundance ratios which are close to relative solar abundances, if the stellar structure is not seriously altered at the same time. Thus a 20–25 $M_\odot$ star, regarded as an average type II supernova progenitor in respect to ejected matter with Z>2, would actually give a solar abundance pattern.

REFERENCES

Arnett, W.D.: 1972, Ap. J. 176, 681
Arnett, W.D.: 1977, Ap. J. Suppl. 35, 145
Arnett, W.D., Wefel, J.P.: 1978, Ap. J. (Letters) 224, L139
Arnould, M.,Nørgaard, H.: 1978, Astron. Astrophys. 64, 195
Fowler, W.A.: 1981, private communication
Fowler, W.A., Caughlan, G.R., Zimmerman, B.A.: 1975, Ann. Rev. Astron.
      Astrophys. 13, 69
Fuller, G.M., Fowler, W.A., Newman, M.J.: 1982, Ap. J. Suppl. 48, 279
Kettner, K.U., Becker, H.W., Buchmann, L., Görres, J., Kräwinkel, H.,
      Rolfs, C., Schmalbrock, P., Trautvetter, H.P., Vlieks, A.: 1982,
      Z.Physik A308, 73
Langanke, K., Koonin, S.E.: 1983, submitted to Nucl. Phys. A
Seelmann-Eggebert, W., Pfennig, G., Münzel, H., Klewe Nebening, H.: 1981
      "Chart of the Nuclides", Kernforschungszentrum Karlsruhe
Thielemann, F.-K.: 1980, Ph.D. thesis, TH Darmstadt and MPI München,
      unpublished
Thielemann, F.-K., Arnett, W.D.: 1983, in preparation
Weaver, T.A., Woosley, S.E.: 1980, Ann. N.Y. Acad. Sci. 336, 335
Weaver, T.A., Zimmerman, G., Woosley, S.E.: 1978, Ap. J. 225, 1021
Woosley, S.E., Holmes, J.A., Fowler, W.A., Zimmerman, B.A.: 1978, Atomic
      Data Nuc. Data Tables 22, 371
Woosley, S.E., Arnett, W.D., Clayton, D.D.: 1973, Ap. J. Suppl. 26, 231
Woosley, S.E., Weaver, T.A.: 1982, in "Essays in Nuclear Astrophysics",
      (Cambridge University Press), p. 377

DISCUSSION

Iben: It was my understanding from an earlier talk by Nomoto,
that type I SN carbon deflagnation models produce elements in just
the region in which type II models underproduce elements as a
consequence of explosive oxygen burning (using the old $^{12}C(\alpha,\gamma)^{16}O$
cross section). Would you comment?

Thielemann: Carbon deflagnation models of type I supernovae seem to
be able to reproduce observed abundance patterns and will therefore
contribute to the abundance of intermediate mass nuclei. But even
detailed network calculations show only large overabundances for
even Z nuclei (mainly α-nuclei). Thus type II supernovae are still
very important to explain solar abundances in this mass range.

r-PROCESS PRODUCTION IN LOW MASS STARS

J.J. Cowan
Department of Physics & Astronomy, University of Oklahoma,
A.G.W. Cameron
Harvard-Smithsonian Center for Astrophysics, and
J.W. Truran
Department of Astronomy, University of Illinois

ABSTRACT

Using steady flow calculations, we find that the solar system r-process abundance distribution could have been formed in astrophysical environments with neutron number densities ranging from 10^{20} to 10^{21} cm^{-3}. Assuming mixing between the H-rich envelope and the helium core, such neutron number densities can be produced by ^{13}C during helium core flashes in low mass stars. These results indicate that low mass stars may have been responsible for synthesizing many of the solar system r-process nuclei.

1. INTRODUCTION

The rapid neutron capture process (r-process) is not well understood. Most of the stable nuclear isotopes above iron are made in this or the slow neutron capture process (s-process); roughly half of all of these isotopes are made in the r-process (see Cameron 1982). There have been two major areas of uncertainties in studying the production of neutron-rich heavy nuclei in the r-process. First, there are uncertainties in the nuclear physics involved. Unlike the s-process, the r-process produces many radioactive nuclei that are far from the so-called valley of beta stability. In practical terms, that means that we have very little hope of measuring the nuclear properties (e.g., masses, beta decay rates, neutron-capture cross sections, etc.) of these radioactive nuclei. Instead, we must rely on theory to predict the properties of these isotopes. Normally the theories adjust their parameters to fit the behavior of nuclei near the valley of beta stability, and then these parameters are used to extrapolate the behavior of the very radioactive nuclei. Clearly such a procedure will result in uncertainties in the nuclear properties of the r-process nuclei.

The second major area of uncertainty has to do with the astrophysical environments in evolved stars. Observational evidence suggests that some processes in evolved stars are responsible for the synthesis

151

C. Chiosi and A. Renzini (eds.), Stellar Nucleosynthesis, 151–159.
© *1984 by D. Reidel Publishing Company.*

of heavy elements. But in evolved stars there are many phases not well
understood, such as mixing. Also, it is not known what mass range of
stars will ignite in supernova explosions or even how the supernova
explosion is triggered.

The important question is what types of astrophisical environment (or environments) could produce the r-process nuclei. Originally,
it was thought that only explosive environments were capable of producing
the extreme conditions that were thought to be necessary to produce
the r-process elements. The central cores of supernovae were suggested
by several groups as a site for the r-process (e.g., Seeger, Fowler,
and Clayton 1965; Cameron, Delano, and Truran 1970; Kodama and Takahashi
1973; Schramm 1973; Hillebrandt, Takahashi, and Kodama 1976). While
these models were promising, none of them could reproduce the observed
solar system r-process abundances using physically plausible conditions.
This led to many other suggested sites for the r-process including
supernova shocks (Colgate 1971), magnetized bubbles in rotating magnetized
stellar cores (LeBlanc and Wilson 1970; Meier et al. 1976) and the
helium zone of stars subjected to supernova shock wave heating (Truran,
Cowan, and Cameron 1978; Hillebrandt and Thielemann 1977).

All of these models examined high mass stars under explosive
conditions. Recent advances in nuclear physics and the development of
more sophisticated stellar evolutionary models have provided impetus to
examine whether r-process nuclei could be produced under nonexplosive
conditions in low mass stars. We recently studied the production of
r-process nuclei during helium core flashes (Cowan, Cameron, and
Truran 1982). The neutron source in this case was assumed to be 2% or
more by mass of ^{13}C in a degenerate helium-rich gas. The ^{13}C would have
to be produced in such a core by mixing between the H-rich envelope
and the ^{12}C in the core; it must be emphasized that such a process is
not presently understood or known to take place. The abundances of the
heavy elements were assumed to be solar. As a result of helium ignition
in the degenerate gas a thermal runaway occurs. Our approach was to
examine various physical conditions typical of helium core flashes in
low mass stars and to calculate the abundances of r-process nuclei that
were produced. To test the models we then compared our calculated
r-process abundance curve with the observed solar system curve. For
certain physical conditions we found that the calculated curve approximated the solar system distribution. While the fit was not excellent,
it was suggestive that r-process nuclei could be synthesized in helium
core flashes.

These thermal runaway models, however, suffered from some
deficiencies. The beta decay rates used in our early models had been
based on the gross theory of beta decay (Takahashi and Yamada 1969).
More recent work by Klapdor and Oda (1980) had indicated that the beta
decay rates predicted by the gross theory were approximately ten times
too low for neutron-rich nuclei. We therefore scaled the gross beta
decay rates consistent with the results of Klapdor and Oda for the heavy,
neutron-rich nuclei. Although this method was an improvement over the

gross rates, it was still an approximation. Our original thermal
runaway models also used neutron capture rates that were based on techniques that we felt could be improved. Finally, we used a simple approximation to describe the time dependent behavior of the temperature and
density.

2. CALCULATIONS

        In an attempt to understand better how r-process nuclei are
synthesized in thermal runaways in the helium cores of low mass stars,
we have made some further numerical experiments. For these models we
have assumed a steady flow condition. In a steady flow, the rate of
inflow of nuclei into the bottom of a network (in our case at ^{28}Si) will
equal the rate of outflow of nuclei at the top. Such a condition will
result after a sufficient time and if fission cycling is ignored. In a
steady flow pattern all of the nuclei in the network will approach a
steady state abundance. It should be noted, however, that the steady
flow approximation does not in any way assume an $(n,\gamma) \rightleftarrows (\gamma,n)$ equilibrium
(also known as the waiting point approximation). It has been shown that
for conditions typical of helium burning, the waiting point approximation
is not valid for most values of Z (Cameron, Cowan, and Truran 1983).
The advantage of using the steady flow procedure is that it allows us
to solve the numerical equations one row of Z at a time due to the
progressive nature of the flow (i.e., each row beta decays to the next
row of higher Z). The calculations are then straight-forward and take
little computer time. This procedure is of course not a dynamic model
but the important nuclear dependences in the r-process can be determined.

        For these new models we have incorporated new nuclear data.
The beta decay rates for the 6,033 nuclei in our network were calculated
using the techniques described by Klapdor et al. (1981). These rates
tend to be between one and ten times the rates based on the gross theory
(Cameron et al. 1983). Our code also allows for beta decay followed by
the emission of up to three delayed neutrons. The neutron capture cross
sections were also recalculated. The radiation widths and the level
density parameters in the cross sections were recalculated to give better
fits to the measured 30 Kev neutron capture cross sections (see Cameron
et al. 1983 for further details).

        The results of some of our new models are shown in Figure 1.
All three cases shown represent steady flow calculations in environments
typical of helium core burning. In each case the calculated abundances
(lower curve), after all the beta decays are completed, are compared
with the observed solar system r-process abundances (upper curve) as
given by Cameron (1982). The calculations were performed at a temperature of 3×10^8 K although the results are insensitive to temperature as
photodisintegration rates are small for the conditions studied. The
neutrons are again produced by ^{13}C during the helium core flash. The
upper set of curves shows the calculated r-process abundances for a
neutron number density, N_n, of 10^{19} cm^{-3} in comparison with the solar

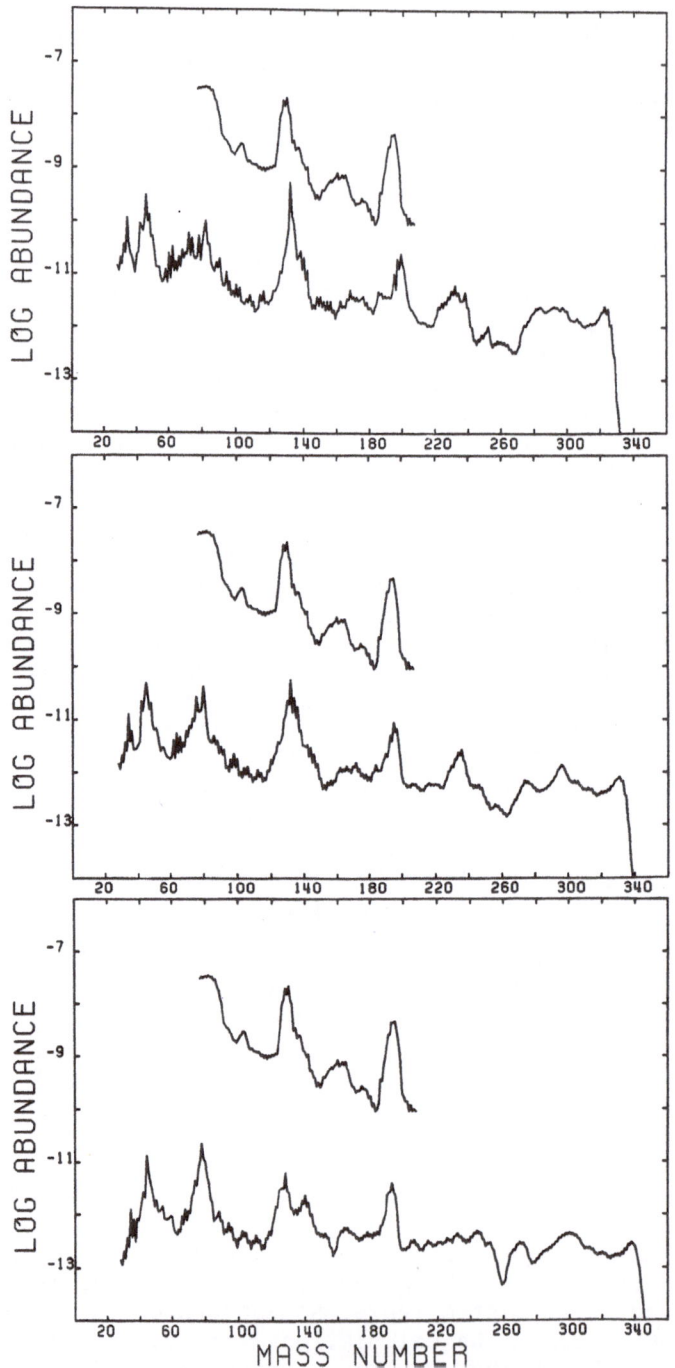

Figure 1. R-process yield curves for neutron number densities of 10^{19} (top), 10^{20} (middle), and 10^{21} cm^{-3} (bottom). The upper curve in each panel is the observed r-process abundance curve (Cameron 1982), and the bottom curve is the calculated steady flow abundance curve. The vertical normalization of the calculated curve is arbitrary.

system curve. The vertical normalization of the curve is arbitrary.
The middle set of curves shows the calculated curve for a neutron number
density of 10^{20} cm^{-3} and the bottom set of curves shows the case for a
neutron number density of 10^{21} cm^{-3}. Since there is no fission in the
models the calculated abundances of nuclei with mass number, A, above
approximately 240 are probably not meaningful.

It is clear from the figures that the position and the shape
of the calculated r-process peaks depend upon the neutron number density.
For $N_n=10^{19}$ cm^{-3} the calculated peaks at A∿130 and A∿195 occur at too
high a mass number and do not line up exactly with the observed r-process
peaks. On the other hand, in the case of $N_n=10^{20}$ shown in the middle
panel of Figure 1, the calculated peak at A∿130 does line up very well
with the observed peak. The agreement is also good for the peak at
A∿80. It should be noted, however, that the observed peak at this mass
number is not well defined in shape since the abundances on the low mass
side of the peak are really upper limits. Thus the observed solar
system r-process peaks at A∿80 and A∿130 can be reproduced in our steady
flow models with a neutron number density of 10^{20} cm^{-3}. The r-process
peak at A∿195, however, is best reproduced with a neutron number density
of 10^{21} as shown in the bottom panel in Figure 1. It may be therefore
that the solar system r-process nuclei were produced in a range of astrophysical environments. Most of the r-process nuclei could be produced
in an environment with the number density near 10^{20} cm^{-3} for a time of
order of seconds. However, in some environments the number density
would rise to 10^{21} cm^{-3} for a fraction of a second.

3. DISCUSSION

It is interesting to note that with the inclusion of the new
nuclear data (i.e., the new beta decay rates and the new neutron capture
cross sections) we find that the calculated r-process peaks line up at
the same mass number as the observed solar system peaks. One of the
primary reasons for performing this series of numerical experiments was
to investigate the dependence of the r-process claculations on the
nuclear parameters. In our earlier dynamic calculations using older
nuclear data we had found good agreement between the calculated and
observed r-process peaks, but there had been some slight offset in mass
number. That offset is not evident in the steady flow calculations using
the new nuclear data.

Our results also indicate that the solar system r-process
abundances could have been produced with relatively low neutron number
densities. The values that we find of 10^{20}-10^{21} cm^{-3} are orders of magnitude less than what was considered in the classical r-process under
$(n,\gamma) \gtrless (\gamma,n)$ equilibrium. This is a very important result and as a
consequence of this, the r-process nuclei need not be produced in an explosive environment (i.e., a supernova). Instead, the conditions that we
have investigated (temperatures of 3-5x10^8 K and ρ∿10^5 gcm^{-3}) are typical
of helium core flashes in low mass stars.

To produce the r-process nuclei in our models requires ^{13}C as a neutron source. We have found that to produce the total solar system r-process abundance curve requires several per cent by mass of ^{13}C in the helium core. We have not performed detailed model calculations to show how ^{13}C is produced in the helium core. It is quite evident that mixing between the H-rich envelope and the ^{12}C in the core is required to produce the ^{13}C. There have been several studies that suggest that such mixing might occur as a result of helium core flashes in low mass stars (see, for example, Deupree and Cole 1983). However, it is evident that such mixing cannot take place unless the stellar core is sufficiently heated and then expanded so that the temperature falls enough not to burn hydrogen rapidly during the mixing process. The core thermal runaway leading to copious neutron production must then be a second or later major helium core flash in the evolution of the star.

In trying to understand how and where the r-process nuclei are produced, our approach has been independent of any particular stellar evolutionary model. As a consequence of this, we have been able to examine a wide range of physical conditions as opposed to a narrow range characteristic of one particular model. Our approach is also independent of model biases and instead relies only on the input physics. The steady flow calculations, in particular, have enabled us to isolate certain conditions that can produce the solar system r-process nuclei. In fact, however, the solar system abundances were not produced in a steady flow process, but rather a dynamic process. It is clear from the curves in Figure 1 that the calculated abundance curves do not have the same slope as the solar system abundance curve. Thus while the steady flow is a good approximation, only a dynamic calculation can properly take into account all of the relevant physics. A series of dynamical calculations is currently in progress by the authors (Cowan, Cameron, and Truran 1983) to determine the abundances of r-process nuclei during helium core flashes. These new models employ the new beta decay rates and the neutron capture cross sections described earlier. These studies, along with continued stellar evolutionary studies of mixing, will help in confirming the helium cores of low mass stars as a site for the production of the r-process elements.

If low mass stars are responsible for synthesizing r-process nuclei, only a small amount of matter must be ejected per star to reproduce the total galactic r-process abundances. If we assume that all stars with mass $M \leq 2\ M_\odot$ experience degenerate He core flashes and produce r-process nuclei, then each star must inject back into the interstellar medium only $8\times10^{-7}\ M_\odot$ of strictly r-process material to satisfy galactic nucleosynthesis requirements. This means that a total of only approximately $10^{-3}\ M_\odot$ of stellar material need be ejected per star. Thus, low mass stars could be responsible for producing the total observed level of galactic r-process abundances as well as reproducing the solar system r-process abundances, and yet still not appear chemically peculiar in the later stages of evolution.

This work has been supported in part by the National Science

Foundation under grants AST 82-14964 at the University of Oklahoma, AST 81-19545 at Harvard University, and AST 80-14964 at the University of Illinois, and by the National Aeronautics and Space Administration under grant NGR 22-007-272 at Harvard University.

REFERENCES

Cameron, A.G.W.: 1982, in "Essays in Nuclear Astrophysics," ed. C.A.
        Barnes, D.D. Clayton, and D.N. Schramm (Cambridge: University
        Press), pp. 23-43.
Cameron, A.G.W., Delano, M.D., and Truran, J.W.: 1970, CERN, 70-30, p. 735.
Cameron, A.G.W., Cowan, J.J., Klapdor, H.V., Metzinger, J., Oda, T., and
        Truran, J.W.: 1983, Astrophys. Space Sci., 91, p. 221.
Cameron, A.G.W., Cowan, J.J., and Truran, J.W.: 1983, Astrophys. Space
        Sci., 91, p. 235.
Colgate, S.A.: 1971, Ap.J., 163, p. 221.
Cowan, J.J., Cameron, A.G.W., and Truran, J.W.: 1982, Ap.J., 252, p. 348.
Cowan, J.J., Cameron, A.G.W., and Truran, J.W.: 1983, in preparation.
Deupree, R.G., and Cole, P.W.: 1983, Ap.J., in press.
Hillebrandt, W., Takahashi, K., and Kodama, T.: 1976, Astron. Astrophys.,
        52, p. 63.
Hillebrandt, W., and Thielemann, F.K.: 1977, in "Mitt. Astron.
        Gesellschaft," 43, p. 234.
Klapdor, H.V., and Oda, T.: 1980, Ap.J. (Letters), 242, p. L49.
Klapdor, H.V., Oda, T., Metzinger, J., Hillebrandt, W., and Thielemann,
        F.K.: 1981, Z. Phys. A-Atoms and Nuclei, 299, p. 213.
Kodama, T., and Takahashi, K.: 1973, Phys. Letters, 43B, p. 167.
LeBlanc, J.M., and Wilson, J.R.: 1970, Ap.J., 161, p. 541.
Meier, D.L., Epstein, R.I., Arnett, W.D., and Schramm, D.N.: 1976,
        Ap.J., 204, p. 869.
Schramm, D.N.: 1973, Ap.J., 185, p. 293.
Seeger, P.A., Fowler, W.A., and Clayton, D.D.: 1965, Ap.J. Suppl., 11,
        p. 121.
Takahashi, K., and Yamada, M.: 1969, Progr. Theor. Phys., 41., p. 1470.
Truran, J.W., Cowan, J.J., and Cameron, A.G.W.: 1978, Ap.J. (Letters),
        222, p. L63.

DISCUSSION

        MATTHEWS: I have two questions. One is - what about the helium driven r-process in massive, e.g., 25 $M_\odot$ stars. The other question is - what about normal thermal pulses in intermediate-mass stars. At the pulse peak there is a substantial neutron density for short time intervals.
        COWAN: In regard to your first question, we have performed calculations for the helium-driven r-process under conditions typical of massive stars. The helium-driven r-process does produce r-process isotopes in this case, but it does not produce the solar system r-process curve unless there has been extensive prior s-processing and even then only for fairly low temperatures (T less than about 400 million degrees).

As for your second question, the thermal pulses in intermediate mass stars will probably provide neutron number densities intermediate between the s- and r-process. Again, you might produce some r-process isotopes, but not the solar system distributions.

EDMUNDS: How long does the process go on for? Also, isn't fission from the end of the chain entering at lower A important in reality (i.e., steady flow is not really applicable), or is this only important if a given flash gets a fair amount of material to the fission point?

COWAN: The burning is very quiet for a couple of thousand seconds and then there is a rapid thermal runaway over a period of a few seconds. We followed the flash and the following quiescent phase for about ten thousand seconds. As to fission, if the steady flow conditions had reproduced the total spectrum of solar system r-process abundances then indeed fission would have contributed two nuclei from the top of the network for every one nuclei produced by beta decay from the bottom of the network. However, because the actual slope of the solar system abundance curve is so much steeper than the steady flow curve, the actual flow of nuclei out of the top of the network will be very small compared to the flow from the bottom. This indicates that the steady flow conditions were not responsible for producing the observed r-process nuclei except in approximations to local regions, and that fission would not be important in the steady flow calculations. Fission recycling might be important for conditions under which $(n,\gamma) \rightleftarrows (\gamma,n)$ equilibrium existed. Those conditions correspond to temperatures and neutron number densities that are far higher than conditions that we are examining.

GALLAGHER: Given that there seem to be more possible sites and processes which can lead to near solar system very heavy element abundance distributions, is it then true that the solar system element distribution is the result of the averaging of similar yields by a variety of processes? If so, then does this represent a change in viewpoint from earlier studies in which, as I recall, different parts of the elemental mass spectrum were identified with specific production sites.

COWAN: Maybe; one of the main problems, of course, is that we do not understand how the r-process nuclei are made. Our work does, however, indicate that the solar system r-process curve might not have been produced by one environment. Instead, there may have been several different environments with different neutron number densities. The higher neutron number densities would be primarily responsible for producing the heavier r-process nuclei while the lighter r-process nuclei would result from lower neutron exposures. If a variety of sites under a variety of neutron exposures was responsible for producing the solar system abundances, it may be difficult to untangle them because of all of the free parameters.

RENZINI: (to Nomoto and Woosley) Could any r-process elements be produced in the region experiencing explosive carbon burning in SNI models?

WOOSLEY: (answer to A. Renzini) Some "r-process" nucleosynthesis is expected from explosive carbon burning in the deflagrating models as the peak temperature in the burning front falls to $<2 \times 10^9$ K. In carbon this process produces only nuclei just above the iron group as discussed by Howard et al. in the Ap.J. in the early 1970's. If the carbon is capped by a helium layer containing s-processed nuclei then there may be an r-process at $T_9 \sim 1$ as Cowan, Thielemann and collaborators have described. It might be a useful exercise to calculate this synthesis in carbon layers again using Cowan's big code and an enhanced s-process seed (not considered by Howard et al.). An adiabatic/exponential expansion parameterization would be as good as any.

NOMOTO: (answer to A. Renzini) No r-process nucleosynthesis takes place in the carbon deflagration wave. However, a strong precursor shock wave ahead of the deflagration wave passes through the helium layer where the r-process elements exist. This may contribute to r-process synthesis. Hillebrandt is suggesting that the dynamic r-process is expected to take place in the ejecta from a 9 $M_\odot$ star.

SYNTHETIC H-R DIAGRAMS AS AN OBSERVATIONAL TEST OF STELLAR EVOLUTION THEORY

G. J. Mathews
Lawrence Livermore National Laboratory
and
S. A. Becker and W. M. Brunish
Los Alamos National Laboratory

Synthetic H-R diagrams are constructed from a grid of stellar models. These are compared directly with observations of young clusters in the LMC and SMC as a test of the models and as a means to determine the age, age dispersion, and composition of the clusters. Significant discrepancies between the observed and model H-R diagrams indicate the possible influences of convective overshoot, large AGB mass-loss rates, and the best value for the mixing length parameter.

1. INTRODUCTION

As we have already heard at this workshop, the Magellanic Clouds are an important arena for the study of stellar evolution. One reason for this is that they have experienced a recent burst of star formation activity during the last 100 million years and therefore contain numerous young clusters whose H-R diagrams are well populated in the various evolutionary phases of intermediate-mass ($3M_\odot < M < 10M_\odot$) stars. Such stars are of interest in the study of stellar evolution and nucleosynthesis as progenitors of planetary nebulae, asymptotic giant branch (AGB) stars, and even as possible progenitors of carbon-deflagration or core-bounce supernovae.

In this work we construct synthetic H-R diagrams from theoretical evolutionary tracks as a function of mass and composition to directly compare with the observations. Such comparisons serve a two-fold purpose. For one, they are a sensitive means to estimate the age and composition of the observed clusters. Perhaps more importantly, such comparisons allow for a determination of the quality of the stellar models and highlight the areas in which new input physics is required.

2. CONSTRUCTION OF SYNTHETIC DIAGRAMS

The basis for the construction of the theoretical H-R diagrams is a three-dimensional grid in M, Y, and Z of evolutionary models mostly

161

C. Chiosi and A. Renzini (eds.), Stellar Nucleosynthesis, 161–167.
© *1984 by D. Reidel Publishing Company.*

from Becker (1981). AGB evolution to the first thermal pulse is interpolated from Becker and Iben (1979, 1980). The remaining AGB evolution
is extrapolated as described in Becker and Mathews (1983). A few additional tracks were computed to complete the grid (Becker and Brunish
1983), and more massive tracks are from Brunish and Truran (1982a, b).
These tracks span a range in mass from 3 to 40 $M_\odot$, and helium mass
fractions from 0.20 to 0.36, as well as metallicities from 0.001 to
0.030. Thus, by interpolation (Becker and Mathews 1983) we are able to
infer the evolution for essentially any mass or composition likely to
be encountered in young Magellanic-Cloud clusters. For the first time
a simultaneous age and composition fit can be attempted.

The synthetic HR diagrams are constructed from a random population of
stars with a gaussian dispersion in ages and an appropriate initial
mass function. (We chose IMF $\propto m^{-2.3}$). The mass and metallicity-
dependent color-magnitude conversions of Kurucz (1979) are applied for
regions where applicable, and the spectral-class-dependent conversions
of Flower (1977) are used for temperatures and luminousities outside
the range of the Kurucz tables. We estimate a theoretical uncertainty
of about 0.1 magnitudes in V and B-V in these conversions from the
difference between the two tables when compared at solar metallicity.

3.   RESULTS

Figures la-b are examples of theoretical diagrams as a function of age
(15, 30, 60, 90, 120, 240 x 10^6 yrs) for two different compositions.

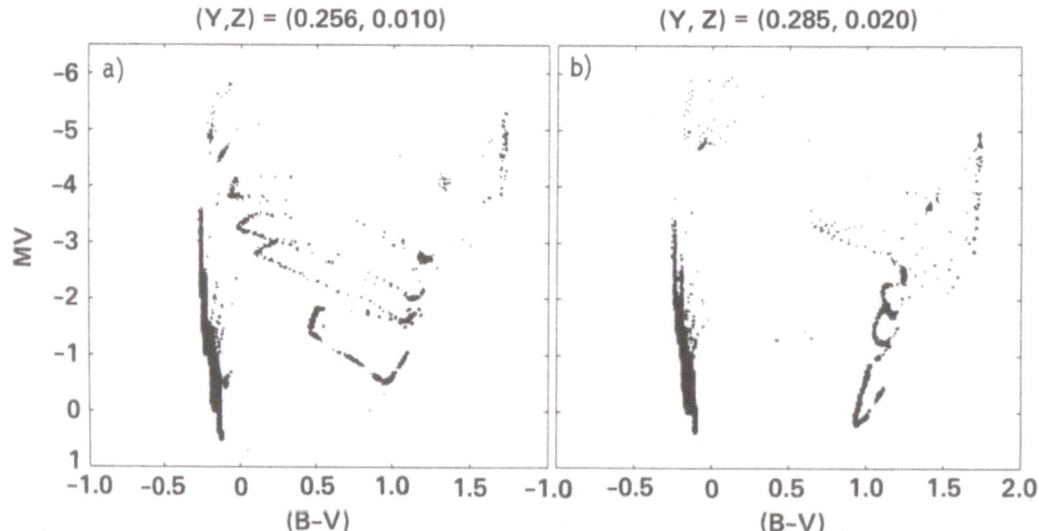

Figure 1. Synthetic H-R diagrams as a function of age for different
compositions.

We assume that Y is correlated with Z according to the relation given in Lequeux et al. (1979). The most dramatic changes in the diagrams of figures la-b are in the location of the core helium burning loop (CHBL) which roughly decreases in brightness with increasing age, and reddens with increasing metallicity and helium abundance. These connections between age and luminousity, and composition and color can be understood for the most part in terms of the influences on the evolutionary tracks from the stellar mass, opacity, and mean molecular weight (Iben 1967).

Figure 2 is a summary of the locations of the blue tip of the CHBL as a function of age for various compositions. Overlayed on this are the locations of the CHBL for eight clusters in the LMC with sufficient data to easily identify a CHBL blue tip (Alcaino 1975). It is interesting to note that nearly all of these clusters lie approximately along a line corresponding to a single composition of (Y,Z) = (0.27, 0.015). The SMC contains fewer clusters with a recognizable CHBL. They tend to indicate a lower metallicity

Comparisons between synthetic and observed H-R diagrams for three of the best studied LMC clusters are given in Figures 3-5. The synthetic clusters are normalized to the same number of stars in the observed ranges of V and (B-V). An artificial dispersion (Robertson 1974) has been added to the theoretical points for faint stars to simulate the dispersion in the observations. These clusters span a range in ages and turnoff masses from fairly young clusters like NGC 1854 ($\sim$30 x 10^6 yrs) with an estimated turnoff mass of 8-9 $M_\odot$, to older clusters like NGC 1866 with an estimated turnoff mass of 5$M_\odot$.

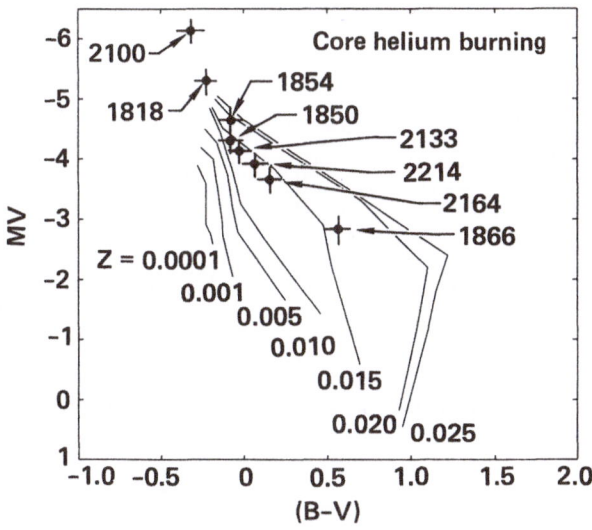

Figure 2. Summary of the location of the tip of the CHBL as a function of age along lines of constant composition.

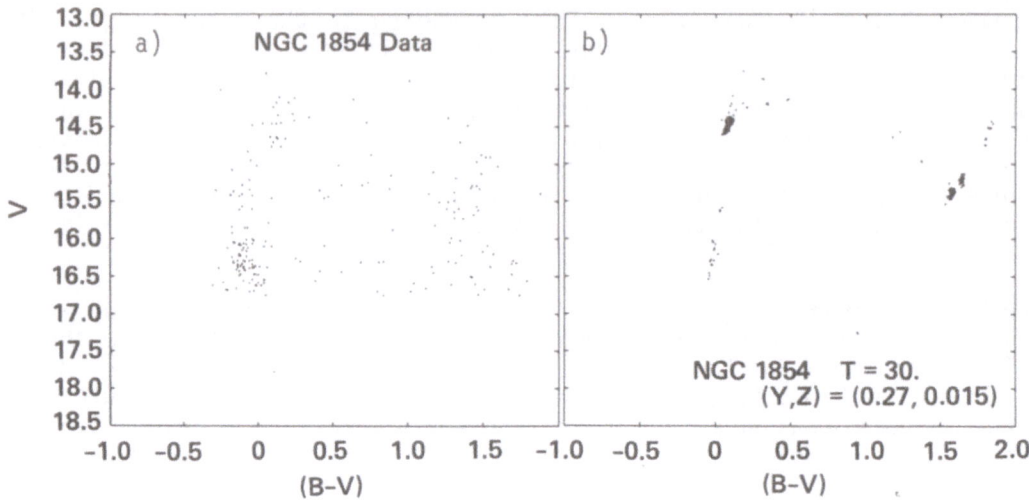

Figure 3. Comparison of observed and model H-R diagrams for the LMC cluster NGC 1854.

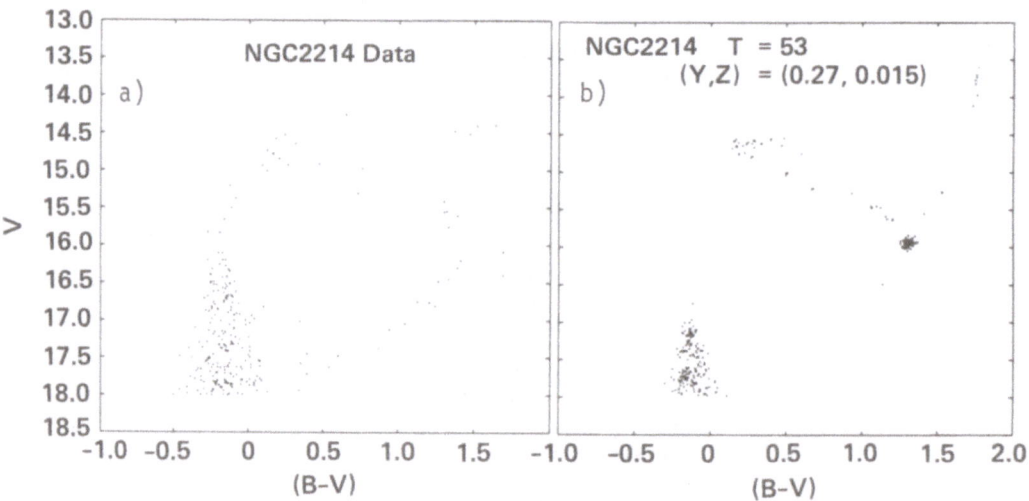

Figure 4. Comparison between observed and theoretical H-R diagrams for the LMC cluster NGC 2214.

4.   INTERPRETATION

The basic features of the observed diagrams are well reproduced in Figures 3b-5b but there are some significant discrepancies which may

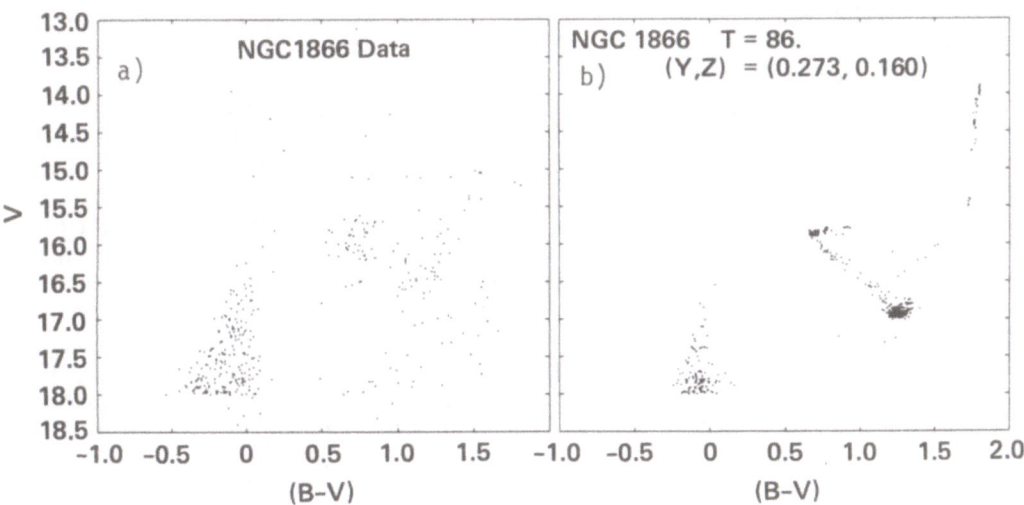

Figure 5. Comparison between observed and model H-R diagrams for the LMC star cluster NGC1866.

indicate the need for refinements in the input physics to the stellar models. These descrepancies are visible in all of the synthetic clusters and are best apparent in the comparison between theory and observations for NGC 1866. (Figs. 5a-b) for which the largest number of stars have been studied.

4.1. Populations

The first discrepancy which we note is the apparent deficiency of main sequence stars relative to red giant stars in the model clusters shown in Figures 3b-5b. This is particularly true for the brightest blue stars which actually correspond in the models to post main-sequence stars near the ignition of shell hydrogen burning. The implication is that either the calculated lifetimes of these stars are too short (particularly for the post main sequence), or that the red-giant lifetimes are too long in the stellar models. One possible explanation for this discrepancy is that the main-sequence lifetimes have been underestimated in the models by the neglect of convective overshoot. This additional mixing could introduce the necessary lengthening of the main-sequence lifetime (Maeder and Mermilliod 1981), but the effect of convective overshoot on the red-giant lifetimes must be considered. Among other possible interpretations for this discrepancy, the possible contributions from rotation, binary evolution, and uncertainties in the opacities should also be considered.

4.2. AGB Evolution

A second striking deviation between the observed and model H-R diagrams is the absence in the data of the asymptotic giant branch which is predicted in the models. The AGB evolution for the model clusters has been interpolated from the calculations of Becker and Iben (1979, 1980), based on the Reimers (1975) mass loss rate. These tracks predict an extended AGB evolution to high luminousities with eventual termination by explosive carbon ignition in an electron-degenerate core, i.e. a carbon-deflagration supernova. The data, on the other hand, appear to reach a limiting absolute visual magnitude of about -4 (assuming a distance modulus of 18.6 for the LMC). In NGC 1866 (Figure 5a), and to some extent in the other clusters, there is even evidence for stars turning off from the AGB toward becoming central stars of planetary nebulae. This is apparent in NGC 1866 as a string of stars with $V \sim 15$. Apparently the mass-loss rate exceeds the assumed Reimers estimate, or an envelope instability is reached early in the AGB evolution which leads to the ejection of the outer layers. Thus, even clusters with a turnoff mass of 8-9 $M_\odot$ (like NGC 1854 in Figure 3a) seem to exhibit white-dwarf formation rather than an extended AGB. This is consistent with recent searches for white dwarfs in young local galactic clusters which imply an upper mass limit for white-dwarf progenitors of about $8 M_\odot$ (Reimers and Koester 1983; Koester and Weidemann 1983).

From the luminousity of the observed post AGB stars we estimate a core mass of $\sim 0.7 M_\odot$. The ejecta mass divided by the time spent on the AGB then implies a mass loss rate of $M/M_\odot > 10^{-6}$ yr^{-1} for a $5 M_\odot$ star.

4.3. Mixing Length to Pressure Scale Height

We note another discrepancy which is that the AGB evolution for the model clusters is too red. The model tracks have been adjusted to a ratio of mixing length to pressure scale height of 1.0. The data seem to indicate that a value of 1.5 is preferable (assuming that the color-magnitude conversions are correct in this region of the H-R diagram).

4.4. Age Dispersion

Finally we point out that in these studies an upper limit of 5 x 10^6 yrs. could be derived for the dispersion in ages of stars in a cluster by the very fact that the CHBL is an identifiable collection of stars at about the same color and brightness. Large dispersions in age tend to produce a large dispersion in the CHBL which is not compatable with the observations (see Becker and Mathews 1983).

5.   ACKNOWLEDGEMENT

Work performed under the auspices of the U. S. Department of Energy by the Lawrence Livermore National Laboratory under contract number W-7405-ENG-48. The authors greatfully acknowledge useful discussions and encouragement from Professor J. W. Truran.

REFERENCES

Alciano, G. 1975, Astron. Ap. Suppl., 21, 279.
Becker, S. A.: 1981, Ap. J. Suppl. 45, p.475.
Becker, S. A. and Brunish, W. M.: 1983, (to be submitted to Ap. J.).
Becker, S. A. and Iben, I. Jr.: 1979, Ap. J. 232, p.831.
_____.: 1980, Ap. J. 237, p.111.
Becker, S. A. and Mathews, G. J.: 1983, Ap. J. 270, p.155.
Brunish, W. M. and Truran, J. W.: 1982a, Ap. J. 256, p.247.
_____.: 1982b, Ap. J. Suppl. 49, p.447.
Flower, P. J.: 1977, Astron. Ap. 54, p.31.
Iben, I. Jr.: 1967, Ann. Rev. Astron. Ap. 5, p.571.
Kurucz, R. L.: 1979, Ap. J. Suppl. 40, p.1.
Lequeux, S. et al.: 1979, Astron. Ap. 80, p.155.
Maeder, A. and Mermilliod, J. C.: 1981, Astron. Ap. 93, p.136.
Reimers, D.: 1975, Mem. Soc. Roy. Sci. Liege, Ser. 6. 8, p.369.
Reimers, D. and Koester, D.: 1983, Astron. Ap. 116, p.341.
Weidemann, V. and Koester, D.: 1983, Astron. Ap. 121, p.77.
Robertson, J. W.: 1974, Ap. J. 191, p.67.

DISCUSSION

Bond--How much of this effect can be explained by variation of the IMF?

Mathews--Very little. To reproduce the ratio of red giants to main seqence stars would require an IMF $\sim m^{-14.3}$, which is unreasonably steep. Furthermore even this IMF would not account for the paucity of post main sequence stars. The reason for this insensitivity to the IMF is because a very narrow mass range (less than $1M_\odot$) is contributing to the red-giant population

Torres-Piembert--You used $\Delta Y \sim 3\Delta Z$ to compute the properties of the clusters, and certainly the clusters seem to be compatable with only one composition. I wonder what would happen if you used a fixed value for Y and varied Z only. Then would you find the same agreement or not?

Mathews--The location of the core helium burning loop depends both upon the helium abundance (via the mean molecular weight) and the metallicity (via the opacity). Therefore a change in the metallicity can be compensated by an opposite change in Y to keep the CHBL in the same position. On the other hand, stars with higher opacities have longer main sequence lifetimes. Therefore, the red giant to main sequence ratio could be reproduced by a low ($\sim$primordial) helium abundance, and a high ($\sim$solar) metallicity. This seems to me to be an exotic composition, however, and probably would not correct for the low number of post main-sequence stars.

EVIDENCE FOR A LARGE MAIN SEQUENCE WIDENING: A PLEA FOR OVERSHOOTING
AND HIGHER HEAVY-ELEMENT OPACITY

Bertelli G.,[1,3] Bressan A.,[2] Chiosi C.[1,2]

1) Institute of Astronomy, Padova, Italy

2) International School for Advanced Studies, Trieste, Italy

3) Fellow of the National Council of Research (C.N.R.)

Summary
It is shown that a moderate increase of the opacity due to heavy elements
in models of massive stars, incorporating convective overshoot and mass
loss by stellar wind, can remove the well known discrepancy between theoretical expectation and observed frequencies of luminous stars in the
HR diagram. These models in fact extend their main sequence band far beyond the classical limit.

1. Introduction
The HR diagrams of luminous (massive) stars in various regions of our own
Galaxy and in external galaxies, like the two Magellanic Clouds, consti-
tute a very useful tool for testing models of massive stars, properly
accounting for a variety of physical processes (mass loss, convective
overshoot, etc.), and for studying the dependence of these latter on local
properties, the metallicity in particular, of the interstellar medium.
The most popular samples of observational data for luminous stars in young
galactic clusters and associations are by Humphreys (1981, and references
therein) and Mermilliod (1976, 1981). The comparison of the relative number
of stars in different regions of the HR diagram with the expectation from
standard theoretical models reveals that too many stars fall outside the
main sequence band. Unless severe selection effects are present in the
star counts for the earliest spectral types (O and B), everything occurs
as if the main sequence band were much wider than predicted, apparently
extending to the spectral type A0. Although this discrepancy has long been
around in the literature (Stothers and Chin, 1977; Chiosi et al.,1978;
Cloutman and Whitaker, 1980), only recently the problem has been completely focussed and fully appreciated (Bressan et al., 1981; Meylan and Maeder,
1982; Bertelli et al., 1983). Various possible causes of main sequence
extension have been indicated over the past few years. These are: i)
mass loss by stellar wind, which under suitable values for the mass loss
rate may extend the main sequence band of stars initially less massive
than about 60 $M_\odot$ up to the spectral type B0.5 (de Loore et al., 1977;
Chiosi et al., 1978). However the lifetime spent by those models beyond
the classical limit (O9.5) turned out to be very short and in turn very

169

few stars to be observable. ii) effect of the stellar wind on the hydrostatic radius (de Loore et al., 1982), which in presence of high rates of
mass loss would shift the models to lower effective temperatures, thus
making wider the main sequence band. However, under the current rates of
mass loss in the range of mass of interest here, this effect turned out
to be marginal (Bertelli et al., 1983). iii) overshoot from the inner
convective core, which combined with mass loss is found to extend the
main sequence band to the spectral type B1 (Bressan et al., 1981). The
lifetime of those models in the region outside the standard limit was
about 20% of the core H-burning lifetime. iv) additional mixing by some
process such as shear flow instabilities and/or meridional circulation
has been suggested by Meylan and Maeder (1981). However, numerical models
incorporating the above effect have not yet been calculated. We suspect
that their behaviour should be quite similar to that of models with convective overshoot. As a conclusion, it seems that no one classical model
is able to satisfy the observational demand. In this paper, we endeavour
to show that *increasing the opacity due to heavy elements by a factor 2
to 4* in models with overshoot (mass loss during the main sequence phase
will turn out to play a marginal role) may significantly improve upon the
disagreement. The proposed increase is not as ad hoc as it may appear,
because heavy element opacities in the region between 10^5 K° to 2×10^6 K°
may plausibly have been underestimated by a factor of 2 to 3 as recently
pointed out by Simon (1982). Furthermore, such an increase has been shown
to remove the mass anomalies in both the double mode and bump Cepheid regimes, and perhaps to energize β Cephei variables (Simon, 1982). Finally,
it is worth recalling that opacities showing a strong enhancement in the
above temperature range already exist in the literature (Carson, 1976),
although they have not received widespread acceptance.

2. Star Counts in the HR Diagram

The relative number of stars in each spectral type is determined from the
M_b vs Log T_e diagram for supergiant stars of Humphreys (1978) with the
following selection criteria. Only stars in the luminosity range $-7 > M_b$
> -9 and within 2.5 Kpc from the sun are considered. This region of the
HR diagram is by far the most popolous one, whereas the low luminosity
cutoff somehow represents the completeness limit of the Humphreys catalogue. The spatial limit is imposed by the comparison of Humphreys' list of
O type stars with Garmany's et al. (1982) catalogue of stars of the same
type. This latter in fact is fairly complete up to this distance. With
these criteria, we count 138 O type stars in Humphreys (1978) against 280
in Garmany et al. (1982). Furthermore, as this area of the HR diagram also
contains most of the WR stars, these have been taken into account when
performing the star counts. In fact, WR stars may significantly alter the
relative star frequencies, if according to current scenarios (Chiosi et
al., 1978; Maeder, 1982) they have progenitors in this range of luminosity
(initial mass) and are generated via mass loss in the post red supergiant
scheme. The number of WR stars falling within the distance of 2.5 Kpc from
the sun is derived from Van der Hucht et al. (1981). The results of the
counts are given in Table 1 for the case of Humphreys' list alone, and
Humphreys' plus Garmany's et al. catalogues. The first column in Table 1

Table   1
(Star   Counts)

Sp	O - B1	B2 - B9	A	F - G	K - M	WR	Case
N	290	49	10	5	30	20	Humphreys
N/N$_t$	0.73	0.12	0.02	0.01	0.07	0.05	
N	432	49	10	5	30	20	Humphreys +
N/N$_t$	0.79	0.09	0.02	0.01	0.05	0.04	Garmany et al.

gives the number of stars within the main sequence band. Only the most fa-
vourable grouping, namely that fixed by models with overshoot of Bressan
et al. (1981) (but classical opacity) is shown for the sake of simplicity. It is well evident that in both cases about 20% of stars fall beyond
the limit of the main sequence band, contrary to the theoretical expectation of less than 10%. If the limit for the main sequence band of standard
constant mass and mass losing models (O9.5 and B0.5, respectively) are
used, the above percentages would increase to 66% and 36% (Humphreys) or
to 49% and 27% (Humphreys plus Garmany et al.) in the order.

3. Models with Enhanced Opacity
Previous models of massive stars with opacity by heavy elements much
higher than the standard value - Carson (1976) and Cox and Stewart
(1970) respectively - are by Stothers (1976) and Stothers and Chin (1977,
1978). These models however are not useful to our purpose at least for
the three following reasons: i) the use of Carson's (1976) opacities has
been much debated as they perhaps overestimate the true opacity in the
temperature range 10^5 K$^\circ$ to 2×10^6 K$^\circ$. ii) the models do not take into
account convective overshoot, which on the contrary appears to be impor-
tant. iii) when mass loss by stellar wind during the main sequence phase
is considered, too high rates of mass loss have been used, which are not
supported by current observational information (Garmany et al., 1981;
Lamers, 1981). In the light of this, we thought worth recomputing models
of massive stars, in which the opacity enhancement in the temperature
range of interest is considered as a free parameter. The opacity is given
by the following relationship

$$\kappa = \kappa_{CS}\left[1 + \chi \exp(- 10 (5.8 - \text{Log } T)^4) \right] , \qquad (1)$$

where κ_{CS} stands for the Cox and Stewart opacity, whereas χ is a free
parameter. The opacity of eq. (1) has been used to compute models with
various assumptions as far as mass loss, overshoot and χ are concerned.
In particular, the convective overshoot is treated according to the formulation of Bressan et al. (1981), which contains the mixing length λ (
ratio of the maximun distance travelled by convective elements to the
pressure scale height) as a parameter. When mass loss is taken into account, the rate is fixed by the location of the models in the HR diagram

and the underlying evolutionary stage. More precisely, the mass loss rate $\dot{M}_B$ during the early stages of central H-burning is given by the well known relation of Castor et al. (1975). The rate of mass loss in the red supergiant region is taken from current observational determinations (Dupree, 1981; Linsky, 1981), $\dot{M}_R = 1 \times 10^{-5}$ $M_\odot$/yr. More advanced stages, if occuring at high effective temperatures showing no H at the surface, are computed using the mass loss rate of WR stars. $\dot{M}_{WR}$ is in the range 0.5×10^{-5} $M_\odot$/yr to 3×10^{-5} $M_\odot$/yr (Barlow, 1982; Bieging et al., 1982). Various evolutionary sequences for an initial 20 $M_\odot$ star with chemical composition X = 0.7 and Z = 0.02 are computed for different combinations of $\dot{M}_B$, $\dot{M}_R$, $\dot{M}_{WR}$, λ and χ . An extensive presentation of the results can be found in Bertelli et al. (1983). It suffices here to mention that i) the main sequence band systematically broadens at increasing χ till when a significant fraction of the core H-burning may even occur in the red supergiant region; ii) in models with overshoot, a small increase of the opacity gives the same broadening of the main sequence that would otherwise be obtained only with an high enhancement of the heavy element opacity (Stothers and Chin, 1977; 1978). Out of those numerical results, we present here only the case of the 20 $M_\odot$ star computed with $\dot{M}_B \simeq 0$ $M_\odot$/yr, $\dot{M}_R = 1 \times 10^{-5}$ $M_\odot$/yr, $\dot{M}_{WR} = 0.5 \times 10^{-5}$ $M_\odot$/yr, $\lambda = 1$, $\chi = 4$ and evolved till the stage of core He exhaustion. This sequence in fact offers the most interesting features to the problem under consideration. Table 2 summarizes the lifetimes spent in the various spectral types.

Table 2

(Theoretical Lifetimes) [*]

Sp	0 - B1	B2 - B9	A	F - G	K - M	OB*	WR
t	7.67	1.14	0.1	0.1	1.0	0.3	0.8
Phase	H	H	H	H	H	H	He
%	0.69	0.10	0.01	0.01	0.09	0.04	0.06

*) in units of 10^6 yr

4. Discussion and Conclusions

Looking at the data of Table 2, it is soon evident that these new models almost entirely eliminate the disagreement between theory and observation. They spent in fact about 68% and 12% of their total life in the spectral range 0 to B1 and B2 to B9 respectively as required by the star counts of Table 1. Another interesting feature is that about 9% and 4% of the core H-burning lifetime is spent by the models while appearing as red supergiant and 0 type stars with low H- content at the surface. This latter stage is indicated with OB* in Table 2. Whether these OB* stages can be identified with OBN and/or Of stars of low luminosity, is a point which deserves further investigation. When the H-rich envelope is completely lost by mass loss, this stage is approximately reached at the end of the core H-burning phase, the so-called WR phase is supposed to begin. Contrary to other pre-

vious model computations (Chiosi et al., 1978; Maeder, 1982) the whole
core He-burning lifetime is now available to the WR phase. This , much
better than ever before, can account for the relative percentage of WR
stars with respect to blue and red supergiants. More details on the problem of formation and statistics of WR stars can be found in Bertelli et
al. (1983). Finally, the new models with overshoot and moderate increase
in the heavy element opacity can easily explain the gradient across the
galactic plane in the percentage of WR stars relative to blue and red
supergiants. This gradient has been attributed by Maeder et al. (1980) to
the effect of metallicity on the mass loss rates $\dot{M}_B$ and $\dot{M}_R$. The metal
content is in fact known to increase toward the galactic centre. We suggest on the contrary that *the systematic increase in the opacity via the
increase in the metallicity is the dominant parameter, even though a minor effect on the mass loss rates cannot be excluded.* As a major conclusion, we argue that opacity changes of the order we have used in models
with overshoot and mass loss by stellar wind, this latter being important
during the post main sequence phases, can reconcile theory and observation,
and explain the major properties of supergiant and WR stars at the same
time.

This work has been financially supported by the National Council of
Research of Italy (C.N.R.-G.N.A.).

References
Barlow, M.J., 1982, in Wolf Rayet stars:Observation, Physics, Evolution,
        IAU Symp. n. 99, ed. C. de Loore and A. Willis, D. Reidel Publ.
        Comp., Dordrecht, Holland, p. 149
Bertelli, G., Bressan, A., Chiosi, C., 1983, Astron. Astrophys. submitted
Bressan, A., Bertelli, G., Chiosi, C., 1981, Astron. Astrophys., 102, 25
Bieging, J.H., Abbott,D.C., Churchwell, E.B., 1982, in Wolf Rayet stars:
        Observation, Physics, Evolution, IAU Symp. n. 99, ed. C. de Loore
        and A. Willis, D. Reidel Publ. Comp., Dordrecht, Holland, p. 87
Carson, T.R., 1976, Ann. Rev. Astron. Astrophys., 14, 95
Castor, J.I., Abbott, D.C., Klein, R.I., 1975, Astrophys. J., 195, 157
Chiosi, C., Nasi, E., Sreenivasan, S.R., 1978, Astron. Astrophys., 63, 103
Cloutman, L.D., Whitaker, R.W., 1980, Astrophys. J., 237, 900
Cox, A.N., Stewart, J.N., 1970, Astrophys. J. Suppl., 19, 243
de Loore, C., de Grève, J.P., Lamers, H.J.G.L.M., 1977, Astron. Astrophys.
        61, 251
de Loore, C., Hellings, P., Lamers, H.J.G.L.M., 1982, in Wolf Rayet stars:
        Observation, Physics, Evolution, IAU Symp. n. 99, ed. C. de Loore
        and A. Willis, D. Reidel Publ. Comp., Dordrecht, Holland, p. 53
Duprèe, A.K., 1981, in Effects of mass loss on stellar evolution, IAU
        Coll. n. 59, ed. C. Chiosi and R. Stalio, D. Reidel Publ. Comp.,
        Dordrecht, Holland, p. 87
Garmany, C.D., Conti, P.S., Chiosi, C., 1982, Astrophys. J., 263, 777
Garmany, C.D., Olson, G.L., Conti, P.S., van Steenberg, M., 1981,
        Astrophys. J., 250, 660
Humphreys, R.M., 1978, Astrophys. J. Suppl., 38, 309
Humphreys, R.M., 1981, in The most massive stars, ESO workshop, ed. S.
        D'Odorico, D. Baade, K. Kjar, p. 5

Lamers, H.J.G.L.M., 1981, Astrophys. J., 245, 593
Linsky, J.L., 1981, in Effects of mass loss on stellar evolution, IAU
        Coll. n. 59, ed. C. Chiosi and R. Stalio, D. Reidel Publ. Comp.,
        Dordrecht, Holland, p. 187
Maeder, A., 1982, in Wolf Rayet stars: Observation, Physics, Evolution,
        IAU Symp. n. 99, ed. C. de Loore and A. Willis, D. Reidel Publ.
        Comp., Dordrecht, Holland, p. 405
Maeder, A., Lequeux, J., Azzopardi, M., 1980, Astron. Astrophys., 90, L17
Mermilliod, J.C., 1976, Astron. Astrophys. Suppl., 24, 159
Mermilliod, J.C., 1981, Astron. Astrophys. Suppl., 44, 467
Meylan, G., Maeder, A., 1982, Astron. Astrophys., 108, 148
Simon, N.R., 1982, Astrophys. J., 260, L87
Stothers, R., 1976, Astrophys. J., 209, 800
Stothers, R., Chin, C.W., 1977, Astrophys. J., 211, 189
Stothers, R., Chin, C.W., 1978, Astrophys. J., 225, 939
Van der Hucht, K.A., Conti, P.S., Lundstrom, I., Stenholm, B., 1981,
        Space Sci. Rev., 28, 227

DISCUSSION

TORRES-PEIMBERT: I am worried about the completeness of the sample to
start with. It is probably complete for large masses, but I think part
of the problem may be that the sample for enclosed low mass stars is very incomplete.
CHIOSI: This in fact is the reason why we have performed star counts only
for stars brighter than M_b = -7, that is to exclude or to reasonably li-
mitate the effect of relatively low mass stars. In any case we agree with
you in that completeness is one of the major problems to worry about.
RENZINI:How much secure is the claim that catalogues are "reasonably"
complete? I remember to have seen HR diagrams in which the zero age main
sequence is practically depopulated.
CHIOSI: The sample of O type stars of Garmany et al. (1982) has been
found to satisfactorily complete within 2.5 Kpc from the sun. This is the
reason why we have rescaled all star counts based on the Humphreys list
to the number of O type stars sorted out of the Garmany et al. catalogue
for the distance and luminosity intervals that concern us. Other possible causes of zero age main sequence depopulation were discussed by Garmany et al. even though I think they cannot completely solve the problem.
BOND: Massey, Conti and Garmany have a "complete" sample of O star binaries. The probability of finding an O star in a close binary is just above 50%. In agreement with Iben's point, this should imply al least 50%
of WR stars might be expected to be associated with close binaries of
all O stars above a given mass because WR stars. How can this be reconci-
led with van der Hucht's estimate of WR binarism?
CHIOSI: The percentage of binary WR stars in the total population in the
solar vicinity is not yet very well established. However, I like to stress
the point that the problem of a wider main sequence would still remain
even if the percentage Of single WR's is lowered by a factor of two. On
the other hand binarism alone cannot account for the high number of B
type stars compared to earlier spectral types.

ON THE STRUCTURE OF THE UPPER HRD OF HUMPHREYS

C. Doom
Astrophysical Institute
Vrije Universiteit Brussel

1. INTRODUCTION

The observational HR diagram of luminous galactic stars, compiled by Humphreys (1978) is a benchmark in the study of these objects. The diagram was considered as resulting from continuous star formation, and has been used by many authors to put constraints on the evolution of massive stars (Bertelli et al., Maeder: this conference; Bressan et al., 1981; Doom, 1982a). However, there is a puzzling characteristic of this HRD: there are too many stars outside the main sequence band.

Bertelli et al. (1983) have offered an explanation of this effect by changing the opacities in 20 to 60 M_0 models. Their evolutionary tracks cover a spectral range from O to M during core hydrogen burning. Their basic hypothesis (the changed opacities) is much debated.

In the next Section, we will discuss the structure of the HR diagram of Humphreys (1978). We will then show that there is a natural explanation of the excess of late type supergiants in terms of time dependent star formation in OB associations.

2. THE STRUCTURE OF HUMPHREYS' CATALOGUE

The catalogue of Humphreys consists exclusively of luminous stars in OB associations. The catalogue is almost complete up to the Zero Age Main Sequence (ZAMS) for O stars, and is consequently almost complete for all stars with an initial mass, larger than some 15 M_0.

In this HRD, there is a sharp gradient in the stellar density in the region $\log L/L_0 \simeq 5.2$, around $\log T_e = 4.25$. This gradient indicates that the main sequence band in this region extends up to $\log T_e = 4.3$ to 4.2.

There is an absence of supergiants with

C. Chiosi and A. Renzini (eds.), Stellar Nucleosynthesis, 175–178.
© *1984 by D. Reidel Publishing Company.*

$\log L/L_o \gtrsim 5.8$ and $\log Te \lesssim 4.2$.
These constraints have been used by Doom (1982a,b)
to determine the amount of overshooting in 10 to 100 M_o
stars. But even in this case, there remains an excess of
stars outside the main sequence.

3. TIME DEPENDENT STAR FORMATION IN OB ASSOCIATIONS

For each association in the Humphreys catalogue we
have constructed the following diagram: for each star in the
association we determined the age t and the initial, ZAMS
mass, M_i. All stars of the association are then put on a
M_i vs. t diagram. These diagrams show the following
features:

We find stars with $10 M_o \lesssim M_i \lesssim 60 M_o$. We do not
consider the stars with $M_i < 15 M_o$ for reasons of completeness.

There is a clear correlation between the initial
mass and the age of stars within a single association: the
less massive stars ($M_i \simeq 10\text{-}25 M_o$) appear to be systhematically _older_ than the more massive stars ($M_i \simeq 40\text{-}60 M_o$).

Consequently, there is a considerable lack of stars
with $M_i = 10\text{-}25 M_o$ near the ZAMS.

For a given M_i, there is a continuous spread of the
stars over a certain time interval. For stars with $M_i \lesssim 25 M_o$
this interval is situated partly within the main sequence
lifetime (t_{ms}), partly out of the main sequence lifetime.
For more massive stars, this interval is situated fully
within the main sequence lifetime.

This effect influences the structure of the HRD of
a separate OB association in the following way.

There is a serious lack of O stars near the ZAMS.

For $\log L/L_o \lesssim 5.7$ there are much stars outside the
main sequence. This is obviously caused by the fact that for
$M_i \lesssim 25 M_o$, the stars are clustered around $t = t_{ms}$ in the
M_i vs. t diagram, and not within the main sequence lifetime.

We believe that these effects are caused by a time
dependent star formation in OB associations. In order to
explain the effect in terms of stellar rotation, we have
to assume unusual large rotational velocities, and only in
stars with $M_i \lesssim 25 M_o$. The effects, seen in the M_i vs. t
diagram cannot be due to observational selection effects:
Humphreys should have missed some 70% of all O stars in
each association in order to account for the lack of young
O stars in the sample.

4. CONSEQUENCES FOR THE STRUCTURE OF HUMPHREYS' HRD

The effects discussed in Section 3 are found in 90% of all associations present in Humphreys' catalogue. Therefore, the systhematic effects, found in the separate associations, will also be found in the overall HRD, namely:
- a lack of O stars near the ZAMS;
- too much stars outside the main sequence, compared to what would be expected from continuous star formation.

5. CONCLUSIONS

The HR diagram of Humphreys (1978) cannot be considered as a sample of stars, formed by continuous star formation. Instead, the sample is composed of OB associations in which the more massive stars are systhematically younger than the less massive stars. Using this fact, the structure of this HR diagram can be perfectly explained, using models with mass loss and overshooting, and the classical assumptions on opacities.

An extended paper on this work is in preparation by Doom, De Grève and de Loore

This research is supported by the National Foundation of Collective Fundamental Research of Belgium under contract N° 2.9009.79 .

REFERENCES

Bressan, A.G.; Bertelli, G.; Chiosi, C.: 1981: Astron. Astrophys. 102, 25
Doom, C.: 1982a: Astron. Astrophys. 116, 303
Doom, C.: 1982b: Astron. Astrophys. 116, 308
Humphreys, R.M.: 1978: Astrophys. J. Suppl. Ser. 38, 309
Bertelli, G.; Bressan, A.G.; Chiosi, C.: 1983: Astron. Astrophys. (in press)

DISCUSSION

<u>Gallagher</u>: Supergiants are intrinsically very luminous, and near log Te = 4.2 bolometric corrections drop rapidly. As a result these stars are optically much more luminous and easy found. Won't these problems affect your statistics?

<u>Doom</u>: I believe that the sample is complete down to M_i = 15-16 M_o. This implies that the catalogue should be complete above the 15 M_o track in the HRD. So, near the ZAMS, all O stars are needed (log L/L_o = 4.2), but for B1-B2 stars, the sample only has to be complete above log L/L_o = 4.7, since the luminosity of a 15 M_o star rises sharply during its evolution. The completeness of the O stars was confirmed by comparing Humphreys' catalogue to the catalogue of galactic O stars of Garmany et al. (1982).

SESSION IV: SUPERNOVA PRECURSORS AND EXPLOSIVE NUCLEOSYNTHESIS

I. IBEN, Jr. and A. V. TUTUKOV: An Essay on Possible Precursors of
        Type I Supernovae

K. NOMOTO: Nucleosynthesis in Type I Supernovae: Carbon Deflagration
        and Helium Detonation Models

K. NOMOTO: Type II Supernovae from 8-10 $M_\odot$ Progenitors

A. YAHIL: The Energetics of Type II Supernovae

S. E. WOOSLEY, T. S. AXELROD and T. A. WEAVER: Nucleosynthesis in
        Stellar Explosions

AN ESSAY ON POSSIBLE PRECURSORS OF TYPE I SUPERNOVAE[†]

Icko Iben, Jr. and Alexander V. Tutukov*
University of Illinois at Champaign-Urbana

I. PROPERTIES TO BE EXPLAINED

Although it is a fascinating problem that has attracted attention
for many years, the origin of type I supernovae (= SNeI) has continued
to defy convincing quantitative solution. The absence of clearcut hydrogen and helium lines at maximum light may be interpreted as evidence that presupernovae are highly evolved stars which have lost
their original hydrogen-rich envelopes, probably as a consequence of
mass exchange in close binaries. SNeI are also characterized by uni-
formity of light curves, spectra, and ejection velocities [$(11 \pm 2) \times 10^3$ km s^{-1}]. Another important characteristic of SNeI is that they
are the only type of supernova which occur in elliptical galaxies,
where now all dying stars have masses $\lesssim 0.8$ $M_\odot$ and ages exceeding
10^{10} yr.

An estimate of the frequency of SNeI in our Galaxy can be achieved by interpolation within observationally founded estimates of the
frequency of occurrence of SNeI in external galaxies (Tammann 1978).
As part of an extended discussion on which this summary is based, Iben
and Tutukov (1983a, hereinafter ITa) infer that approximately one SNI
explosion occurs for every one hundred stars of primordial mass near
0.8 M_0 which are reaching the end of their nuclear burning evolution.
Since, in our Galaxy, about one low mass star completes its nuclear
burning evolution in any given year, the Galactic frequency of SNeI is
on the order of 10^{-2} yr^{-1}.

*On leave from the Astronomical Council of the USSR Academy of Sciences,
Moscow.

[†]Supported in part by the USA National Science Foundation Grant AST 81-
15325.

181

C. Chiosi and A. Renzini (eds.), Stellar Nucleosynthesis, 181–204.
© *1984 by D. Reidel Publishing Company.*

II. SNeI AS A RESULT OF ACCRETION ONTO AN ELECTRON-DEGENERATE DWARF

The observational requirements make it natural to suppose that
SNeI events are produced by electron-degenerate dwarfs which accrete
matter either from a companion filling its Roche lobe or by capture
from a strong wind emitted by a companion which does not fill its
Roche lobe. One of the first concrete models of this type -- a carbon-oxygen dwarf accreting matter from a red giant companion -- was
proposed over ten years ago by Whelan and Iben (1973). In spite of a
significant expansion in our understanding of the evolution of close
binary stars (see, e.g., Webbink 1979, Tutukov and Yungelson 1979) the
CO dwarf plus red giant companion model has continued to enjoy wide
currency.

An attractive feature of the model is that, since an explosion
initiated by carbon burning will not occur until the accreting CO
dwarf has nearly achieved the Chandrasekhar mass (~ 1.4 $M_\odot$), all immediate SN precursors of this type are essentially identical; this iden-
tity provides a natural explanation for the observed uniformities
among SNeI. The effective lack of hydrogen, which has (by assumption)
been converted into helium, and the lack of helium, which has (also by
assumption) been converted into carbon and oxygen (except in an extremely thin skin ($< 10^{-5}$ $M_\odot$) near the surface of the precursor dwarf),
account for the lack of hydrogen and helium lines in the SNI spectrum
at maximum light. One must, of course, beware that the absence of
telltale lines of H and He could simply be a matter of unfavorable
conditions for exciting these lines and not an indication of the
absence of the elements themselves.

A consideration of the energy budget provides another argument in
favor of the exploding CO dwarf model. The conversion of 1.4 $M_\odot$ of
carbon and oxygen (say, 50 percent each) into ^{56}Ni releases approximately 22 x 10^{50} erg of nuclear energy. About 5 x 10^{50} erg is required
to unbind the dwarf, only about 10^{50} erg comes out as photons, and the
rest of the released energy goes into spewing the once degenerate matter outward at average velocities of 11,000 km s^{-1}, remarkably consistent with observed velocities. Additional favorable attributes of the
exploding CO dwarf model are that (1) the observed light curve is a
natural consequence of the time characteristics of the Ni-Co-Fe decay
chain coupled with the characteristics of the photon diffusion process
in an outward flowing medium (Colgate and Petschek 1980, Chevalier
1981) and that (2) matter in the outer portions of the exploding star
is not completely incinerated, so that elements such as Si and Ca can
survive at finite abundances (Ivanova et al 1977, Mazurek et al 1977,
Chechetkin et al 1980, Woosley et al 1980, Weaver et al 1980, Woosley,
Axelrod, and Weaver 1983, Nomoto 1984).

The one, possibly fatal flaw in the "traditional" model involving
a CO dwarf and a red giant companion filling its Roche lobe is that,
in order to achieve mass accretion rates which will permit the CO
dwarf to grow in mass as far as the Chandrasekhar limit, it is nec-

essary to "fine tune" the characteristics of the primordial binary to such a degree that, although finite, the likelihood of occurrence of appropriate systems is extremely small relative to the observed frequency of SNeI (see section IV). Considerable space in ITa is devoted to demonstrating this fact.

Surprisingly, a slight variant of the traditional scenario, again involving a degenerate CO dwarf, but this time calling upon capture by the dwarf of matter from a wind emitted by a red supergiant AGB star which does not fill its Roche lobe, appears at first sight to have a much larger probability of occurrence, but a careful examination of permissible accretion rates may considerably limit the number of systems which can ultimately evolve to explosive conditions.

Still other variants of the traditional model replace the red giant or supergiant donor by a low mass main sequence or near main sequence companion. Observational counterparts of potential supernova precursors of this type are the cataclysmic and cataclysmic-like binaries. The observational fact that many cataclysmic systems experience nova outbursts, coupled with the observationally based and theoretically supported inference that most of the matter accreted between nova events is lost in the nova event, makes it evident that not all (if indeed any of) such systems will evolve to an explosive state of SNI magnitude. Nevertheless, we are at present not in a position to deny that some fraction of the $\sim 3 \times 10^{-3}$ cataclysmic systems that are formed each year may eventually give birth to SNeI (ITa).

Two major stumbling blocks in the way of all scenarios that rely on the accretion of hydrogen on a degenerate dwarf are the low ignition temperature and the high energy content of hydrogen. These characteristics lead both to nova outbursts (if $\dot{M}$ is too small) and to expansion and formation of a common envelope (if $\dot{M}$ is too large). One way to avoid the accompanying stringent limitation on those values of $\dot{M}$ which will permit the dwarf to grow steadily in mass is to invoke accretion of fuels with higher ignition temperatures and less energy content. In the context of our current understanding of single and binary star evolution, this means that the mass donor must be an electron-degenerate dwarf composed either of helium or of carbon and oxygen. Furthermore, since degenerate dwarfs typically have radii no larger than one or two earth radii, we are led to contemplate the probability of formation of two <u>very</u> close degenerate dwarfs, each of which can have evolved only from a precursor system of semimajoraxis larger than several times the solar radius. As it turns out, the most likely immediate supernova precursors are CO dwarfs accreting carbon and oxygen from a companion dwarf which is composed typically also of carbon and oxygen.

Another set of models involving an accreting dwarf replace the CO dwarf with one composed of oxygen, neon, and magnesium (an ONeMg dwarf, for short) and some appropriate mass donor. The ONeMg dwarf will ignite oxygen explosively if its mass approaches the Chandrasekhar mass, but

electron captures on Mg and other elements will cause the core of the
dwarf to collapse to form a neutron star (not a black hole) and, at
last, only a small fraction of the outer portion of the dwarf will be
expelled. Since only a portion of the total nuclear energy liberated
(also on the order of 10^{51} erg in this case) goes into expelling matter, it is not possible to estimate the expansion velocity of this
matter for comparison with the observations. Even so, the prediction
that the ejected matter is considerably less massive than ~ 1 $M_\odot$ is at
variance with estimates of the mass in the ejectum of the Tycho SNI
(Seward et al 1983), and the prediction of a condensed neutron star
remnant is at variance with the lack of evidence for such remnants at
the center of extended SNeI remnants (e.g., Helfand 1980). Apart from
this, the estimated frequency of formation of appropriate systems containing an ONeMg dwarf is far below the observed SNI occurrence frequency (ITa).

One final system which may lead to an explosive state with the
release of $\sim 10^{51}$ erg of nuclear energy consists of a degenerate dwarf
composed of helium and a mass-donating companion which is either a
star with a hydrogen-rich envelope (main sequence or red giant) or
another, lighter, helium dwarf. Nomoto and Sugimoto (1977) have shown
that, for particular choices of effective growth rates of the helium
dwarf, a dwarf-disrupting explosion will occur when the mass of the
dwarf reaches (0.65–0.8) $M_\odot$ and the burning will continue all the way
to the formation of iron peak elements. This means that about 15 x
10^{17} erg gm^{-1} of nuclear energy is released. Since the binding energy
is less than about 10^{17} erg gm^{-1}, the typical velocity of ejected matter will be on the order of 16,000 km s^{-1} and this is somewhat larger
than inferred from the observations. Furthermore, the fact that the
matter is completely incinerated is at variance with the presence of
elements such as Ca and Si in SNeI spectra. However, there are indications (Arnett 1983) that a slight amount of rotation may alter the
character of the explosion in such a way that matter is not completely
incinerated and, since the interaction between a dwarf and its accretion disk is expected to make large rotation rates the rule, rather
than the exception (Iben and Tutukov 1983b, hereinafter ITb), this
last potential variance with the observations may be an academic one.
Finally, most models of exploding helium dwarfs produce light curves
which are "narrower" at the peak by about a factor of two than observed light curves (see, e.g. Woosley et al 1980, Nomoto 1984). In contrast, those scenarios which terminate with the deflagration of carbon
produce light curves of the observed peak half width.

In summary, a CO dwarf accreting matter from another CO dwarf is,
at the present writing, the most likely theoretical candidate for a
typical SNI precursor system in that it can account for the uniformity, the energetics, the spectrum and the photospheric composition at
maximum light, and the light curve (see Branch 1984, Woosley et al
1984). Further, as we shall show in section IV, the estimated formation frequency of appropriately close and massive double CO degenerates is comparable to the observed SNI rate.

As a caveat, however, we emphasize that, as it is a function of initial mass and accretion rate and as its character possibly depends quite sensitively on the rotation rate, the nature of the explosion of a helium dwarf is not as well explored or understood as that of CO dwarfs, all of near Chandrasekhar mass, and we suggest deferring final judgement as to whether or not binaries consisting of two degenerate helium dwarfs might contribute a substantial fraction of observed SNeI.

III. FACTORS TO CONSIDER IN SEARCHING FOR LIKELY SNeI PRECURSORS AND GENERAL IMPLICATIONS OF THE BINARY HYPOTHESIS

1. Limitations on Accretion Rate

Accretion onto a degenerate dwarf will lead to an inexorable increase in the mass of the dwarf only if the dwarf mass, the average accretion rate, and the compositions of the dwarf and of the accreted matter are within specific ranges. Consider first a degenerate CO dwarf accreting carbon and oxygen. One might suppose that the rate of gravitational potential energy release of the accreted matter must be less than the Eddington luminosity, $L_{Edd} \sim 4\pi cGM/\kappa \sim 6.55 \times 10^4 (M/M_\odot)$ $L_\odot$. With $L_{acc} \sim GM\dot{M}/R \sim 3.16 \times 10^9 (M/M_\odot) (R/10^{-2} R_\odot)^{-1} \dot{M}(M_\odot \text{ yr}^{-1})$, we have $L_{acc}/L_{Edd} \sim 4.82 \times 10^4 (R/10^{-2} R_\odot)^{-1} \dot{M}(M_\odot \text{ yr}^{-1})$. For example, if the mass of the CO core is $M_{CO} \sim 1.25 M_\odot$, the radius of the white dwarf is given by $R_{cold} \sim 0.005 R_\odot$ if the dwarf is cold and $L_{acc} > L_{Edd}$ if $\dot{M}$ $\gtrsim \dot{M}_{Edd} \cong 10^{-5} M_\odot \text{ yr}^{-1}$. However, why should the effective accretion rate onto the accreting star be limited to $\dot{M}_{Edd}$ for a compact dwarf? The most probable outcome of accretion at rates larger than the formal Eddington limit for a cold dwarf is simply a heating and expansion of surface layers until $GM\dot{M}/R' \sim L_{Edd}$, where R' is the new equilibrium radius, and thus there is, in principle, no restriction on the rate of growth of the accreting CO dwarf.

Does the nature of the ultimate explosion of merging CO dwarfs depend on the rate of merging? It is known that, for effective accretion rates in the range $10^{-9} < \dot{M}(M_\odot \text{ yr}^{-1}) < \dot{M}_{AGB \text{ star}} \sim 10^{-6} (M_{CO}/M_\odot -$ 0.5), both the central density ρ_c^{ig} at carbon ignition (Ergma and Tutukov 1976) and the central density ρ_c^{def} when a deflagration is initiated (Couch and Arnett 1975) decrease with increasing accretion rate. For example, when $\dot{M} \sim \dot{M}_{AGB}$, $\rho_c^{ig} \sim 2 \times 10^9$ gm cm^{-3} and $\rho_c^{def} \sim 4 \times 10^9$ gm cm^{-3} and, when $\dot{M} \sim (10^{-9}-10^{-8}) M_\odot \text{ yr}^{-1}$, $\rho_c^{ig} \sim 4 \times 10^9$ gm cm^{-3} and $\rho_c^{def} \sim 9 \times 10^9$ gm cm^{-3} (Iben 1982a). The convective Urca process is responsible for delaying deflagration to densities larger than ignition densities.

The fact that ρ_c^{def} varies with $\dot{M}_{effective}$ has important consequences for the nucleosynthetic products spewed out by the model SNI
(e.g., Woosley et al 1984, Nomoto 1984). For example, if $\rho_c^{def} \sim 9$ x
10^9 gm cm^{-3}, the ratio of ^{58}Fe to ^{56}Fe appearing in the final ejectum
is perhaps a factor of ten larger than the solar system ratio. If
$\rho_c^{def} \sim 4$ x 10^9 gm cm^{-3}, the ratio of ^{54}Fe to ^{56}Fe in the final ejectum
is larger than the solar system ratio by perhaps a factor of four
(Woosley 1983). Only if $\rho_c^{def} \gtrsim 2$ x 10^9 gm cm^{-3}, corresponding to,
say, $\dot{M} \gtrsim (3$ x 10^{-6}-$10^{-5})$ $M_\odot$ yr^{-1}, are the iron isotopes produced in
the solar system distribution. Thus, if SNeI are the major producers
of solar system iron isotopes and if double degenerate CO dwarfs are
responsible for most SNeI, then we have the intriguing result that the
rate of growth of the presupernova dwarf must exceed $\dot{M}_{AGB}$, and perhaps
even approach $\dot{M}_{Edd}$. However, since there are uncertainties in the nuclear physics, including uncertainties in electron capture rates which
affect calculated isotopic ratios, we cannot as yet insist absolutely
on this point. Furthermore, there exists the possibility that, for
accretion rates approaching $\dot{M}_{Edd}$, carbon will be ignited off center and,
instead of developing into a deflagration, a series of shell flashes
will occur until the degree of degeneracy in the core is reduced and
the CO core is converted into an ONeMg core. Continued mass accretion
may then ultimately lead to neutron star formation.

Consider next the accretion of pure helium onto a degenerate CO
dwarf. Not all of the relevant parameter space has been explored and
therefore we can only touch upon a fraction of the possibilities. For
example, if $M_{CO} \gtrsim 1.15$ M_0, and $\dot{M} \lesssim 10^{-9}$ $M_\odot$ yr^{-1}, accretion will continue uneventfully until $M_{CO} \sim 1.4$ $M_\odot$ and then the central ignition of
carbon at densities larger than $\rho_c^{ig} \gtrsim 4$ x 10^9 gm cm^{-3} will lead to the
formation of a deflagration wave which stops short of converting all
matter into iron peak elements (Fujimoto 1980, Taam 1980b, Nomoto 1982,
1984). If $10^{-9} \lesssim \dot{M}$ $(M_\odot$ yr$^{-1}) \lesssim 4$ x 10^{-8} and $M_{CO} \sim (0.5$-$1.0)$ $M_\odot$, helium
will accumulate without burning until helium detonates, setting off a
carbon detonation; all matter in the initially layered dwarf is converted into iron peak elements and the star is completely disrupted.
If 4 x $10^{-8} \lesssim \dot{M}$ $(M_\odot$ yr$^{-1}) \lesssim \dot{M}_{steady}^{He} \sim 1.3$ x 10^{-6} $(M_{CO}/M_\odot)^{3.57}$, finite
amplitude helium flashes will occur, converting helium into carbon and
oxygen, and the CO core will grow in mass until, if M_{CO} reaches ~ 1.4
$M_\odot$, a deflagration will be set off as the central density reaches
$\rho_c^{def} \gtrsim 4$ x 10^9 gm cm^{-3}. For accretion rates larger than $\dot{M}_{steady}^{He}$,
helium will burn quiescently (without flashing) at the base of the
helium layer and, for accretion rates only slightly larger than this,
the helium layer will expand to giant dimensions (presumably incorporating the donor as part of the new, giant helium envelope) and the
system will adopt the characteristics of an R CrB star (see also ITb).
The rate at which the CO core will grow is given by $\dot{M}_{CO}$ $(M_\odot$ yr$^{-1}) \sim$

6×10^{-6} $(M_{CO}/M_\odot - 0.5)$. Since R CrB stars lose mass at fairly high
rates, the most probable final state will be a single, cold degenerate
CO dwarf of mass less than 1.4 $M_\odot$. If by chance the CO core is able
to reach the Chandrasekhar limit, a central deflagration or off center
ignition will occur, and the subsequent evolution of the core will
proceed in essentially the same fashion as when CO matter is accreted
rapidly onto a degenerate CO dwarf. However, the giant helium envelope will modify the flow of energy from the explosion and the light
curve will be more characteristic of a SNII event than of a SNI event.

The third situation to examine is accretion of helium onto a degenerate helium dwarf (accretion of carbon and oxygen onto a helium
dwarf will never occur since always $M_{CO} > M_{He}$ and therefore $R_{CO} < R_{He}$).
The maximum mass of a degenerate helium dwarf in an initial system is
~ 0.5 $M_\odot$ (the mass at which a single low mass red giant experiences
core helium flashes which converts it into a core helium burning hori-
zontal branch star). Accretion at rates $\dot{M}_{RG} \lesssim \dot{M} \lesssim \dot{M}_{Edd}$, where $\dot{M}_{RG}$
$(M_\odot \text{ yr}^{-1}) \sim 10^{-5.36}$ $(M_{He}/M_\odot)^{6.6}$ is the rate at which the helium core
in a red giant grows, should lead to a series of core helium flashes
when M_{He} reaches ~ 0.5 $M_\odot$. For $\dot{M} < \dot{M}_{RG}$, complete exploration of parameter space for initial M_{He} and $\dot{M}$ has not as yet been accomplished,
but indications are that, when $\dot{M} < \dot{M}_{RG}/\text{few}$, a detonation supernova may
result if accretion can be sustained until M_{He} reaches, say, (0.65-
0.8) $M_\odot$ (Nomoto and Sugimoto 1977).

We come finally to the accretion of hydrogen-rich matter onto a
degenerate dwarf. Accretion on a degenerate CO dwarf at rates in the
range ~ $10^{-9} \lesssim \dot{M}$ $(M_\odot \text{ yr}^{-1}) \lesssim 1.3 \times 10^{-7}$ $(M_{CO}/M_\odot)^{3.57} = \dot{M}^H_{steady}$ results
in a series of relatively weak hydrogen shell flashes and in a growth
of the dwarf at an effective rate which is a function of the semimajoraxis but is, in any case, always at least as large as $\dot{M}/2$ (Iben
1982b). Accretion at rates larger than $\dot{M}^H_{steady}$ leads to the formation
of a common envelope and no effective accretion. If $\dot{M} \gtrsim 4 \times 10^{-8}$
$M_\odot \text{ yr}^{-1}$, then helium shell flashes are also relatively weak, $\dot{M}_{effective}$
$\gtrsim \dot{M}/2$, and the CO dwarf will grow until, if M_{CO} approaches 1.4 $M_\odot$, a
carbon deflagration event will take place. However, if $10^{-9} \lesssim \dot{M}$
$(M_\odot \text{ yr}^{-1}) \lesssim 4 \times 10^{-8}$, the helium zone above the initial CO core will
grow until first a helium detonation and then a carbon detonation will
develop, transmitting matter all the way to the iron peak and completely
disrupting the star.

2. Drivers of Mass Transfer in Precursor Systems

Among double degenerate dwarfs, the only known mechanism for
bringing the components close enough together so that the lighter of
the two can fill its Roche lobe is gravitational wave radiation (GWR).
The time scale for reducing the system semimajoraxis A is proportional

to the fourth power of A (e.g., Landau and Lifshitz 1962) and, for typically expected initial masses, evolution to Roche-lobe filling will not occur in less than $\sim 10^{10}$ yr unless the initial value of A is less than $\sim 3\ R_{\odot}$. Once Roche-lobe contact is made, then mass transfer will occur on a dynamic time scale (Tutukov and Yungelson 1979) and one expects that the initially less massive dwarf will be transformed almost immediately into a "heavy" disk or rapidly rotating envelope about the initially more massive dwarf. One may expect that, in some cases, accretion from the disk onto the dwarf may be quite rapid, with L_{acc} initially exceeding L_{Edd}, and that the dwarf may expand until the outer surface of the dwarf and the inner edge of the ring are in "deep" contact.

In the case of a system composed initially of a helium dwarf and a CO dwarf, if the eventual accretion of helium onto the CO dwarf is sufficiently rapid, helium will burn quiescently at the accretion interface and the composite star will swell to giant dimensions, taking on the appearance of an R CrB star, as already discussed, and either become a SN if the CO core mass can reach $\sim 1.4\ M_{\odot}$ or end as a degenerate CO dwarf. In the case of rapid accretion of helium onto a helium dwarf, the ultimate result is the ultimate conversion of two helium dwarfs into a single CO dwarf of mass less than $1\ M_{\odot}$.

In summary, close enough systems of two CO degenerates or of CO plus helium degenerates of total mass $> 1.4\ M_{\odot}$ will experience the deflagration phenomenon, whereas close systems of two helium dwarfs (total mass always less than $\sim M_{\odot}$) may evolve relatively quietly into CO dwarfs.

In cataclysmic and cataclysmic-like precursors, no known mechanism can bring the potential donor of hydrogen-rich matter into contact with its Roche lobe in less than the Hubble time unless the initial semimajoraxis A is less than $\sim 10\ R_{\odot}$. For values of A in the range $3 \lesssim A/R_{\odot} \lesssim 10$, it is thought that a magnetic stellar wind (MSW) acts to abstract angular momentum from the system and thus causes a shrinkage in A, whereas, for $A \lesssim 3\ R_{0}$, GWR can do the job. Once Roche-lobe contact has been made, the driver which keeps the (near) main sequence star in contact with its Roche lobe is the swelling caused by nuclear transformations in the interior.

In the case of precursor systems consisting of a dwarf and a red giant, it is the growth of the red giant in response to the growth of its electron-degenerate core which achieves and maintains Roche lobe contact. One final system of interest involves a degenerate dwarf and an AGB star which does not fill its Roche lobe. As a consequence of a rapid expansion during thermal pulses, if the AGB star were to fill its Roche lobe, a common envelope which strips the entire envelope from the CO core would occur almost immediately and terminate further mass transfer. On the other hand, AGB stars produce robust winds, and capture by the dwarf of matter from the wind could cause its mass to grow to an interesting size. The driver for mass transfer in this case is thus a stellar wind.

3. Some Observational Consequences of the Binary Hypothesis

There are a number of features of binary scenarios for SNI production that have other consequences of observational significance.
For example, all of the scenarios have in common that the degenerate
dwarf which eventually accretes enough matter to explode has been
formed as a consequence of a common envelope stage. Since the observed SNI frequency is about one percent of the formation rate of all low
mass stars, most of which are expected to form planetary nebulae even
as single stars, we might predict that one percent of all observed
planetary nebulae have at their centers a close binary which will
ultimately evolve into a SNI. However, detailed calculations of the
behavior of model central stars that emerge from the common envelope
state (Iben and Tutukov 1983c) show that primordial systems of long
period ($P_{orb} \gtrsim 1$ yr) will produce remnants which are "lazy" (Renzini
1979) in the sense that the "subdwarf" component does not become hot
enough to ionize the ejected common envelope before this envelope has
dispersed. Although short period systems will produce observable PNe,
it is nevertheless probably best to view those planetary nebulae that
do contain very close double stars (e.g., Grauer and Bond 1983) as
evidence for the reality of the common envelope process and not primarily as possible precursors of SNeI.

Another feature common to all scenarios is the liberation of
gravitationl potential energy at far ultraviolet wavelengths. If we
suppose that this energy is emitted continuously and that all pre-
supernova accreting systems are similar, then the average UV luminosity of the Galaxy due to the release of gravitational potential energy
by presupernovae is simply $\langle L_{acc} \rangle = (GM\Delta M/R) \nu_{SNI} = 3.16 \times 10^7 \; L_\odot$
$(M\Delta M/M_\odot^2) \; (R/10^{-2} \; R_\odot)^{-1}$, where ΔM is the average amount of mass accumulated by the dwarf and M and R are its average mass and radius, re-
spectively. Adopting $M \sim 1.2 \; M_\odot$, $R \sim 0.005 \; R_\odot$, and $\Delta M \sim 0.3 \; M_\odot$, we
have $\langle L_{acc} \rangle \sim 2.3 \times 10^7 \; L_\odot$. If a typical source emits UV radiation at
the rate $L_{UV} \sim 10^4 \; L_\odot$, then there should be ~ 2300 such sources in the
sky. These are not particularly exciting numbers, but we must remem-
ber that only a small fraction, say 10 percent, of all similar systems
are able to evolve all the way to the explosive state. Thus, accreting
degenerate dwarfs may altogether number $\sim 2 \times 10^4$ and produce $L_{acc} \sim$
$2 \times 10^8 \; L_\odot$ in the far UV; this exceeds the contribution of central stars
of planetary nebulae ($L_{PN} \sim 10^8 \; L_\odot$) and should therefore be taken into
account, along with planetary nebula nuclei, in spectrum synthesis models of the radiation from external galaxies which contain no young OB
stars.

A more careful examination of the two major classes of scenarios
-- merging double degenerates and hydrogen-accreting degenerates --
shows that there are additional observational consequences that dif-
ferentiate between them.

If all presupernovae were degenerate dwarfs accreting hydrogen, the nuclear energy released in converting hydrogen into helium would exceed the rate at which gravitational energy is released by a very large factor: $\langle L_{NUC}^{preSNeI} \rangle \sim 6 \times 10^{18}$ erg gm^{-1} ΔM $\nu_{SNI} \sim 10^9$ $L_\odot (\Delta M/M_\odot)$ $\sim 3 \times 10^8$ $L_\odot$, say. Adding in the other similar but unsuccessful systems, we would predict the existence of about 2×10^5 bright sources, emitting a total UV luminosity of about $L_{UV} \sim 3 \times 10^9$ $L_\odot$, making hydrogen accreting dwarfs far more important than planetary nebulae as sources of Galactic UV radiation. Possibly the most likely of the hydrogen accretion scenarios involves capture by a CO dwarf of matter from a wind emitted by an AGB companion which does not fill its Roche lobe. Some symbiotic stars may be observational counterparts. But, since there are only about 10^4 symbiotics in the sky (Boyarchuk 1975), it would appear that symbiotic stars could account for at most 10 percent of all SNeI, even if all symbiotics were to evolve to the SNI state.

In the case of double degenerates, the formation of a heavy disk is expected to occur on a dynamic time scale, say $\sim 10^2$ s, and thus one expects gravitational potential energy to be released in a mighty burst at the rate $L_{burst} \sim (GM\Delta M/R)$ 10^{-2} s^{-1} $\sim 3.16 \times 10^7$ $\langle L_{acc} \rangle \sim 7 \times 10^{14}$ $L_\odot$! The burst energy will of course not be immediately converted into photon energy but will first be transformed into the energy of bulk motions and into heat, and finally leak out over an interval long compared to 100 s. One might expect the formation of a strong shock front at an outwardly growing accretion interface that causes the disk to expand outward and form an extended envelope. In fact, there is every reason to suppose that the collapse-bounce sequence will bear a qualitative resemblance to a SN outburst. However, there will also be substantial quantitative differences: (1) the duration of the light curve at peak will not be controlled by the time-scale for a nuclear decay process, as is thought to be the case in SNeI explosions, and (2) the gravitational potential energy released ($\sim 10^{50}$ erg) is on the order of the binding energy of a degenerate dwarf of mass $\sim (0.3-0.7)$ $M_\odot$, so outward velocities at light curve maximum will be well below escape velocities from the central dwarf ($v < $ few $\times 10^3$ km s^{-1}).

An additional consequence of all double degenerate scenarios is the production of gravitational waves. If we assume that all SNeI have double degenerates as precursors and that only about ten percent of all double degenerates become SNeI, then we estimate at the Earth a flux of energy in the form of gravitational waves of $\sim 10^{-3}$ erg cm^{-2} s^{-1} at a frequency of ~ 0.1 s^{-1} (ITa).

IV. ESTIMATES OF REALIZATION FREQUENCIES

1. Preliminary

Having concluded that presupernova systems are binaries containing at least one electron-degenerate dwarf and a companion that is close enough to transfer matter onto this dwarf, we now focus on establishing concretely the parameters of those primordial systems which might be expected to evolve into a presupernova configuration and to deduce from these parameters the probabilities with which "successful" primordial systems of various types are formed.

We suppose that (ITa) the "realization frequency" of each "scenario" may be approximated by

$$\nu \sim q \ (0.2 \ \Delta\log A) \ (\int_{M_A}^{M_B} \frac{dM}{M^{2.5}}), \tag{1}$$

where $q = M_2/M_1$ is the ratio of secondary mass M_2 to primary mass M_1 in the primordial system, A is the semimajoraxis of the primordial system, and M_A and M_B are the limits on the mass of the primordial primary. When the primordial components are of comparable mass, we take $q = 1$.

The approximation represented by equation (1) is admittedly extremely crude. Among the many uncertainties are the absence of a quantitative theory which translates primordial masses and semimajoraxes of primordial systems into masses and semimajoraxes of successor systems. The transformation of primordial M's and A's to successor M's and A's frequently involves the development of a common envelope (Sparks and Stecher 1974, Paczynski 1976, Taam et al 1978), which serves the double role of forcing mass loss from the binary system and bringing the binary components together and, at this juncture, we are forced to make (hopefully reasonable) guesses as to the outcome. So, even if 0.2 Δlog A were to properly assess that part of a realization frequency due to primordial component separation, the appropriate values of A are uncertain. Further, although the probability of primordial binary formation as a function of primary mass and semimajoraxis appears to be reasonably well established (Popova et al 1983), the appropriate values of M_A and M_B are uncertain due to the uncertainty in the quantitative aspects of the transformation to successor masses.

The distribution of unevolved binaries over the mass ratio q is also rather poorly known, especially when q is less than ~ 0.3. Nevertheless, in some successful scenarios, primordial ratios smaller than 0.3 are required and we are forced to live with the uncertainty. Fortunately, the (formally) most probable scenarios involve primordial masses which are comparable and the choice $q \sim 1$ is not unreasonable.

2. Double Degenerates

We shall illustrate the concepts involved in tracing the evolution from primordial to immediate presupernova systems by examining in
some detail those scenarios which lead to double degenerate dwarfs at
an "initial" separation small enough that gravitational wave radiation
will bring the two dwarfs close enough for merging in a time less than
$\sim 10^{10}$ yr.

Schematics describing the four scenarios we shall discuss explicitly are provided in Figures 1-4. Component masses and radii, system
semimajoraxes and orbital periods, and time scales for the various
phases are shown.

We consider first a scenario which will produce initially "close"
presupernova systems consisting of two CO dwarfs which will merge on a
time scale shorter than $\sim 2 \times 10^8$ yr. The logical steps involved are
shown in schematic form in Figure 1. If we begin with a system of
comparable component masses in the range $\sim$ (5-7) $M_\odot$ and at a separation between 70 $R_\odot$ and 460 $R_\odot$, this system will pass through two common envelope stages in which the mass losing star produces in each
case a bare non degenerate helium core which will, as a consequence of
helium burning, be converted into an electron degenerate CO dwarf of
mass in the range (0.7-1.0) $M_\odot$. The choice of 5 $M_\odot \lesssim M_1, M_2 \lesssim 7 M_\odot$ is
determined by the desire to achieve $0.7 \lesssim M_{1R}/M_\odot \lesssim 1.0$ and the precise
limits are subject to modification as a result of careful computations.

To estimate the semimajoraxis A_f of the intermediate system formed during the first common envelope stage, we adopt the parameterization of Tutukov and Yungelson (1979):

$$\frac{M_1^2}{A} = \alpha \frac{M_2 M_{1R}}{A_f} , \tag{2}$$

where M_{1R} is the mass of the bare helium core left by the primary and
α is a free parameter which we shall choose as 1. The physical content of equation (2) is simply that the energy required to disperse
the common envelope ($M_{common} \sim M_1 - M_{1R} \sim M_1$) is comparable to the reduction in the binding energy of the primordial system during the common envelope phase. Using equation (2) with $\alpha = 1$ we estimate that,
when the primary fills its Roche lobe after constructing a helium core
of mass (0.7-1) $M_\odot$, the common envelope which is formed reduces the
semimajoraxis of the intermediate system to $A_f \sim$ (10-65) $R_\odot$. In a
time of the order of a few times 10^6 yr the helium core is converted
into a degenerate CO dwarf.

A second common envelope stage occurs when the primordial secondary swells beyond its Roche lobe. We use an equation analogous to
(2):

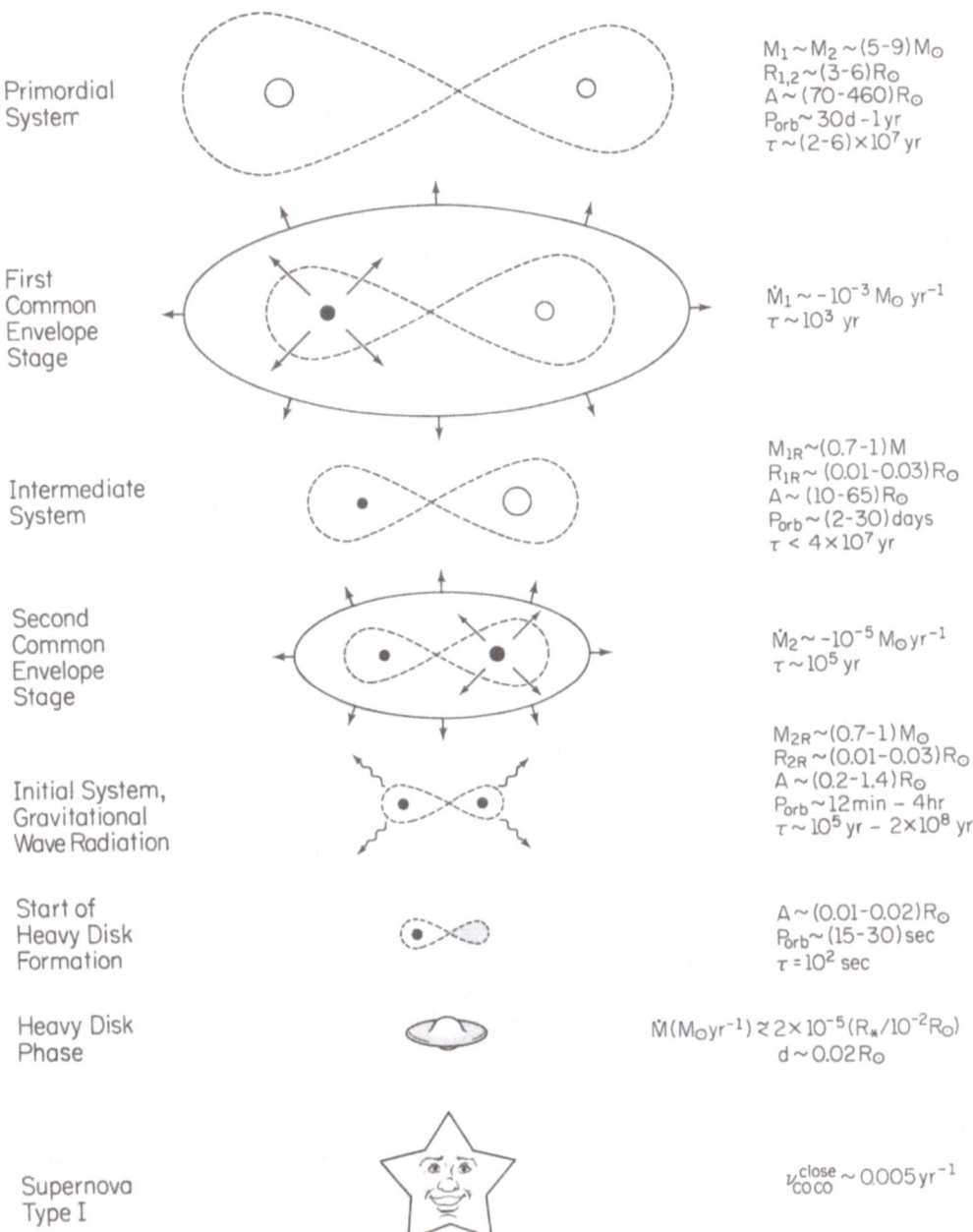

Primordial
System

$M_1 \sim M_2 \sim (5-9) M_\odot$
$R_{1,2} \sim (3-6) R_\odot$
$A \sim (70-460) R_\odot$
$P_{orb} \sim 30d - 1 yr$
$\tau \sim (2-6) \times 10^7 yr$

First
Common
Envelope
Stage

$\dot{M}_1 \sim -10^{-3} M_\odot yr^{-1}$
$\tau \sim 10^3 yr$

Intermediate
System

$M_{1R} \sim (0.7-1) M$
$R_{1R} \sim (0.01-0.03) R_\odot$
$A \sim (10-65) R_\odot$
$P_{orb} \sim (2-30) days$
$\tau < 4 \times 10^7 yr$

Second
Common
Envelope
Stage

$\dot{M}_2 \sim -10^{-5} M_\odot yr^{-1}$
$\tau \sim 10^5 yr$

Initial System,
Gravitational
Wave Radiation

$M_{2R} \sim (0.7-1) M_\odot$
$R_{2R} \sim (0.01-0.03) R_\odot$
$A \sim (0.2-1.4) R_\odot$
$P_{orb} \sim 12 min - 4 hr$
$\tau \sim 10^5 yr - 2 \times 10^8 yr$

Start of
Heavy Disk
Formation

$A \sim (0.01-0.02) R_\odot$
$P_{orb} \sim (15-30) sec$
$\tau = 10^2 sec$

Heavy Disk
Phase

$\dot{M}(M_\odot yr^{-1}) \gtrsim 2 \times 10^{-5} (R_*/10^{-2} R_\odot)$
$d \sim 0.02 R_\odot$

Supernova
Type I

$\nu_{CO CO}^{close} \sim 0.005 yr^{-1}$

Figure 1. Two Degenerate CO Dwarfs (Close Systems)

JP-1155

$$\frac{M_2^2}{A_f} = \alpha \frac{M_{1R}M_{2R}}{A_{ff}} , \qquad\qquad (3)$$

where M_{2R} is the mass of the bare helium core left by the primordial secondary, to determine the semimajoraxis A_{ff} of the "initial" pre-supernova system. In this way we find $0.2 \lesssim A_{ff}/R_\odot \lesssim 1.4$. The helium core rapidly transforms into a CO dwarf of mass $(0.7-1.0)$ $M_\odot$.

Using the formalism of Landau and Lifshitz (1975), we find that the timescale for the reduction of A_{ff} to a value such that the lighter of the two degenerate dwarfs fills its Roche lobe varies from $\sim 2 \times 10^8$ yr ($A_{ff}^{initial} \sim 1.4$ $R_\odot$) to $\sim 10^5$ yr ($A_{ff}^{initial} \sim 0.2$ $R_\odot$). Thus, systems of this sort do not account for SNeI in elliptical galaxies. However, the transformation scheme which we have adopted is highly uncertain, and we must be careful to view this result only as a qualitative suggestion.

Now, using $q \sim 1$, $\Delta\log A \sim \log (460/70)$, $M_A \sim 5$, and $M_B \sim 7$ in equation (1), we estimate a realization frequency for "close" primordial systems which, after two common envelope stages, produce SNeI consisting of two degenerate CO dwarfs to be $\nu_{COCO}^{close} \sim 3 \times 10^{-3}$ yr^{-1}. There is currently, for lack of concrete numerical experiments to prove otherwise, no real reason to restrict the primordial mass of the primary to a value less than ~ 7 $M_\odot$. The choice of $M_1 \sim (7-9)$ $M_\odot$ leads to $M_{1R} \gtrsim 1$ $M_\odot$ and a helium star of this mass will expand during helium burning, with the consequence that an intermediate common envelope stage will occur after the first one. Preliminary experiments (Iben and Tutukov 1983c) suggest that this will not result in a significant reduction in M_{1R}. The enlarged range in permissible M_1, coupled with a reduction in the mimimum required value for M_{2R} (and hence of M_2), increases the realization frequency of the close double degenerate CO dwarf scenario to $\nu_{COCO}^{close} \sim (5-6) \times 10^{-3}$ yr^{-1}.

Examples of real primordial systems which may evolve according to this "close" scenario are Π Andromeda and HD 778322. The orbital characteristics of Π And are: $M_1 \sim M_2 \sim 9.7$ $M_\odot$, $P \sim 143.6$ days, and $A \sim 343$ $R_\odot$, while those of HD 778322 are $M_1 \sim 6.5$ $M_\odot$, $M_2 \sim 6$ $M_\odot$, $P \sim 12.5$ days, and $A \sim 58$ $R_\odot$ (Batten 1968).

The other class of primordial systems which may lead to explosive double degenerate CO dwarf systems differs qualitatively from the first only in that the first common envelope stage is delayed until the primary has become an AGB star with a fully developed electron-degenerate CO core. In this case we choose $A \gtrsim 460$ $R_\odot$, $M_A \sim 5$, $M_B \sim 9$ (maximum mass of single stars that produce degenerate CO cores), and $q \sim 1$. An upper limit on primordial A is established by insisting that $A_{ff} \lesssim 3.5$ $R_\odot$, a choice which ensures that gravitational wave radiation will drive the two CO dwarfs to the explosive state in less

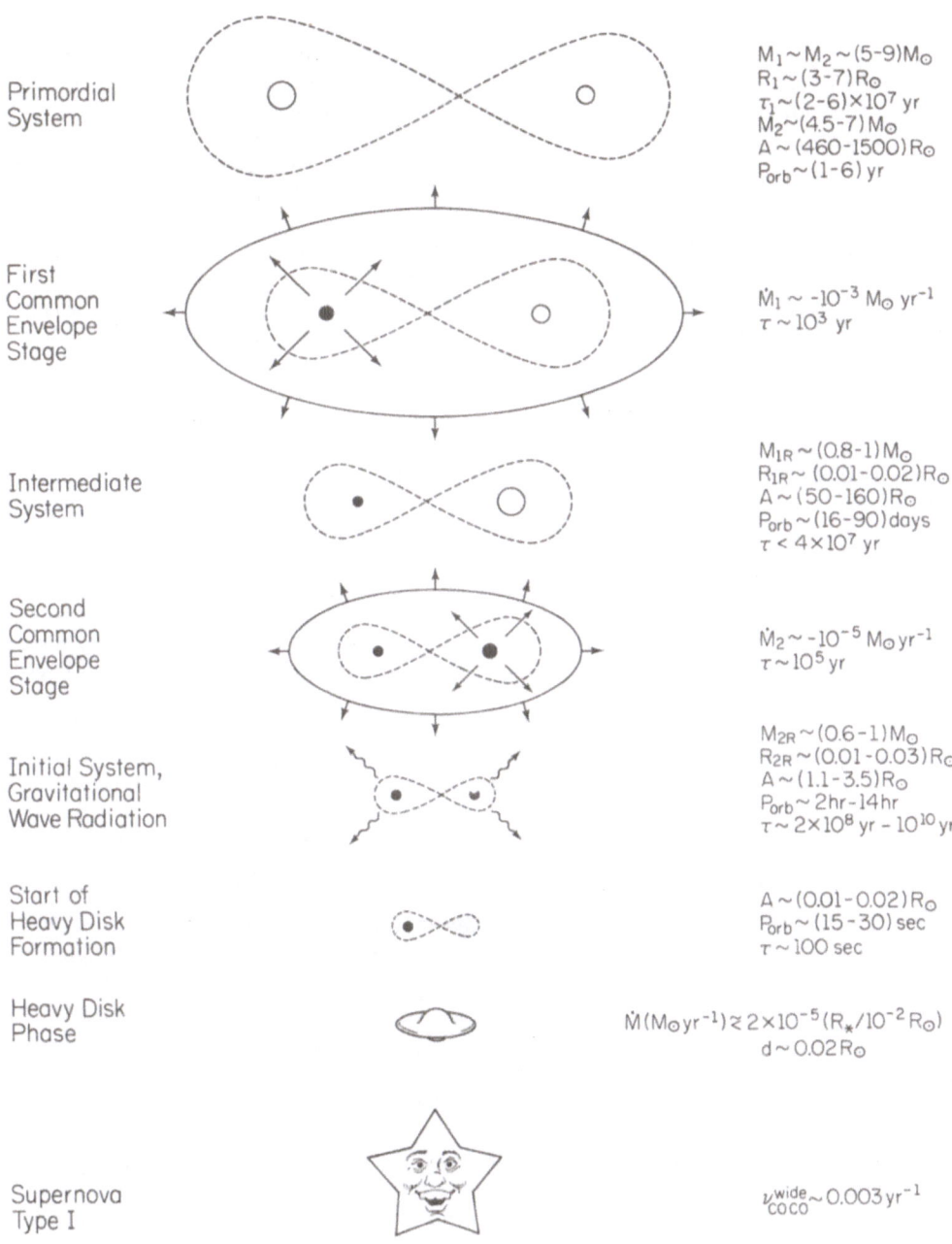

Figure 2. Two Degenerate CO Dwarfs (Wide Systems)

JP-1156

than 10^{10} yr.

Setting $M_{1R} \sim (0.8\text{--}1.0)\ M_\odot$, $M_{2R} \sim (0.6\text{--}1.0)\ M_\odot$ (and therefore M_2 between 4.5 $M_\odot$ and 9 $M_\odot$), we derive from equations (2) and (3) that $A \lesssim 1500\ R_\odot$. Then with $\Delta \log A \sim \log (1500/460) \sim 0.5$, we have from equation (1) that the formation frequency of initial presupernova systems which are made up of "wide" pairs of CO dwarfs is $\nu_{COCO}^{wide} \sim (3\text{--}4) \times 10^{-3}$ yr^{-1}. The evolution of the mass and orbital characteristics of representatives of this scenario are shown in Figure 2.

Out of the 1041 spectroscopic binaries studied by Popova et al (1982) there are two double line spectroscopic binaries with $460 < A/R_\odot < 1500$ and $5\ M_\odot < M_1, M_2 < 9\ M_\odot$. These two may evolve according to the "wide" scenario.

Altogether, we have a total formation frequency of $\nu_{COCO}^{all} \sim (8\text{--}10) \times 10^{-3}$ yr^{-1}. The fact that (in the framework of our transformation scheme) only the primordially wide systems can delay reaching the explosive state for $\sim 10^{10}$ yr means that, per unit galactic luminosity, the frequency of SNeI in elliptical galaxies should be smaller than in later types, a trend which shows up in the observations (Tammann 1980). In fact, the frequency of SNeI per unit luminosity decreases by about a factor of two in going from Sb spirals to E0 ellipticals (Tammann 1980) and this agrees remarkably well with our estimate that, in regions of active star formation, the rate at which wide CO + CO systems are formed is about one half to one third of the rate at which all CO + CO systems are formed.

Another double degenerate configuration to consider consists of a helium dwarf and a CO dwarf (see Figure 3). Choosing $M_1 \sim (6\text{--}9)\ M_\odot$, $M_{1R}^{CO} \sim (0.9\text{--}1.2)\ M_\odot$, $A > 460\ R_\odot$, $M_2 \sim (2\text{--}4)\ M_\odot$, $M_{2R}^{He} \sim (0.25\text{--}0.5)\ M_\odot$, and insisting that $A_{ff} \lesssim 2.8\ R_\odot$ (to ensure a total lifetime less than 10^{10} yr), we estimate that $\nu_{COHe}^{wide} \sim 0.6 \times 10^{-3}$ yr^{-1}. In a similar fashion we find that, beginning with a primordial system with $A \lesssim 460\ R_\odot$, the frequency of realization of an appropriate initial configuration is $\nu_{COHe}^{close} \sim 0.7 \times 10^{-3}$ yr^{-1}, so that the total realization frequency of potentially explosive systems is on the order of $\nu_{COHe}^{all} \sim 10^{-3}$ yr^{-1}. Once again, only wide systems can survive for 10^{10} yr, so that the contribution of CO + He systems (if finite) also decreases by about a factor of two in going from ellipticals to spirals.

An example of a primordial system which may evolve according to the "close" CO + He scenario is HD 208095, which has the characteristics $M_1 \sim 5\ M_\odot$, $M_2 \sim 2.8\ M_\odot$, $P \sim 17.3$ days, and $A \sim 61.5\ R_\odot$ (Batten 1968).

We next examine the probability of forming explosive systems composed of two degenerate helium stars. If we choose $M_1 \sim M_2 \sim (3.2\text{--}3.8)\ M_\odot$ and a primordial A between 60 $R_\odot$ and 460 $R_\odot$, both the primary and

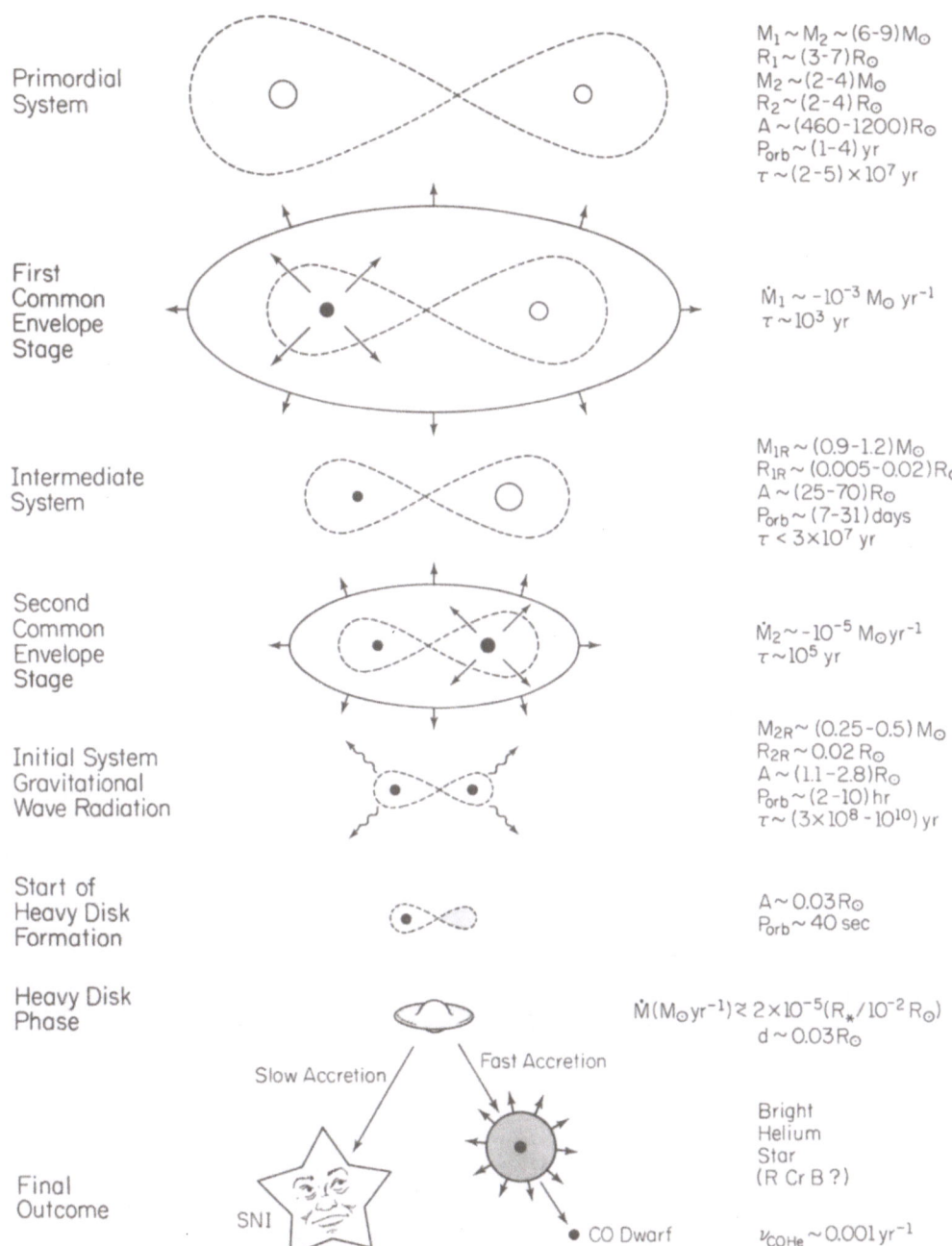

Figure 3. CO and He Degenerate Dwarfs (Close Systems)

JP-1157

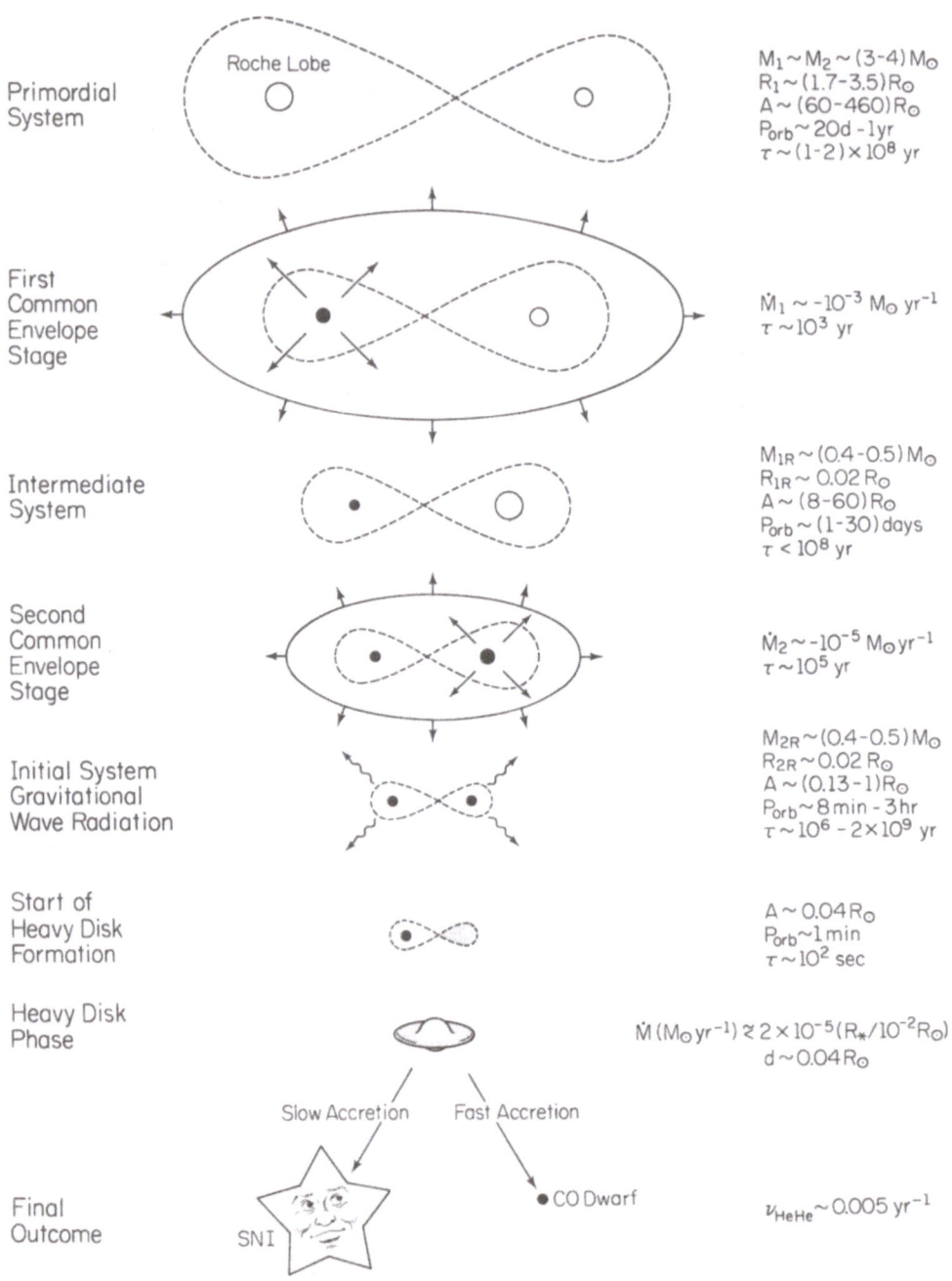

Figure 4. Two Degenerate Helium Dwarfs

JP-1154

the secondary will fill their Roche lobes for the first time when they have developed helium cores of mass $M_{1R} \sim M_{2R} \sim (0.4-0.5)$ $M_\odot$. These cores are too small (in mass) to sustain helium burning and they quickly cool off to become electron-degenerate helium dwarfs. The orbital parameters derived by using equations (2) and (3) are shown in Figure 4.

Setting $q \sim 1$, $M_A = 3.2$ $M_\odot$, $M_B = 3.8$ $M_\odot$, and $\Delta\log A = \log (460/60) \sim 0.9$ in equation (1), we obtain $\nu_{HeHe} \sim 5 \times 10^{-3}$ yr^{-1} as the probability of forming presupernova systems made up of two helium dwarfs. Just as in the case of close double degenerate CO dwarfs, the initial separation of double helium degenerate dwarfs is so small that evolution to the merging state cannot be delayed as long as 10^{10} yr, and such systems (in the framework of our transformation scheme) can therefore not account for SNeI in elliptical galaxies.

If the rate at which matter from the heavy disk formed by the transformation of the less massive of the two dwarfs is accreted onto the more massive one at a rate larger than $\sim 10^{-6}$ $M_\odot$ yr^{-1}, then the ultimate result of merging may be quiescent burning into a single degenerate CO dwarf and we infer that double degenerate helium dwarfs may not contribute to the observed SNI frequency even in spiral galaxies.

Examples of primordial systems which may evolve into close double degenerate helium dwarfs are DI Hercules with $M_1 \sim 5.2$ $M_\odot$, $M_2 \sim 4.2$ $M_\odot$, $P \sim 10.6$ days, and $A \sim 48$ $R_\odot$ (Popper 1980) and Θ Aquilae with $M_1 \sim M_2 \sim 3.5$ $M_\odot$, $P \sim 17.1$ days, and $A \sim 59$ $R_\odot$ (Batten 1968).

3. A Degenerate Dwarf and A (Near) Low Mass Main Sequence Companion

The initial presupernova system is assumed to have a semimajor-axis between 2.5 $R_\odot$ and 10 $R_\odot$. The lower limit is chosen to achieve a detached system after mass loss via a common envelope and the upper limit is demanded by observational-theoretical arguments which suggest that there is no mechanism to bring the initial system closer together for larger separations. Cataclysmic variables with $A < 10$ $R_\odot$, which are prime examples of the systems under consideration, are known to be losing angular momentum (e.g., Patterson 1983) and a magnetic stellar wind (MSW) is held responsible (e.g., Tutukov 1983).

We first consider initial systems which are composed of a degenerate CO dwarf and a main-sequence star. Once again, there are two channels for producing the CO dwarf: by common envelope formation after the primary has grown a degenerate CO core (wide primordial systems) and by common envelope formation after the primary has grown a non-degenerate helium core of mass greater than 0.5 $M_\odot$ (close primordial systems).

The requirement that $A < 10$ $R_\odot$ means that M_1^2/M_2 must be fairly large (see equation 2). For primordially wide systems, the choice M_1

$\sim$ (5-9) $M_{\odot}$ gives $M_{1R} \sim$ (0.7-1.0) $M_{\odot}$, and the further choice of $M_2 \sim M_{\odot}$ (much larger than this leads to too rapid mass transfer) gives $A_f \sim$ A/60, so that, with $A_f < 10 R_{\odot}$, we have $A < 600 R_{\odot}$. The lower limit on A for wide systems follows from the fact that the AGB phase is not reached until $R_{star} \gtrsim 230 R_{\odot}$, giving $A \gtrsim 460 R_{\odot}$. Placing $\Delta \log A \sim \log$ (600/460) = 0.14, $M_A = 5$, $M_B = 9$, and $q \sim 1/7$ in equation (1), we have $\nu_{COMS}^{wide} \sim 1.4 \times 10^{-4}$ yr^{-1}.

For primordially close systems we choose $M_1 \sim$ (3.8-9) $M_{\odot}$, giving $M_{1R} \sim$ (0.5-1.0) $M_{\odot}$, and the requirement that $2.5 < A_f/R_{\odot} < 10$ translates into $80 < A/R_{\odot} < 320$ and therefore $\Delta \log A \sim \log$ (320/80) = 0.6. Finally, with $M_2 \sim M_{\odot}$ and $q \sim 1/5$, we obtain $\nu_{COMS}^{close} \sim 1.6 \times 10^{-3}$ yr^{-1}. The total realization frequency of potentially explosive cataclysmic systems made up of a degenerate CO dwarf and a low mass main sequence star is thus $\nu_{COMS}^{all} \sim 2 \times 10^{-3}$ yr^{-1}.

Next we explore the formation frequency of possible presupernova cataclysmics in which the degenerate dwarf is composed of helium. Again there are two channels. The first has the primordial primary fill its Roche lobe only after it has formed a degenerate helium core of mass $M_{1R} \sim$ (0.4-0.5) $M_{\odot}$. This channel requires that $M_1 < 2.3 M_{\odot}$ and that Roche lobe overfilling does not occur until the radius of the primary exceeds $\sim 80 R_{\odot}$, but the necessity to achieve $A_f < 10 R_{\odot}$ also means that $A \lesssim 22 M_1^2/M_2$. We fulfill all of these requirements by choosing $M_2 \sim 0.5 M_{\odot}$, $M_1 \sim$ (2-2.3) $M_{\odot}$, and $A \sim$ (80-200) $R_{\odot}$. Then, with $q \sim 0.25$, $\Delta \log A \sim 0.4$, $M_A \sim 2$, and $M_B \sim 2.3$, we have $\nu_{HeMS}^{wide} \sim 9 \times 10^{-4}$ yr^{-1}.

The second channel supposes that the primordial primary fills its Roche lobe after developing a helium core of mass $\sim$ (0.4-0.5) $M_{\odot}$, but also before reaching the giant branch. This means that $M_1 \sim$ (3.2-3.8) $M_{\odot}$. The requirement that $2.5 < A/R_{\odot} < 10$ is easily met so we may set $\Delta \log A = \log$ (10/2.5) = 0.6. Then, with $M_2 \sim$ (0.5-1.0) $M_{\odot}$, so that $q \sim 0.2$, we have from equation (1) that $\nu_{HeMS}^{close} \sim 7 \times 10^{-4}$ yr^{-1}. Altogether we have $\nu_{HeMS}^{all} \sim 1.5 \times 10^{-3}$ yr^{-1}. Adding in the systems with CO degenerate dwarfs we have that the realization frequency of potentially explosive cataclysmic systems is $\nu_{cataclysmic} \sim$ (3-4) $\times 10^{-3}$ yr^{-1}.

A final set of scenarios of this general type are those which lead to a degenerate CO or helium dwarf and a star of primordial mass $M_2 \gtrsim 1.5$ which is near the main sequence and which, on filling its Roche lobe transfers matter to the dwarf on a thermal time scale. We call such systems "cataclysmic-like". Since the only difference in primordial parameters between these systems and cataclysmic systems is that the primordial mass ratio q is slightly larger in the former systems, we estimate that $\nu_{cataclysmic-like} \sim \nu_{cataclysmic} \sim 3-4 \times 10^{-3}$ yr^{-1}.

We have chosen primordial orbital parameters in such a way that the total mass of the initial system is large enough to ensure that a supernova-like explosion may occur, provided the accretion rate is in a range which will permit the degenerate dwarf to grow. However, cataclysmic systems experience nova outbursts and a combination of observational and theoretical studies (see, e.g. Faber and Gallagher 1979, MacDonald 1983) suggests that if $\dot{M}_{accrete} \lesssim 10^{-9}$ $M_\odot$ yr^{-1}, the nova explosion removes essentially all of the matter accreted between outbursts. These stars may also suffer from anexoria nervosa, losing more mass during an outburst than they have accreted between outbursts (MacDonald 1983).

It is known that, for cataclysmic systems, $\dot{M}_{accrete}$ decreases on average with decreasing semimajoraxis and that for $P_{orb} \lesssim 3^h$, $\langle \dot{M}_{accrete} \rangle$ becomes less than $\sim 10^{-9}$ $M_\odot$ yr^{-1} (Patterson 1983). It is clear then, that only a fraction of all cataclysmic variables can evolve into explosive systems of SNI magnitude, but what that fraction is we are not able at present to estimate.

4. A Degenerate Dwarf and A Red (Super)Giant

We come at last to an estimate of the frequency of formation of initial systems of the sort originally proposed by Whelan and Iben (1973). The derivation is rather complicated and we content ourselves here with simply stating the results. Considerations of the stability of the mass transfer process in the potential presupernova system of CO dwarf plus Roche-lobe filling red giant require that the mass of the red giant be less than ~ 0.61 times the mass of the dwarf. To satisfy this limitation and still achieve a total mass of the accreting white dwarf plus transferrable matter equal to 1.4 $M_\odot$, it is necessary to begin with a primary of mass $M_1 \sim (8-9)$ $M_\odot$ which produces a bare CO core of mass $M_c \sim (1.3-1.4)$ $M_\odot$. With $M_2/M_{1R} \lesssim 0.61$, we have that $M_2 \lesssim 0.85$ $M_\odot$ and therefore q in equation (1) is ~ 0.1. Since the core mass and radius of an AGB star are related by R α $(M_{CO}-0.5)^{0.68}$, we have that $\Delta \log A \sim 0.07$. Then, with $M_A = 8$ and $M_B = 9$, equation (1) gives $\nu_{CORG} \sim 0.1 \times 0.014 \times 0.005$ yr$^{-1} < 10^{-5}$ yr^{-1}. Thus, the specific model visualized by Whelan and Iben does not at present appear to be able to account quantitatively for the observed SNI frequency.

It does not help to replace the red giant donor by a Roche-lobe filling asymptotic giant branch (AGB) star. Such stars experience thermal pulses and, for brief periods, swell beyond the size they maintain between pulses. During the first such swelling which overfills the Roche lobe, a common envelope will be formed and the core of the AGB star will be stripped of its entire remaining hydrogen rich envelope. There is then no further mass to transfer and the system rapidly evolves into an inert double CO white dwarf system.

One final variant which has not yet been discredited involves an AGB donor which does not fill its Roche lobe (to avoid termination by a thermal pulse). Luminous single AGB stars are known to emit winds at fairly substantial rates ($|\dot{M}_{wind}| \gtrsim 10^{-6}$ $M_\odot$ yr^{-1}) and at relatively small velocities ($\sim$ 10 km s^{-1}). If the degenerate CO dwarf in such a system could capture matter from this wind with an efficiency of, say, 10 percent, then perhaps (0.1–0.2) $M_\odot$ of matter could be transferred to the CO dwarf without being blown off in nova outbursts. We estimate that the frequency with which "successful" initial systems of this sort are formed could be, under the most favorable circumstances, as large as ν_{COAGB} $\sim$ 4 x 10^{-3} yr^{-1}.

V. SUMMARY AND CONCLUSIONS

We have estimated the rates of formation of various primordial binary systems which may possibly evolve into immediate pre-SNeI systems and have found that the most attractive scenario begins with primordial binaries consisting of two intermediate mass stars of mass $\sim$ (5–9) $M_\odot$ that are at a separation A in the range 70 $\lesssim$ A/R$_\odot$ $\lesssim$ 1500. These systems evolve into binaries consisting of two degenerate CO dwarfs at an initial separation A$_{ff}$ less than $\sim$ 3 R$_\odot$. As a consequence of the radiation of gravitational waves, the two dwarfs are drawn closer together until (in a time between 10^8 and 10^{10} yr, depending on A$_{ff}$) the lighter of the two dwarfs fills its Roche lobe and is transformed on a dynamic time scale into a heavy disk that then transfers matter to the more massive primary on a time scale that is probably on the order of 10^5–10^6 yr. This scenario is able to account for an impressive number of SNeI characteristics: overall frequency of occurrence, decrease in frequency of occurrence as one goes from spiral galaxies to ellipticals, appropriate expansion velocities, peak half width of light curves, chemical abundances inferred from spectral distributions (in particular the absence of lines of H and He), and lack of evidence for compact remnants.

Nevertheless, other scenarios cannot be unequivocally excluded, and even in the scenario which leads to the merging of two degenerate CO dwarfs there are crucial problems which must be solved before this scenario can be considered fully established. These problems include: (1) the nature of the dynamical transition of the lighter of the two CO dwarfs into a heavy disk about the more massive dwarf, (2) the rate at which mass is transferred from the disk onto the central dwarf, (3) the manner in which the initially cold central dwarf responds to the (probably large) accretion rate, and (4) the nature of the thermonuclear explosion that occurs when carbon is ignited.

ACKNOWLEDGEMENTS

The emphasis placed here on the scenario involving two electron-degenerate CO dwarfs was inspired by the sum total of the remarks made by other participants at this workshop, both informally and in lectures, and by the recognition of the ease with which this scenario is able to account for many of the observed properties of SNeI. We would, in particular, like to thank Dave Branch, Ken Nomoto, and Stan Woosley for their magnificent presentations and to also thank them and the other participants who made the workshop a most rewarding and educational experience. Thanks also to Masayuki Fujimoto for guiding us through the literature dealing with the thermonuclear responses of degenerate dwarfs to accretion at various rates.

REFERENCES

Arnett, D.W. 1983. private communication.
Batten, A.H. 1980. Ann. Domin. Observ., 13, 1.
Boyarchuk, A.A. 1975. in Variable Stars and Stellar Evolution, Proc. I.A.U. Symposium No. 67, eds. V.E. Scherwood and L. Plaut, (D. Reidel, Dordrecht-Holland: Boston, USA), p. 377
Branch, D. 1984, in Stellar Nucleosynthesis, eds. C. Chiosi and A. Renzini (Dordrecht: Reidel), this volume.
Chechetkin, V.M., Gershtein, S.S., Immshennik, V.S., Ivanova, L.N., and Kholpov, M.Yu 1980. Ap. Space Sci., 67, 61.
Chevalier, R. 1981. Ap. J., 246, 267.
Colgate, S.A. and Petschek, A.G. 1980. in Type I Supernovae, ed. J.C. Wheeler (Austin: Univ. of Texas), p. 42.
Couch, R.G. and Arnett, W.D. 1975. Ap. J., 196, 791.
Ergma, E. and Tutukov, A.V. 1976. Acta Astron., 26, 69.
Faber, S.M. and Gallagher, J.S. 1979. Ann. Rev. A. and Ap., 17, 135.
Fujimoto, M. Y. 1980. in Type I Supernovae, ed. J.C. Wheeler (Austin: Univ. of Texas), p. 155.
Fujimoto, M.Y. and Sugimoto, D. 1979. in White Dwarfs and Variable Degenerate Stars, eds. H.M. Van Horn and V. Weidemann (New York: U. of Rochester), p. 285.
Fujimoto, M.Y. and Sugimoto, D. 1982. Ap. J., 257, 291.
Grauer, A.D. and Bond, H.E. 1983. Ap. J., 271, 259.
Helfand, D.J. 1980. in Type I Supernovae, ed. J.C. Wheeler (Austin: Univ. of Texas), p. 20.
Hills, J.G. 1975. Astr. J., 80, 809.
Iben, I. Jr. 1982a. Ap. J., 253, 248.
Iben, I. Jr. 1982b. Ap. J., 259, 244.
Iben, I. Jr. and Tutukov, A.V. 1983a. U. of Illinois preprint IAP 83-16.
Iben, I. Jr. and Tutukov, A.V. 1983b. in Proc. of the Santa Cruz Workshop on High Energy Transients.
Iben, I. Jr. and Tutukov, A.V. 1983c. in progress.
Ivanova, L.N., Imshennik, V.S., and Chechetkin, V.M. 1977. Sov. Astr., 21, 374.
Landau, L. and Lifshitz, 1975. The Classical Theory of Fields, (Pergamon Press).
MacDonald, J. 1983. Ap. J., 267, 732.
Mazurek, T.J., Meier, D.L., and Wheeler, J.C. 1977. Ap. J., 213, 518.
Nomoto, K. 1984, in Stellar Nucleosynthesis, eds. C. Chiosi and A. Renzini (Dordrecht: Reidel), this volume.
Nomoto, K. 1982. Ap. J., 257, 780.

Nomoto, K. and Sugimoto, D. 1977. Publ. Astr. Soc. Japan, 29, 765.
Paczynski, B. 1976. in Proc. of the IAU Symp. No. 73, Structure and
    Evolution of Close Binary Systems, eds. P. Eggleton, S. Mitton,
    J. Whelan (Dordrecht: Reidel), p. 75.
Patterson, J. 1983. in The Evolution of Cataclysmic Binaries and Low
    Mass X-ray Binaries, eds. J. Patterson and D.Q. Lamb (Cambridge:
    Harvard U. Press).
Popper, D.M. 1980. Ann. Rev. A. and Ap., 18, 115.
Renzini, A. 1979. in Stars and Star Systems, ed. B.E. Westerlund
    (Dordrecht: Reidel), p. 155.
Seward, F., Gorenstein, P., and Tucker, W. 1983. Ap. J., 266, 287.
Shklovski, I.S. 1981. Soviet Scientific Reviews/Section E V 1. ed.
    R.A. Sunyaev (Harwood Academic Publishers), p. 177.
Sparks, W.M. and Stecher, T.P. 1974. Ap. J., 188, 143.
Taam, R.E. 1980a. Ap. J., 237, 142.
Taam, R.E. 1980b. Ap. J., 242, 749.
Taam, R.E., Bodenheimer, P., and Ostriker, J.P. 1978. Ap. J., 222,
    684.
Tammann, G.A. 1978. Ann. New York Acad. Sci., 302, 61.
Tutukov, A.V. 1983. in The Evolution of Cataclysmic Binaries and Low
    Mass X-ray Binaries, eds. J.Patterson and D.Q. Lamb (Cambridge:
    Harvard U. Press), in press.
Tutukov, A.V. and Yungelson, L.R. 1979. Acta Astron., 23, 665.
Weaver, T.A., Axelrod, T.S., and Woosley, S.E. 1980. in Type I Super-
    novae, ed. J.C. Wheeler (Austin: Univ. of Texas), p. 113.
Webbink, R.F. 1979. in White Dwarfs and Variable Degenerate Stars,
    eds. H.M. Van Horn and V. Weidemann (New York: U. of Rochester),
    p. 426.
Webbink, R.F., Rappaport, S., and Savonije, G.J. 1983. Ap. J., 270,
    678.
Whelan, J. and Iben, I. Jr. 1973. Ap. J., 186, 1007.
Woosley, S.E. 1983. private communication.
Woosley, S.E., Weaver, T.A., and Taam, R.E. 1980. in Type I Super-
    novae, ed. J.C. Wheeler (Austin: Univ. of Texas), p. 96.
Woosley, S.E., Axelrod, T.S., and Weaver, T.A. 1984, in Stellar
    Nucleosynthesis, eds. C. Chiosi and A. Renzini (Dordrecht:
    Reidel), this volume.

NUCLEOSYNTHESIS IN TYPE I SUPERNOVAE:
CARBON DEFLAGRATION AND HELIUM DETONATION MODELS

Ken'ichi Nomoto
Dept. of Earth Science and Astronomy, College of Arts and
Sciences, University of Tokyo, Meguro-ku, Tokyo 153
and
Max-Planck-Institut für Physik und Astrophysik
Institut für Astrophysik, Garching b. München

ABSTRACT

    The evolution of accreting white dwarfs in close binary systems is
studied from the onset of accretion through the thermonuclear explosion.
Relatively rapid accretion onto C+O white dwarfs leads to a carbon
deflagration supernova which disrupts the star completely. Explosive
nucleosynthesis in the deflagration wave produces 0.5-0.6 $M_\odot$ ^{56}Ni in the
inner part of the star; this amount of ^{56}Ni is sufficient to power the
light curve of Type I supernovae by the radioactive decays. In the outer
layers of the star, the deflagration wave synthesizes appreciable amount
of intermediate mass elements such as Ca, Ar, S, Si, Mg, and O; this is
consistent with the spectra of Type I supernovae near maximum light.
Thus the carbon deflagration model can account for many of the observed
features. Moreover, the nuclear products in this model are quite
complementary to nucleosynthesis in Type II supernovae.
    On the other hand, slow accretion results in the detonation
supernova explosion, namely, a helium detonation in helium white dwarfs
and double or single detonations in C+O white dwarfs. This type of
explosion produces almost exclusively iron peak elements and thus cannot
account for the observed features of Type I supernovae.
    Possible progenitors of Type I supernovae are discussed, namely,
white dwarfs undergoing hydrogen burning near the surface and double
white dwarf systems undergoing transfer of helium or C+O onto the more
massive white dwarf.

1. INTRODUCTION

    Type I supernovae (SN I) are quite distinct from Type II supernovae
(SN II) because of hydrogen-deficiency in their spectra near maximum
light. SN I are observed in elliptical galaxies as well as in other type
of galaxies and are not concentrated in spiral arms (see, e.g., Trimble
1982 for a review). These facts have led to the idea that the
progenitors of SN I are white dwarfs (WDs) (Finzi and Wolf 1967) or
helium stars with relatively small mass (Wheeler 1978).

C. Chiosi and A. Renzini (eds.), Stellar Nucleosynthesis, 205–237.

A recent breakthrough in understanding the nature of SN I has been brought about by the success of the radioactive decay model (^{56}Ni → ^{56}Co → ^{56}Fe) in reproducing the characteristic SN I light curves (Arnett 1979; Colgate et al. 1980; Chevalier 1981; Axelrod 1980b; Weaver et al. 1980; Sutherland and Wheeler 1984). This model is supported by the identification of the features in the late time spectra of SN 1972e with emission lines of Fe (Meyerott 1980· Axelrod 1980a,b) and decaying Co (Axelrod 1980a,b). The amount of ^{56}Ni required to power the maximum light is estimated to be 0.2-1 M$_\odot$ depending on the Hubble constant (see Wheeler 1982 for a review).

The first hundred day's spectra of SN I 1981b have been compared with synthetic spectra (Branch et al. 1982, 1983); the maximum-light spectrum is well interpreted by the presence of Ca, Si, S, Mg, O and possibly Co in the outer layer of SN I at the expansion velocity of v_{exp} ∿ 10^4 km s^{-1}.

These results suggest the following composition structure of SN I; the inner layer, exposed at later times, contains 0.2-1 M$_\odot$ ^{56}Ni while the outer layer, observed at maximum light, is composed mainly of intermediate mass elements Ca-Si-O. This picture is consistent with the composition structure of the remnant of SN 1006 suggested from the IUE observation by Wu et al. (1983). X-ray emission from the Tycho and Kepler supernova remnants also indicates the existence of Si and S-rich outer shells (Becker et al. 1980), although the observation of little or no enhancement of the iron abundance is still a problem (e.g., Shull 1982).

The failure to detect point X-ray sources or a synchrotron nebulosity in the SN I remnants suggests that no neutron star is left after an SN I explosion, because the upper limit of the X-ray luminosity is lower than the lower limit to the luminosity expected from the cooling neutron stars (Nomoto and Tsuruta 1981) and from the nebular emission observed around pulsars (Helfand 1983). The neutron star might exist but be cooled down more rapidly than expected due to the effects of exotic particles such as quarks and pion-condensates. However, the observed flux from the point source in RCW 103 (Tuohy and Garmire 1980), which is very likely a thermal black body radiation from the neutron star (Tuohy et al. 1983), is quite consistent with the "standard" cooling curve where the exotic cooling is not included (Nomoto and Tsuruta 1981). Therefore it is very likely that SN I do not leave neutron stars behind.

Many of these observed features of SN I are consistent with the thermonuclear explosion model of accreting white dwarfs, especially, carbon deflagration supernovae (see reviews by Nomoto 1981, Wheeler 1982, and Branch 1983). From the point of view of stellar nucleosynthesis, carbon deflagration supernovae may synthesize most of the iron-peak elements and a significant fraction of intermediate mass elements (Si-Ca) in the Galaxy as shown by Nomoto et al. (1976). However, the contribution of SN I to galactic nucleosynthesis has been poorly studied. In the next section, therefore, the carbon deflagration model is

described in some detail with particular emphasis on the explosive
nucleosynthesis in the deflagration wave.

Since the evolution of accreting WDs depends on several parameters
of the binary system, carbon deflagration is not a unique mechanism to
explode white dwarfs: for a certain range of accretion rate, $\dot{M}$, and WD
composition, helium detonation is initiated. The hydrodynamics and
nucleosynthesis of the detonation-type supernovae are discussed in § 3.
Comparison with observations of SN I is less satisfactory, since the
detonation-type explosion cannot account for the presence of Si-Ca at
$v_{exp} \sim 10^4$ km s^{-1}.

Despite the plausibility of the carbon deflagration model, the
evolutionary origin of accreting WDs which could plausibly lead to carbon
deflagration is still a controversial subject. In § 4, therefore,
possible evolutionary paths of close binary systems up through the
explosion of a WD component are discussed.

2. CARBON DEFLAGRATION SUPERNOVAE IN ACCRETING WHITE DWARFS

Hoyle and Fowler (1960) first presented the idea that SN I are
triggered by the thermonuclear runaway of carbon burning in electron-
degenerate cores. This idea was refined to the carbon detonation
supernova model in the carbon-oxygen (C+O) cores of intermediate mass (4-
8 $M_\odot$) stars (Arnett 1969). Recently many people have argued that stars
less massive than $\sim$ 8 $M_\odot$ become white dwarfs rather than supernovae by
loss of the hydrogen-rich envelope (e.g., Weidemann and Koester 1983).
However, the carbon flash model for SN I has been revived in the case of
mass accreting WDs. Recent calculations have shown that the evolution of
accreting C+O WDs, under certain conditions of the binary system,
actually reaches the point of carbon ignition which grows into the carbon
deflagration supernovae (see Nomoto 1981 for a review). Whether this
model may violate certain observational constraints is open to question.

Here we describe the whole evolutionary processes of WDs from the
onset of accretion through the explosion and nucleosynthesis based on the
most detailed carbon deflagration model by Nomoto and Thielemann (1984).

2.1 Heating of White Dwarfs by Mass Accretion

Two initial models were chosen; i.e., a 1 $M_\odot$ C+O core (hereafter
referred to as "core") of an asymptotic giant branch (AGB) star and a 1
$M_\odot$ WD which has been cooled down for 5.8 x 10^8 yr. Compositions were
assumed to be $X(^{12}C) = 0.475$, $X(^{16}O) = 0.5$, and $X(^{22}Ne) = 0.025$. Initial
values of central temperature, T_C, and density, ρ_C, are summarised in
Table 1. The accretion rates assumed were $\dot{M} = \dot{M}_{AGB} = 8.5$ x 10^{-7} (M/$M_\odot$-
0.52) $M_\odot$ yr^{-1} for the core and $\dot{M} = 4$ x 10^{-8} $M_\odot$ yr^{-1} for the WD. As will
be discussed in §§ 3-4, the accretion of hydrogen-rich matter at a rate
of $\dot{M}_{AGB} > \dot{M} \gtrsim 4$ x 10^{-8} $M_\odot$ yr^{-1} results in a steady or weakly unstable
hydrogen and helium shell burnings which increase the mass of C+O WDs.

Table 1

	Initial Models		...	Ignition Models		
	ρ_c(g cm^{-3})	T_c (K)	...	$\rho_{c,ig}$	$M/M_\odot$	E_{Bind}(ergs)
Core ...	2.9×10^7	1×10^8	...	1.5×10^9	1.366	5.0×10^{50}
WD ...	3.4×10^7	1×10^7	...	2.6×10^9	1.378	5.3×10^{50}

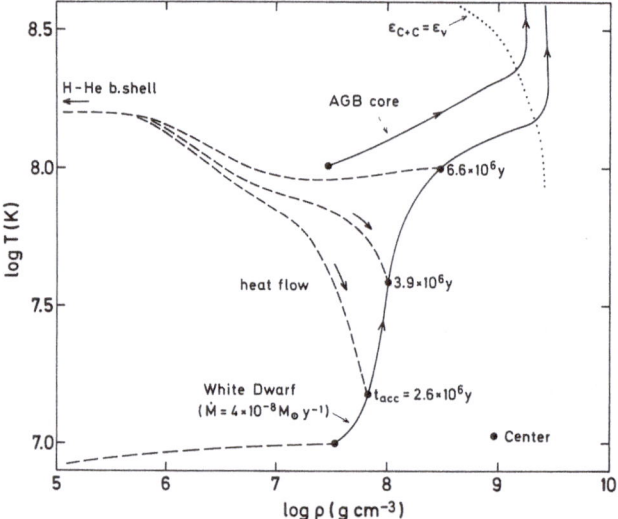

Figure 1: Accretion onto the white dwarf and the core growth of the AGB star. The evolution of (ρ_c, T_c) is shown by the solid lines. Dashed lines are the structure lines of the white dwarf where heat flows from the surface into the interior. The dotted line is the ignition line of carbon burning.

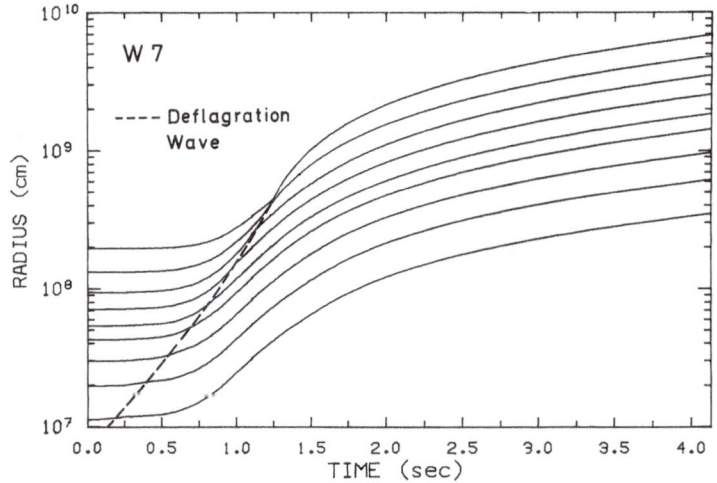

Figure 2: Propagation of the carbon deflagration wave (dashed line) and the associated expansion of the Lagrangian shells for case W7. These shells correspond to $M_r/M_\odot$ = 0.007, 0.03, 0.10, 0.25, 0.41, 0.70, 1.00, 1.28, and 1.378, from the interior to the exterior.

Thus the accretion was followed assuming steady hydrogen-helium (H-He) double burning shells at the outer edge of the WD.

Figure 1 shows the temperature, T, and density, ρ, plane. A change in the ρ-T structure line of the WD during the accretion is shown by the dashed lines, and the evolutionary tracks corresponding to the central conditions, i.e., (ρ_c, T_c) for both the core and the WD are shown by the solid lines. As mass is accreted, both ρ_c and T_c increase by compression. At the same time, heat conduction from the outer layers into the central cold layer raises T_c. In the present models, the H-He burning shells are the dominant heat sources. Moreover, the accreted materials contract and release gravitational energy which also heats up the outer layer; the resultant heat flow into the central layer is comparable to the effects of shell burning for the high $\dot{M}$ considered here. As a result, T_c increases to reach a point where the neutrino loss is significant; afterwards the evolution of (ρ_c, T_c) is controlled by the balance between compressional heating and neutrino loss, which depends only on $\dot{M}$. Similar results have been obtained for colder WDs with a crystallized central region. The WD is first melted by the heat inflow from the surface layer and the evolution of (ρ_c, T_c) merges into the same track as is determined only by $\dot{M}$ in Fig. 1. (If the initial mass of the WD is more massive than, say, $\sim$ 1.2 $M_\odot$, the evolution of (ρ_c, T_c) in later phases will depend on the initial condition.)

When the evolutionary path of (ρ_c, T_c) reaches the dotted line corresponding to $\varepsilon_{C+C} = \varepsilon_\nu$ (where ε_{C+C} and ε_ν denote the nuclear energy generation rate and the neutrino energy loss rate, respectively), carbon is ignited at the center. Because of the strong electron degeneracy, carbon burning is unstable to a flash and T_c increases rapidly as seen in Fig. 1. The central density and the mass of the core and the WD at the carbon ignition, summarized in Table 1, are important quantities for the subsequent evolution and nucleosynthesis. The strong screening effect due to degenerate electrons (Ichimaru and Utsumi 1983a,b) significantly enhances the C+C reaction rate so that the ignition curve in the (ρ, T) plane does not depend sensitively on the temperature for T < 2×10^8 K. Because of the strong density dependence of ε_{C+C}, the ignition density, $\rho_{c,ig}$, does not depend much on $\dot{M}$; even for $\dot{M}$ < 10^{-9} $M_\odot$ yr^{-1} (test calculation), $\rho_{c,ig} \lesssim 3.5 \times 10^9$ g cm^{-3} which is significantly lower than the previous value of $\rho_{c,ig} \sim 10^{10}$ g cm^{-3} for such a slow accretion and/or low temperature WDs (Canal and Shatzmann 1976; Duncan et al. 1976; Ergma and Tutukov 1976; Nomoto 1982a).

The above results imply that, unless the initial mass of the WD is as large as M > 1.2 $M_\odot$, the WD mass and $\rho_{c,ig}$ at carbon ignition are fairly indepedent of initial condition of the WDs and $\dot{M}$. This is consistent with the uniformity of SN I.

2.2 Hydrodynamic Behavior of Carbon Deflagration Supernovae

The central carbon flash grows into thermonuclear runaway, which incinerates the material into nuclear statistical equilibrium (NSE).

However, the released nuclear energy, $\sim 3 \times 10^{17}$ erg g^{-1}, is only 20 % of
the Fermi energy of degenerate electrons so that the thermal overpressure
is too weak to form a carbon detonation wave (cf. Ivanova et al. 1974;
Buchler and Mazurek 1975; Nomoto et al. 1976; Mazurek et al. 1977). The
spherical geometry in the central region also damps the shock wave. It
should be noted that the carbon detonation models (Arnett 1969; Bruenn
1971) assumed the formation of a detonation wave.

After the thermonuclear runaway in the central region, the explosive
carbon burning front propagates outward on the timescale for convective
energy transport across the front, because the density inversion at the
front is unstable to the Rayleigh-Taylor instability (see the 2D
calculation by Muller and Arnett 1982 for development of the R-T
instability). A burning front which propagates at a subsonic velocity
with respect to the unburnt material is called a deflagration wave (DFW)
in contrast to a detonation wave which propagates at supersonic speed
(e.g., Courant and Friedrichs 1948).

To simulate the propagation of a convective carbon DFW with a 1D
hydrodynamic code, the time-dependent mixing theory of convection by Unno
(1967) was employed. The mixing length, ℓ, was taken to be $\ell = \alpha \, H_p$
where H_p is a pressure scale height and α is a parameter (cf., Sutherland
and Wheeler 1984 for a different mixing prescription).

Two cases with $\alpha = \ell/H_p = 0.6$ (C6) and 0.8 (C8) for the core and
three cases with $\alpha = 0.6$ (W6), 0.7 (W7), and 0.8 (W8) for the WD were
calculated. Here the result of case W7 will be discussed in some detail.
Figure 2 shows the propagation of the DFW and the associated expansion of
the Lagrangian shells. Changes in the profiles of temperature, density,
and velocity are shown in Figs. 3-5.

The propagation velocity, v_{DF}, of the DFW with respect to the
unburnt material is slow in the early stages; e.g., $v_{DF} \sim 0.08 \, v_s$ for
stage 2, where v_s is the sound velocity of the burnt material. As the
DFW propagates outwards, v_{DF} gets larger because of the increasing
density jump across the front. However, at stage 7, v_{DF} is still as slow
as $\sim 0.3 \, v_s$. Thus it takes 1.2 sec for the DFW to reach the shell at M_r
$= 1.3 \, M_\odot$ (Fig. 2) which is 6 times longer than the propagation of a
detonation wave (Arnett 1969).

During a slow propagation of the DFW, the WD gradually expands (Fig.
2), and the density and temperature decrease (Figs. 3-4). Such an
expansion weakens the explosive nuclear burning at the DFW and eventually
quenches the carbon burning when the DFW reaches $M_r = 1.3 \, M_\odot$ where $\rho \sim$
10^7 g cm^{-3}.

The DFW compresses the material ahead of it and forms a precursor
shock wave as seen from the density and velocity profiles (Figs. 4-5).
This precursor shock is not so strong as to ignite carbon; in other
words, the DFW does not grow into a detonation. However, when the DFW
arrives at the outer layer of around $M_r \sim 1.3 \, M_\odot$, the precursor shock is

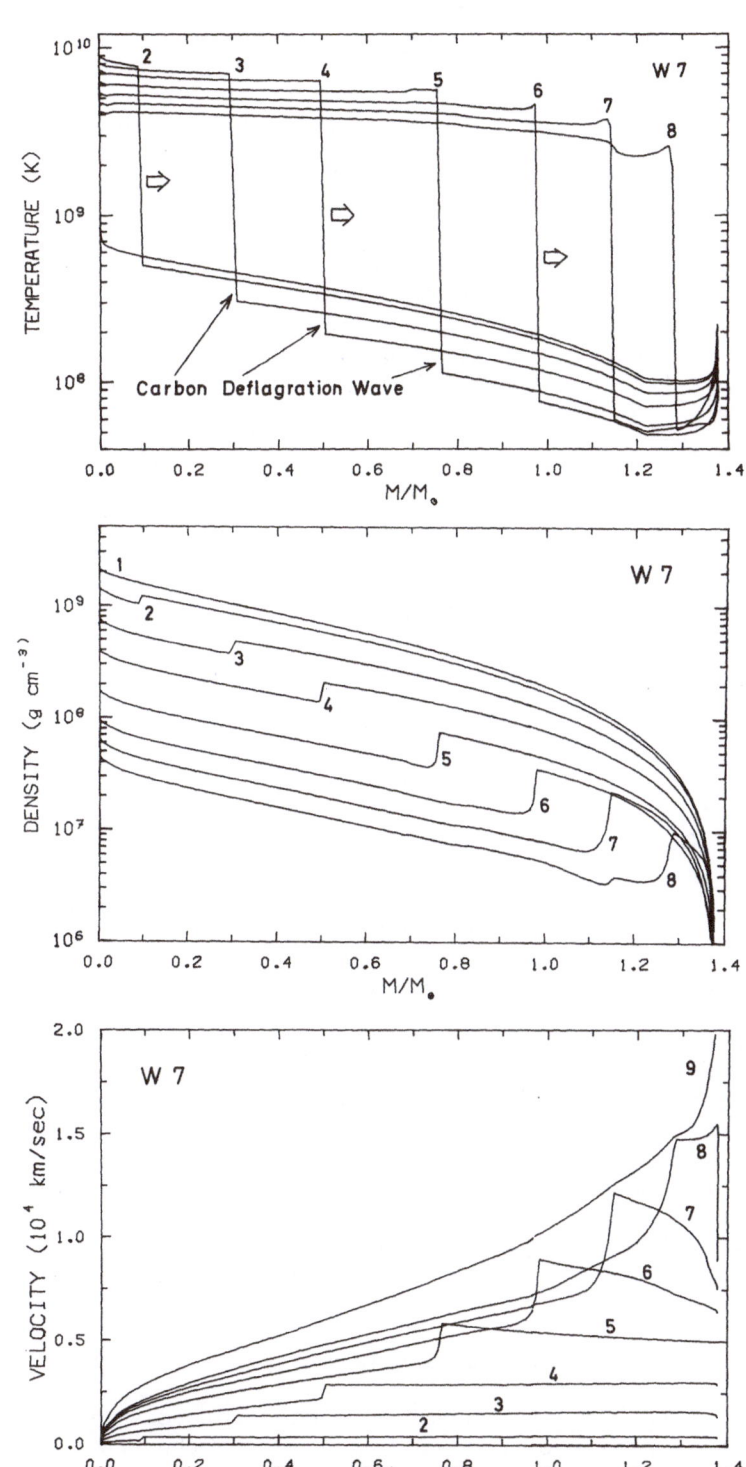

Figure 3:
Change in the temperature profile against M_r during the propagation of the carbon deflagration wave (case W7). Stage numbers 1-9 correspond to t(sec) = 0.0 (#1), 0.60 (#2), 0.79 (#3), 0.91 (#4), 1.03 (#5), 1.12 (#6), 1.18 (#7), 1.24 (#8), 3.22 (#9), respectively.

Figure 4:
Same as Fig. 3 but for the density profile.

Figure 5:
Same as Fig. 3 but for the velocity profile.

strengthened appreciably due to the steep density gradient near the
surface. Whether the shock ignites carbon depends on the propagation
speed of the DFW and also the pre-shock temperature. For the present
models, carbon ignition in the outer layer does occur for case C8 but
does not for other cases.

The energetics of the deflagration are summarized in Table 2; the
total nuclear energy release exceeds the initial binding energy of the WD
and of the core so that the star is disrupted completely with no compact
remnant star. Because of the low $\rho_{c,ig}$, neutrino energy loss associated
with electron captures is as small as $\sim 10^{49}$ ergs, and thus negligible.
The explosion energy is $\sim 10^{51}$ ergs which is in good agreement with SN I.
As seen from the velocity profile at stage 9 (Fig. 5), the expansion
velocity of 10000-14000 km s^{-1} deduced from the spectra at maximum light
corresponds to $M_r = 0.9$-1.2 $M_\odot$.

2.3 Explosive Nucleosynthesis in the Carbon Deflagration Wave

The WD material undergoes explosive burning of carbon, neon, oxygen,
and silicon at the passage of the DFW. The nuclear products of such
explosive burning depends mainly on the peak temperature, T_p, and the
corresponding density, ρ_p, at the DFW. The final composition structure
is shown in Fig. 6 for case W7 and Fig. 7 for case C6.

a) NSE Layer

For the inner layer at $M_r < 0.7$ $M_\odot$, $\rho_{p,7} > 9$ and $T_{p,9} > 6$ so that
the nuclear reactions are rapid enough to incinerate the material into
NSE elements. Here $\rho_{p,7} \equiv \rho_p/10^7$ g cm^{-3} and $T_{p,9} \equiv T_p/10^9$ K. NSE
abundances for 450 species were obtained from a table (Yokoi and Nomoto
1984, updated version of Yokoi, Neo, and Nomoto 1979), and the
neutronization in the NSE layer was calculated by using electron capture
rates of Fuller et al. (1982) whose rates on NSE elements are larger than
the rates of Mazurek et al. (1974) at least by a factor of 2.

As the white dwarf expands, the interior temperature decreases and
the NSE composition moves to iron-peak elements. The abundances were
assumed to be frozen at $T_9 = 3$, where $T_9 \equiv T/(10^9$ K). The resultant
composition structure is shown in Fig. 6 where the region interior to the
dashed line ($M_r < 0.7$ $M_\odot$) is the NSE region. The central layers are
composed of neutron-rich iron-peak elements (^{58}Fe, ^{56}Fe, ^{54}Fe) because of
electron captures at the high density; at the center, the number of
electrons per baryon is $Y_e = 0.45$ for case W7 and 0.46 for case C6. In
the lower density region at $M_r > 0.1$ $M_\odot$, Y_e is larger and so ^{56}Ni is the
dominant product.

b) Explosive Nucleosynthesis in Partial Burning Layers

For the outer layers at $M_r \gtrsim 0.7$ $M_\odot$, detailed nucleosynthesis is
calculated until the actual freezing occurs at $T_9 \sim 1$. When the DFW
reaches such outer layers, the density at the front is significantly

Table 2. Energetics of Explosion

Case	...	C6	C8	...	W6	W7	W8
$\dot{M}$	...		$\dot{M}_{AGB}$	...		4×10^{-8} $M_\odot yr^{-1}$	
$\alpha(\equiv \ell/H_p)$	...	0.6	0.8	...	0.6	0.7	0.8
E(neutrino)*	...	0.018	0.020	...	0.033	·0.035	0.037
E(nuclear)*	...	1.4	1.9	...	1.5	1.8	2.0
E(kinetic)*	...	0.91	1.4	...	0.99	1.3	1.5
$M(^{56}Ni)/M_\odot$	...	0.47	0.64	...	0.49	0.58	0.65

*   Energies in units of 10^{51} ergs.

Table 3. Abundances of Major Products

		Case W7		...	Case C6	
		mass $(M_\odot)$	$\langle X_i/^{56}Fe\rangle$**	...	mass $(M_\odot)$	$\langle X_i/^{56}Fe\rangle$**
Cr-Ni	...	0.78		...	0.66	
^{58}Ni	...	0.054	2.2	...	0.040	2.0
^{56}Fe	...	0.611	1	...	0.485	1
^{54}Fe	...	0.145	3.9	...	0.106	3.6
^{40}Ca	...	0.033	1.0	...	0.028	1.1
^{36}Ar	...	0.021	0.49	...	0.017	0.51
^{32}S	...	0.082	0.41	...	0.067	0.42
^{28}Si	...	0.155	0.46	...	0.130	0.48
^{24}Mg	...	0.023	0.088	...	0.017	0.081
^{20}Ne	...	0.011	0.018	...	0.001	0.002
^{16}O	...	0.140	0.036	...	0.255	0.082
^{12}C	...	0.032	0.018	...	0.166	0.12

** $\langle X_i/^{56}Fe\rangle \equiv (X_i/X(^{56}Fe)) / (X_i/X(^{56}Fe))_\odot$

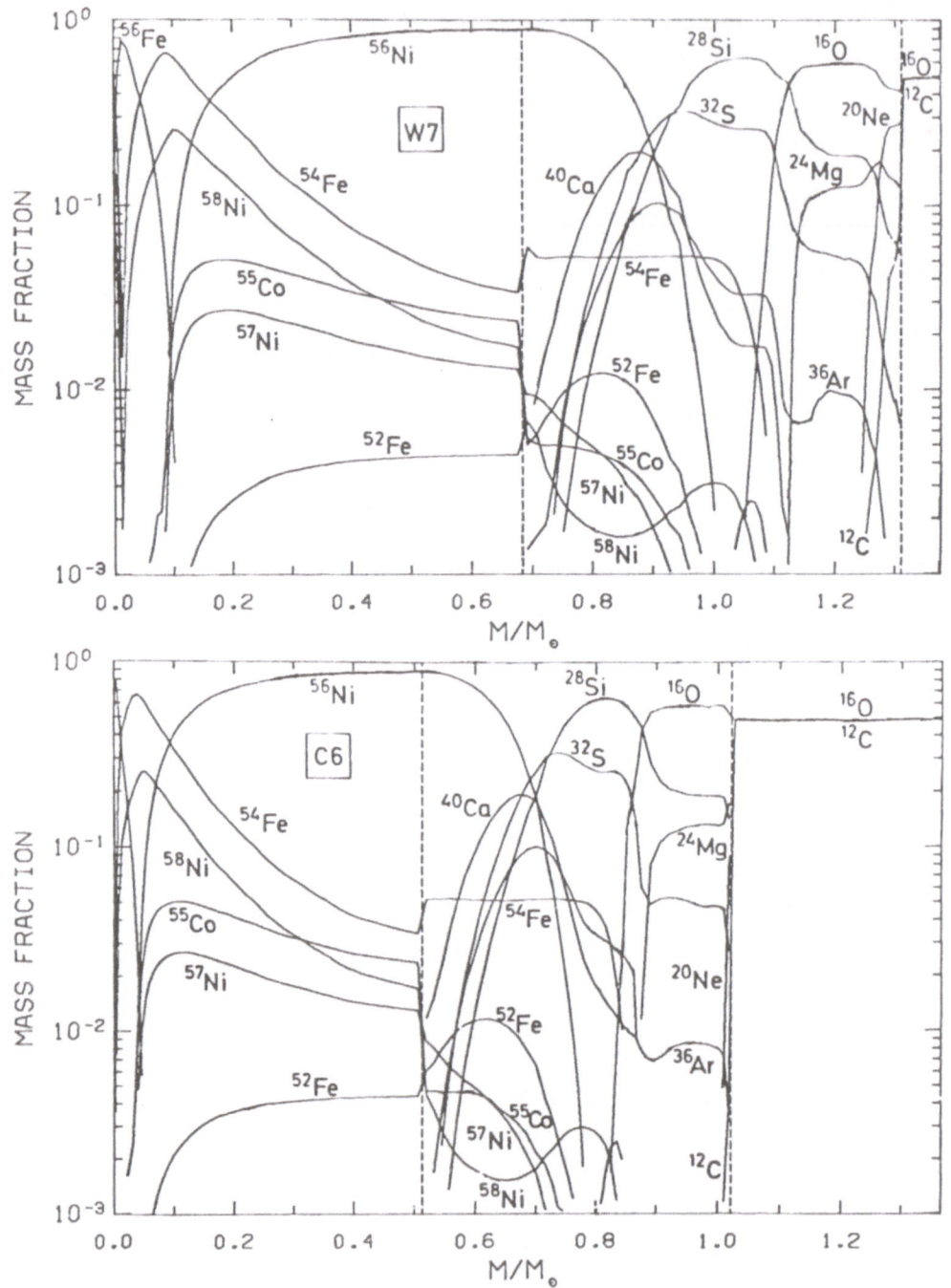

Figures 6 and 7: Composition structures of the deflagrating white dwarfs for cases W7 (Fig. 6) and C6 (Fig. 7). At $M_r \lesssim 0.7\ M_\odot$ (Fig. 6), the WD undergoes incineration into NSE. In the intermediate region at $0.7 < M_r/M_\odot \lesssim 1.3$, the WD undergoes partial explosive burning. In the outer layer ($M_r > 1.3\ M_\odot$), the DFW is quenched.

lower than the initial value at t = 0 as a result of expansion. The peak temperature attained during explosive burning, T_p, is lower for lower density because of the larger heat capacity. Moreover, the temperature and density quickly decrease after the passage of DFW because of the accompanying rarefaction wave and the rapid expansion of the WD. These effects slow down the nuclear reactions so that the material undergoes burning but is not incinerated into NSE composition. The resultant composition structure is shown in Fig. 6.

Near the shell at $M_r \sim 0.7$ $M_\odot$, the reaction network calculation demonstrates that the NSE composition consisting mainly of ^{56}Ni is realized. For the outer layers at $0.7 < M_r/M_\odot < 0.9$, however, T_p is lower, i.e., $6 \gtrsim T_{p,9} > 5$ and the corresponding density is $9 > \rho_{p,7} \gtrsim 4$. The reactions are not rapid enough to process the material into ^{56}Ni; in other words this layer undergoes the partial Si-burning whose products are ^{28}Si, ^{32}S, ^{36}Ar, ^{40}Ca, ^{54}Fe, ^{56}Ni, etc.

For $0.9 \lesssim M_r/M_\odot < 1.1$, $5 > T_{p,9} \gtrsim 4$ and $4 > \rho_{p,7} \gtrsim 2.5$. Explosive carbon and oxygen burning produces ^{28}Si, ^{32}S, and ^{36}Ar, but T_p is too low for Si burning to proceed. Thus this layer contains Si-peak elements.

For $1.1 \lesssim M_r/M_\odot < 1.25$, $4 > T_{p,9} > 3$ so that this layer undergoes explosive carbon and neon burning which produces ^{16}O, ^{24}Mg, ^{28}Si, etc.; oxygen burning is too slow to proceed appreciably. Finally, in the quenching phase of the DFW, only the carbon burning products appear for $1.25 < M_r/M_\odot \lesssim 1.3$. In the outer layer at $M_r > 1.3$ $M_\odot$, the original C+O remains unburnt.

The hydrodynamical calculation was done with an α-network including 14 species (He-Zn). After obtaining the pressure change as a function of time for each shell, i.e., $P(M_r, t)$, detailed nucleosynthesis was calculated by solving the reaction network with 205 species together with the energy conservation equation and $P = P(M_r, t)$ (Thielemann and Nomoto 1984). The energy generation rates and the bulk of the composition based on the large and small networks are consistent. The composition structures in Figs. 6-7 are based on the large network.

2.4 Abundances in the Ejecta

Both the WD and the core are disrupted completely so that all the synthesized materials in Figs. 6-7 are ejected into space. The mass and the ratio of $\langle X_i/^{28}Si \rangle \equiv (X_i/X(^{28}Si))/(X_i/(^{28}Si))_\odot$ for several important species are summarized in Table 3 for both cases W7 and C6. The integrated abundances of the stable isotopes relative to solar values are given in Fig. 8 (W7). There the ratio is normalized to the ^{56}Fe abundance. Case C6 shows an almost identical pattern except for the smaller abundances of ^{54}Cr and ^{58}Fe and the larger ^{16}O abundance. Figure 9 shows the abundances of elements relative to solar, which is normalized to Si (case W7).

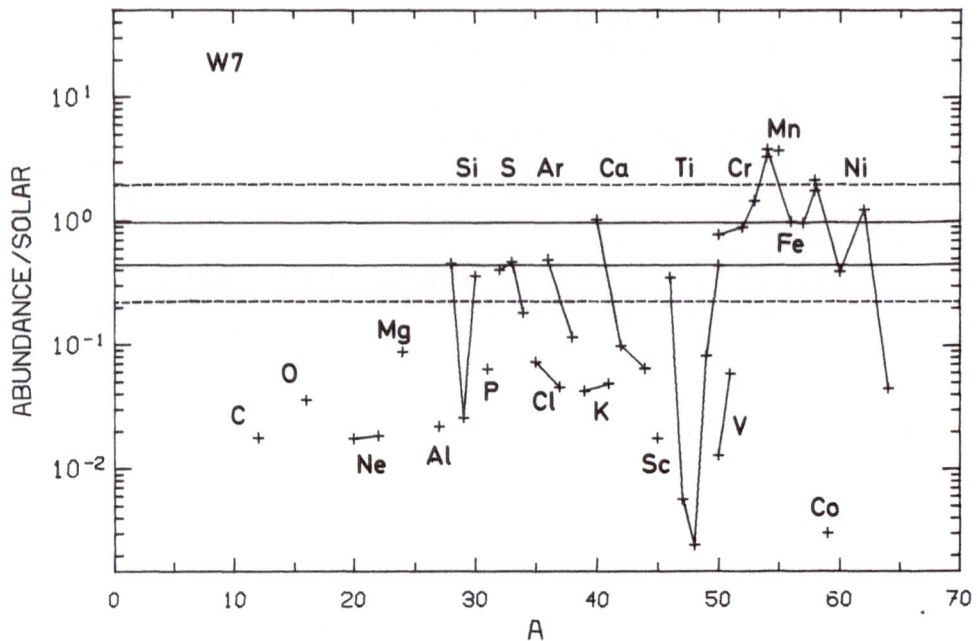

Figure 8: Nucleosynthesis in the carbon deflagration model W7. The
abundances of stable isotopes relative to the solar values are shown.
The ratio is normalized to ^{56}Fe.

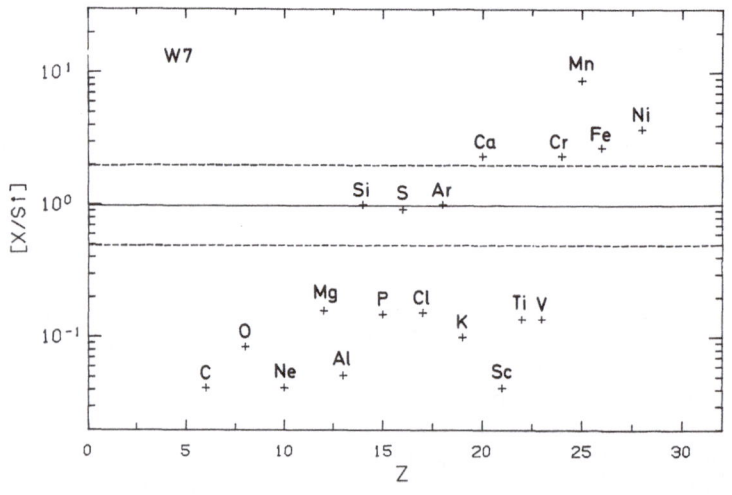

Figure 9:
The abundances of
elements relative
to the solar
values (W7). The
ratio is
normalized to Si.

a) Iron peak elements

The masses of the iron peak elements ejected are 0.86 $M_\odot$ (W7) and 0.66 $M_\odot$ (C6). Among them, ^{56}Ni amounts to 0.5-0.6 $M_\odot$ which is enough to power the light curve of SN I by the decays into Co and Fe.

As seen by the isotope ratios in Figs. 8 and 9, the abundances of the iron peak elements are generally in good agreement with solar values as has been found for the carbon detonation model (Arnett et al. 1971; Bruenn 1971) except for ^{54}Cr, ^{55}Mn, and ^{54}Fe; these species are enhanced relative to ^{56}Fe by a factor of 3-4 (W7). However, their abundances are subject to uncertainties in the model and thus may not be a serious trouble.

The ^{54}Cr abundance is sensitive to $\rho_{c,ig}$; for $\rho_c = 1.5 \times 10^9$ g cm^{-3} (C6), $<^{54}Cr/^{56}Fe> = 0.22$ which is smaller than 3.3 for $\rho_c = 2.6 \times 10^9$ g cm^{-3} (W7) by a factor of 15. This suggests that the ^{54}Cr abundance may give such a constraint as $\rho_{c,ig} < 2.3 \times 10^9$ g cm^{-3} which correspond to $\dot{M} > 10^{-7}$ $M_\odot$ yr^{-1}.

The abundance of ^{55}Mn, which is a decay product of ^{55}Co, is quite sensitive to the assumed value of freezing temperature, T_f, of NSE; if $T_{f,9} = 2$ is assumed, $X(^{55}Co)$ is ~ 5 times smaller than for the case of $T_{f,9} = 3$. In fact, the network calculation for the shell at $M_r \sim 0.7$ $M_\odot$ shows that $X(^{55}Co)$ decreases with decreasing temperature until freezing at around $T_9 \sim 2.2$ and $X(^{55}Co)$ is 3 times smaller than in the adjacent NSE layer where $T_{f,9} = 3$ was assumed (Fig. 6).

The ratio of $^{54}Fe/^{56}Fe$ is not decreased by changing T_f. This ratio depends on the distribution of the neutron excess, $\eta = 1-2Y_e$, in the NSE layer, which, in turn, depends on the ignition density and on the competitive processes of electron captures and the propagation of the DFW: for lower $\rho_{c,ig}$ (higher $\dot{M}$) and faster DFW, η is smaller in the central region. Therefore, the distribution of η or Y_e is sensitive to the treatment of DFW. In view of the uncertainties involved in both the presupernova evolution and the theory of convection, the overabundance of ^{54}Fe may not be so severe. More accurate calculation of the electron capture rates for the NSE composition might also be important, since a factor of 3 change of rates significantly affects $^{54}Fe/^{56}Fe$ ratio (Bruenn 1971; Yokoi and Nomoto 1984).

The most natural way to avoid the excess neutronization is to initiate a carbon deflagration at lower ρ_c than C6, which corresponds to $\dot{M} > 10^{-6}$ $M_\odot$ yr^{-1}. Such a case is expected to occur in the double WD scenario for SN I progenitors (see § 4).

b) Si-Ca

The abundances of Si-Ca are important for comparison with the early time spectra of SN I (Branch 1983). Both models (W7 and C6) eject ^{40}Ca with the abundance ratio of $<^{40}Ca/^{56}Fe> \sim 1$, and Si, S, and Ar with the ratio of $\langle X_i/^{56}Fe \rangle \sim 0.4\text{-}0.5$. Moreover, the relative abundances of Si, S, Ar are remarkably close to the solar values. On the other hand, the abundances of odd Z elements in Si-Ti range, i.e., P, Cl, K, and Sc, are as small as $\langle X_i/^{28}Si \rangle \sim 0.04\text{-}0.14$ probably because of relatively small η assumed in the model.

It is noteworthy that Si-peak elements are synthesized in the layer with relatively high density, i.e., $\rho_{p,7} = 2.5\text{-}4$. This is significantly higher than the $10^5 - 10^6$ g cm^{-3} in the oxygen-rich layer of a 25 $M_\odot$ star where Si-peak elements are produced (Woosley and Weaver 1982) and so the relative production of Si peak elements is enhanced.

c) s- and r-process elements

The carbon deflagration supernova is a possible source of s- and r-process elements. During the phase of mass accretion onto the white dwarf, helium shell flashes recur many times when enough helium is accumulated below the hydrogen-burning shell. As demonstrated by Sugimoto et al. (1978), the helium shell flash develops a convective zone which mixes hydrogen into the helium layer. Subsequent reactions of $^{12}C(p,\gamma)$ $^{13}N(p\ \beta^+)$ $^{13}C(\alpha,n)$ ^{16}O could produce neutrons to synthesize s-process elements. Even without such a mixing of hydrogen, the peak temperature during the helium shell flashes is high enough to produce neutrons via $^{22}Ne(\alpha,n)$ ^{25}Mg reaction for M > 1 $M_\odot$ (Iben 1975; Sugimoto and Nomoto 1975; Iben 1981). Therefore, if hydrogen or helium accretes onto the WD, the outer layer of the WD contains some s-process elements at the time of explosion.

The carbon DFW produces a precursor shock wave as discussed in § 2.2. When the DFW reaches near the WD surface, the precursor shock grows strong because of the steep density gradient. If the temperature at the passage of the shock wave becomes high enough for $^{22}Ne(\alpha,n)^{25}Mg$ reaction to generate neutrons, the r-process could operate on the seed s-process elements (Truran et al. 1978; Thielemann et al. 1979; Cowan et al. 1980). The amount of s- and r-process elements produced by this mechanism deserves further quantitative investigation.

d) Contribution of SN I to Galactic Nucleosynthesis

If we adopt a carbon deflagration supernova model like case W7 as a SN I model, SN I have significant contribution to the galactic nucleosynthesis in the Si-Ni range. Following Tinsley (1980), we may make a crude estimate of the abundance of elements, X_i, ejected from SN I by the yield $X_i \equiv m_i r/s$, where m_i is the mass of species i ejected by each SN I, r is the SN I rate, and s is the net star formation rate. If we adopt values of s = 5 $M_\odot$ pc^{-2} Gyr^{-1} (Tinsley 1980) and r = (0.01-0.03) pc^{-2} Gyr^{-1} (Tammann 1982), X_i = (0.002-0.006) $(m_i/M_\odot)$. Model W7 gives X(Fe) = 0.0017-0.0051 which is marginally consistent with the solar abundance if we adopt the lower limit for the SN I rate. This implies that most of the iron peak elements in the Galaxy may originate from SN I. If the SN I rate is higher, a part of the Fe from SN I may form grains or escape from the Galaxy.

Assuming $r = 0.01$ pc^{-2} Gyr^{-1} which corresponds to 10^{-2} yr^{-1} for the Galaxy (Tammann 1982), model W7 gives $X(^{56}Fe)/\odot \sim 0.91$, $X(^{40}Ca)/\odot \sim 0.94$, $X(^{36}Ar)/\odot \sim 0.45$, $X(^{32}S)/\odot \sim 0.37$, $X(^{28}Si)/\odot \sim 0.42$, $X(^{24}Mg)/\odot \sim 0.080$, $X(^{16}O)/\odot \sim 0.033$, and $X(^{12}C)/\odot \sim 0.017$, respectively. Therefore SN I could also produce a significant fraction of ^{40}Ca, and Si-peak elements in the Galaxy.

It has been pointed out that Si-Ca are underproduced relative to ^{16}O in the explosion of a 25 $M_\odot$ star, which may be a typical site of massive star nucleosynthesis (Woosley and Weaver 1982). Woosley and Weaver (1982) proposed that elements in the range of S-Ti could be produced by stars more massive than 25 $M_\odot$ or very massive population III stars. Arnett and Thielemann (1984) show that a part of these problems are removed with an enhanced $^{12}C(\alpha,\gamma)$ ^{16}O rate in He-burning. Here I would like to make an alternative proposal that at least a significant fraction of Si-Ca originate from SN I. Nucleosynthesis in the deflagration model for SN I is quite complementary to the SN II in the sense that products of SN I tend to fill the gap of Si-Ni region in SN II products and also with respect to the isotopic abundance distribution within one elements (Thielemann and Nomoto 1984).

e) γ-radioactivities

Besides ^{56}Ni and ^{56}Co, SN I produces some γ-radioactivities, i.e., $^{22}Na(2\times10^{-7}$ $M_\odot)$, $^{26}Al(4\times10^{-6}$ $M_\odot)$, ^{44}Ti $(5\times10^{-5}$ $M_\odot)$, $^{48}V(2 \times 10^{-7}$ $M_\odot)$, $^{60}Fe(1 \times 10^{-6}$ $M_\odot)$, and $^{57}Co(2\times10^{-3}$ $M_\odot)$ are ejected from the model W7.

2.5 Comparison with Observations

As briefly summarized in §1, the late time spectra show some evidence for the existence of Fe and probably Co in SN I. Quantitatively, the radioactive model for light curves requires 0.2-1 $M_\odot$ ^{56}Ni to power the maximum light. The carbon deflagration supernova models presented here eject 0.5-0.7 $M_\odot$ ^{56}Ni which is consistent with this requirement. Moreover, theoretical light curves for this type of models show a very good fit to the observations (Chevalier 1981; Arnett 1982; Sutherland and Wheeler 1984).

The early time spectra show the existence of O, Mg, Si, Ca, and possibly Co in the outer layer which is expanding at $\sim 10^4$ km s^{-1} (Branch 1983). The observations suggest that these elements are not stratified but somewhat mixed. In the present models, although a mixing process after the passage of the DFW was not included in the calculation, convective mixing in the outer layers is expected to occur because of the decaying nature of the DFW: The incinerated NSE region is almost isothermal and thus convectively stable. On the other hand, material in the outer layers at $M_r > 0.8$ $M_\odot$ is processed by the decaying DFW and there the nuclear energy release gets smaller as the DFW decays. Therefore, such layers would be convectively unstable because entropy

decreases outward. The extent of mixing needs more investigation, but
some ^{56}Ni could be brought to the surface layer and mixed with O-Si-Ca as
shown in the model by Nomoto (1980).

As seen from the almost final velocity profile at stage 9 in Fig. 5,
v_{exp} = (1-1.4) x 10^4 km s^{-1} in the layers at M_r = 0.9-1.2 $M_\odot$. Therefore,
if the mixing takes place very efficiently in the outer layers, a mixture
of O, Mg, Si, Ca, Ni, and Co could be observed at $v_{exp} \sim 10^4$ km s^{-1},
which is consistent with observations. Branch et al. (1982) estimated
that abundance ratios of O-Ca are roughly consistent with the solar
values under the LTE assumption. If we assume complete mixing in the
outer layer at M_r > 0.7 $M_\odot$ for models W7 and C6, abundance ratios among
Si-peak elements are close to the solar values, but O and Mg abundances
are smaller than the solar value by a factor of $\sim$ 5 relative to Si. The
latter ratios may be within the uncertainties in the observed values
(Branch et al. 1982).

The present model is also consistent with the IUE observations of
the remnant of SN 1006 and X-ray observations of SN I remnants which
suggest the existence of some Si-peak elements in the outer layers of the
ejecta.

3. HELIUM DETONATION SUPERNOVAE

The carbon deflagration supernovae occur when the accretion rate of
hydrogen-rich or helium-rich matter onto the WD is as high as $\dot{M}$ > 4x10^{-8}
$M_\odot$ yr^{-1} or when there is a direct transfer of C+O from the companion
star. If the accretion rate is lower, a different type of thermonuclear
runaway triggers an explosion, namely, a helium detonation.

Such a dependence of the explosion mechanism on the accretion rate
has been revealed by extensive numerical calculations: i,e., the
accretion of helium onto He WDs (Nomoto and Sugimoto 1977) and C+O WDs
(Fujimoto and Sugimoto 1982; Taam 1980; Nomoto 1982a). These
calculations have covered a parameter range of $\dot{M}$ = 10^{-7} - 10^{-10} $M_\odot$yr^{-1}
and M_0 = 0.4 - 1.4 $M_\odot$ where M_0 denotes the initial mass of the WD.

Here we discuss the reason why the rate of accretion is crucial to
determine the fate of WDs, in other words, the physics of accretion. The
accretion onto the WD is specified by a set of parameters ($\dot{M}$, M_0, L_0)
where L_0 denotes the initial luminosity of the WD. As will be discussed
in §3.1, the accretion sets up the conditions at nuclear ignition,
namely, the WD mass, M, (or ρ_c) for central ignition or another set of
parameters (M, ΔM) for the off-center ignition where ΔM is the mass above
the burning shell. The development of the flash and the hydrodynamical
outcome depend only on (M, ΔM) or M almost independently of the history
of accretion (§3.2). The variation of the composition (He or C+O) of
the accreted matter and of the WD are also important factor to yield a
variety of explosion. (§§ 3.3-3.4). (See Sugimoto and Nomoto 1980 for a
review).

3.1 Thermal Effects of Accretion

As matter is accreted by the WD, the interior material contracts owing to the increase in the weight of the overlying matter and then gravitational energy is released; in other words, the material is compressed causing an increase in the temperature. The compression rate due to accretion, $\lambda_\rho \equiv (\text{dln } \rho/dt)_m$ is not uniform in the WD and is conveniently expressed as $\lambda_\rho = \lambda_\rho(M) + \lambda_\rho(q) \equiv (\partial\text{ln } \rho/\partial\text{ln } M)_q(\text{dln } M/dt) - (\partial\text{ln } \rho/\partial\text{ln } q)_M(\text{dln } M/dt)$, where the mass fraction, $q \equiv m/M$, for a given Lagrangian shell of m decreases as M increases (Nomoto 1982a). The term $\lambda_\rho(M)$ is due to the increase in ρ_c as a result of mass increase and is almost spatially uniform since the change in the ρ-distribution against q is quite homologous. In contrast $\lambda_\rho(q)$, which shows the compression as the matter moves inward in q-space, is much larger than $\lambda_\rho(M)$ near the surface because of the steep density gradient, but is negligible for q < 0.8 (see Fig. 2 of Nomoto 1982a). If the compression is adiabatic, therefore, the temperature of the accreted matter increases first near the surface.

However, the entropy of the accreted matter is lost by radiative heat transport which makes the WD interior almost isothermal on a time scale of τ_h. Typically $\tau_h \sim 10^7-10^8$ yr in the interior but it is much shorter near the surface.

These two competitive effects of compression and radiative heat transport determine the temperature distribution in the WD. In view of such a competition, the accretion is classified into three types according to $\dot{M}$ (Sugimoto et al. 1979).

1) Hot Accretion for $\dot{M} \gtrsim \dot{M}_{RG}$: The accreted matter can not be compressed into the interior but simply piles up to form a red-giant-size envelope. This is because $\lambda_\rho^{-1} < \tau_h$ near the surface so that the matter cannot lose its entropy (e.g., Nomoto et al. 1979). If the stable nuclear burning shell exists near the bottom of the accreted envelope, $\dot{M}_{RG}$ is given by the rate at which the shell burning processes the accreted material into heavier elements.

2) Warm Accretion for $\dot{M}_{RG} > \dot{M} \gtrsim \dot{M}_{warm}$: Here $\lambda_\rho^{-1} < \tau_h$ near the bottom of the accreted matter so that an appreciable amount of original entropy is retained within the accreted matter. As a result the temperature peak within the WD appears in the middle part of the accreted layer while the deep interior remains cool; i.e., a temperature inversion appears.

3) Cold Accretion for $\dot{M}_{warm} > \dot{M}$: τ_h is short enough for the WD to reach a thermally relaxed, i.e., almost isothermal state. For low $\dot{M}$, the temperature in the WD can even decrease. In the long run, the temperature reaches a steady value which is determined by the balance between the two effects and thus the temperature is higher for higher M.

The increase in ρ and T in the accreting WD eventually leads to ignition of nuclear burning. The ignition location (center or off-

center) depends on the distributions of composition and temperature which
are set by accretion. In this way, M (or ρ_c) or (M, ΔM) at the ignition
is determined from ($\dot{M}$, M_0, L_0); among the latter parameters, M_0 and L_0
are less important as discussed in §2.1. Examples of such a
transformation from ($\dot{M}$, M_0) to (M, ΔM) are given in Fig. 1 of Nariai and
Nomoto (1979). In general, both M and ΔM at the ignition are smaller for
higher $\dot{M}$ because the ignition density is lower for higher temperature.
In many cases, the nuclear burning ignited in this way is unstable to a
flash because of the strong electron-degeneracy and/or the plane-like
geometry if the burning shell lies near the surface.

3.2 Development of a Flash and Initiation of a Detonation Wave

Development of the nuclear flash depends on how the pressure at the
burning shell responds to the energy input. For example, if the pressure
is constant, the temperature increases up to a limiting value of T =
$(3P/a)^{1/4}$ = 3.0 x 10^9 (P / 2 x 10^{23} dyn $cm^{-2})^{1/4}$ K at vanishing density
(Sugimoto and Nomoto 1975). If the pressure drops quickly, the resultant
decrease in temperature quenches the nuclear burning. On the other hand,
if the expansion is slow, the temperature can reach T_d above which the
flash grows into the formation of a deflagration/detonation wave. Such a
response of the pressure is described by gravothermal specific heat which
is a combination of usual thermodynamical specific heat and a term due to
hydrostatic readjustment of the star upon entropy input (Sugimoto et al.
1981; Sugimoto and Miyaji 1981).

Assuming hydrostatic equilibrium and an adiabatic temperature
gradient in the convective layer, the peak temperature, T_{peak}, during the
flash can be obtained as a function of M for the central flash (Sugimoto
and Nomoto 1980, p.184) and a function of (M, ΔM) for the off-center
flash (Sugimoto et al. 1979; Fujimoto and Sugimoto 1982). Then critical
values of M or (M, ΔM) are determined above which T_{peak} exceeds T_d. The
central helium (carbon) flash for M > 0.65 $M_\odot$ (1.3 $M_\odot$) leads to the
dynamical event. The He-shell flash grows into runaway if M and/or ΔM
are large enough to yield ρ > 2 x 10^6 at the shell. Once thermonuclear
runaway starts, explosive nucleosynthesis proceeds. If the temperature
exeeds T_{NSE} $\sim$ 3 x 10^9, material is incinerated into NSE composition
(Truran et al. 1967). From the energy conservation equation, T_9 $\sim$ 3
corresponds to ρ_{NSE} $\sim$ 2 x 10^6 g cm^{-3} for helium and $\sim$ 1 x 10^7 g cm^{-3} for
C+O; for lower density, the heat capacity of the radiation field is too
large for the temperature to reach T_{NSE}. In most of the actual
situation, the burning layer expands so that the density and temperature
drops rapidly. Therefore a higher density is required for the iron peak
elements to be synthesized (see the example of carbon deflagration).

The next question is whether the thermonuclear runaway results in
the formation of a detonation wave that is a burning front ignited by a
precursor shock and propagating at a supersonic velocity. An important
quantity to determine the hydrodynamics is q/u_0 (where q is the nuclear
energy release by incineration and u_0 is the initial internal energy of
the matter) since an overpressure $\Delta P/P_0$ produced by the incineration is

roughly proportional to q/u_0. As seen in Fig. 2 of Nomoto (1982b), q/u_0 is larger for lower ρ; for example $q/u_0 = 6$ and 1.2 for $\rho = 10^7$ and 2×10^9 g cm^{-3}, respsctively, for carbon incineration. For helium incineration, q is larger than for carbon by a factor of 1.7. Moreover, $T_d \sim 3 \times 10^8$ K for helium while $T_d \sim 1.5 \times 10^9$ K for carbon at $\rho = 10^7$ g cm^{-3}. Therefore, the shock wave ahead of the burning front is stronger and thus the formation of a detonation wave is easier for lower density, and, in particular, for helium.

Other important factors which control the initiation and propagation of the detonation/deflagration wave are the pressure gradient and the spherical geometry in the central region (Ōno 1960). The pressure gradient strengthens the outgoing shock wave while the sphericity damps it. For an inward shock wave, the effects are the opposite. We need a detailed computation to see if the detonation wave is formed.

According to the shock tube analysis by Mazurek et al. (1977), a carbon detonation is unlikely to be formed for $\rho > 5 \times 10^7$. Therefore, the carbon flash in C+O WDs does not form a detonation but forms a subsonic carbon deflagration wave. On the other hand, a helium detonation wave may always be formed by helium flash in a WD interior.

3.3 Accretion of Helium onto Helium White Dwarfs

Accretion of helium includes two cases of accretion. One is a case where the H-rich matter accretes to ignite H-burning near the surface of the WD. A steady burning or a relatively weak flash gradually builds up a He layer. The other case is a direct transfer of helium from the companion star if it has evolved to a He WD or a relatively low mass He star (see §4).

For the former case, $\dot{M}$ is governed by the H-shell burning so that $\dot{M} < \dot{M}_{RG}$. For the latter case of direct accretion, $\dot{M}$ can exceed $\dot{M}_{RG}$. The hot accretion of helium will fôrm an extended He envelope as demonstrated for the H-accretion (Nomoto et al. 1979). The warm and cold accretions have been explored by Nomoto and Sugimoto (1977) for several cases of $\dot{M}$ starting from an initial mass of 0.4 $M_\odot$. They obtained $\dot{M}_{warm} \sim 3 \times 10^{-8}$ $M_\odot \text{yr}^{-1}$.

a) <u>Warm Accretion leading to a Weak Shell Flash or a Double Detonation Supernova</u>: For $\dot{M} = 4 \times 10^{-8}$ $M_\odot \text{yr}^{-1}$, a temperature inversion in the accreted matter persists until the He-flash is ignited at the outer shell of $q = 0.86$ when $M = 0.66$ $M_\odot$. The ignition density of $\rho = 3 \times 10^5$ g cm^{-3} is too low to induce a dynamical event. Subsequent heat transport from the burning shell and the continuing accretion would ignite helium at the center. If M has exceeded 0.65 $M_\odot$ at the central ignition, the flash grows into a supernova explosion.

For $\dot{M}$ slightly lower than 4×10^{-8} $M_\odot \text{yr}^{-1}$, the density at the ignited shell may exceed 2×10^6 g cm^{-3}. Then the shell flash would produce double detonation waves of helium which propagate both outward

and inward. As will be shown in §3.4, the double detonation waves burn
the material mostly into NSE and disrupt the star completely.

b) Cold Accretion leading to a He Detonation Supernova: The He WD is
well thermally relaxed due to heat diffusion and so undergoes helium
ignition at the center. Ignition density depends only slightly on the
initial condition but it depends mainly on $\dot{M}$, because the evolutionary
path of (ρ_c, T_c) near the ignition is determined by the balance between
compression and the neutrino loss (see §2.1).

     As discussed in §3.1, slower accretion results in the ignition at
higher density (larger mass): Numerical example shows that the accretion
with $\dot{M}$ = (2-1) x 10^{-8} $M_\odot$ r^{-1} correspond to the ignition at M = (0.78 -
0.99) $M_\odot$ and ρ_c = 7 x 10^6 - 2 x 10^7 g cm^{-3}. Even for the extremely slow
accretion and thus low temperature, however, the ignition condition is
limited to ρ_c < 2.5 x 10^8 and M < 1.25 $M_\odot$ because of the nonresonant
nature of the 3α reaction at low temperature (Cameron 1959; Nomoto 1982a)
and the effect of strong screening (Ichimaru and Utsumi 1983b).

     For these cases of cold accretion, the central flash grows to form a
helium detonation wave because M > 0.65 $M_\odot$. As has been actually
demonstrated for M = 0.78 $M_\odot$ (Nomoto and Sugimoto 1977), the detonation
wave incinerates most of the WD material into NSE composition and the WD
is disrupted completely.

3.4 Accretion of Helium onto C+O White Dwarfs

     In this case, the physics of accretion is the same as of accretion
onto He WDs. However, the difference in the composition of the WD and
the accreted matter plays an important role.

     The critical rate of $\dot{M}_{RG}$ for the C+O WD of mass, $\dot{M}_{CO}$, is given by
the growth rate of the C+O core in He stars (Uus 1971), namely, $\dot{M}_{RG}$ = 7.2
x 10^{-6} ($M_{CO}/M_\odot$-0.6) $M_\odot yr^{-1}$. The hot accretion with $\dot{M} \gtrsim \dot{M}_{RG}$ forms an
extended helium envelope. Subsequent evolution with a common helium
envelope depends on the mass-donating companion star: If the companion
is a He WD, the system could become a R CrB-like He star (Webbink 1984;
Iben and Tutukov 1984) which may or may not become an SN I depending on
the total mass of the system. If the companion is a He star having a C+O
core, double C+O WDs would be left after the formation of a planetary
nebula.
     Another critical rate is $\dot{M}_{warm} \sim$ 3 x 10^{-8} $M_\odot yr^{-1}$. For the warm
accretion, the helium shell flash is ignited in the middle layer of the
accreted envelope because of a temperature inversion. For the cold
accretion, the He shell flash is ignited at the bottom of the He envelope
if M_{CO} is small enough. This is in contrast to the accretion onto He WDs
where the cold accretion leads always to the central ignition.
Accordingly the outcome of accretion depends not only on $\dot{M}$ but also on
M_{CO}. Thus it is more useful to classify the result in the ($\dot{M}$, M_{CO}) plane
as given in Fig. 10. This figure is an updated version of Fig. 8 of
Nomoto (1982a) by taking into account the new strong screening (Ichimaru

and Utsumi 1983b). Here the ignition densities of helium and carbon are 2.5×10^8 and 3.5×10^9 g cm^{-3}, respectively. For carbon burning at T < 3×10^7 K, the ignition density is extraporated from the strong screening regime to pycnonuclear regime (S. Ichimaru, private communication).

a) <u>Weak He Shell Flashes leading to a Carbon Deflagration</u> ($\dot{M} \gtrsim 4 \times 10^{-8}$ M$_\odot$yr^{-1}): The ignition density is lower than $\rho_d \sim 2 \times 10^6$ g cm^{-3}. The flash is mild and does not form a detonation wave (Taam 1980; Fujimoto and Sugimoto 1982). During the continuing accretion, such mild flashes recur many times to increase the mass of the C+O WD. Eventually, the carbon burning is ignited at the center and grows into a carbon deflagration supernova.

b) <u>Double Detonation Supernovae</u>: For $4 \times 10^{-8} > \dot{M}(M_\odot yr^{-1}) \gtrsim 10^{-9}$, the ignition density is in the range of $(2 \times 10^6 - 10^8)$ g cm^{-3}. The helium shell flash grows into the thermonuclear runaway which incinerates the material into NSE. The resultant large overpressure forms double shock waves which propagate both outward and inward. The outgoing shock wave is strong enough to initiate a helium detonation. The strength of the inward shock wave that propagates into the C+O core is also large enough to form a carbon detonation wave (Nomoto 1980, 1982b; Woosley et al. 1980). This is the case only for $\rho < 10^8$ g cm^{-3}.

In short, the helium flash for this range of $\dot{M}$ forms the double detonation waves, namely, a helium detonation wave that propagates outward and a carbon detonation wave that propagates inward. The double detonation waves incinerate most of the WD material into NSE. The released nuclear energy is large enough to disrupt the star completely and no compact star is left.

c) <u>Single Detonation Supernovae</u>: For the case with slower accretion rate ($\dot{M} < 1 \times 10^{-9}$ M$_\odot$yr^{-1}), the outcome depends on both $\dot{M}$ and M_{CO} as seen in Fig. 10. Since the temperature of the WD is so low as T < 4×10^7 K, the material at the bottom of the helium layer does not undergo the helium ignition until the density there exceeds $\sim 1 \times 10^8$ g cm^{-3}. As a result, in the low $\dot{M}$ and large M_{CO} domain in Fig. 10 (roughly $M_{CO} \gtrsim 1.1$ M$_\odot$), carbon is ignited at the center prior to the helium ignition.

For small M_{CO}, helium ignition occurs prior to carbon ignition, which we discuss here first (case C of Nomoto 1982b). As in the case of the double detonation, the helium shell flash forms double shock waves. The outgoing shock wave forms a helium detonation wave; it propagates toward the surface and incinerates most of helium into NSE. On the other hand, the inward shock fails to initiate a carbon detonation; the energy release due to helium incineration is as small as $q/u_0 < 1$ at $\rho > 1 \times 10^8$ g cm^{-3}, while $q/u_0 > 2-3$ is required to initiate a carbon detonation (Mazurek et al. 1977).

The single helium detonation leads to the ejection of most of the envelope material in the form of ^{56}Ni and would give rise to a relatively weak supernova explosion. Two possible cases are resulted from such an

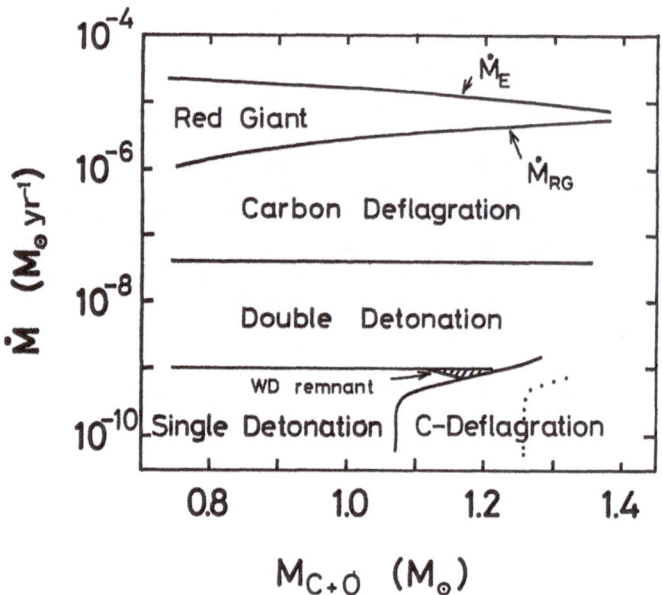

Figure 10: Possible models of SN I in C+O WDs, which depend on the accretion rate of helium, $\dot{M}$, and the initial mass, M_{CO}. $\dot{M}_E$ is the Eddington limit. The hot accretion ($\dot{M} \gtrsim \dot{M}_{RG}$) forms a red giant-like envelope. For $\dot{M} < 10^{-9}$ $M_{\odot}yr^{-1}$, both single detonation and carbon deflagration is possible depending on M_{CO}, where the solid and dotted lines correspond to the carbon ignition density of 3.5×10^9 and 1×10^{10} g cm^{-3}, respectively. The single detonation results in either the total disruption or the explosion leaving a WD remnant (for shaded region).

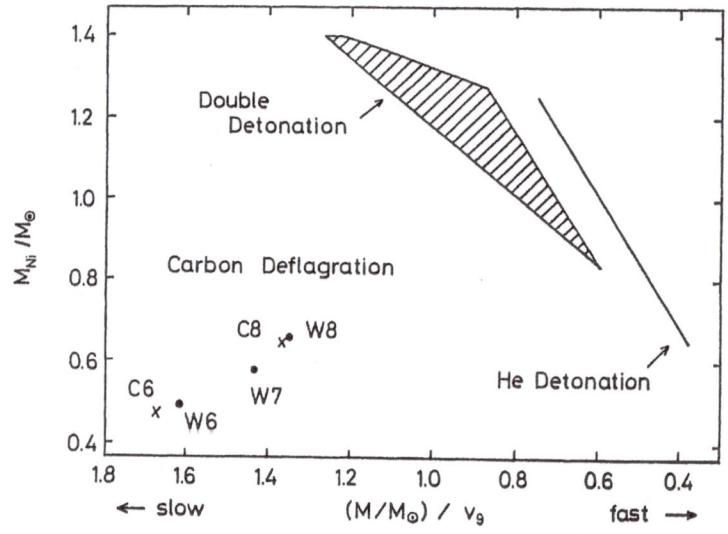

Figure 11: The brightness ($\propto M_{Ni}$ and the speed class (indicated by M/v) of the light curves due to radioactive decays. Possible ranges for various types of explosion are indicated, i.e., He detonation in He WDs, double detonations in C+O WDs, and several carbon deflagration models.
$v_9 \equiv v/(10^9 cm\ s^{-1})$.

explosion, namely, formation of a white dwarf remnant or total disruption.

White Dwarf Remnant: If $M_{CO}/M \gtrsim 0.8$-0.9 and $M \gtrsim 1.35$ $M_\odot$ at the helium ignition, the total energy of the C+O core is negative so that a white dwarf remnant is left behind the supernova explosion (case C of Nomoto 1982b). The corresponding region of the white dwarf remnant in ($\dot{M}$, M_{CO}) plane is shown in Fig. 10. Since the ejected mass is rather small, the binary system would not be disrupted. Then the WD remnant would continue to accrete matter from the companion star and so recurrent supernova explosions are possible.

Total Disruption: For most cases of single detonation, the C+O core is not bound (e.g., case E of Nomoto 1982b). Thus the white dwarf is disrupted completely. After the detonated envelope (^{56}Ni) is ejected with a velocity in excess of $\sim 10^4$ km s^{-1}, C+O material expands slowly with $v \sim 10^3$ km s^{-1} (Sutherland and Wheeler 1984). The amounts of ejected ^{56}Ni and unburnt C+O are in the range of 0.2 - 0.8 $M_\odot$ and 1.2 - 0.56 $M_\odot$, respectively (Nomoto 1982b).

d) Carbon Deflagration for $\dot{M} < 10^{-9}$ $M_\odot yr^{-1}$: In the region with low $\dot{M}$ and large M_{CO} in Fig. 10, the mass of the He envelope is too small for the density to reach 2.5×10^8 g cm^{-3} at the bottom when carbon is ignited at the center ($\rho_c \sim 3.5 \times 10^9$ g cm^{-3}).

If the central temperature is as low as $T_c < 3 \times 10^7$ K, the central region of the WD is in solid state. The carbon deflagration wave would propagate on the timescale to melt the overlying solid by penetrating convection (Mochkovitch 1980) or on the conduction timescale if the convection cannot melt the solid (Canal and Isern 1979). The final outcome may not be much different from the model W7 in § 2, unless the ignition density is so high as $\sim 10^{10}$ g cm^{-3}. (See Isern et al. 1983 for possible effects of carbon-oxygen separation.)

3.5 Comparison with Observations

a) Light Curves

Helium detonation supernovae of various types eject a large amount of ^{56}Ni. Decays of ^{56}Ni and ^{56}Co into ^{56}Fe can power the supernova light curve. Numerical calculations for the detonated He WD and the double detonation C+O WDs (Chevalier 1981; Weaver et al. 1980) show good fits to the observed light curves of SN I within the uncertainties involved in the opacity. The single detonation model does not seem to show a good fit (Sutherland and Wheeler 1984).

For the detonation-type models, there are variations among the ejected mass of ^{56}Ni, M_{Ni}, and explosion energy, E. The possible range of M_{Ni} are 1) $(0.56$-$1.25)$ $M_\odot$ for the He WDs, 2) $(0.8$-$1.4)$ $M_\odot$ for the double detonation models in C+O WDs, 3) $(0.2$-$0.8)M_\odot$ for the single detonation with an unbound C+O core, 4) $(0.08$-$0.24)$ $M_\odot$ for the single detonation learving a C+O WD remnant.

In order to see the possible variation of the light curves among above models, it is useful to employ the analytic solution for light curves based on the ^{56}Ni decays (Arnett 1982). According to it, the post-maximum decline is slower (faster) for the longer (shorter) effective diffusion time, $\tau_m \equiv (\kappa M/v)^{1/2}$, where M is the total ejected mass, κ is the opacity of an expanding matter, and $v \propto (E/M)^{1/2}$ is the velocity scale. The maximum luminosity is approximately proportional to M_{Ni}.

Figure 11 of a (M_{Ni} - M/v) plane shows a possible range of these quantities for the models of He-detonation of He WDs and the double detonation of C+O WDs. It is noteworthy that a model with a larger amount of ^{56}Ni has a larger M/v (or longer τ_m if κ is the same); in other words, brighter supernovae show slower decline. This tedency is easily interpreted as follows: For the completely incinerated models of He WDs, $M_{Ni} \simeq M$ and $E \propto M$. Thus v is constant and $\tau_m \propto M^{1/2}$ for the constant κ. Therefore more massive He WD has a larger M_{Ni} and longer τ_m. For the double detonation C+O WDs, a model containing a larger fraction of He has a larger E (and v) and thus shorter τ_m. The slower-brighter tendency seems to be consistent with the observed tendency as suggested by Pskovskii (1977) and Branch (1981). On the other hand, the light curves of the carbon deflagration supernovae are quite uniform. If we compare the models with the same $\alpha \equiv \ell/H_p$ but different $\dot{M}$, i.e., C6-W6 and C8-W8, their locations in the (M_{Ni} - M/v) plane are very close with each other. This is due mainly to the fact that the ignition density is not sensitive to $\dot{M}$.

If both types of thermonuclear explosion, i.e., the helium detonation and the carbon deflagration occur as SN I in nature, the deflagration-type is slower and dimmer than the detonation-type. Such a combination gives a tendency just opposite to the Pskovskii-Branch relation. However, whether SN I show such a tendency or rather uniform is controversial (Branch 1984).

b) Chemical Composition

The chemical composition in the ejected matter of the detonation-type supernova is almost exclusinvely iron. The He detonation fails to incinerate the material into NSE and produces O- and Si-peak elements only for $\rho < 2 \times 10^6$ g cm^{-3} (Woosley et al. 1980; Hashimoto et al. 1983). Such a low density layers exist only near the surface so that the mass of intermediate mass elements is as small as 10^{-3} - 10^{-4} $M_\odot$. Moreover, the surface layer has a steep density gradient which accelerates the shock wave. As a result, the expansion velocity of the material after the passage of the shock wave is higher than 2×10^4 km s^{-1}. This is in conflict with the SN I spectrum near maximum light which indicates the existence of intermediate mass elements at $v_{exp} \sim 10^4$ km s^{-1} (Branch et al. 1982). Also the detonation-type model may be inconsistent with the X-ray observations of SN I remnants (Tycho, Kepler, SN 1006) which suggest the enrichment of Si-peak elements in the shocked layer.

4. EVOLUTION OF SN I PROGENITORS

Hydrodynamical behavior of the exploding WDs and associated nucleosynthesis have shown that the carbon deflagration model can naturally account for the basic observational features of SN I such as light curves and early time spectra. On the other hand, detonation-type models (the helium detonation in He WDs, and the single and double detonation in C+O WDs) cannot account for the existence of outer layers containing intermediate mass elements (O-Si-Ca).

Suppose that most of SN I are the carbon deflagration supernovae in the accreting WDs. Then the question is why the carbon deflagration occurs so frequently in nature compared with the detonation-type supernova (Branch 1983). Possible answers are as follows:

1) For C+O WDs, the carbon deflagration is simply more frequent than the double detonations because the deflagration occurs for more rapid accretion (see Fig. 10).

2) The occurence of a detonation-type explosion is rare because hydrogen shell flashes for the slow accretion are so strong that most of the accreted matter is lost from the WD and cannot build up a helium layer with sufficient amount of mass.

3) Most of SN I originate from the double WD system (Iben and Tutukov 1984; Webbink 1984). In this case, accretion rate is so high that the double detonation supernova corresponding to slow $\dot{M}$ is a rare event. Only the C+O WD pair could become SN I and the He-WD pair may not explode for some reason, e.g., too small a total mass.

4.1 Hydrogen Shell Burning in the Accreting White Dwarfs

First let us discuss the case where the hydrogen-rich matter accretes onto the WD. Possible mass donating stars are the Roche lobe filling main-sequence stars, subgiant stars, and red-giant stars. Even the star which does not fill its Roche lobe can donate matter as a stellar wind (Iben and Tutukov 1984). The mass accretion rates from these stars cover a rather wide range. When a certain amount of mass, ΔM_H, is accumulated, a hydrogen shell flash is ignited. Whether the flash leads to ejection of most of the accreted matter or produces a substantial helium layer depends on $\dot{M}$ as follows (Sugimoto and Miyaji 1981; Nomoto 1982a; Fujimoto and Taam 1982).

The hot accretion with $\dot{M} \gtrsim \dot{M}_{RG}$ forms a red-giant-size envelope which becomes a common envelope. Here $\dot{M}_{RG} = 8.5 \times 10^{-7}$ (M/M$_\odot$ - 0.52) M$_\odot$yr^{-1}. Subsequent evolution could form a planetary nebula and leave double WDs which is also a candidated of SN I progenitor.

For $\dot{M}_{RG} > \dot{M} \gtrsim 0.4 \dot{M}_{RG}$, the hydrogen shell burning is stable and steady. Thus the accreted H-rich matter could be processed into helium at the same rate as of accretion. However, such a WD is as bright as AGB stars so that the substantial mass could be lost from the WD as a stellar wind. Therefore the net growth rate of the WD mass could be somewhat smaller than $\dot{M}$.

For $\dot{M} < 0.4 \, \dot{M}_{RG}$, the H-shell burning is unstable to a flash. The flash causes an expansion of the accreted envelope. A part of accreted matter is lost from the system during the phase with large radius and its amount depends on $(M, \Delta M_H)$. For larger M and lower $\dot{M}$ (i.e., larger ΔM_H), the expansion is more rapid and the envelope reaches larger radius because of larger and faster nuclear energy release (e.g., Nariai et al. 1980; Iben 1982). Therefore, the WD mass could be almost prevented from increasing for $\dot{M} < 10^{-8} \, M_{\odot} yr^{-1}$, especially for the case of nova-like explosion. On the other hand, a large portion of ΔM_H can be processed into helium for the accretion as rapid as $\dot{M} \sim 2 \times 10^{-7} - 10^{-8} \, M_{\odot} yr^{-1}$.

Because of such a dependence of the strength of the flash on $\dot{M}$, the accreting WD could evolve into SN I only for $\dot{M} >$ several $\times 10^{-8} \, M_{\odot} yr^{-1}$, which correspond to the region of carbon deflagration in Fig. 10. According to Webbink et al. (1983), the subgiant star with a radius larger than $\sim 50 \, R_{\odot}$ can supply mass at such high rate for sufficiently long time. The WDs undergoing rapid accretion might be observed as some type of symbiotic stars (Paczyński and Rudak 1980; Kenyon and Truran 1984).

4.2 Double White Dwarf System

The hot accretion of hydrogen or helium-rich matter onto the WD does not lead directly to the supernova but could lead indirectly to SN I through the formation of double WDs (Nomoto 1982a). More general scenario for the formation of the double WDs and their possible evolution to SN I have recently been discussed in detail by Iben and Tutukov (1984) and Webbink (1984). Possible combinations of the WDs are He-He, He-CO, and CO-CO. The ONeMg WDs add more variations.

The double WD system originates from double degenerate cores embeded in a common envelope that is formed by a hot accretion in several kinds of binary system. Thus the separation of the double WDs is already small at their birth and gets smaller as angular momentum is lost from the system through gravitaional radiation. When the smaller mass WD fills its Roche lobe, mass transfer of helium or carbon-oxygen to the more massive WD commences. Since the radius of the mass losing WD gets larger as M decreases, the system could be unstable to a dynamical timescale mass transfer. Even if it is stable, $\dot{M}$ may well exceed the Eddington limit because of a rapid decrease in angular momentum (Webbink 1984). Although angular momentum constraint is important for the accretion in the double WDs, such a rapid accretion will eventually form an extended envelope of small mass igniting a shell flash of either helium or carbon near the surface (Webbink 1984; see also §3.1).

If the He-CO system forms an extended massive He envelope, which is suggested to be the origin of R CrB stars (Webbink 1984), further evolution up to supernova explosion will be similar to the He star of 1.5-2 $M_\odot$ studied by Nomoto (1982c). Interior to such a He star, a degenerate CO core grows to initiate a carbon deflagration. Owing to rather rapid increase in the CO core mass, carbon shell flashes are ignited and recur many times. As a result, outer layers of the core contains a large amount of carbon burning products at the initiation of carbon deflagration. Such a composition structure might be significant for comparison with the SN I observations.

Shell flashes of either helium or carbon for the He + He and CO + CO pairs would also synthesize heavier elements in the outer layers of the WD. Heat diffusion from the burning layer and the mass increase would eventually ignites a flash at the center. Further evolution depends on the total mass of the pair. The flash can grow into runaway if the mass of the accreting WD exceeds 0.65 $M_\odot$ for helium and 1.3 $M_\odot$ for carbon. Thus the carbon deflagration would be a likely outcome while the possibility of helium detonation seems to be marginal. In any case, the double detonation supernova corresponding to low $\dot{M}$ must be a rare event. It is noteworthy that rapid accretion would ignite carbon at the density lower than in models shown in §2 so that excess neutronization can be naturally avoided as mentioned in §2.4. The evolutionary scenario of the double WDs is highly speculative and deserves further study.

5. CONCLUDING REMARKS

Many of the observational features of SN I, in particular, near maximum light, can be accounted for by the carbon deflagration model of accreting WDs. Detailed nucleosynthesis calculation by Nomoto and Thielemann (1984) has shown that the carbon deflagration wave synthesizes ∿ 0.6 $M_\odot$ ^{56}Ni which is enough to power the light curve. In the outer layers of the WD, substantial amount of intermediate mass elements such as Ca, Ar, S, Si, Mg, and O are synthesized. This is consistent with the early time spectra of SN I. Such nucleosynthesis in SN I is complementary to the nuclear products from massive stars.

Theoretical models of accreting WDs predict the occurence of different type of explosion for the slow accretion, namely, detonation supernovae initiated by a helium flash. However, the detonation-type explosion produces almost exclusively iron peak elements, which is inconsistent with the observations. Probably the accreting WDs can increase their masses through the explosion only for relatively rapid accreion leading to a carbon deflagration. This is the case for the mass increase through the hydrogen shell burning and also for the accretion in the double WDs.

The thermonuclear explosion disrupts the WD completely leaving no neutron star behind except for the case with a single detonation leaving a WD remnant. On the other hand, the evolutionarry origin of low mass X-ray binary systems and a binary containing the 6.1 ms pulsar might be the collapse of a certain class of accreting WDs to form neutron stars (e.g., van den Heuvel 1981; Helfand et al. 1983). In this regard, possible effects of carbon and oxygen separation in the crystalizing WD have been investigated by Isern et al. (1983) and Mochkovitch (1983) although such a separation is quite hypothetical (S. Ichimaru, private communication). If the separation occurs, solid oxygen core is formed interior to the C+O layers. Most probable outcome of the accretion onto such a WD is the off-center carbon ignition prior to the onset of electron captures on ^{16}O at the center because of rather low ignition density of carbon ($\sim 3 \times 10^9$ g cm^{-3}). Therefore the outcome could be SN I rather than collapse to form neutron stars.

However, collapse of accreting WDs naturally occurs for O+Ne+Mg WDs because electron captures on ^{24}Mg and ^{20}Ne trigger the collapse prior to the explosion (Nomoto et al. 1979; Miyaji et al. 1980). Such O+Ne+Mg WDs form from 8-10 $M_\odot$ stars in close binary systems. Although the frequency of collapse is several order of magnitude smaller than SN I, it is consistent with the statistics of low mass X-ray binaries and binary pulsars (van den Heuvel 1981; Webbink et al. 1983; Iben and Tutukov 1984).

I would like to thank Drs. F.-K. Thielemann and K. Yokoi for the collaboration on the carbon deflagration model, which will be published elsewhere. I am also indebted to Prof. R. Kippenhahn, Drs. W. Hillebrandt, and E. Muller for stimulating discussion and hospitality during my stay at the Max-Planck-Institut and to Profs. D. Sugimoto and J.C. Wheeler for useful comments and discussion. This work is supported in part by the Japanese Ministry of Education, Science, and Culture through the Research Grant No. 58340023.

REFERNCES

Arnett, W.D. 1969, Ap. Space Sci., 5, 180.
_____. 1979, Astrophys. J. (Letters), 230, L37.
_____. 1982, Astrophys. J., 253, 785.
Arnett, W.D., and Thielemann, F.-K. 1984, this volume.
Arnett, W.D., Truran, J.W., and Woosley, S.E. 1971, Astrophys. J., 165, 87.
Axelrod, T.S. 1980a, in Type I Supenovae, ed. J.C. Wheeler (Austin: University of Texas), p.80.
_____. 1980b, Ph.D. Thesis, University of California at Santa Cruz.
Becker, R.H., Holt, S.S., Smith, B.W., White, N.E., Boldt, E.A., Mushotzky, R.F., and Serlemitos, P.J. 1980, Astrophys. J. (Letters), 235, L5.
Branch, D. 1981, Astrophys. J., 248, 1076.
_____. 1983, Ann. N.Y. Acad. Sci., Review paper presented at the XI Texas Symposium on Relativistic Astrophysics.
_____. 1984, this volume.
Branch, D., Buta, R., Falk, S.W., McCall, M.L., Sutherland, P.C., Uomoto, A., Wheeler, J.C., and Wills, B.J. 1982, Astrophys. J., (Letters), 252, L61.

Branch, D., Lacy, C.H., McCall, M.L., Sutherland, P.G., Uomoto, A.,
    Wheeler, J.C., Wills, B.J. 1983, Astrophys. J., 270, 123.
Bruenn, S.W. 1971, Astrophys. J., 168, 203.
Buchler, J.R., and Mazurek, T.J. 1975, Mem. Soc. Roy. Sci. Liege, 8, 435.
Cameron, A.G.W. 1959, Astrophys. J., 130, 916.
Canal, R., and Schatzman, E. 1976, Astron. Astrophys., 46, 229.
Canal, R., and Isern, J. 1979, in IAU Colloquium 53, White Dwarfs and
    Variable Degenerate Stars, ed., H.M. Van Horn and V. Weidemann
    (Rochester: University of Rochester), p.52.
Chevalier, R.A. 1981, Astrophys. J., 246, 267.
Colgate, S.A., Petschek, A.G., and Kriese, J.T. 1980, Astrophys. J.,
    (Letters), 237, L81.
Courant, R., and Friedrichs, K.O. 1948, Supersonic Flow and Shock Waves
    (New York: Interscience).
Cowan, J.J., Cameron, A.G.W., and Truran, J.W. 1980, Astrophys. J.,
    241, 1090.
Duncan, M.J., Mazurek, T.J., Snell, R.L., and Wheeler, J.C. 1976.
    Astrophys. Letters, 17, 19.
Ergma, E.V., and Tutukov, A.V. 1976, Acta Astron., 26, 69.
Finzi, A., and Wolf, W.A. 1967, Astrophys. J., 150, 115.
Fujimoto, M.Y., and Sugimoto, D. 1982, Astrophys. J., 257, 291.
Fujimoto, M.Y. and Taam, R.E. 1982, Astrophys. J., 260, 249.
Fuller, G.M., Fowler, W.A., and Newman, M. 1982, Astrophys. J., 252, 715.
Hashimoto, M., Hanawa, T., and Sugimoto, D. 1983, Publ. Astron Soc.
    Japan, 35, 1.
Helfand, D.J. 1983, in IAU Symposium 101, Supernova Remnants and Their
    X-Ray Emission, ed., J. Danziger and P. Gorenstein (Dordrecht:
    Reidel).
Helfand, D.J., Ruderman, M.A., and Shaham, J. 1983, Nature, 304, 423.
Hoyle, F., and Fowler, W.A. 1960, Astrophys. J., 132, 565.
Iben, I., Jr. 1975, Astrophys. J., 196, 525.
_____. 1981, Astrophys. J., 243, 987.
_____. 1982, Astrophys. J., 259, 244.
Iben, I., Jr. and Tutukov, A.V. 1984, preprint.
Ichimaru, S. and Utsumi, K. 1983a, Astrophys. J., (Letters), 269, L51.
_____. 1983b, Astrophys. J., (submitted).
Isern, J., Labay, J., Hernanz, M., and Canal, R. 1983, Astrophys. J., 273.
Ivanova, L.N., Imshennik, V.S., and Chechetkin, V.M. 1974, Astrophys.
    Space Sci., 31, 497.
Kenyon, S.J. and Truran, J.W. 1984, preprint.
Mazurek, T.J., Meier, D.L., and Wheeler, J.C. 1977, Astrophys. J., 213, 518.
Mazurek, T.J., Truran, J.W., and Cameron, A.G.W. 1974, Astrophys. Space
    Sci., 27, 261.
Meyerott, R.E. 1980, Astrophys. J., 239, 257.
Miyaji, S., Nomoto, K., Yokoi, K., and Sugimoto, D. 1980, Publ. Astron.
    Soc. Japan, 32, 303.
Mochkovitch, R. 1980, Thesis, University of Paris.
_____. 1983, Astron. Astrophys., 122, 212.
Muller, E. and Arnett, W.D. 1982, Astrophys. J., (Letters), 261, L107.
Nariai, K. and Nomoto, K. 1979, in IAU Colloquium 53, White Dwarfs
    and Variable Degenerate Stars, ed. H.M. Van Horn and V. Weidemann
    (Rochester: University of Rochester), p.525.

Nariai, K., Nomoto, K., and Sugimoto, D. 1980, Publ. Astron. Soc. Japan, 32, 473.

Nomoto, K. 1980a, in Type I Supernovae, ed. J.C. Wheeler (Austin: University of Texas), p.164.

_____. 1981, in IAU Symposium 93, Fundamental Problems in the Theory of Stellar Evolution, ed. D. Sugimoto, D.Q. Lamb, and D.N. Schramm (Dordrecht: Reidel), p.295.

_____. 1982a, Astrophys. J., 253, 798.

_____. 1982b, Astrophys. J., 257, 780.

_____. 1982c, in Supernovae: A Survey of Current Research, ed. M.J. Rees and R.J. Stoneham (Dordrecht: Reidel), p.205.

Nomoto, K., Miyaji, S., Sugimoto, D., and Yokoi, K. 1979, in IAU Colloquium 53, White Dwarfs and Variable Degenerate Stars, ed. H. M. Van Horn and V. Weidmann (Rochester: University of Rochester),p.56.

Nomoto. K., Nariai, K., and Sugimoto, D. 1979, Publ. Astron. Soc. Japan, 31, 287.

Nomoto, K. and Sugimoto, D. 1977, Publ. Astron. Soc. Japan, 29, 765

Nomoto, K., Sugimoto, D., and Neo, S. 1976, Astrophys. Space Sci., 39, L37.

Nomoto, K. and Thielemann, F.-K. 1984, to be published.

Nomoto, K. and Tsuruta, S. 1981, Astrophys. J. (Letters), 250, L19.

Ōno, Y. 1960, Prog. Theor. Phys., 24, 825.

Paczyński, B., and Rudak, B. 1980, Astron. Astrophys., 82, 349.

Pskovskii, Y.P. 1977, Sov. Astr.-AJ, 21, 675.

Shull, J.M. 1982, Astrophys. J., 262, 308.

Sugimoto, D., Fujimoto, M.Y., Nariai, K., and Nomoto, K. 1978, in IAU Symposium 76, Planetary Nebulae, ed. Y. Terzian, (Dordrecht: Reidel), p.208.

_____. 1979, in IAU Colloquium 53, White Dwarfs and Variable Degenerate Stars, ed. H.M. Van Horn and V. Weidemann (Rochester: University of Rochester), p.280.

Sugimoto, D. and Miyaji, S. 1981, in IAU Symposium 93, Fundamental Problems in the Theory of Stellar Evolution, ed. D. Sugimoto, D.Q. Lamb, and D.N. Schramm (Dordrecht: Reidel), p.191.

Sugimoto, D., Eriguchi, Y., and Hachisu, I. 1981, Prog. Theor. Phys. Suppl., 70, 154.

Sugimoto, D. and Nomoto, K. 1975, Publ. Astron. Soc. Japan, 27, 197.

Sugimoto, D. and Nomoto, K. 1980, Space Sci. Rev., 25, 155.

Sutherland, P. and Wheeler, J.C. 1984, Astrophys. J., (submitted).

Taam, R.E. 1980a, Astrophys. J., 237, 142.

Tammann, G.A. 1982, in Supernovae: A Survey of Current Research, ed. M.J. Rees and R.J. Stoneham (Dordrecht: Reidel), p.371.

Thielemann, F.-K., Arnould, M., and Hillebrandt, W. 1979, Astron. Astrophys., 74, 175.

Thielemann, F.-K. and Nomoto, K. 1984, to be published.

Tinsley, B.M. 1980, in Type I Supernovae, ed. J.C. Wheeler (Austin: University of Texas), p.196.

Trimble, V. 1982, Rev. Mod. Phys., 54, 1183.

Truran, J.W., Arnett, W.D., and Cameron, A.G.W. 1967, Canadian J. Phys., 45, 2315.

Truran, J.W., Cowan, J.J., and Cameron, A.G.W. 1978, Astrophys. J., (Letters), 222, L63.

Tuohy, I., and Garmire, G.P. 1980, Astrophys. J. (Letters), 239, L107.
Tuohy, I., Garmire, G.P., Manchester, R.N., and Dopita, M.A. 1983,
    Astrophys. J., 268, 778.
Unno, W. 1967, Publ. Astron. Soc. Japan, 19, 140.
Uus, U. 1970, Nauch. Inform. Akad. Nauk USSR, 17, 25.
van den Heuvel, E.P.J. 1981, in IAU Symposium 93, Fundamental Problems
    in the Theory of Stellar Evolution, ed. D. Sugimoto, D.Q. Lamb,
    and D.N. Schramm (Dordrecht: Reidel), p.155.
Weaver, T.A., Axelrod, T.S., and Woosley, S.E. 1980, in Type I
    Supernovae, ed. J.C. Wheeler (Austin: University of Texas), p.113.
Webbink, R.F. 1984, preprint.
Webbink, R.F., Rapaport, S., and Savonije, G.J. 1983, Astrophys. J.,
    270, 678.
Weidemann, V. and Koester, D. 1983, Astron, Astrophys., 121, 77.
Wheeler, J.C. 1978, Astrophys. J., 225, 212.
_____. 1982, in Supernovae: A Survey of Current Research, ed. M.J. Rees
    and R.J. Stoneham (Dordrecht: Reidel), p.167.
Woosley, S.E. and Weaver, T.A. 1982, in Essays in Nuclear Astrophysics,
    ed. C.A. Barnes, D.D. Clayton, and D.N. Schramm (Cambridge
    University Press), p.377.
Woosley, S.E., Weaver, T.A., and Taam, R.E. 1980, in Type I Supernovae,
    ed. J.C. Wheeler (Austin: University of Texas), p.96.
Wu, C.-C., Leventhal, M., Sarazin, C.L., and Gull, T.R. 1983, Astrophys.
    J. (Letters), 269, L5.
Yokoi, K., Neo, S., and Nomoto, K. 1979, Astron. Astrophys., 77, 210.
Yokoi, K. and Nomoto, K.1984, in preparation.

DISCUSSION

Branch: 1) Do you expect that there will be any sodium at all in the
material consisting primarily of oxygen-burning products? 2) Will the
density inversion remain in the final density profile? 3) How much mass
will be moving slower than 2000 km s^{-1}?
Nomoto: 1) Sodium is produced in the decaying deflagration wave at $M_r \sim$
1.3 $M_\odot$, where the carbon burning is not explosive. The mixing between
the oxygen burning products and carbon burning products is likely to
occur because the outer layers are convectively unstable after the
deflagration. 2) The density inversion remains in the present model as
seen in Fig. 4, although it depends on the propagation speed of the
deflagration wave. 3) About 0.05 $M_\odot$.

Woosley: 1) You discussed two possible mechanisms for heating the
central regions of accreting white dwarfs, gravitational compressional
heating and conduction from active hydrogen and helium burning shells,
yet you say both give nearly identical ignition density for a given
accretion rate. Why do these two different mechanisms give such similar
results? 2) You emphasize the possibility of making silicon to calcium
group elements in deflagrating Type I's, yet the iron to silicon ratio
you produce is at least two times solar and, in particular, $^{54}Fe/^{56}Fe$ is

about 4x solar. Normalizing to ^{54}Fe production I claim no more than $\sim$10% of Si-Ca could have originated in this type of explosion.
Nomoto: 1) If the accretion is rapid enough, both heating mechanisms raise the temperature up to $\sim$ 10^8 K. Afterwards the evolutionary track of the central temperature and density is controlled by the neutrino energy loss and the rate of compression. Thus the ignition density does not depend on the early phase of heating but depends only on the accretion rate. 2) The ^{54}Fe/^{56}Fe ratio depends on the neutron excess and thus on the electron capture rates, particularly in the central region. If the accretion rate is higher than considered here, the ignition density is lower and electron captures are slower. If the propagation speed of the deflagration wave is higher near the center, the material is less neutronized. Both cases are possible and their effects lead to smaller ^{54}Fe/^{56}Fe ratio. Other isotope ratios among Fe-Ni are rather satisfactory. If the abundances of Ca-Si group are normalized to ^{56}Fe, their production is significant, i.e., 40-50 %.

Renzini: On s-process elements: How massive is the layer in which they will survive?
Nomoto: Outermost layer of $\sim$ 0.1 M$_\odot$ is ejected without being processed by the deflagration wave.

Mathews: Could you comment on whether a type I supernova can produce a helium burning r-process elements?
Nomoto: A precursor shock ahead of the deflagration wave is found to get very strong in the low density helium layer. The temperature at the shock could be high enough to produce neutrons, although the calculation at this phase is preliminary. Since the helium layer has contained s-process elements as seed elements before the explosion, r-process could operate.

Woosley: While you have probably done the best one can (presently) do a modeling these accreting white dwarf models for SN I, there must remain considerable uncertainty in some of your results (e.g., iron mass synthesized, accretion rate range for a given SN I mechanism, etc.) due to 1) uncertainty in the deflagration wave velocity (and geometry). Using mixing length theory, even time dependent mixing length theory, must be extremely approximate when the convective velocities are approaching the speed of sound. 2) The thermal structure of the envelope of the white dwarf, which sets the central conditions for the detonation may be uncertain. If shells are flashing, it is difficult to estimate what fraction of the energy liberated by nuclear reaction flows into the star and what fraction flows out. 3) The accretion rate may not be constant nor spherically symmetric. This also affects 2). Would you care to comment?
Nomoto: 1) Until a full 3D calculation is done to simulate the propagation of the deflagration wave (2D eddies are still strange), the velocity of the deflagration wave must be regarded as a parameter. We need to set a mixing length, for example, to give a reasonable explosion energy. However, qualitative results of the carbon deflagration model is quite encouraging for further study because it can account for the

production of both ^{56}Ni and Si-Ca group elements in SN I. 2) Heat inflow
from the flashing shell is less important to set the ignition density
once the central temperature becomes as high as $\sim 10^8$ K. This is because
the neutrino loss and the compression are dominant in the central region
and, moreover, heat flows outward from the center near the carbon
ignition stage. Thermal structure of the envelope is important for the
case of rapid accretion which ignites an off-center flash. 3) If the
accretion rate is variable, the rate at the very late stage of accretion
is important to set the central condition of the white dwarf.

Renzini: Having such a variety of situations which may lead to SN I
events, I would be very surprized if nature managed to produce them in
roughly equal numbers. I would rather expect one scenario to be much
more frequent than the others. Everybody would be happy if this were the
C-deflagration case.
Nomoto: If the mass of the white dwarf increases as a result of
recurrent hydrogen flashes, the double detonation supernovae would be
rare events. This is because the double detonations occur for slower
accretion and the slow accretion leads to the hydrogen flash which is so
strong as to eject a large portion of the accreted matter. On the other
hand, rapid accretion $(\sim 10^{-7} M_\odot yr^{-1})$ leads to a weaker flash. Thus the
carbon deflagration may be most frequent.

Branch: What will the $10^{-7} M_\odot yr^{-1}$ accretors look like?
Nomoto: Paczyński and Rudak have suggested that the hydrogen shell
burning in the rapidly accreting white dwarfs may give rise to symbiotic
star-like phenomena.

Gallagher: Although symbiotic stars are indeed potential sites for
extensive accretion onto white dwarfs, they may not produce SN I. The
problem lies with the giant primaries in symbiotic binaries, which are
usually luminous giants. They have extensive H-rich envelope which might
well be entrained in supernova ejecta. Could this lead to a violation of
the H-deficient condition imposed by the analysis of SNe I spectra?
Nomoto: I am not sure if it leads to the excess hydrogen in the spectra
over the upper limit set by the recent observations. (See Branch et al.
1982.)

Yahil: Could the high accretion rate resulting in a common envelope red
giant be the source of planetary nebulae in old systems?
Nomoto: I think so.

Gallagher: Perhaps I can reenforce your point about the potential
signigicance of stellar winds in steady burning objects. Post-maximum
classical novae are thought to have near equilibrium H-burning
luminosities. From their rapid (< 10 yr) turn-offs and observations of
spectra, it appears that mass loss rates are quite high (perhaps $> 10^{-7}$
$M_\odot yr^{-1}$). If this is not due to transient properties of novae, then one
would expect mass loss via stellar winds to be extremely important for
objects in the steady-burning regime due to their similar luminosities to
post-maximum novae.

TYPE II SUPERNOVAE FROM 8-10 $M_\odot$ PROGENITORS

Ken'icni Nomoto
Dept. of Earth Science and Astronomy, College of Arts and
Sciences, University of Tokyo, Meguro-ku, Tokyo 153
and
Max-Planck-Institut für Physik und Astrophysik
Institut für Astrophysik, Garching b. München

ABSTRACT

The evolution of stars in the mass range of 8-12 $M_\odot$ is summarized;
it is significantly different from more massive stars (typically a 25 $M_\odot$
star) because of smaller core mass and stronger electron degeneracy.
Stars of 10-12 $M_\odot$ undergo strong shell flashes of neon, oxygen, and
silicon and the resultant iron core eventually collapse. Stars of 8-10
$M_\odot$ form electron-degenerate O+Ne+Mg cores and undergo collapse induced by
electron captures. The collapse of such less massive progenitors does
lead to the Type II supernova explosion. Nucleosynthesis in these stars
does not contribute much to the galactic nucleosynthesis, but provides
useful information on the Crab Nebula's progenitor.

1. INTRODUCTION

Type II supernovae (SN II) are naturally interpreted as the
explosion of massive stars that have hydrogen-rich envelopes with red-
giant size. Presupernova evolution has been well studied for stars more
massive than 15 $M_\odot$; these stars evolve straightforwardly to form iron
cores which collapse due to nuclear dissociation (e.g., Weaver et al.
1978). On the other hand, the late stage evolution of smaller mass stars
is complicated because the effect of electron degeneracy is significant
(Sugimoto and Nomoto 1980). In particular, stars of 8-10 $M_\odot$ form
electron degenerate O+Ne+Mg cores and undergo collapse indueced by
electron captures on ^{24}Mg and ^{20}Ne (Nomoto 1981; Miyaji et al. 1980).

Hydrodynamical behavior of the collapse also depends on the stellar
mass (Hillebrandt 1983). Most recent calculation has shown that 8-10 $M_\odot$
stars do explode as SN II (Hillebrandt, Nomoto, and Wolff 1984). The aim
of the present paper is to summarize these studies and fill the lack of
our understanding of the evolution of 8-12 $M_\odot$ SN II progenitors and
associated nucleosynthesis with a special emphasis on the 8-10 $M_\odot$ range.

C. Chiosi and A. Renzini (eds.), Stellar Nucleosynthesis, 239–250.

2. PRESUPERNOVA EVOLUTION OF 8-12 $M_\odot$ STARS

    Figures 1-4 show the chemical evolution of stars with masses M = 12
$M_\odot$, 10.4 $M_\odot$, 9.6 $M_\odot$ and 8.8 $M_\odot$, respectively (Nomoto 1983, 1984). These
stars are assumed to have helium cores of masses M_H = 0.25 M, i.e., 3 $M_\odot$,
2.6 $M_\odot$, 2.4 $M_\odot$, and 2.2 $M_\odot$, respectively; a static hydrogen-rich envelope
is fitted to the core at the hydrogen-burning shell and the evolution was
calculated from the helium burning phase. (Hereafter, M_H, M_{He}, and M_C,
and M_O denote the masses interior to the H, He, C, and O burning shell,
respectively).

    As seen in Figs. 1-4, evolution depends sensitively on the stellar
mass because their masses are close to the Chandrasekhar limit, M_{Ch}.
Stars of 10-12 $M_\odot$ undergo recurrent strong shell flashes of neon, oxygen,
and silicon and eventually evolve into the iron core collapse. On the
other hand, stars of 8-10 $M_\odot$ form electron degenerate O+Ne+Mg cores
without neon shell flashes.

1) 12 $M_\odot$ Star (Case 3.0)

    The star evolves through the phases of carbon, neon, and oxygen
burning. The masses M_{He} (= 1.62 $M_\odot$) and M_C (= 1.60 $M_\odot$) exceed M_{Ch} so
that silicon burning can be ignited. During the gravitaional contraction
of the silicon core, a temperature inversion appears because of electron
degeneracy in the central layers and the cooling due to neutrino energy
loss which is larger for higher density. As a result, a silicon flash is
ignited at the shell with M_r = 0.17 $M_\odot$ and forms an iron layer above the
small Si core (Fig. 1). Final stages of evolution are characterized by
the irregular nuclear shell flashes and associated neutrino emission as
seen in Fig. 5 (see also Sparks and Endal 1980). Eventually the star
undergoes the Fe photodisintegration after the central silicon flash.
Mass interior to the active silicon burning shell is 1.4 $M_\odot$.

2) 10.4 $M_\odot$ Star (Case 2.6)

    Core carbon burning proceeds under non-degenerate condition. During
the O+Ne+Mg core contraction, electrons become somewhat degenerate (ψ_c ∿
30) so that a large temperature inversion appears. Since M_{He} (= 1.45 $M_\odot$)
and M_C (= 1.43 $M_\odot$) are larger than the critical mass for neon ignition,
namely 1.37 $M_\odot$ (Boozer et al. 1973; Nomoto 1984), a strong neon flash is
ignited at M_r = 0.88 $M_\odot$. Further evolution is complicated but the star
will eventually collapse due to nuclear dissociation in the iron core
(Woosley et al. 1980). The iron core mass should be smaller than 1.45 $M_\odot$.

3) 9.6 $M_\odot$ Star (Case 2.4)

    The evolution is similar to case 2.6 (10.4 $M_\odot$ star) up to
gravitaional contraction of the O+Ne+Mg core except that carbon ignition
first occurs at the outer shell of M_r = 0.05 $M_\odot$. The important
difference from case 2.6 is that M_{He} (= 1.34 $M_\odot$) is smaller than the
critical mass for neon ignition. Therefore, the O+Ne+Mg core never

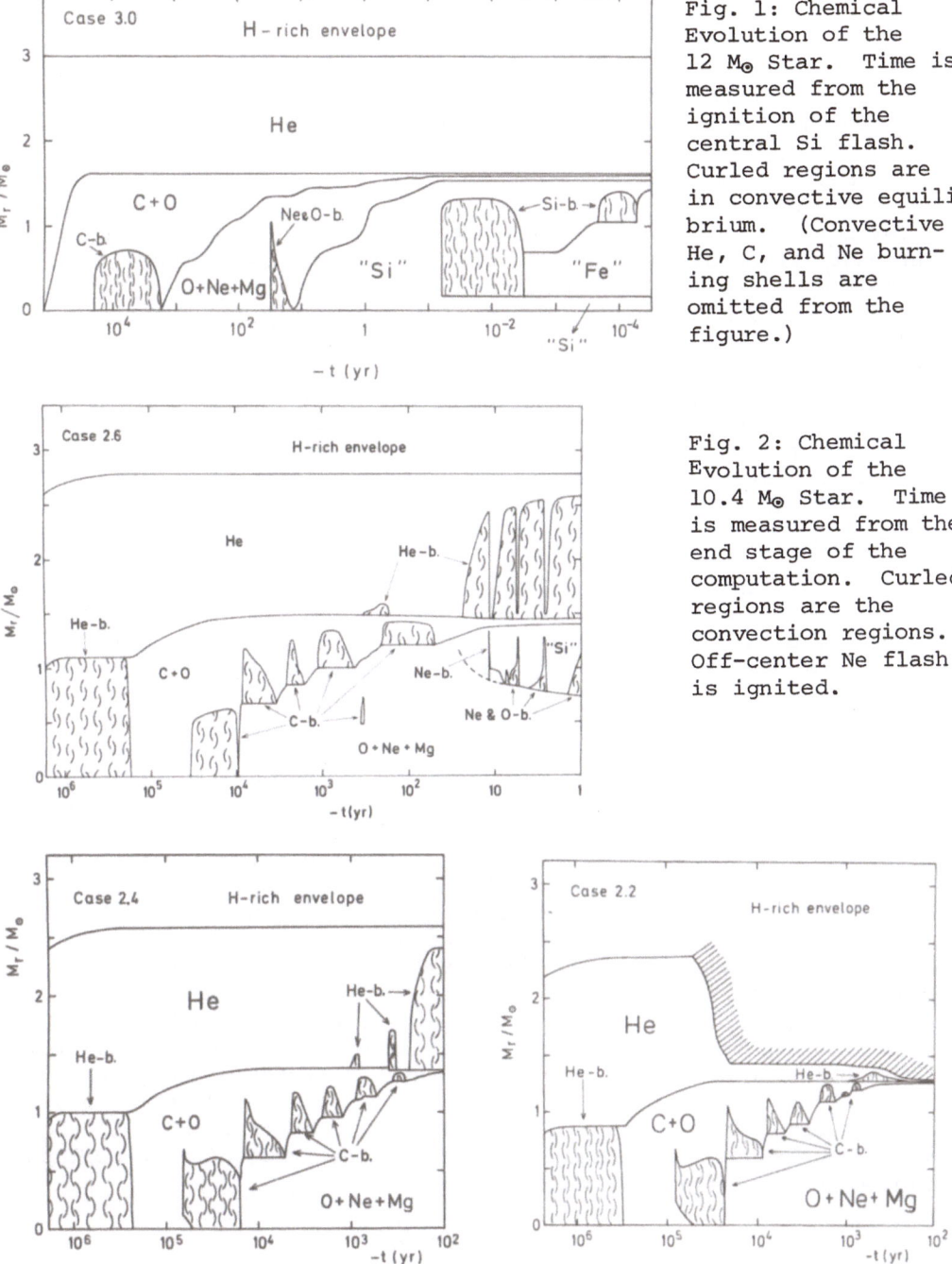

Fig. 1: Chemical Evolution of the 12 M$_\odot$ Star. Time is measured from the ignition of the central Si flash. Curled regions are in convective equilibrium. (Convective He, C, and Ne burning shells are omitted from the figure.)

Fig. 2: Chemical Evolution of the 10.4 M$_\odot$ Star. Time is measured from the end stage of the computation. Curled regions are the convection regions. Off-center Ne flash is ignited.

Fig. 3   Fig. 4

Same as Fig. 2 but for the 9.6 M$_\odot$ star (Fig. 3) and 8.8 M$_\odot$ star (Fig. 4) where the degenerate O+Ne+Mg core is formed. Shaded region indicates the surface convection zone penetrating into the He-layer.

undergoes neon burning and become strongly degenerate. When M_C becomes
1.339 $M_\odot$ being close to M_{He} = 1.343 $M_\odot$, the helium layer expands greatly.
It induces a penetration of the surface convection zone to dredge up the
helium layer. Further evolution up to the electron capture collapse is
similar to the 8.8 $M_\odot$ star described below.

4) 8.8 $M_\odot$ Star (Case 2.2)

     The basic evolutionary feature is similar to case 2.4. The
evolutionary track of the central density, ρ_C, and temperature, T_C, is
shown in Fig. 6 from carbon burning through the early phase of collapse.
The off-center carbon ignition causes a loop of the (ρ_C, T_C) track around
$\rho_C \sim 10^6$ g cm^{-3}. After carbon burning, T_C gradually decreases and a
strongly degenerate O+Ne+Mg core is formed. This is because M_C (= 1.28
$M_\odot$) is smaller than 1.37 $M_\odot$. Owing to the stronger electron degeneracy,
the dredge-up of He layer starts earlier than case 2.4, i.e., at the
early phase of core carbon burning (see also Weaver et al. 1980). Most
of the helium layer is dredged up and H-shell burning is re-ignited.
(More details of this dredge-up phase is described in Nomoto et al.1982.)

3. COLLAPSE OF THE O+Ne+Mg CORE DUE TO ELECTRON CAPTURES

     The evolution of cases 2.2 and 2.4 after the dredge-up of helium
layer is common to 8-10 $M_\odot$ stars and proceeds as follows.

1) Hydrogen-Helium Double Shell Burning Phase

     The core contracts as the masses $M_H \simeq M_{He}$ increase towards M_{Ch}
through the H-He double shell burning phase. As seen in Fig. 7, a carbon
layer is gradually built up, but the effect of compressional heating is
too small to ignite carbon shell burning. During this phase the star
consists of a hydrogen-rich envelope of red-supergiant size, thin helium
layers of 10^{-4} to 10^{-5} $M_\odot$, and a degenerate core with outer carbon layers
and inner O+Ne+Mg parts. Thermal pulses of a helium burning shell
synthesizes s-process elements (Iben 1975). Such a structure is almost
identical to those of 4-8 $M_\odot$ stars having degenerate C+O cores.
Therefore the luminosity is close to that given by Paczynski's (1970)
core-mass to luminosity relation. Also the evolutionary track of (ρ_C, T_C)
merge into the common track of 4-8 $M_\odot$ stars (Paczyński 1970), because it
depends only upon the growth rate of the core (given by the luminosity)
and the neutrino loss rate (Fig. 6). This implies that the evolution
from the double shell burning phase through the stage of core collapse
should be the same for 8-10 $M_\odot$ stars.

2) Rapid Contraction due to Electron Captures

     Once M_H ($\simeq M_{He}$) exceeds 1.375 $M_\odot$, electron captures on ^{24}Mg and ^{20}Ne
start. (Their threshold densities are indicated in Figs. 6-7.) As the
number of electrons per baryon, Y_e, decreases, the core begins to
contract rapidly on the timescale of electron captures. When ρ_C reaches

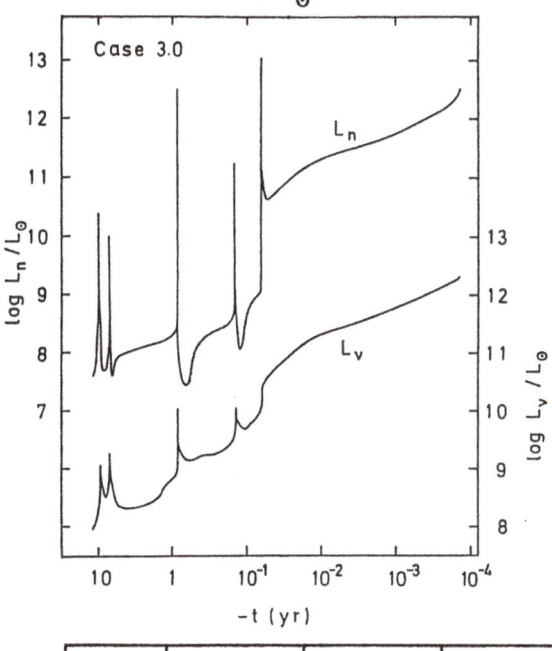

Figure 5:
Evolutionary changes
in the nuclear energy
generation rate, L$_n$,
and the neutrino
luminositym, L$_\nu$, for
the 12 M$_\odot$ star.
Abscissa (t) is the
same as in Fig. 1.
Sharp rises in L$_n$ are
due to shell flashes
of neon, oxygen, and
silicon.

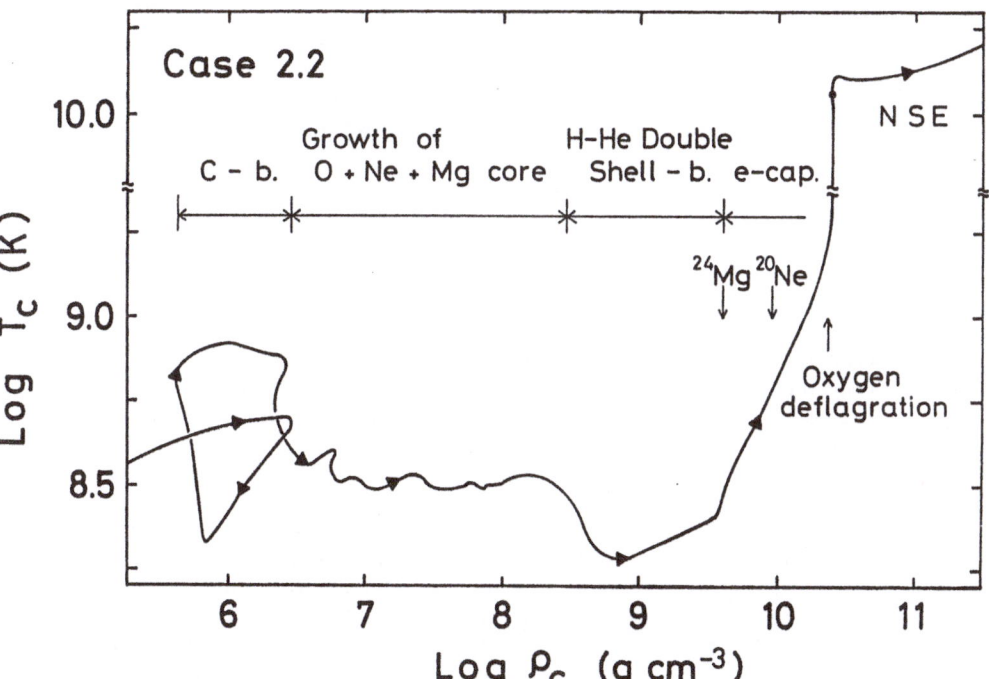

Figure 6: Evolutionary changes in ρ_C and T_C for the
8.8 M$_\odot$ star. Corresponding evolutionary stages from
C-burning through electron capture collapse are
indicated. Threshold densities for electron captures
on ^{24}Mg and ^{20}Ne are also indicated.

2.5×10^{10} g cm^{-3}, oxygen burning is ignited and the material undergoes incineration into nuclear statistical equilibrium (NSE) composition at the combustion front. The resultant energy release is only 4% of the internal energy of the matter, which is too small to initiate a detonation. Thus the oxygen combustion front is almost stationary at a radius of about 170 km as the materials contract. As a result, the mass of the NSE core, M_{NSE}, grows as ρ_c increases (Fig. 7). Although electron captures on NSE elements (mainly on free protons) accelerates the contraction, it is slower than the free fall until the stage with $\rho_c \sim 10^{12}$ g cm^{-3} and $M_{NSE} \sim 0.8$ M$_\odot$.

The central density at which the oxygen combustion starts is subject to the following uncertainty: The electron captures on ^{24}Mg and ^{20}Ne form a semi-convection zone in the present model, because electron captures produce much entropy as well as a gradient of Y_e in the central region. Semi-convection state is secularly overstable with a growth timescale of heat diffusion (Kato 1966; Langer et al. 1983). In the early phase, therefore, electron captures are slow enough for the convection to develop but, later, becomes so rapid as to stop the mixing in the central region. As a result, the oxygen combustion commences at $1 \times 10^{10} \lesssim \rho_c$ (g cm^{-3}) $\lesssim 2.5 \times 10^{10}$; the upper and lower limit to ρ_c correspond to the case where the Schwarzshild criterion for convection is adopted as in the present calculation and the case for the Ledoux criterion (see Miyaji et al. 1980), respectively. If ρ_c is closer to 1×10^{10} g cm^{-3}, the oxygen combustion front would locate at a larger radius so that the explosion would be weaker than presented below.

3) Bounce, Shock Propagation, and Explosion

From the stage (t = 0 ms) with $\rho_c = 2.7 \times 10^{11}$ g cm^{-3}, $T_c = 1.5 \times 10^{10}$ K, s = 1 k/nucleon, and $Y_e = 0.36$, neutrino transport was included in the calculation by Hillebrandt, Nomoto and Wolff (1984). After the stage with $\rho_c \sim 10^{12}$ g cm^{-3}, M_{NSE} exceeds M_{Ch} so that the collapse becomes really dynamical as seen from the change in the radius in Fig. 8. The core makes a bounce at $\rho_c \sim 3 \times 10^{14}$ g cm^{-3} (t = 24 ms) and a shock forms at $M_r = 0.63$ M$_\odot$ (r = 16 km). The mass of the rebound core and the initial energy given to the shock ($\sim 6 \times 10^{51}$ ergs) are $\sim$ 15% smaller than those for more massive stars because of slightly smaller lepton concentration of $Y_\ell \sim 0.35$ in the central region.

Despite that, the shock does not damp as it propagates outward and finally gives rise to an explosion ejecting ~ 0.1 M$_\odot$ material with the energy of $\sim 2 \times 10^{51}$ ergs. Such a shock propagateion is seen from the changes in the distribution of velocity, density, and temperature given in Figs. 9-11.

The reason why the 8.8 M$_\odot$ star does explode while stars more massive than 15 M$_\odot$ do not can be explained by the existence of the oxygen combustion front (Hillebrandt 1983). First the oxygen combustion during the collapse decelerates the infalling material as seen from the velocity profile (Fig. 9) so that the infalling velocity is smaller than in more

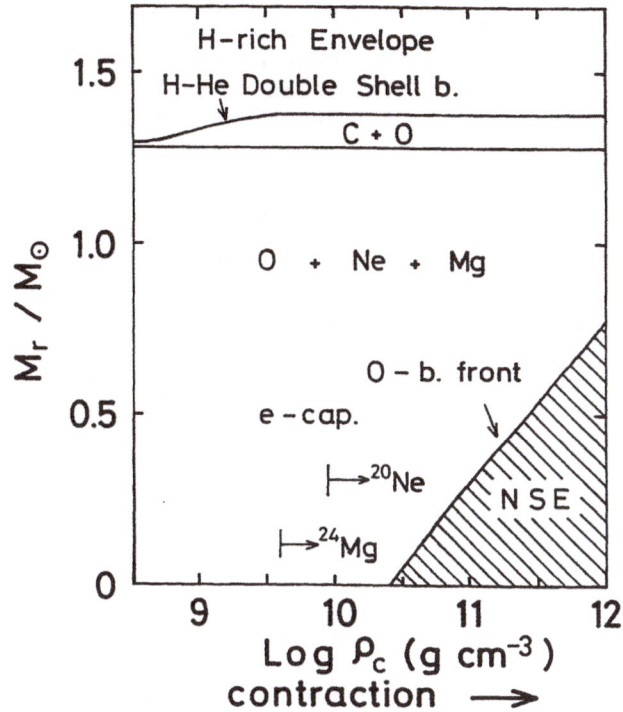

Figure 7:
Evolutionary changes
in the chemical
structure of the 8.8
$M_\odot$ star as a function
of the increasing
central density. C+O
layers grows as a
result of H–He double
shell burnings.
During the rapid
contraction, NSE
layer (shaded) grows
behind the oxygen
combustion front.

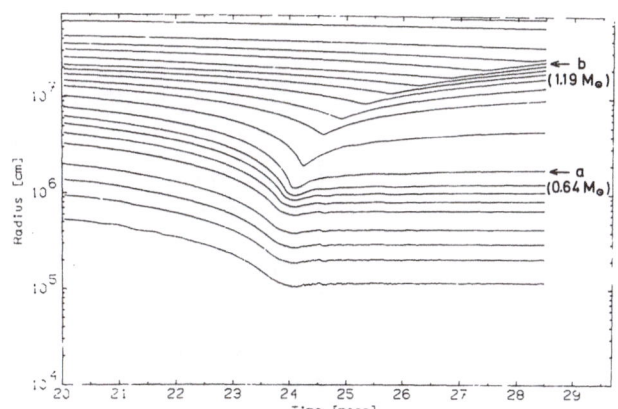

Figure 8: Collapse,
bounce, and shock
propagation for the
8.8 $M_\odot$ star (taken
from Hillebrandt et
al. 1984). Changes
in the radius for
selected mass zones
are shown. Time is
measured from the
stage with $\rho_c = 2.7 \times 10^{11}$ g cm^{-3}.

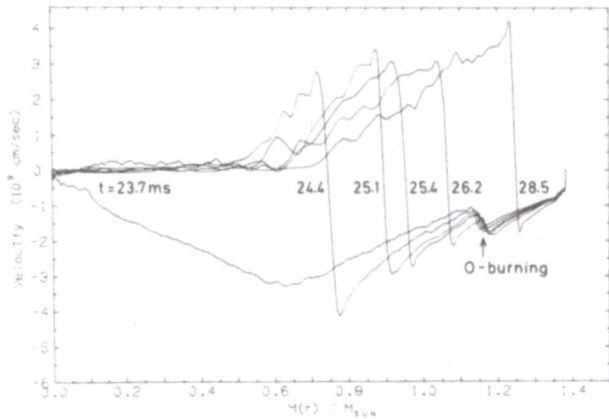

Figure 9:
Changes in the
velocity profile
during the collapse,
bounce, and shock
propagation for the
8.8 $M_\odot$ star
(Hillebrandt et al.
1984). Time is the
same as in Fig. 8.
It is shown that the
infalling matter is
decelerated at the
oxygen burning front.
The shock does not
damp and is
strengthened at the
oxygen burning front.

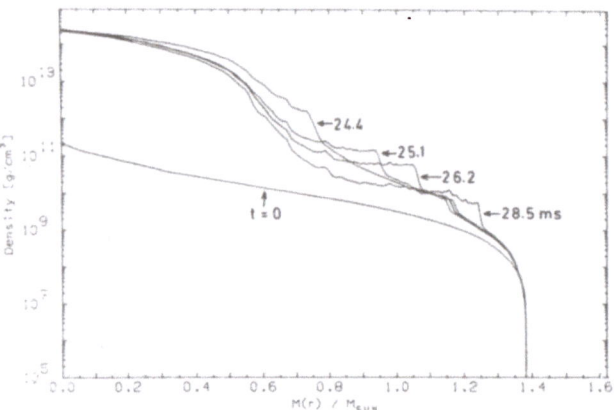

Figure 10:
Same as Fig.9 but for
the density profile.
The steep density
gradient at the
oxygen front ($M_r \sim$
1.2 $M_\odot$) strengthen
the shock wave.

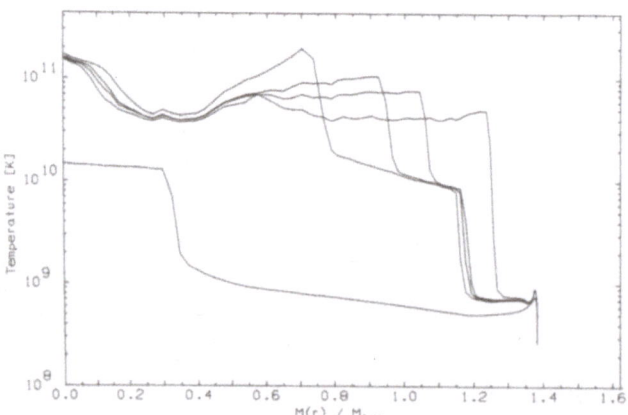

Figure 11:
Same as Fig. 9 but
for the temperature
profile. After
passing through the
oxygen burning front,
entropy behind the
shock becomes as high
as 15 k/nucleon.

massive stars. As a result, the density ahead of the shock is smaller by a factor of 10, and the ram pressure due to the infalling matter is much smaller (Yahil 1984). Second the shock is strengthened when it reaches the oxygen combustion front. At this time, the combustion front has moved to $M_r \sim 1.2\ M_\odot$ ($r = 200$ km) and it forms a rather steep density gradient there corresponding to the velocity gradient accross the front.

Although the calculation by Hillebrandt et al. (1984) was stopped at $t = 28.5$ ms, the explosion of 8-10 $M_\odot$ stars will be observed as SN II because these stars are on the asymptotic giant branch at the time of explosion. The shock wave will travel through the extended hydrogen-rich envelope and heats up the surface to produce Type II light curves. The mean density of the envelope is lower than in more massive stars because of smaller envelope mass. Thus the resultant light curves would be somewhat different from those in stars of $M > 15\ M_\odot$ (e.g., Chevalier 1981).

4. NUCLEOSYNTHESIS IN 8-10 $M_\odot$ STARS

It is common to nucleosynthesis in 8-12 $M_\odot$ stars that the amount of ejected heavy elements are much smaller than those from stars heavier than 15 $M_\odot$, if the mass cut is assumed to be $\sim 1.4\ M_\odot$. This is because M_{He} is smaller than $\sim 1.6\ M_\odot$ in 8-12 $M_\odot$ stars as seen from Figs. 1-4. Therefore these stars do not contribute much to the galactic nucleosynthesis although most of SN II originate from this mass range.

Despite that, chemical abundances of the ejecta from 8-12 $M_\odot$ stars could provide a useful information on the progenitors of young supernova remnants. In fact, the Crab Nebula's progenitor has been suggested to be in the mass range of 8-9.5 $M_\odot$ on the main sequence (Davidson et al. 1982; Nomoto 1982, 1983; Nomoto et al. 1982).

The supernova explosion of 8-10 $M_\odot$ stars eject a core material of $\sim 0.1\ M_\odot$ and a H-He envelope with mass depending on the mass loss prior to the explosion. Although the abundances in the ejected core material has not been calculated, we can estimate them as follows: The temperature profile in Fig. 11 suggests that the temperature behind the shock will be high enough to process the material into NSE when the shock propagates into the outer layers at $M_r \sim 1.25\ M_\odot$. Therefore, the ejected core material is composed mostly of iron-peak elements with negligible amounts of intermediate mass elements (Ca, Ar, S, Si).

The H-He envelope contains some carbon which was originally deposited in the helium layer and then dredged-up together with helium by the surface convection zone. For cases 2.4 - 3.0, i.e., 9.5 $M_\odot < M \lesssim 12\ M_\odot$, 0.03-0.04 $M_\odot$ carbon has been deposited in the convective shell of He burning (shaded regions in Figs. 1-3). For the smaller mass star of case 2.2, on the other hand, the amount of carbon, though enhanced, is as small as 0.007 $M_\odot$ because the dredge-up of helium layer prevents the convective He-burning shell from developing (Nomoto et al. 1982).

Whether any such carbon enrichment can be observed at the stellar surface depends on the stellar mass. If sufficient amount of mass has been lost from the H-rich envelope before the dredge-up ($\dot{M} \sim 10^{-6} M_\odot yr^{-1}$), the carbon to oxygen ratio could exceed unity in the envelope so that the star would be observed as a carbon star. In the extreme case of mass loss, the star would become a helium star of red-giant-size. Such a mechanism to yield carbon stars does not operate for stars with $M \gtrsim 10 M_\odot$ because the dredge-up does not take place.

The carbon abundance in the envelope gives some hints on the Crab Nebula's progenitor (Arnett 1975). Several studies have suggested a main-sequence mass of $\sim 10 M_\odot$ for the Crab's progenitor (Gott et al. 1970; Arnett 1975; Woosley et al. 1980; Nomoto 1982; Hillebrandt 1982). Recent optical, IUE, and near IR observations (Davidson 1979; Davidson et al. 1982; Henry and MacAlpine 1982; Dennefeld and Andrillat 1981) have shown that the Crab Nebula has a helium overabundance of $1.6 \lesssim X_{He}/X_H \lesssim 8$, relatively small oxygen abundance $X_O \sim 0.003$ (see, however, Pequignot and Dennefeld 1983), and carbon to oxygen ratio of $0.4 \lesssim X_C/X_O \lesssim 1.1$ (X denotes the mass fraction). Suppose that the mass of the H-rich envelope has decreased to $\sim 1 M_\odot$ prior to the explosion. Then for the star with M = 8.8 $M_\odot$, mixing of hydrogen and helium layers produces a He-rich envelope with heavy element abundances of $X_C \sim X_N \sim X_O \sim 0.004$. This is consistent with the relatively small carbon abundance of the Crab. On the other hand, the carbon abundance of $X_C \sim 0.04\text{-}0.05$ ($X_N \sim X_O \sim 0.005$) of the He layer in stars with $M > 9.5 M_\odot$ is inconsistent with the Crab even if the dilution due to mixing is taken into account (see Arnett 1975 for more massive stars).

More recently, Henry (1984) reported that Ni/S ratio of the Crab Nebula is ten times larger than the solar value while Fe/S is smaller than the solar (S abundance is normal). He suggested that nickel may be actually overabundant in the Crab and iron may form grains. Qualitatively, this suggestion seems to be consistent with the products from the 8.8 $M_\odot$ model in the sense that only iron-peak elements are enhanced as a result of explosion, though Fe/Ni ratio has not been calculated yet for the model.

5.   COLLAPSE OF O+Ne+Mg WHITE DWARFS

If one of the component star of a close binary system is a 8-10 $M_\odot$ star, it becomes a helium star of 2-2.5 $M_\odot$ as a result of tidal mass loss. This helium star evolves further as described in §§2-3, i.e., it undergoes non-degenerate helium and carbon burning, and develops a degenerate O+Ne+Mg core. Eventually the helium envelope expands to a red-ginant-size and overflows its Roche lobe. (This phase correspond to the dredge-up phase in §2.) Finally an O+Ne+Mg white dwarf is left (Nomoto 1981; Law and Ritter 1983).

When the companion star evolves to transfer a matter, the O+Ne+Mg white dwarf accretes it and eventually collapses due to electron captures. The condition for $\dot{M}$ to allow such a growth of the white dwarf mass is similar to the case of accretion onto C+O white dwarfs (Nomoto 1981). The hydrodynamic behavior of collapse will not be different from the 8.8 M$_\odot$ star; the white dwarf would form a neutron star ejecting a relatively small amount of iron peak elements. Because of the small initial radius, the supernova would be rather dim.

Formation of a neutron star through the collapse of a white dwarf might account for the origin of some low mass X-ray binaries and binary pulsars (van den Heuvel 1981; Helfand et al. 1983; see also Isern et al. 1983 for a possibility of collapse of C+O white dwarfs).

6. CONCLUDING REMARKS

Hydrodynamic behavior of supernova explosion in 8-12 M$_\odot$ stars is different from more massive stars. Relatively small iron core mass helps the explosion of a 10 M$_\odot$ star (Hillebrandt 1982). For 8-10 M$_\odot$ stars, the O+Ne+Mg core collapses and does explode. Thus 8-12 M$_\odot$ stars produce a substantial fraction of SN II. On the other hand, these stars contribute little to the galactic nucleosynthesis. Since evolution of stars in this range is very sensitive to the stellar mass, various types of supernovae would appear, which deserves further study.

It is a pleasure to thank Prof. Kippenhahn, Drs. W. Hillebrandt, F.-K. Thielemann, and R. Wolff for discussion and hospitality during my stay at the Max-Planck-Institut. I am also indebted to Prof. D. Sugimoto and Drs. R. Henry and S. Miyaji for stimulating discussion. This work is supported in part by the Japanese Ministry of Education, Science, and Culture through the Research Grant No. 58340023.

REFERENCES

Arnett, W.D. 1975, Astrophys. J., 195, 727.
Boozer, A.H., Joss, P.C., and Salpeter, E.E. 1973, Astrophys. J., 181, 393.
Chevalier, R.A. 1981, Fund. of Cosmic Phys., 7, 1.
Davidson, K. 1979, Astrophys. J., 228, 179.
Davidson, K. et al. 1982, Astrophys. J., 253, 696.
Dennefeld, M. and Andrillat, Y. 1981, Astron. Astrophys., 103, 44.
Gott, J.R., Gunn, J.E., and Ostriker, J.P. 1970, Astrophys. J. (Letters) 160, L91.
Henry, R.C. and MacAlpine, G.M. 1982, Astrophys. J., 258, 11.
Henry, R.C. 1984, this volume.
Hillebrandt, W. 1982, Astron. Astrophys., 110, L3.
_____. 1983, Ann. N.Y. Acad. Sci., (Review paper at the XI Texas Symposium).
Hillebrandt, W., Nomoto, K., and Wolff, R., 1984, Astron. Astrophys.
Iben, I., Jr. 1975, Astrophys. J., 196, 525.
Isern, J., Labay, M., Hernanz, M., and Canal, R. 1983, Astrophys. J., 273.
Kato, S., 1966, Publ. Astron. Soc. Japan, 18, 374.

Langer, N., Sugimoto, D., and Fricke, K.J. 1983, Astron. Astrophys.
Law, W.Y., and Ritter, H. 1983, Astron. Astrophys., 123, 33.
Miyaji, S., Nomoto, K., Yokoi, K., and Sugimoto, D. 1980, Publ. Astron.
     Soc. Japan, 32, 303.
Nomoto, K. 1981, in IAU Symposium 93, Fundamental Problems in the
     Theory of Stellar Evolution, ed. D. Sugimoto, D.Q. Lamb, and D.N.
     Schramm (Dordrecht: Reidel), p.295.
_____. 1982, in Supernovae: A Survey of Current Research, ed. M.J. Rees
     and R.J. Stoneham (Dordrecht: Reidel), p.205.
_____. 1983, in IAU Symposium 101, Supernova Remnants and Their X-Ray
     Emission, ed. J. Danziger and P. Gorenstein (Dordrecht: Reidel).
_____. 1984, Astrophys. J., 277, (in press).
Nomoto, K., Sparks, W.M., Fesen, R.A., Gull, T.R., Miyaji, S., and
     Sugimoto, D. 1982, Nature, 299, 803.
Paczyński, B. 1970, Acta Astr., 20, 47.
Pequignot, D. and Dennefeld, M. 1983, Astron. Astrophys., 120, 249.
Sparks, W.M., and Endal, A.S. 1980, Astrophys. J., 237, 130.
Sugimoto, D., and Nomoto, K. 1980, Space Sci. Rev., 25, 155.
van den Heuvel, E.P.J. 1981, in IAU Symposium 93, Fundamental Problems
     in the Theory of Stellar Evolution, ed. D. Sugimoto, D.Q. Lamb,
     D.N. Schramm (Dordrecht: Reidel), p.155.
Weaver, T.A., Axelrod, T.S., and Woosley, S.E. 1980, in Type I
     Supernovae, ed. J.C. Wheeler (Austin: University of Texas), p.113.
Weaver, T.A., Zimmerman, G.B., and Woosley, S.E. 1978, Astrophys. J.,
     225, 1021.
Wolff, R. 1983, Ph.D. Thesis, Tech. Universität München.
Woosley, S.E., Weaver, T.A., and Taam, R.E. 1980, in Type I Supernovae,
     ed. J.C. Wheeler (Austin: University of Texas), p.96.
Yahil, A. 1984, this volume.

DISCUSSION

Renzini: Could models in this mass range blow up with a surface N/C ∿ 7
(C/O > ∿ 2) as in SN 1979c?
Nomoto: Carbon abundance is enhanced in the helium layer of thses stars.
The helium-rich layer of the 8.8 $M_\odot$ star has N/C ∿ 1 and C/O ∿ 1. For
stars of M > 9.5 $M_\odot$, the C/N and C/O ratios are larger.

Bedogni: Could the method of zoning (mass distribution, using uniform or
non uniform grid size) affect the numerical solution of the shock wave?
Nomoto: The hydrodynamical calculation was carried out using 185 mass
zones for the 1.38 $M_\odot$ core. Because of such a good spatial resolution,
we didn't need to make rezoning and, in fact, there was very weak
numerical damping for the shock wave propagation.

Li: What is the difference of Wolff's equation of state from that of El
Eid and of Lamb? What role does the EOS play in SN explosion?
Nomoto: The equation of state by Wolff (1983) was obtained from the
Hartree-Fock calculation. His EOS is relatively stiff, which favors the
explosion.

THE ENERGETICS OF TYPE II SUPERNOVAE

Amos Yahil

State University of New York, Stony Brook, NY 11794, USA

ABSTRACT

The energetics of type II supernovae is considered within the gravitational implosion–explosion picture. The total (internal plus gravitational) energies of different parts of the star are calculated, when they are in, or close to, hydrostatic equilibrium. This enables an estimate of the energy transferred to other parts of the star, and ultimately to the supernova ejectum.

I.   INTRODUCTION

It is generally accepted (see, e.g., Woosley, this volume) that the core of a star, which is massive enough to burn carbon nondegenerately, will burn all the way to the iron peak elements. Endoergic reactions and electron capture then cause a pressure reduction, resulting in catastrophic collapse. This collapse process occurs in the iron core, with little or no effect on the outlying mantle, because the time it takes sound waves from the center to reach the edge of the core, ~1 s, is comparable to the time to complete the supernova process.

The collapse of the iron core is now well understood. The nuclei remain intact, because the entropy of the collapsing core remains low during the collapse, s~1, where s is the dimensionless entropy per baryon in units of the Boltzmann constant (Bethe, Brown, Applegate, and Lattimer 1979). Neutrino losses during collapse are also minimal, both in number and energy, with the electron fraction dropping to only $Y_e \cong 0.39$, before the neutrinos become trapped at a density ~10^{12} g cm^{-3} (Cooperstein and Wambach, private communication). Consequently, the

C. Chiosi and A. Renzini (eds.), Stellar Nucleosynthesis, 251–261.

adiabatic index remains just under 4/3 (Lamb, Lattimer, Pethick, and
Ravenhall 1978), and the collapse proceeds unhindered, until the center
of the star reaches nuclear matter density ($\sim$3x10^{14} g cm^{-3}).

The hydrodynamics of the collapse is well described by self-similar
collapse models, in which the equation of state is approximated by a
simple polytropic law p=Kρ^{γ}, where γ is an effective adiabatic index,
and K is constant both in space and time (Goldreich and Weber 1980;
Yahil and Lattimer 1982; Yahil 1983). The original iron core breaks
up into an inner core, which collapses homologously, and an outer core,
which cannot keep up with the homologous collapse, since the infall
velocity can not exceed the free-fall velocity. A suitable demarcation
line between the inner and outer cores during infall is the point of
maximum infall velocity. The mass within this radius is about 10%
higher than the Chandrasekhar mass appropriate to the composition and
entropy of the inner core, or 0.8-0.9 M_{Θ}. (In order to calculate the
Chandrasekhar mass it is first necessary to fit the equation of state to
a polytrope. The effective adiabatic index will, in general, be
different from 4/3, because of the variable Y_{e}, and effects of the
Coulomb lattice energy of the nuclei. For a definition of the
Chandrasekhar mass for polytropes with $\gamma \neq 4/3$, see Yahil 1983, and
below.)

Instead of collapsing subsonically and homolologously like the
inner core, the outer core is infalling supersonically at approximately
half the free-fall velocity, with a Mach number $\sim$2-3, and a density
profile which is approximately a power law, $\rho \propto r^{-2/(2-\gamma)}$, i.e.,
approximately $\rho \propto r^{-3}$ for $\gamma \cong 4/3$. The constant of proportionality in the
expression for the density depends critically on γ: the closer it is to
4/3, the more tenuous the outer core becomes. We shall see below that
the ram pressure of the outer core ahead of the shock plays a crucial
role in the containment of the supernova shock. The value of the
effective adiabatic index γ is therefore critical.

When the center of the star reaches nuclear matter density, the
equation of state suddenly stiffens, and the collapse is quickly brought
to a halt. The information, that the center has stopped collapsing,
first propagates out as a sound wave. A shock forms only when the
relative velocity of the infalling material and the shock front reaches
the sound speed. Even then, the shock is at first weak, with a small
density jump, and only strengthens as it propagates into lower density.
The density gradient at the edge of the inner core is particularly
steep, and the entropy behind the shock quickly rises there, exceeding
s$\cong$3, the point at which nuclei dissociate. Roughly speaking,
therefore, the demarcation between the inner and outer cores,
established during infall, remains after rebound, in the sense that the

entropy of the inner core is unchanged, or increases only slightly,
while the outer core is strongly shocked.

The source of energy of a type II supernova is thought to be
gravitational, being extracted following the implosion and rebound of
the inner core (Colgate and White 1966; but see Hoyle and Fowler
1960; Woosley and Bodenheimer 1983; Burrows and Lattimer 1983). The
problem of a type II supernova is therefore that of nonequilibrium
energy transfer. The progenitor is a bound star, whose total energy is
negative. After a successful explosion, the outer part of the star is
ejected with positive energy, while a more tightly bound residue is left
behind, ultimately to become a neutron star. (We avoid the term remnant
to prevent confusion with supernova remnants in the interstellar
medium.) The total energy of the residue is equal to the original total
energy of the progenitor (usually negligible), minus the ejected energy,
and energy lost via neutrinos. A successful ejection of a supernova
with positive energy can therefore occur if and only if the absolute
value of the (negative) total energy of the residue exceeds the energy
lost via neutrinos. Our purpose here is to try to estimate the total
energy of the residue, by calculating the energy transferred from the
inner to the outer core during rebound, and to find out what use is made
of that energy.

For a variety of reasons, numerical hydrodynamical calculations
over the past decade have given ambiguous answers to the most
fundamental question: does the implosion-explosion mechanism succeed in
ejecting a type II supernova? A major difficulty is that the energy of
the supernova is only about 1% of the gravitational energy of the
residue, and so the calculation is very sensitive to small errors.

In this paper we therefore adopt a 'pedestrian' approach, in the
sense that we forego numerical hydrodynamical calculations, in favor of
simpler calculations. The technique is to oversimplify the problem
enough, so that it reduces to a set of ordinary differential equations,
instead of the hydrodynamical partial differential equations. These
ordinary differential equations can then be solved to any desired degree
of accuracy. The usefulness of such calculations is necessarily
limited, because of the approximations employed. They are, however,
free of the many numerical pitfalls to which the full blown
hydrodynamical calculations are prone, and they do provide a conceptual
framework, within which the numerical results can be better understood.

II.   TOTAL ENERGY OF INNER CORE

The basic characteristic of the unshocked inner core is that its

positive internal energy and negative gravitational energy almost
balance, resulting in a total energy that is much smaller than either
one. Thus, while the internal and gravitational energies of the inner
core are of the order of 100 foes, the total energy of the inner core is
only a few foes (one foe equals ten to the fifty one ergs).

The exact net total energy depends, to first order, only on the
mass of the inner core and on the chemical potentials of the leptons
(Yahil and Lattimer 1982; Lattimer, Burrows, Yahil, and Baron 1983).
In particular, it does not depend on the nuclear part of the equation of
state. The enclosed total energy as a function of enclosed mass is
first positive, and increases with enclosed mass. This is because at
small radii the positive specific internal energy is large, but the
negative specific gravitational energy, $-GM(r)/r$, is small. The
enclosed total energy then reaches a maximum, and begins to decrease,
because the specific internal energy decreases at lower density, but the
specific gravitational energy increases in absolute value. If the mass
of the inner core is sufficiently large, a point is reached at which the
total enclosed energy is zero. We define the mass enclosed within this
point to be the Chandrasekhar mass. This definition is a generalization
of the original one given by Chandrasekhar (1939) for a polytrope of
index $\gamma=4/3$. (Some ambiguity arises in the case of a realistic equation
of state, because the zero point of the nuclear binding energy is
arbitrary. We prefer to use the binding energy of ^{56}Fe as the zero
point, since this best relates the energy to the progenitor.)

If the mass of the inner core is larger than the Chandrasekhar
mass, then the total energy of the inner core is negative. That means
that the sum of the energy lost via neutrinos and the energy of the
outer core is positive. There is therefore a possibility for a
supernova ejection, depending on neutrino losses, and on how the
remaining positive energy, if any, is distributed in the outer core.

In reality, as shown by numerical calculations (e.g., Van Riper and
Lattimer 1982), and by self-similar collapse models (Yahil 1983), the
mass of the inner core is ~10% higher than the Chandrasekhar mass, and
the inner core is bound. Yahil and Lattimer (1982) show that, to first
order, the total energy of the inner core can be written as

$$E_{ic} = - \alpha\mu_e Y_e N_0 (M_{ic} - M_{Ch}) \qquad (1)$$

where M_{ic} is the mass of the inner core, M_{Ch} is the Chandrasekhar mass,
μ_e is the chemical potential of the electrons, N_0 is Avogadro's number,
and $\alpha \cong 0.5$ is a numerical constant. The important point is that the
total energy is independent of the nuclear part of the equation of
state, and depends only on the leptonic characteristics and the central

density (approximately equal to nuclear matter density). Inserting
characteristic numbers into equation (1), gives a total energy ~-5 foe,
if the mass of the inner core exceeds the Chandrasekhar mass by 10%.

   The above estimate of the total energy of the inner core is subject
to a number of modifications, including effects of neutrinos, entropy,
and the finite pressure at the edge of the inner core (Burrows and
Lattimer 1983; Lattimer et al. 1983). First, the total energy of the
inner core really depends on the polytropic constant K, which is a
function of the lepton fractions of both the electrons and the
neutrinos, and of the entropy. The electron abundance Y_e should
therefore be replaced by the total lepton fraction Y_1, and the chemical
potential of the electrons μ_e by an effective chemical potential μ_1,
which is the chemical potential of zero entropy electrons giving rise to
the same polytropic constant K. These are straighforward modifications
to equation (1), and the important change is only in the Chandrasekhar
mass, and not in the other variables.

   The effect of the external pressure at the edge of the inner core
is also small. Owing to the incompressibility of the inner core, the
external pressure is unable to move its edge very much, and therefore
does not do much compressional work. Lattimer et al. (1983) find that
the ratio $4\pi R^3 P/\Delta E=6$; so even if $4\pi R^3 P=10$ foe (a rather high value),
then ΔE is only 1.7 foe.

   A major error can be made if the Chandrasekhar mass is calculated
inaccurately. The supernova works on a small energy 'profit margin', so
to speak. If the mass of the inner core were equal to the Chandrasekhar
mass, it would have, by definition, zero total energy. The mass of the
core is higher than the Chandrasekhar mass, but only by a few percent.
Hence, a mistake of a few percent in the calculation of the
Chandrasekhar mass can halve the estimate of the total energy, or even
cause a change in its sign. Such a mistake was made in preparing for
this conference, and an erroneous result was consequently presented in
the talk. Taking the neutrino fraction to be 0.08 instead of 0.09
resulted in an underestimate of the Chandrasekhar mass by 4.3% (for
$Y_e=0.32$), and an overestimate of the total energy of the inner core.

   An additional source of error in equation (1) is due to the rise in
entropy at the edge of the inner core. Equation (1) was derived
assuming the inner core to be isentropic (same K everywhere). While the
shock becomes strong only at the edge of the inner core, as defined in
Section I, it begins as a weak shock further inside, and the material in
the outer part of the inner core suffers a slight rise in entropy. This
increases the effective Chandrasekhar mass yet again. More
devastatingly, it can reduce α by a factor of 2 or so.

The end result of all these effects is that the estimate based on equation (1) was over-optimistic by a factor of 2-3. For the 25 $M_\odot$ progenitor recently published by Weaver, Woosely and Fuller (1983), the total energy is really only about -2 foes, far less comfortable than the -(5-6) foes imagined earlier.

III. SHOCK PROPAGATION

In Section II we saw that during rebound the inner core transfers energy to the outer core. The problem is to identify what the outer core does with the energy it receives. Since the shock causes entropy to be irreversibly changed, some of that energy is responsible for a permanent change in the structure of the star. Presumably, this results in a supernova.

Except for extremely powerful explosions, however, the supernova will not eject the entire shocked outer core. The so-called 'mass cut', i.e., the demarcation line between the ejected material, and that which falls back in, is in realistic cases expected to be outside the unshocked inner core, and well into the shocked outer core. In fact, since the electron fraction is significantly smaller than $Y_e=0.5$, most of the outer core should not be ejected. Otherwise, the interstellar medium would become more abundant in neutron-rich iron-peak elements than is observed. Moreover, the observational evidence is consistent with the notion that all neutron stars have a gravitational mass of 1.4 $M_\odot$ (Taylor and Weisberg 1982; Rappaport and Joss 1983), and therefore the mass cut is at 1.55 $M_\odot$ (the gravitational binding energy accounts for the 10% difference).

The purpose of this paper is to estimate whether or not the shock manages to eject a supernova at all. One way to estimate this is by repeating the previous calculation of the total enclosed energy, but this time for the combined residue of the supernova, made up of the unshocked inner core and the shocked outer core behind the mass cut. Unfortunately, neither the mass cut, nor the profiles of entropy and composition are known. For the entropy and composition profiles, recourse need be made to results of numerical hydrodynamics, or to parametric studies. The mass cut is as yet unknown, and must be left as a free parameter.

We begin by considering the entropy and composition profiles, Fig. 1, obtained in a specific numerical hydrodynamical calculation (Burrows, Yahil, and Lattimer, in preparation), using the 25 $M_\odot$ progenitor of Weaver, Woosely, and Fuller (1983). The entropy and

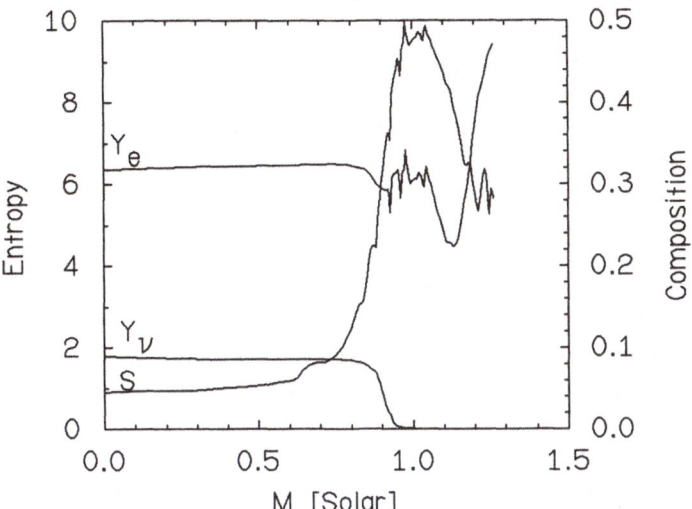

Fig. 1: Entropy and composition profiles after rebound in the core of a 25 $M_\odot$ star.

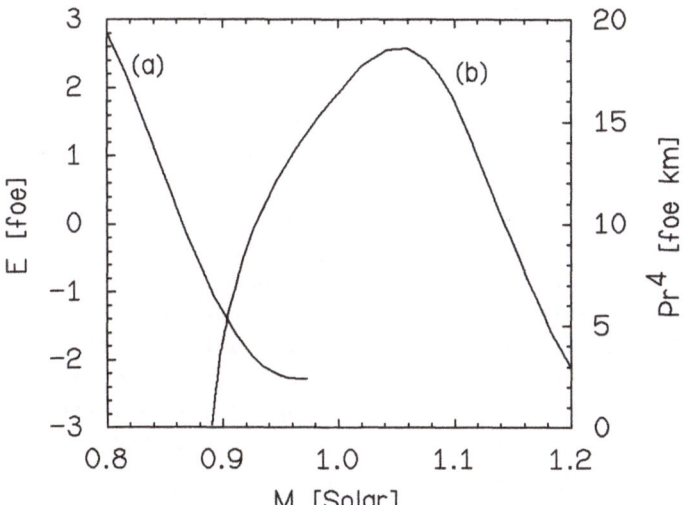

Fig. 2: (a) The total energy as a function of mass for stars having the entropy and composition profiles of Fig. 1. Note that there are no solutions for M>0.97 $M_\odot$.
(b) The surface ram pressure parameter Pr^4 needed to give such stars a total energy of −1 foe. The ram pressure provided by the infalling outer core ahead of the shock is 16 foe km, which is able to contain the shock for M>1.1 $M_\odot$.

composition profiles, together with the total mass, completely define
the structure of the residue. It is therefore possible to calculate the
total energy as a function of the mass cut for these entropy and
composition profiles. (In practice, in order to avoid iterative
calculations, the equation of hydrostatic equilibrium is solved,
starting at the origin, for a range of central densities, and the mass
of the residue is a derived quantity.)

The result of this calculation is shown in Fig. 2. Two
interesting phenomena are seen. First, the total energy as a function
of mass is indeed seen to follow equation (1) in the unshocked inner
core, with the slope dE/dM decreasing as one progresses into the higher
entropy region. Secondly, for $M>0.97\ M_\odot$, where the entropy rises to
really high values, $s\sim 10$, there are no more hydrostatic solutions. This
latter phenomenon is similar to an effect seen in massive giants
(Schonberg and Chandrasekhar 1942), and is the subject of a separate
paper (Yahil, van den Horn, and Baron 1983). Briefly, if the entropy in
the shocked outer core rises high enough, then the sound speed exceeds
the escape velocity, more precisely $Pr/\rho GM>(\gamma-1)/\gamma$, where $\gamma=d\ln P/d\ln\rho$.
When this happens, hydrostatic equilibrium can only be established by
piling up a Jeans mass above this high entropy shell. The Jeans mass,
appropriate to the density and pressure of the high entropy shell, is
much larger than the residue masses under consideration. There is
therefore a mass gap, for which there are no hydrostatic solutions.

By now the perplexed reader is wondering how a residue can exist at
all. The answer, of course, is that, if a supernova is to be ejected,
then the entropy and/or composition profiles must change between the
time when the snapshot in Fig. 1 was taken, and the time at which the
residue settles after ejecting a supernova. The time interval between
these two events may be long, ~ 1 s if Wilson (1983) is correct in
arguing that the shock is only resuscitated by the deleptonization
neutrino flux from the unshocked inner core. Two avenues are now open
to the 'pedestrian' investigator. Either an attempt is made to
calculate those long term changes in the entropy and composition
profiles, or an understanding must be achieved of the interim period.
Here we take the latter approach.

A supernova in its early stages is very different from its later
phases, when it is ploughing into the envelope of the star or into the
interstellar medium: as it propagates into the outer core, material
behind the shock quickly comes into hydrostatic equilibrium. It is
therefore possible to consider the material behind the shock as a
residue, just as we did before. The only difference is that there is a
finite pressure at the surface of this residue. We can now rephrase the
question of total energy as follows. Estimates of the neutrino losses

when the shock emerges through the 'neutrinosphere', and the outer core
deleptonizes, are $\Delta E \sim 1$ foe (Burrows and Mazurek 1982; Bethe, Yahil,
and Brown 1982). The losses during infall are smaller (Brown, Bethe,
and Baym 1982). The total energy of the 'residue' behind the shock
should therefore be ~ 1 foe, as long as the slow deleptonization of the
inner core can be neglected. (The time scale for the deleptonization of
the inner core is ~ 1 s, see Burrows, Mazurek, and Lattimer 1981). The
question is what surface pressure is needed in order to have a
hydrostatic residue with a total energy ~ 1 foe, and is this surface
pressure available?

Physically, if the shock stalls, the surface pressure of the
residue must be provided by the infalling outer core ahead of the shock.
Since this, as yet unshocked, outer core is moving supersonically, the
surface pressure is the ram pressure ρv^2 of the infalling material.
(Some allowance must be made for material right behind the shock, which
is not yet in hydrostatic equilibrium. Consequently the surface
pressure of the real hydrostatic 'residue', somewhat behind the shock,
need not be equal to the pressure right behind the shock. In practice,
the difference between the two is small, and we ignore it.)

From the self-similar collapse model (Yahil 1983) we know that
$\rho \propto r^{-3}$, and $v^2 \propto r^{-1}$. It is therefore natural to describe the ram pressure
of the infalling material in terms of the radius independent quantity
$\rho v^2 r^4$, and the surface pressure of the residue in terms of Pr^4. Fig. 2
shows the Pr^4 required in order to give a residue with the entropy and
composition profiles of Fig. 1 a total energy of -1 foe. The actual
ram pressure provided by the outer core in the first tens of
milliseconds after rebound is $\rho v^2 r^4 = 16$ foe km. We thus see from Fig. 2
that, except for the mass range $1.0 < M/M_\odot < 1.1$, the infalling outer core
is indeed able to contain the supernova, at least as long as it
maintains its ram pressure.

The above analysis suggests that the supernova is unable to break
through. The shock moves forward through the mass range between 1.0 $M_\odot$
and 1.1 $M_\odot$, but then stalls, and is contained by the infalling outer
core ahead of it. This is indeed what the numerical hydrodynamical
calculations show.

What next? We know from the self-similar collapse model that in
the postcatastrophe phase, i.e. sufficiently long after rebound, the
infalling outer core goes into free-fall. At this stage the velocity at
any given radius is about double what it was at rebound, but the density
is much smaller. The ram pressure therefore falls with time. If it
were not for neutrino transport, and its associated change in the
entropy and composition profiles, we could say with certainty that over

a time scale $\lesssim 1$ s the ram pressure would drop sufficiently for the shock to move out again.

The first thought is that neutrino losses can only harm the supernova. The success or failure of the supernova would seem to be determined by a race between the dropping ram pressure of the infalling outer core ahead of the shock and neutrino losses behind the shock. In this scenario a successful supernova, whose neutrino losses are manageable, would move out slowly, on a time scale determined by the rate at which the ram pressure drops. This time scale is proportional to $r^{1/(2-\gamma)}$, i.e., approximately proportional to $r^{3/2}$. The shock velocity would therefore be proportional to $r^{-1/2}$.

The above scenario is not what the recent numerical hydrodynamical calculation by Wilson (1983) shows. In his calculation the shock indeed stalls for hundreds of milliseconds, but when it starts moving again, it moves very rapidly, with a constant, if not accelerating, shock velocity. This behavior need not be a numerical artifact, as some may suspect. Parametric studies of the requisite ram pressure needed to contain the shock (in preparation) show that it is very sensitive to the entropy profile. For some critical configurations, a small change in the entropy profile can result in large changes in the containment ram pressure. The reason, not yet fully understood, may be related to the mass gap seen in the case of the zero surface pressure solutions. It is therefore tempting to speculate that neutrino transport may, in fact, help the shock, instead of hindering it. If neutrino transport changes the entropy profile through such a critical point, where the containment ram pressure suddenly becomes greater than the one provided by the outer core, then a rapid resuscitation of the shock may be possible. This scenario is now under investigation.

This research was supported in part by a USDOE grant DE-AC02-80ER10719 at the State University of New York, and by an Ernest E. Fullam Award.

REFERENCES

Bethe, H.A., Brown, G.E., Applegate, J., and Lattimer, J.M.: 1979, Nucl. Phys. A324, 487.

Bethe, H.A., Yahil, A., and Brown, G.E.: 1982, Astrophys. J. Lett. 262, L7.

Brown, G.E., Bethe, H.A., and Baym, G.: 1982, Nucl. Phys. A375, 481.

Burrows, A., and Lattimer, J.M.: 1983a, Astrophys. J. 270, 735.

——————————————————————————: 1983b, preprint.

Burrows, A., and Mazurek, T.J.: 1982, Astrophys. J. 259, 325.

Chandrasekhar, S.: 1939, 'An Introduction to the Study of Stellar Structure', Chicago: University of Chicago Press.

Colgate, S.A., and White, R.H.: 1966, Astrophys. J. 143, 626.

Goldreich, P., and Weber, S.V.: 1980, Astrophys. J. 238, 991.

Hoyle, F., and Fowler, W.A.: 1960, Astrophys. J. 132, 565.

Lamb, D.Q., Lattimer, J.M., Pethick, C.J., and Ravenhall, D.G.: 1978, Phys. Rev. Lett. 41, 1623.

Lattimer, J.M., Burrows, A., Yahil, A., and Baron, E.: 1983, preprint.

Rappaport, S., and Joss, P.C.: 1983, in 'Accretion Driven Stellar X-Ray Sources', W.H.G. Lewin and E.P.J. van den Heuvel (eds.), Cambridge: Cambridge University Press, in press.

Schonberg, M., and Chandrasekhar, S.: 1942, Astrophys. J. 96, 161.

Taylor, J.H., and Weisberg, J.M.: 1982, Astrophys. J. 253, 908.

Van Riper, K.A., and Lattimer, J.M.: 1981, Astrophys. J. 249, 270.

Weaver, T.A., Woosley, S.E., and Fuller, G.: 1983, preprint.

Wilson, J.R.: 1983, preprint.

Woosley, S.E., and Bodenheimer, P.: 1983, Astrophys. J. 269, 281.

Yahil, A.: 1983, Astrophys. J. 265, 1047.

Yahil, A., and Lattimer, J.M.: 1982, in 'Supernovae: A Survey of Current Research', M.J. Rees and R.J. Stoneham (eds.), Dordrecht: Reidel, pp. 53-70.

Yahil, A., van den Horn, L.J., and Baron, E.: 1983, preprint.

NUCLEOSYNTHESIS IN STELLAR EXPLOSIONS

S. E. Woosley
Board of Studies in Astronomy and Astrophysics
University of California, Santa Cruz 95064,
and
Special Studies Group, Lawrence Livermore National Laboratory
University of California, Livermore CA 94550,

Timothy S. Axelrod
T-Division, Lawrence Livermore National Laboratory
University of California, Livermore CA 94550,

and

Thomas A. Weaver
Special Studies Group, Lawrence Livermore National Laboratory
University of California, Livermore CA 94550

ABSTRACT

The final evolution and explosion of stars from 10 $M_\odot$ to 10^6 $M_\odot$ are reviewed with emphasis on factors affecting the expected nucleosynthesis. We order our paper in a sequence of decreasing mass. If, as many suspect, the stellar birth function was peaked towards larger masses at earlier times (see e.g., Silk 1977; but also see Palla, Salpeter, and Stahler 1983), this sequence of masses might also be regarded as a temporal sequence. At each stage of Galactic chemical evolution stars form from the ashes of preceding generations which typically had greater mass. A wide variety of Type I supernova models, most based upon accreting white dwarf stars, are also explored using the expected light curves, spectra, and nucleosynthesis as diagnostics. No clearly favored Type I model emerges that is capable of simultaneously satisfying all three constraints.

1. SUPERMASSIVE STARS (M > 10^5 $M_\odot$)

The most massive stars to form early in the Universe might have had masses on the order of the Jeans' mass at the time of radiation decoupling, the Big Bang, $\sim 10^5$ to 10^6 $M_\odot$, coincidentally the mass of a typical globular cluster (Dicke and Peebles 1968). Alternatively such supermassive objects might have formed at a later time, and with a greater metallicity, from the relaxation of a dense stellar cluster (Sanders 1970). Stars having such great mass are *general relativistically unstable* (Iben 1963; Fowler 1966). After spending only a few thousand years contracting as gravity provides the radiation leaving their surface (Kelvin-Helmholtz stage),

C. Chiosi and A. Renzini (eds.), Stellar Nucleosynthesis, 263–293.

the stars reach a critical radius, $\lesssim 10^{14}$ cm for a star of 5×10^5 $M_\odot$, where the first order relativistic corrections to Newtonian gravity render the star, which has γ very nearly equal to $4/3$, unstable to continuing collapse. Were it not for the presence of unburned nuclear fuel this collapse would continue indefinitely until the star became a black hole. Because the star *does* contain hydrogen, however, it is possible for nuclear reactions during the implosion to release sufficient energy to power a gigantic explosion.

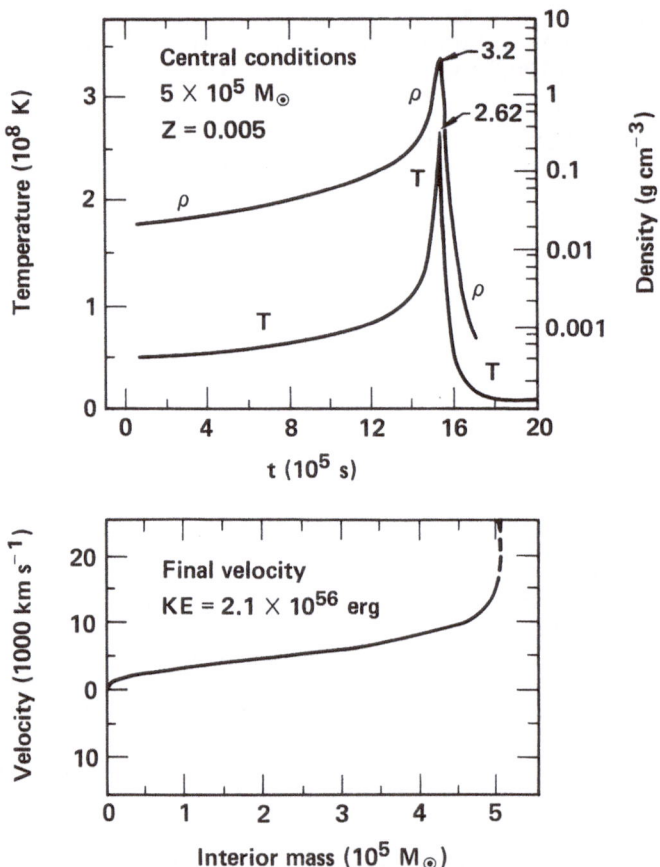

Figure 1. Thermodynamic conditions and final expansion velocity in a supermassive stellar explosion. Peak values of central temperature (in units of 10^8 K) and density (g cm^{-3}) are indicated. The velocity curve is dashed in the outermost layers where the mass zoning was not adequate for an exact calculation. The kinetic energy is given at infinity.

Fuller, Woosley, and Weaver (1982,1983) have recently constructed numerical models of supermassive stars in the range 10^5 to 10^6 $M_\odot$ using a one-dimensional implicit hydrodynamics code (KEPLER; Weaver, Zimmerman, and Woosley 1978) revised to include first order post-Newtonian corrections to gravity and nuclear burning by the rp-process (Wallace and Woosley 1981). They find, similar to Fricke (1973, 1974); Appenzeller and Fricke (1972); and Appenzeller and Tscharnuter (1973), but contrary to Ober (1981), that non-rotating stars in this mass range will collapse to black holes with no reversal of implosion unless the initial metallicity of the star is already a substantial fraction of solar (i.e., $Z \gtrsim 0.005$). The results are quite sensitive to the metallicity since the rate of hydrogen burning is proportional to the abundances of catalytic nuclei in the CNO group. Lower burning rates do not release sufficient energy during the infall to reverse the implosion. Supermassive stars having mass 1×10^5 ($Z = 0$), 2.5×10^5 ($Z = 0$), 5×10^5 ($Z = 0, 0.002, 0.005, 0.01$), and 10×10^5 ($Z = 0, 0.006, 0.01$) solar masses were studied. The 10^5 $M_\odot$ star burned hydrogen stabling, i.e., it did not experience the general relativistic instability prior to hydrogen depletion in the center of the star and it did not collapse. All other models with the exception of three produced black holes. The three exceptions were two 5×10^5 $M_\odot$ stars with metallicity 0.005 and 0.01, and one 10^6 $M_\odot$ star with metallicity 0.01. These models completely disrupted with kinetic energies of $\sim 10^{56}$ to 10^{57} erg. The central temperature and density history during the implosion/explosion of one 5×10^5 $M_\odot$ model is shown in Figure 1. In the 10^6 $M_\odot$ explosion the luminosity, which during the Kelvin-Helmholtz stage was near 10^{44} erg s^{-1}, i.e., the Eddington value, rises to $\gtrsim 10^{45}$ erg s^{-1} and stays that bright for approximately 10 years. Apparently if such gigantic stars ever do form they will make their presence known!

The chief nucleosynthetic consequences of these explosions are the production of helium from hydrogen and trace amounts of ^{15}N and ^{7}Li (Wallace and Woosley 1981; Norgaard and Fricke 1976). The helium mass fraction in both 5×10^5 $M_\odot$ explosions rises by 3% and in the 10^6 $M_\odot$ explosion, it rises by 16%. One should keep in mind, however, that were it not for a near solar concentration of heavy elements these stars would never have exploded in the first place. Thus non-rotating supermassive stars (heavier than 10^5 $M_\odot$) played absolutely no role in producing pre-Galactic helium.

Quite different results may be expected if these stars are endowed with rotation (Fowler 1966; Fricke 1974). In particular, higher central temperatures may be reached by objects that have lower initial metallicity (zero ?) yet succeed in exploding because of the centrifugal potential. At these higher temperatures the rp-process can produce heavy elements up to the iron group and beyond (Wallace and Woosley 1981). Fuller et al intend to model rotating supermassive stars during the forthcoming year by introducing a one-dimensional "psuedo-rotational" potential into the same KEPLER code used to study the non-rotating models. Full two-dimensional calculations should eventually be attempted.

2. VERY MASSIVE STARS ("HYPERNOVAE")

Stars forming with initial masses in the range roughly 100 to 300 $M_\odot$ (without rotation) experience an instability that bears certain similarities to the supermassive stars discussed above. Following stages of relatively stable hydrogen and helium burning (during which *pulsational* instabilities exist that may drive rapid mass loss, Stothers and Simon 1970; Talbot 1971; Papaloizou 1973) the stellar core enters a thermodynamic regime that favors the creation of large quantities of electron-positron pairs. Creation of these pairs takes energy that might otherwise have provided pressure support and thus the star collapses (Barkat, Rakavy, and Sack 1967), again with γ close to 4/3. Nuclear fusion during the collapse liberates energy from carbon, neon, and oxygen burning, but silicon burning *requires energy*. This is because, at the low densities characteristic of such massive stars, silicon burning produces ^{54}Fe + 2p, not ^{56}Ni, and the net reaction sequence is *endoergic*. If sufficient energy is liberated during the collapse by carbon, neon, and oxygen burning then the star blows up; but if the implosion continues well into silicon burning (and rotation is negligible), a black hole forms. Consequently, non-rotating stars having mass greater than about 300 $M_\odot$ (and less than 10^5 $M_\odot$) end up as black holes. Below 100 $M_\odot$ the pair instability is not encountered and the star undergoes stable carbon, neon, oxygen, and silicon burning, culminating in iron core collapse (see next Section).

In the range 100 to 300 $M_\odot$ (these being rough estimates of the mass of the star *at its birth on the main sequence* prior to mass loss), stars end their lives in gigantic explosions called "hypernovae". Such stars have been recently modeled by Woosley and Weaver (1982a); Bond, Arnett, and Carr (1982, 1983; see also Bond, this volume); and Ober, El Eid, and Fricke (1983). The evolution of a 200 $M_\odot$, Pop III (i.e. Z = 0) star studied by Woosley and Weaver is typical. This star burns hydrogen at a central temperature of $\sim 1.2 \times 10^8$ K and density ~ 20 g cm^{-3} for ~ 2.5 million years (note that the main sequence lifetimes of stars in this mass range are approximately constant owing to the Eddington limitation on their luminosities and the fact that they are convective throughout most of their mass). The luminosity of the 200 $M_\odot$ star on the main sequence is $\sim 2.0 \times 10^{40}$ erg s^{-1} and the effective emission temperature (neglecting mass loss) is 80,000 K. Because the luminosity is so close to Eddington and the core pulsationally unstable, extensive mass loss may well modify the surface properties leading to softer, perhaps even infra-red emission. It is interesting that hydrogen burning in this star is mediated by the CNO-cycle despite the fact that the star was born with no metals. Prior to hydrogen ignition the star contracts, heats up and actually burns a trace of helium to nitrogen ($\gtrsim 10^{-9}$ by mass) before igniting hydrogen burning (see also Ezer and Cameron 1971). At one-half hydrogen depletion the nitrogen mass fraction is $\sim 10^{-8}$.

Following hydrogen exhaustion, the 200 $M_\odot$ star burns helium for 265,000 years at a central temperature near 2.5×10^8 K and density 250 g cm^{-3}. During this time the radius of the star expands greatly to $\sim 10^{14}$ cm. At the end of helium burning, the inner 112 $M_\odot$ is comprised of oxygen and carbon in a roughly 3 to 1 ratio with traces of neon and magnesium. As the star contracts and heats in its center to $\sim 10^9$ K, seeking to ignite the $^{12}C + ^{12}C$ reaction, it encounters the pair instability and begins to collapse rapidly (although not super-sonically and not in free fall). The central temperature rises to 3.9×10^9 K and density to 2.2×10^6 g cm^{-3} as carbon burns in the inner ~ 80 $M_\odot$, neon in the inner ~ 70 $M_\odot$, and oxygen in the inner ~ 20 $M_\odot$. The energy released by this burning is adequate to explode the star with total kinetic energy 2.6×10^{52} erg. For 140 days the light output of this hypernova is $\sim 10^{44}$ erg s^{-1}, or about 10 times that of a typical Type II supernova (see Woosley and Weaver 1982a for light curve).

The nucleosynthesis from this explosion is summarized in Table 1 with results normalized to solar system abundances (Cameron 1973) relative to ^{16}O, an abundant product of the explosion. The lack of iron production is a common trait of this class of models owing to the endoergic nature of silicon burning. It is interesting that these explosions are such efficient producers of heavy elements (more than 1/2 of the initial mass ends up in elements heavier than helium) that the explosion of only one or two could account for the abundances of elements from carbon through calcium found in a typical globular cluster of stars (10^6 $M_\odot$, Z $\sim 10^{-4}$). Ten thousand such explosions could produce a metallicity *throughout the Galaxy* comparable to that observed in the most metal-deficient stars (eg. CD-38°-245, Bessel and Norris 1981; see also Bond 1981). If this did in fact occur at very early times, one might expect to find, in the most metal deficient stars, large overabundances of the elements from carbon through calcium relative to iron (compared to the sun). Observations indicate that such a correlation may indeed exist (Sneden,

Lambert, and Whitaker 1979; Butler, Dickens and Epps 1978; Bessel and Norris 1981). Unless some unforeseen circumstance prohibited star formation in this mass range or led to dispersal of the star prior to the pair instability, it is hard to see how the nucleosynthetic contribution from stars in this mass range could have been anything other than very important.

Table 1: Nucleosynthesis in a 200 M$_\odot$ Pop III Star

Species	^{4}He	^{12}C	^{16}O	^{20}Ne	^{24}Mg
Mass ejected (M$_\odot$)	70	11	57	4.5	5.5
Compared to sun	0.03	0.45	1	0.44	1.7

Species	^{28}Si	^{32}S	^{36}Ar	^{40}Ca	^{56}Fe
Mass ejected (M$_\odot$)	20	6.8	0.85	0.66	0.08
Compared to sun	4.6	2.7	1.4	1.4	0.01

Other important implications for stars in the mass range > 100 M$_\odot$, especially for providing "missing mass", helium synthesis, and microwave background are discussed by Bond elsewhere in this volume (see also Bond, Arnett, and Carr 1982, 1983). We would only briefly comment on work in progress by Stringfellow, Woosley, and Bodenheimer regarding the evolution of very massive stars *including rotation*. As noted above the fate of non-rotating stars having main sequence mass greater than ~ 300 M$_\odot$ is a black hole. Stringfellow recently examined a 300 M$_\odot$ *helium core* (corresponding to a main sequence mass of $\gtrsim 500$ M$_\odot$) using a two-dimensional explicit hydrodynamics code (Bodenheimer and Woosley 1983). The 300 M$_\odot$ core was first evolved to the point of pair instability using a one-dimensional *implicit* hydrodynamics code (KEPLER) then remapped onto the two-dimensional grid at a time when its central temperature had reached 2×10^9 K. Continued evolution in *one* dimension led to the formation of a black hole. This same evolution to black hole was replicated in the 2-D code if the star was not rotating. Stringfellow found, however, that the implosive evolution is greatly modified by the introduction of only a small amount of rotation. A rotational energy corresponding to 0.4% of the gravitational binding energy at the beginning of the calculation (i.e., when the central temperature was 2×10^9 K and the binding energy 7.4×10^{53} erg), or a ratio of centrifugal force to gravity of only 1%, leads to a reversal of the implosion and a gigantic explosion. During this explosion matter at the center of the star reaches 12 billion degrees and the central density rises to 4×10^7 g cm^{-3}, much greater than in any of the one-dimensional calculations. A similar calculation employing 5 times as much rotational energy also explodes but reaches a somewhat lower temperature (5.9×10^9 K). Unfortunately the nuclear physics included in these two-dimensional calculations (only oxygen burning) is not adequate for such extreme thermodynamic conditions. In particular, photodisintegration, a very important endoergic process that can diminish the prospects for an explosion, was left out. Stringfellow is currently rewriting his program to include photodisintegration and electron capture and we should have reliable results in the near future. An interesting possibility is mass "bifurcation", i.e., the central portions of the star may continue to collapse while outer layers are ejected by a combination of nuclear burning and rotation (see also Bodenheimer and Woosley 1983). Important implications loom for both nucleosynthesis and the "missing mass" problem. The higher explosion temperatures accessible with the inclusion of rotation may allow iron group nuclei to be synthesized and ejected. Ejection of only a small fraction of the stars mass as heavy elements could eliminate the residual black hole as an important component of the missing mass.

3. MASSIVE STARS (10 to 100 $M_\odot$)

Stars in the range 10 to $\sim$ 100 $M_\odot$ end their lives by producing iron cores of roughly a Chandrasekhar mass which collapse owing to a combination of photodisintegration reactions (induced by the high core temperature) and electron capture (resulting from both the high core temperature and density). In the limited, although numerically significant mass range 8 to 10 $M_\odot$, the late evolution may be diverse and complex. This results, in part, from an inverted temperature gradient set up by efficient neutrino cooling in a star whose central pressure is only marginally sensitive to temperature and, in part, from the semi-degenerate nature of the nuclear burning itself (Weaver and Woosley 1979; Woosley, Weaver, and Taam 1980; Miyaji et al 1980; Nomoto, 1980a, 1981, 1983, this volume; Hillebrandt 1982a). Here we will restrict our attention to the (comparatively) simpler evolution of more massive stars.

Much has been written about the evolution of stars in this mass range and, especially, the core bounce/neutrino transport mechanism which may (or may not, depending seemingly upon the year in which one lives!) lead to their explosion. We shall not, in this limited space, even attempt to review the extensive literature dealing with this subject. The interested reader may consult reviews by Arnett (1980); Trimble (1982); Wheeler (1981); Hillebrandt (1982b,1983); Bowers and Wilson (1982ab); and Brown, Bethe, and Baym (1982). Here we will briefly discuss some recent results from our ongoing attempt to numerically simulate the complete evolution of a grid of stellar masses 10, 15, 20, 35, 50, and 100 $M_\odot$ (of constant mass) from birth on the main sequence through their death as Type II supernovae.

In 1978 Weaver, Zimmerman, and Woosley published presupernova models of 15 and 25 $M_\odot$ stars. These were subsequently followed through simulated core bounce and explosion (Weaver and Woosley 1980a) and yielded luminosity, radius, and effective temperature histories in good agreement with observations of typical Type II supernovae. Furthermore, the nucleosynthesis calculated by Weaver and Woosley (1980a) and, subsequently in greater detail by Woosley and Weaver (1982b), for the 25 $M_\odot$ star agreed reasonably well with solar system proportions of isotopes from ^{12}C to ^{60}Ni. There were notable deficiencies, however, as noted by the authors. The abundances of species in the silicon to titanium range were too small by roughly a factor of 5 compared to other important isotopes such as ^{16}O; ^{23}Na was *overabundant* by roughly the same factor; and several interesting species: ^{13}C, ^{14,15}N,^{19}F, ^{48}Ca, and ^{47}Ti, were hardly produced at all. Some of these deficient productions were not particularly worrisome, eg. ^{13}C and ^{14}N can probably be produced adequately in lower mass stars, ^{15}N by novae, ^{48}Ca by the occasional ejection of highly neutronized core material, etc., but others, eg. the deficient production of silicon through titanium, all of which are *explosively* synthesized during shock wave passage, led us to suggest that solar abundance set requires a component of nucleosynthesis from stars *more massive than* 25 $M_\odot$. This is a proposition to which we still adhere for several reasons. First, as discussed in the previous section, stars of mass $\gtrsim$ 100 $M_\odot$ are copious producers of silicon through calcium (Table 1). Unless nature has purposefully avoided their creation, hypernovae should form an important component of the nucleosynthetic matrix. Second, as one goes to higher mass stars the density gradient around the core becomes less steep. A shallower density gradient implies that the shock wave generated by core bounce (in stars less massive than 100 $M_\odot$ but more massive than 25 $M_\odot$) will heat more matter to peak temperatures in the range 3 to 4 billion degrees, i.e., the requisite range for explosive oxygen burning, and produce proportionately greater amounts of silicon through calcium. Finally, it is unrealistic to expect the nuleosynthesis of all stars in the 10 to 100 $M_\odot$ range to be so similar that element production in any one mass (eg. 25 $M_\odot$) is anything other than an extremely qualitative representation of the entire ensemble. An accurate calculation of *iron group* synthesis, for example, is an especially difficult problem as it relies on knowing both the strength of the shock and the location of the "mass cut" (last stellar zone to fall back onto the neutron star) in stars of different total mass. Therefore, quantitative predictions of iron synthesis are likely to remain highly uncertain for some time to come.

During the last several years we have computed the evolution of a number of other massive stars up to the point of silicon ignition. These include, in addition to the 15 $M_\odot$ and 25 $M_\odot$ models of Weaver, Zimmerman and Woosley (1978), stars of mass 10, 20, 35, 50, and 100 $M_\odot$. We stopped these calculations at silicon ignition because the nuclear physics in the code at that time, especially the rates for and coupling of hydrostatic neutronization processes to the calculation, was not wholly adequate for an accurate representation of the final evolutionary stages (the same goes for the 15 and 25 $M_\odot$ stars that we *did* publish; see below). Even so, we were able to determine several interesting quantities that are not expected to be greatly modified during silicon core burning or by the ejection process. These include the helium core mass at the end of the stars life; the mass of heavy ($Z > 2$) elements in the core (from which we can subtract roughly 1.5 $M_\odot$ to obtain a rough estimate of the total mass of heavy elements ejected by the star); and the nucleosynthesis of light isotopes ($Z \lesssim 12$). Given the helium abundance dredged up by convection into the envelope, one can also estimate $\Delta Y/\Delta Z$, the change in helium mass fraction divided by the change in metallicity, for these stars. The nucleosynthesis of light isotopes was reported in Woosley and Weaver (1982a). The remaining quantities are summarized in Table 2. The metallicity, i.e., that fraction of the initial stellar mass finally ejected in the form of elements heavier than helium, can be approximated in the mass range 15 to 100 $M_\odot$ by $Z = 0.5$ - 6.3 $M_\odot/M$.

Table 2: Bulk Nucleosynthesis in Massive Stars

Mass ($M_\odot$)	Helium Core ($M_\odot$)	Metal Core ($M_\odot$)	Y_{env}*	$\Delta Y/\Delta Z$
10	2.6	1.5	0.30	large
15	4.5	3.0	0.33	0.70
20	7.4	5.0	0.36	0.40
25	9.5	7.1	0.40	0.30
35	17.0	13.0	0.41	0.08
50	24.0	21.0	0.51	0.14
100	47.0	47.0	0.57	0.05

*Initially the helium mass fraction in envelope was 0.28

More recently Weaver, Woosley, and Fuller (1982, 1983; WWF) *have* revised the stellar evolution code so that the electron capture processes that commence during oxygen and silicon burning *are* properly modeled. Modifications include weak

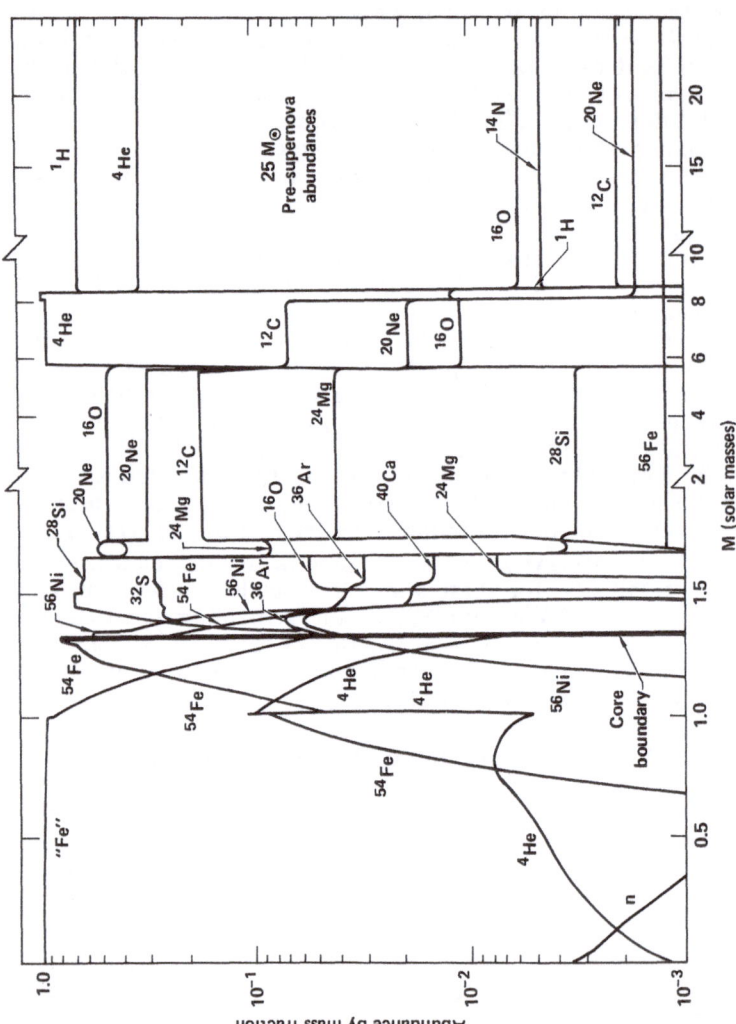

Figure 2. Abundances in the new 25 $M_\odot$ pre-supernova star (WWF). The composition is sampled at a time when the collapse velocity has first reached a value of 1000 km s^{-1} in any zone. Note the smaller iron core mass as compared to earlier calculations. Horizonal abundance lines are, in general, indicative of regions that are, or once were, convective. An exception is the curve labeled "Fe" which is the sum of the mass fractions of all iron group species more neutron-rich than ^{54}Fe.

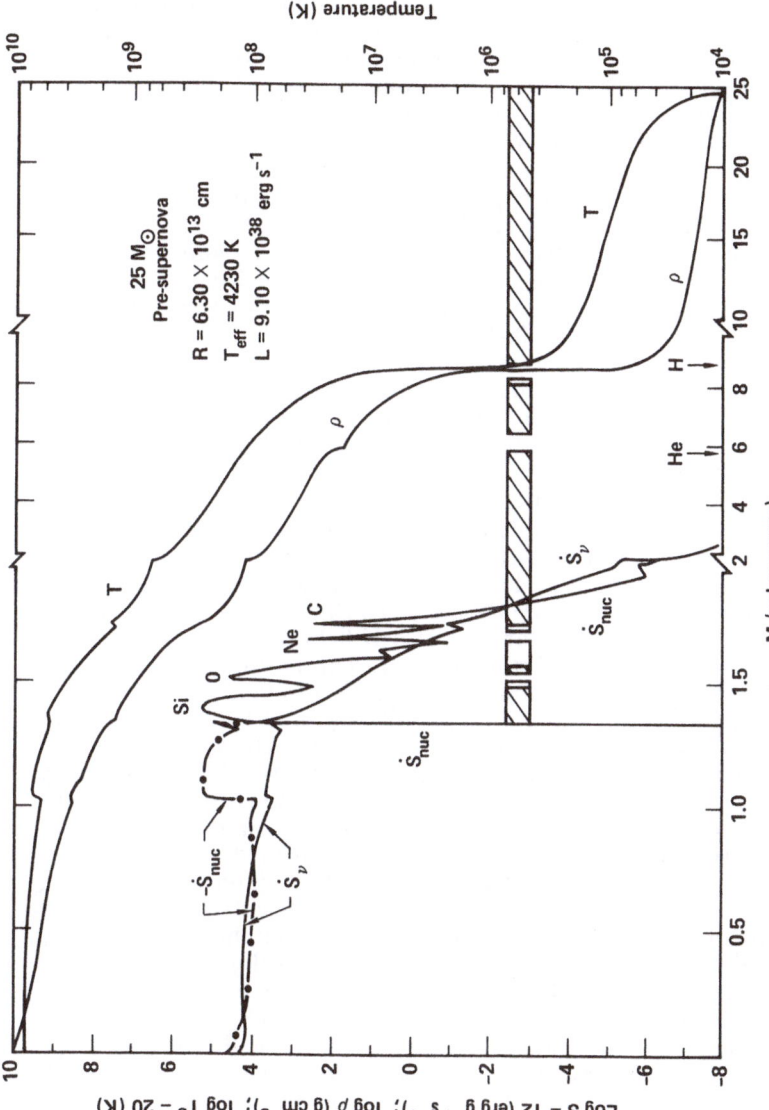

Figure 3. Thermodynamic conditions in the new 25 $M_\odot$ presupernova star (WWF). In the region external to the iron core the nuclear energy generation rate is positive and the neutrino loss rate negative. Internal to the core both are negative, with neutrino losses primarily coming from electron capture. Collapse here (at the same time as Figure 2) is a consequence of both photodisintegration and electron capture. Cross-hatched and open rectangles denote regions of convective and semi-convective instability respectively. Hydrogen and helium shells are burning at rates that are off-scale as indicated by arrows. Curves for ρ and T are parallel in regions where the density scales as T^3. The stellar radius, effective emission temperature, and photon luminosity are also given.

interaction rates from Fuller, Fowler, and Newman (1982ab); a new treatment of quasiequilibrium and "cluster equilibrium" (Woosley, Arnett, and Clayton 1973) that calculates the rate of neutronization (principally due to electron capture) during core and shell oxygen burning as well as during silicon burning and core collapse; new sequences of nuclear reactions for silicon burning in a composition that has already become so neutron-rich that ^{30}Si is a major constituent; minor corrections to nuclear screening factors; initial abundances from Cameron (1973) rather than Cox and Stewart (1970ab); and finer time step and zoning criteria (roughly a factor of 2 in each). Using the improved code, WWF have calculated new pre-supernova models of 20 and 25 $M_\odot$ stars. Other masses, eg. 10, 11, 15, 50, and 100 $M_\odot$ are in progress.

The chief differences in the new pre-supernova models (Figures 2 and 3) are smaller iron cores at the time of collapse; smaller values of Y_e, the electron mole number; and lower values of entropy (see WWF for plots of Y_e and entropy in the new models). In 20 and 25 $M_\odot$ pre-supernova models the iron core masses are now 1.41 and 1.35 $M_\odot$ (as opposed to a value of 1.61 $M_\odot$ in the old 25 $M_\odot$ star); central values of Y_e are 0.422 and 0.423; and central values of entropy are 0.69 and 0.73 respectively. The smaller cores and lower entropies should favor the prospects for explosion by the core bounce mechanism (eg. Hillebrandt 1982b; 1983). Additionally, the density gradients surrounding the iron cores of the new models are less steep. This implies, all else being equal, improved nucleosynthesis of elements from silicon through calcium. We plan to follow the (continued) *explosive* evolution of these new models in the near future.

Perhaps the most exciting recent development in studies of massive star evolution are the calculations of Wilson (1983) showing the *delayed* explosion of the collapsed core following a long period of shock stagnation. *All* the models Wilson has calculated now explode *when the calculation is continued to sufficiently late times*. These include the "old" 10 $M_\odot$ star (see also Hillebrandt 1982a); the new 25 $M_\odot$ star (WWF); and *the old 15 $M_\odot$ star* (Weaver, Zimmerman, and Woosley 1978). Explosions occur at time ranging from 0.25 s (10 $M_\odot$) to 0.55 s (25 $M_\odot$); produce total explosion energies of from 1.8 to 4.5 $\times 10^{50}$ erg; and leave neutron star masses (not corrected for neutrino losses) of 1.48 $M_\odot$ (10 $M_\odot$), 1.76 $M_\odot$ (15 $M_\odot$), and 1.66 $M_\odot$ (25 $M_\odot$). Note that the 25 $M_\odot$ core accretes roughly 0.3 $M_\odot$ before finally exploding. The mechanism for these explosions involves the slow diffusion of neutrinos out of the core which capture on free nucleons behind an almost stationary accretion shock. The heating from this capture plus the diminished ram pressure as regions of decreasing density pass through the accretion shock (and the scattering of neutrinos on pairs produced as the temperature rises and density drops, a lesser effect) ultimately combine to change the standing accretion shock into an outwardly propagating wave capable of ejecting the stellar mantle and envelope. Considerable numerical uncertainty exists in these calculations owing to the necessity of running thousands of time steps using explicit hydrodynamics and continuous mass rezoning. It will be very interesting to see the results of future calculations, perhaps employing implicit hydrodynamics and mixed Eulerian/LaGrangian mass zoning. Also interesting are the implications this all has for "leptonic convection" (Epstein 1979) at these late times and the possibility of core overturn (Colgate and Petschek 1979; Smarr *et al* 1981).

TYPE I SUPERNOVAE

There is general consensus that Type I supernovae (SNI) are the explosions of relatively low mass stars, at least compared to Type II's, and that the stars involved are void of substantial hydrogenic envelopes. The former contention is supported by the lack of association of SNI's with the spiral arms of (spiral) galaxies within which they occur (Maza and van den Bergh 1976) and the occurrence of SNI in elliptical galaxies where the rate of massive star formation is very small. The latter is consistent with the absence of hydrogenic lines in the SNI spectrum (unlike SNII) and the lack of a well defined "plateau" phase in the light curve (Falk and Arnett 1973; Weaver and Woosley 1980a). Most currently favored models are based upon white dwarfs accreting mass in cataclysmic variable systems (Wheeler and Hansen 1971; Whelan and Iben 1973; Mazurek 1973; Nomoto and Sugimoto 1977; Nomoto 1980ab, 1981, 1982ab, this volume; Taam 1980ab; Chevalier 1981; Weaver and Woosley 1980b; Weaver, Axelrod, and Woosley 1980, (WAW); Woosley, Weaver, and Taam 1980; Woosley and Weaver 1981; Woosley, Taam, and Weaver 1983,(WTW); Iben and Tutokov 1983) although a limited range of stellar masses between 8 $M_\odot$ and 10 $M_\odot$ may also contribute to the SNI rate (WAW) if such stars lose their hydrogen envelopes and ignite a degenerate thermonuclear runaway.

The fate of the accreting white dwarf as well as the physical mechanism for its explosion is critically dependent upon its mass, accretion rate, and the composition of the accreted matter (see Iben and Tutokov, 1983, for detailed discussion of various evolutionary scenarios expected to lead to SNI). Some combinations lead to common nova outbursts. If the thermonuclear model for novae as commonly formulated is correct (cf., Starrfield, Sparks, and Truran 1974; Starrfield, Truran, and Sparks 1981; Mcdonald 1983), newly accreted material must mix with the carbon-oxygen substrate during the inter-flash period. Since the nova itself does not appear likely to produce fresh carbon during its outburst, this implies that the white dwarf mass does not grow but may, in fact, shrink. Either way, the critical mass for a SNI is not attained. For some range of accretion rates and dwarf masses, though, the hydrogen shell flashes are expected to be weak (Fugimoto and Taam 1982; WTW; Nomoto this volume) and the mass of the dwarf *does* grow as a layer of degenerate helium accumulates between the hydrogen burning shell and the carbon-oxygen core (see Nomoto and Sugimoto 1977 for the interesting case of *helium* accretion onto a *helium* white dwarf). Providing that the accretion process is not truncated, the white dwarf has three possible fates: 1) off-center ignition with the accreted helium layer consumed in an outwardly propagating *detonation* wave and (most of) the carbon-oxygen core left behind as a lower mass white dwarf remnant; 2) off-center ignition with *dual-detonation waves* propagating inwards and outwards leading to the complete nuclear incineration and disruption of the star; or 3) a centrally ignited *deflagration*, with runaway occurring just short of the Chandrasekhar mass, that burns only the inner portion of the white dwarf but also leads to its complete disruption. We shall briefly discuss representative models of each variety.

3.1 Detonations and Deflagrations

WTW have recently computed the complete evolution of three accreting white dwarfs, all of which experience "off-center" detonation (cases 1 and 2 above). The results should be representative of a broad range of accretion rates and dwarf masses. Specifically examined were carbon-oxygen white dwarfs of mass 1.2, 0.8, and 0.5 $M_\odot$ accreting helium (i.e., assuming hydrogen shell flashes to be weak or the donor star to be pure helium at rates of 10^{-9}, 2×10^{-9}, and 10^{-8} $M_\odot$ y^{-1} respectively. We shall refer to these calculations as Models 1, 2, and 3. All three models ignited helium burning under conditions that were extremely degenerate and, in all cases, this ignition was achieved prior to the star attaining the Chandrasekhar mass. The total masses of the white dwarfs at helium ignition were 1.43 $M_\odot$, 1.34 $M_\odot$, and 1.12 $M_\odot$ respectively for Models 1, 2, and 3. Model 3 has been discussed previously (Weaver and Woosley 1980b; Woosley, Weaver, and Taam 1980; WAW). The results of Model 2 are summarized in Figure 4.

The temperature and density as helium first begins to run away in Model 2 are 6.1×10^7 K and 7.5×10^7 g cm^{-3} respectively. Owing to the extreme degeneracy, the runaway proceeds all the way to the iron group, chiefly ^{56}Ni at this relatively low

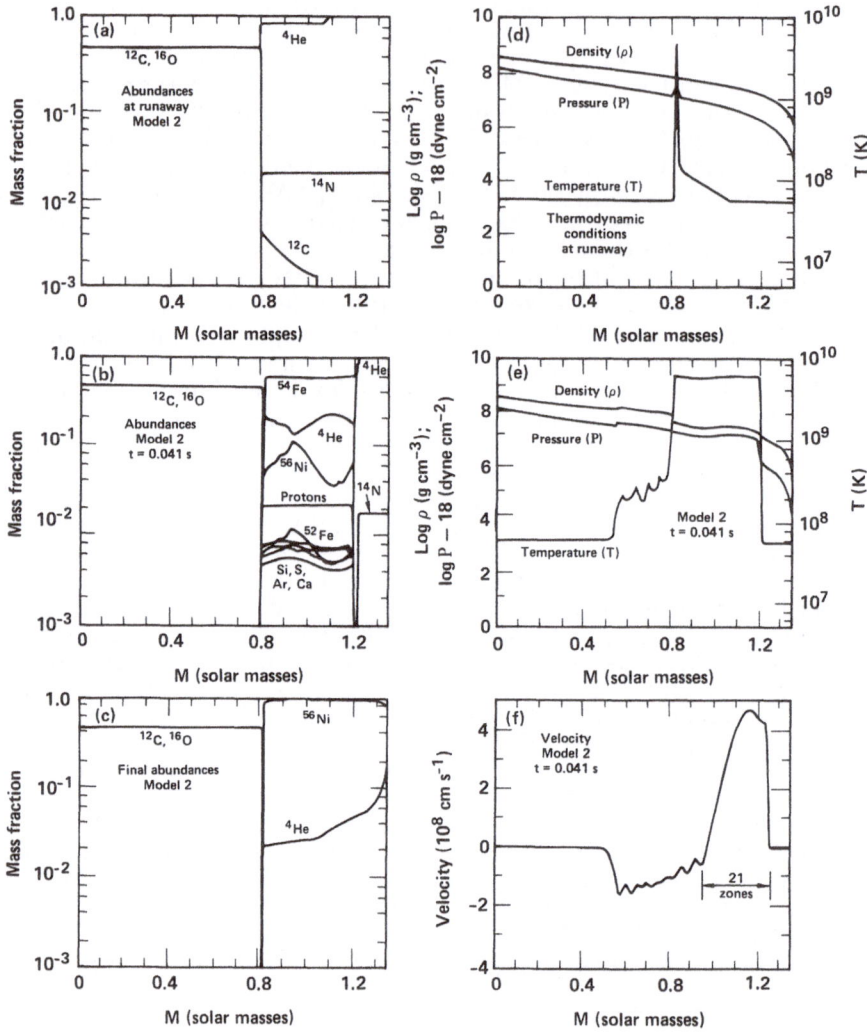

Figure 4. A sequence of figures showing the explosive evolution of "off-center" detonation Model 2. The composition at the time of runaway, as defined by that time when the temperature first reaches 5×10^9 K at the base of the accreted layer (d) is shown in a. A small amount of helium has burned to ^{12}C before the runaway ensues. A short time later (41 ms; b, e, and f), a detonation wave has burned 0.4 $M_\odot$ of the outlying material to iron group species. As expansion leads to cooling the ^{54}Fe shown in b recombines with protons to produce ^{56}Ni as shown in c. The large overpressure and velocity in the outgoing shock and small overpressure and velocity of the inwardly proagating wave are seen in e and f. The carbon oxygen core never ignites. A portion is left behind as a white dwarf remnant of the supernova explosion.

value of neutron excess, before substantial expansion occurs. The nuclear reactions release about 1.5×10^{18} erg g^{-1} and raise the local pressure by about a factor of 5. During the latter stages of the runaway the nuclear time scale becomes extremely short, much shorter than the sound crossing time for the burning region. The short burning time and large overpressure lead to expansion that, in the unburned material, is supersonic. Moving outwards the shock wave experiences a decreasing density gradient as well as a fuel (helium) that has a low ignition temperature. This allows the outward moving shock to propagate as a successful detonation wave, converting all but the outermost portions of the helium shell into iron. The inbound wave meets with less success. It must move up a density gradient and attempt to ignite a fuel (carbon) that has both a higher flash temperature and lower specific energy. As Figure 4 shows, the inward wave does *not* succeed in igniting the carbon-oxygen core (as it *does* in Model 3; see Woosley, Weaver and Taam 1980), but simply dies out and becomes an acoustic wave which compresses and heats the core only slightly. Consequently, that layer which was helium is converted into iron and ejected with high velocity while most of the carbon-oxygen core is left behind as a bound remnant. Not all of the original white dwarf remains, however, only about 0.54 M$_\odot$ of the initial 0.8 M$_\odot$. This is because the total energy (internal minus gravitational) of the new, lower mass white dwarf is less than the total energy of the core of equivalent mass prior to removal of its outer layers and the difference is available for ejecting material. A total of 0.79 M$_\odot$ is ejected by Model 2 with kinetic energy 1.15×10^{51} erg. This ejected mass includes 0.26 M$_\odot$ of carbon and oxygen and ~ 0.50 M$_\odot$ of ^{56}Ni. Model 1 also leaves behind a white dwarf remnant, 1.16 M$_\odot$ in this case, and ejects a correspondingly smaller amount of mass, 0.27 M$_\odot$, which is mostly ^{56}Ni. The total kinetic energy is also much smaller, 3.8×10^{50} erg. Model 3, as mentioned previously, leaves no remnant and generates an explosion of 2.2×10^{51} erg. The light curves, spectra, and nucleosynthesis from these explosions are discussed in subsequent sections.

We have also recently studied a number of models characterized by centrally ignited *deflagration*. Model 4 is the end point of the evolution of a 2 M$_\odot$ helium core, which might typify a main sequence star of ~ 9 M$_\odot$ that has lost its hydrogen envelope prior to the reduction of the helium core mass by convective dredge-up (WAW). This core spends 2.9 million years burning helium in a convective core of 0.73 M$_\odot$. Typical conditions during this time (evaluated at 1/2 helium depletion) are central temperature 1.6×10^8 K, central density, 2.7×10^3 g cm^{-3}, radius, 2.3×10^{10} cm, luminosity, 1.2×10^{37} erg s^{-1}, and effective emission temperature, 75,000 K. During the next 300,000 years helium burns in a shell as the carbon-oxygen core grows to 1.1 M$_\odot$ (60% carbon, 40% oxygen). The envelope of the helium star also expands to $\sim 1.3 \times 10^{12}$ cm and the luminosity rises to $\sim 1.3 \times 10^{38}$ erg s^{-1}. Neutrino losses become quite important during this stage, comparable to the surface luminosity, and a central temperature inversion develops. At this point, *i.e.*, when the CO-core is 1.1 M$_\odot$, carbon burning ignites non-degenerately, but off-center at ~ 0.35 M$_\odot$. Typical conditions in the convective carbon burning shell are 7×10^8 K and 3×10^5 g cm^{-3}, while the temperature at the center of the star is only $\sim 4 \times 10^8$ K. Over the next 50,000 years the carbon burning shell moves inwards, eventually reaching the center of the star. During this time the boundary of the CO-core (*i.e.*, the helium burning shell) moves out to 1.22 M$_\odot$ and, over the next 15,000 years, to 1.3 M$_\odot$. At this time the inner 1.1 M$_\odot$ consists of a neon-magnesium core (60% neon, 25% magnesium, and 15% oxygen) with the region 1.1 to 1.3 M$_\odot$ composed of carbon and oxygen.

Here the helium shell burning became (perhaps unrealistically) violent. We purposefully maintained coarse zoning in the helium layer in order not to spend prohibitive amounts of computer time following thin shell instabilities. As the core grew from 1.3 M$_\odot$ to the Chandrasekhar value, successive helium shells burned one at a time as they added to the core. Owing to the degenerate conditions, high temperatures, up to 1.5 to 2.0×10^9 K, were attained as these zones experienced carbon, neon, and in some cases, even oxygen burning following their addition. This stage of the evolution is probably not realistic and should be computed with much finer zoning before it is taken seriously. The composition from 1.3 to 1.43 $M_\odot$ must therefore be regarded, for the time being, as artificial.

Even so, the evolution of the *central core* should be correctly represented and, for purposes of exploring the generic class of deflagrating models, this should be adequate. It is, after all, the central density at runaway, and, to a lesser extent, the composition of the inner solar mass of material, that influences the outcome. The central density determines the mass and binding energy of the (nearly Chandrasekhar) core as well as the rate of electron capture once the explosion gets underway. The composition determines the specific energy released as burning proceeds to the iron group. Of course the exact composition of the outer layers will be important spectroscopically.

The conditions at runaway in Model 4, *i.e.*, when central nuclear energy generation first exceeded neutrino losses, were 2.9×10^8 K and 2.3×10^9 g cm^{-3}. As the runaway developed, the central density reached a peak value of 3.7×10^9 g cm^{-3} before burning finally led to expansion. Burning in Model 4 was propagated as a *convective deflagration*, which is to say, ignition proceeded at a subsonic rate with overlying material ignited by the thermal transport of energy by convection from underlying material that was already burning. Numerical simulation of such a phenomenon using mixing length theory is highly uncertain at best, even in one dimension (see Muller and Arnett 1982 for a two-dimensional attempt). The temperature gradients are so extreme (super-adiabatic) that the convective velocity is near sonic. Peak temperature in the burning front is $\sim 7 \times 10^9$ K compared to only a few $\times 10^8$ K a short distance ahead of the front. Furthermore the time scale for nuclear burning is orders of magnitude shorter than the Courant crossing time for any reasonable zoning. Thus it is practically impossible to resolve the burning front on a numerical grid. This leads to a computational *impasse* which, if great care is not taken, may propagate the burning at an artificial velocity. This occurs, for example, if the luminosity from one zone to the next is allowed to increase at an arbitrary rate set only by the (artificial) temperature gradient which may itself change on a *very* short time scale. As that gradient becomes large, the "convective luminosity" may become enormous heating the overlying zone rapidly to the flash point in a time short compared to what actual mixing could have achieved, even at the sound speed. After flashing, that zone can ignite the next one in a similar fashion and a chain reaction occurs that can erroneously propagate the burning at a supersonic velocity even though the overpressure is not, in reality, enough to propagate a detonation wave.

We avoided this artificial propagation by restricting the rise in luminosity in a given time step to be no greater than the luminosity at the previous step multiplied by $exp(\Delta t/\tau_{conv})$ where τ_{conv} is the convective crossing time for the zone (thickness of the zone divided by the convective velocity which was, itself, limited to subsonic values). Obviously, this prescription is somewhat arbitrary and, for that reason, we regard the velocity at which the deflagration propagates as essentially a free parameter (limited, of course, by the speed of sound). Besides the central density at runaway, it is the most important parameter in the deflagrating models since it determines how much material will burn to iron and how quickly the star starts to expand and quenches the burning. For our calculations this critical parameter turned out to be roughly 1/3 the speed of sound (or about 2000 km s^{-1}). We note, in anticipation of subsequent discussions, that models which employ larger propagation velocities will synthesize more iron, develop greater expansion velocities, and

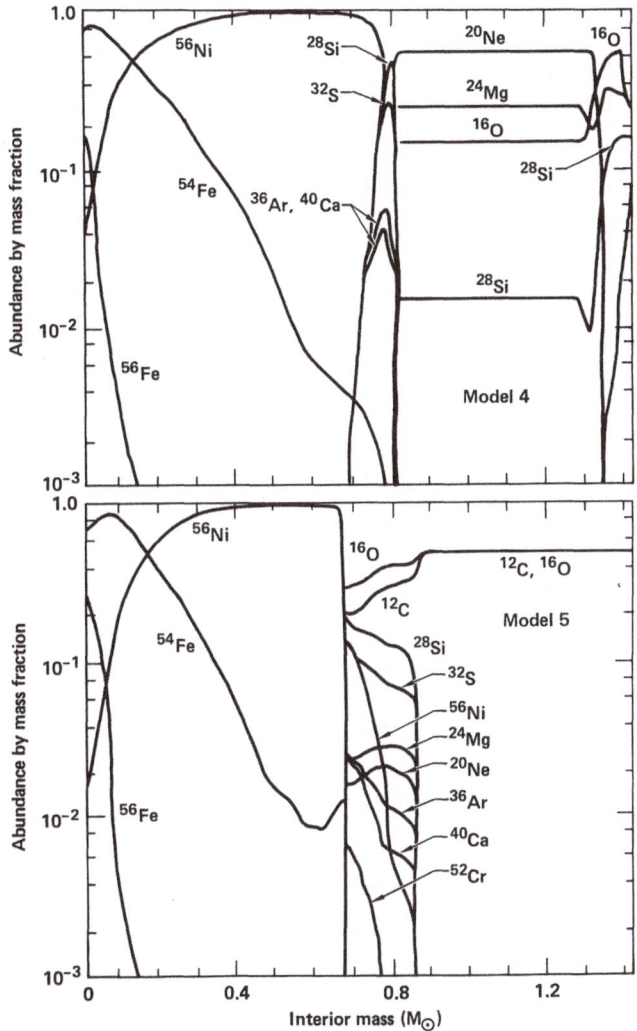

Figure 5. Final compositions of deflagrating Models 4 and 5. A shell of the products of explosive oxygen burning surrounds the iron core in both cases. This is a result of the expansive quenching of the deflagration. Large quantities of ^{54}Fe are present in the cores of both models. The composition of the outer 0.15 M$_\odot$ of Model 4 may be artificial.

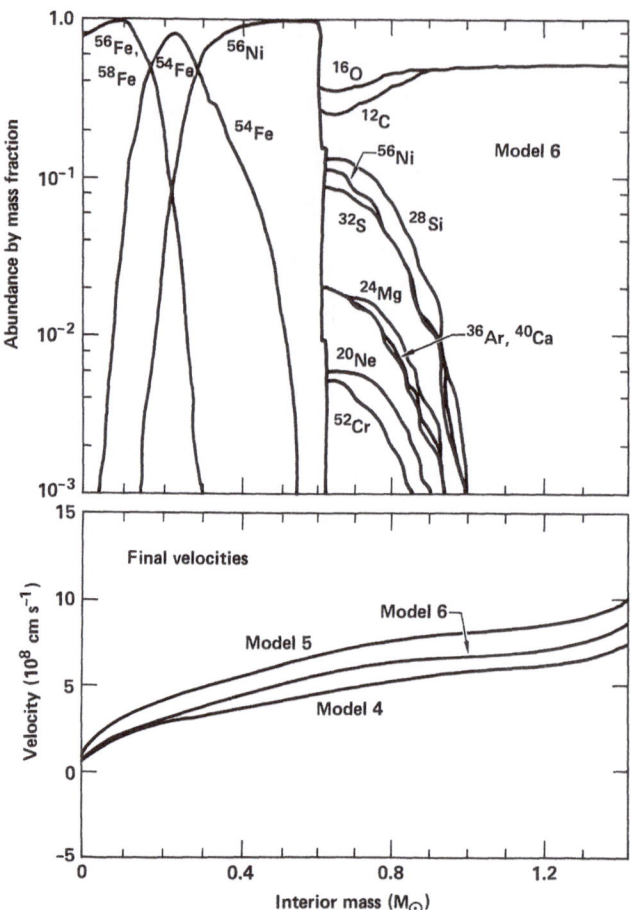

Figure 6. Final composition of deflagrating Model 6 and terminal velocities for all 3 deflagrating models. Owing to its large central density, Model 6 produces excessive quantities of neutron-rich isotopes. The velocity curves reflect the central densities at runaway (Model 6 more tightly bound than Model 5) and the composition (less specific energy from neon burning in Model 4 than carbon and oxygen burning in Models 5 and 6).

experience less electron capture. The final composition and velocity for Model 4 are given in Figures 5 and 6. The total kinetic energy at infinity was 3.5×10^{50} erg.

We also prepared two additional models to simulate deflagration in accreting carbon-oxygen white dwarfs. Models 5 and 6, composed initially of 50% carbon and 50% oxygen throughout their mass, were parametrized by their central densities at the time of runaway. We leave the task of relating these densities to actual evolutionary scenarios and, especially, accretion rates, to others (*cf.* Nomoto 1982ab, Iben 1978ab, 1982; Taam 1980b; Iben and Tutukov 1983). In our calculations ignition was achieved by forming white dwarfs of nearly Chandrasekhar mass having initial central temperature 2×10^8 K and adding carbon and oxygen to their surfaces until they exploded. The preignition temperature should not be critical except as it partly determines the runaway *density*. In Model 5 explosion occurred at a peak central density 2.4×10^9 g cm^{-3} and in Model 6 explosion occurred at a peak density of 7.0×10^9 g cm^{-3}. Final compositions and velocities are shown in Figures 5 and 6. The kinetic energies at infinity were 6.2×10^{50} erg and 4.8×10^{50} erg for Models 5 and 6 respectively.

In the final composition of all 3 deflagrating models one sees a tenth of a solar mass or so where burning at intermediate temperatures ($5 \times 10^9 \gtrsim T \gtrsim 2 \times 10^9$ K) synthesizes elements from neon to calcium. This burning transpires as the deflagration dies out, quenched by the expansion of the white dwarf. An interesting effect, poorly modeled in the present study, is the mixing that may occur for several seconds following the explosion owing to Raleigh-Taylor instability. The central deflagration naturally produces a density inversion which is maintained by its higher temperature. Partial "overturn" of the region just outside the iron core is a possible consequence.

Another effect, seen in some calculations, but which we regard as "bogus" is classical convection occurring at very late times. The luminosity input by nickel decay is greatly super-Eddington and, while the envelope remains optically thick (several days) and dominated by radiation pressure, it is formally unstable to overturning convection. Allowing this convection to proceed at sonic velocity in the co-moving frame could stir freshly synthesized nuclei throughout most of the exploding star. It is not proper, however, to apply traditional convection theory in a medium which is freely streaming and in which gravity has become a negligible force. The luminosity of a supernova (Type I or II) is *advectively* transported, i.e. carried by expanding matter and released at the optically thin photosphere. The total energy in the radiation is, at times later than a few seconds, an inconsequential fraction of the streaming kinetic energy of the ejecta and, as such, it can have negligible effect on its motion. We turned convection off in our calculations at times later than a few seconds.

3.2 Light Curves

It has long been realized that the light output of a Type I supernovae *cannot* be a consequence simply of the high temperature and kinetic energy created during the explosion. Some late time energy input is required that is not degraded by adiabatic expansion. To see this, one must only realize that the expanding white dwarfs of the previous section will not become optically thin until their radii have increased to roughly 10^{15} cm. This estimate follows from an initial density of $\sim 10^8 - 10^9$ g cm^{-3} and an opacity of a few tenths cm^2 g^{-1} (WAW), typical of the large ensemble of Doppler-broadened lines in the expanding plasma. Since the initial radius was roughly a million times smaller than this, any initial internal energy deposited by the explosion is degraded to a trivial value. In Type II supernovae this problem is circumvented by starting with a configuration that already has a large radius, *i.e.*, the low density envelope of a super-giant star.

The solution to this problem in SNI has also been realized for a long time, namely that the light curve of Type I's is a result of the delayed energy input by the decay of radioactive nuclei produced in the explosion (Borst 1950; Burbidge et al 1957), principally the nucleus ^{56}Ni ($\tau_{\frac{1}{2}} = 6.1\ d$) and its daughter ^{56}Co ($\tau_{\frac{1}{2}} = 78.5\ d$) (Pankey 1962; Colgate and McKee 1967). It is also important to note that the decay of ^{56}Co proceeds 19% of the time though a mode that produces positrons. Thus even at late times when the supernova remnant becomes transparent to γ-radiation the kinetic energy of the positrons still deposits energy ($\sim 4\%$ of the total decay energy) and can power the late time light curve (Arnett 1979; Axelrod 1980ab). It is also important that the time history of the emission will not be given simply by the decay rate of the radioactive nuclei but will be modulated by both the efficiency of energy deposition by γ-rays, a factor that is density dependent, and by a time dependent fraction of deposited energy that managed to escape without being degraded by adiabatic expansion (WAW). The correct expression for the light output as a function of time is

$$L(t) = M_{56}\dot{S}(t - \tau_l)f_{dep}(t - \tau_l)f_{esc}(t),$$

where M_{56} is the mass of freshly synthesized ^{56}Ni, $\dot{S}$ is the nuclear energy generation rate from radioactive decay, f_{dep} is the fraction of the radioactive decay energy deposited in the remnant (minimum value 4%), f_{esc} is the fraction of previously deposited γ- and positron energy that avoids decompression to escape as optical (or near optical) radiation, and τ_l is the retarding time to approximately reflect the mean lag time between energy deposition and escape (or decompression). Analytic expressions that may be related to any supernova model are given for all of these quantities by WAW. It is interesting that the results are most sensitive to i) the mass of ^{56}Ni synthesized, ii) the expansion velocity (i.e., the kinetic energy of the explosion), and iii) the density structure of the expanding supernova. All these would be quite similar for exploding white dwarfs of nearly Chandrasekhar mass, which might explain why the class of Type I light curves is such a homogeneous sample.

The light curves for Models 1, 2 and 3 are given in Figure 7. A detailed comparison of the light curve for Model 3 and the observational data for supernova 1972e was given by WAW and showed excellent agreement. Model 2, with an appropriate but reasonable adjustment of supernova date and distance to NGC5253, gives almost as good a fit. Light curves for the deflagrating models have not yet been computed, but are expected to closely resemble Models 2 and 3 except that for Model 4, the 2 $M_\odot$ helium star, there may be important modifications to its light curve brought about by the presence of $\sim$0.5 $M_\odot$ of low density helium surrounding the exploding core. Model 1 is too fast and cannot be made to fit the data at all well. The reasons for the dim, rapid display in Model 1 are obvious. Too little ^{56}Ni is synthesized and the ejecta move too fast, hence become thin to γ-radiation too quickly. Models 1 and 2 may, in fact, be even "faster" in their optical light curves than Figure 7 suggests. As we shall see in the next section, the emission from these explosions is dominated by lines of highly ionized iron that may, shortly after peak, emit principally in the ultraviolet. Thus there may be a substantial bolometric correction.

It is also important to recall that Models 1 and 2 leave white dwarf remnants. The calculations of WTW therefore suggest the existence of a class of SNI characterized by i) comparatively fast, dim light curves (hence rendering their detection less likely); ii) ultraviolet emission and "hot" spectra (see next section); and iii) white dwarf stars present in their remnants, perhaps even still bound in the binary that produced the explosion. No event of this type has been unambiguously identified although the models WTW constructed should correspond to events that occur in nature.

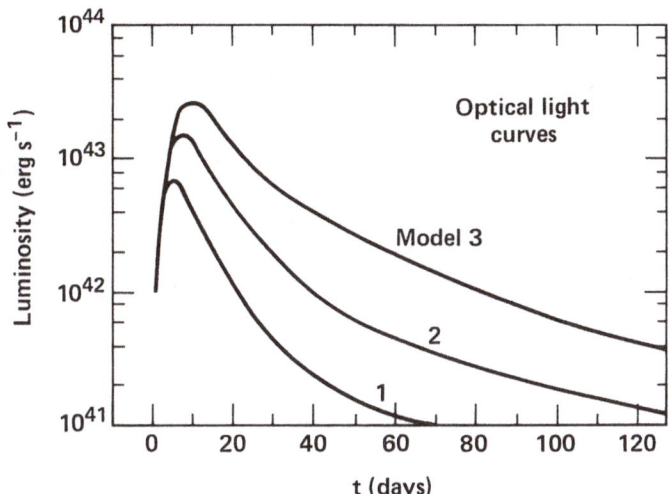

Figure 7. Optical (bolometric) light curves for the 3 "off-center" detonation models as calculated by WTW using the prescription of WAW. The light curves are most sensive to the mass of ^{56}Ni ejected; the density of the expanding supernova; and the gradient in that density. Model 1 is an extremely "fast", dim Type I explosion because the mass of ^{56}Ni produced is small and the expansion velocity great. The factor of 4 variation in peak luminosity should roughly span the allowable parameter space for Type I models. See WTW for γ-line fluxes corresponding to these light curves.

3.3 Type I Spectra

As was first recognized by Meyerott (1978; 1980ab), the presence of fast particles from radioactive decay in the SNI plasma has important implications for the spectrum produced. In particular, ionization states may result which are far from LTE. Meyerott's work and the observation by Kirshner and Oke (1975) that the late time SNI spectrum bears a close resemblance to that of a pure, Doppler-broadened spectrum of Fe II led Axelrod (1980ab) to undertake a numerical calculation of the spectra resulting from a uniform expanding shell of ^{56}Ni. The numerical model self-consistently determined the temperature and ionization state of the nebula and accounted for a variety of spectral effects due to the rapid expansion of the nebula. Comparison of the results with the observed spectrum of SN1972e showed a close approximation was achieved if the ^{56}Ni had a mass in the range 0.3 to 1.2 $M_\odot$ (depending on an uncertain distance to NGC5253) and an expansion velocity of roughly 7000 km s^{-1}. Of most direct importance for the ^{56}Ni hypothesis, the calculated spectrum contained a feature near 6000 Å due to ^{56}Co which decayed with time and a corresponding decaying feature was also present in the observations. This formed the first *direct* evidence for the presence of freshly synthesized radioactive ^{56}Ni in supernovae.

This picture has been recently clarified by new calculations which account for the fact that the SNI spectrum is strongly influenced by the density profile of the ejecta. This occurs because the ionization balance is very sensitive to density with both the average degree of ionization and the electron temperature decreasing as density increases. The density profile is also reflected in the spectrum through its effect on the transport of far ultraviolet radiation which results from radiative recombination. We have included these effects in calculating the spectra form Models 3 and 5 (Figures 8 and 9). These spectra include only the contribution from those regions of the models which are dominantly ^{56}Ni and would be modified by the inclusion of the overlying "atmosphere". During the nebular phase, however, that atmosphere is expected to have only a small effect (Axelrod 1981). The difference between the spectra of Models 3 and 5 is substantial and might therefore, unlike their light curves, be employed to discriminate which model is appropriate for a given explosion. The spectrum calculated for Model 3 is not in agreement with that of SN1972e nor with any other observations we could find in the literature. It is too "hot". That is, the high expansion velocity causes the density to decrease to too low values at too early times thus leading to a preponderance of high ionization states of iron (eg., Fe V and Fe VI) and excessive Doppler broadening. Spectra for Models 1 and 2 would be even worse (The spectrum of Model 2 has been computed but is not shown. It is dominated by lines of Fe V, VI, and even VII). In fact, an accurate calculation of Models 1 and 2 and the outer regions of Model 3 would require higher temperatures and ionization states than are currently accessible by the code and would probably give an even hotter spectrum. Thus the generic class of detonating (carbon-oxygen) white dwarfs is in difficulty spectroscopically. Such models also pose a problem to the spectrum at early times (Branch 1980ab; 1982; Branch et al 1982) in that the velocities of elements such as Ca and Si, which form p-Cygni absorbtion lines at early times, tend to be far too high to agree with observations. There may also be inadequate calcium and silicon synthesis to account for the strength of observed absorption features. Possible resolutions to this apparent dilemma involve the two-dimensional geometry of the off-center detonation, currently under study by Fryxell and Woosley, or simply that this class of SNI models is relatively infrequent and our observational data base too small.

On the other hand the computed spectrum for Model 5 *is* in good agreement with SN1972e, in fact, given uncertainties in the atomic data, the agreement is *excellent*. Many details of the spectrum are congruent in both theory and observation lending credence to the modeling process in general. The spectrum for Model 4 (not shown) has also been computed (without including the overlying atmosphere) and is similar to that for Model 5, although a slightly "cooler". Model 6 is expected to be similar. Moreover, the deflagrating models offer a natural site for the synthesis of silicon and calcium seen as absorption features in the observed spectrum (but not included in the present spectral synthesis calculations). Model 5 makes ~ 0.02 M$_\odot$ each of ^{28}Si and ^{32}S and ~ 0.003 M$_\odot$ each of ^{36}Ar and ^{40}Ca. Model 4 makes ~ 0.08 M$_\odot$ of ^{28}Si, ~ 0.02 M$_\odot$ of ^{32}S, ~ 0.003 M$_\odot$ of ^{36}Ar, and ~ 0.004 M$_\odot$ of ^{40}Ca. Unfortunately, the calcium in Model 5 is moving at a maximum velocity of 8,000 km s^{-1}, which is too slow according to Branch (1980ab; 1982). Indeed, all things considered, it would be better if the deflagrating models had acquired greater kinetic energy and a more rapid expansion (see especially the next section on nucleosynthesis). This might occur if the deflagration wave has greater velocity, and as we argued that that velocity is essentially a free parameter, such may well be the case. The deflagration of a *helium* dwarf would also give a greater expansion velocity because of the greater specific energy available from burning helium to iron. There will, of course, always be *some* calcium, as well as traces of other heavy elements, present primordially in the outermost layers which are ejected with the greatest velocity (roughly 11,000 km s^{-1} in Model 5). Thus it is important to know *quantitatively* just what fraction of

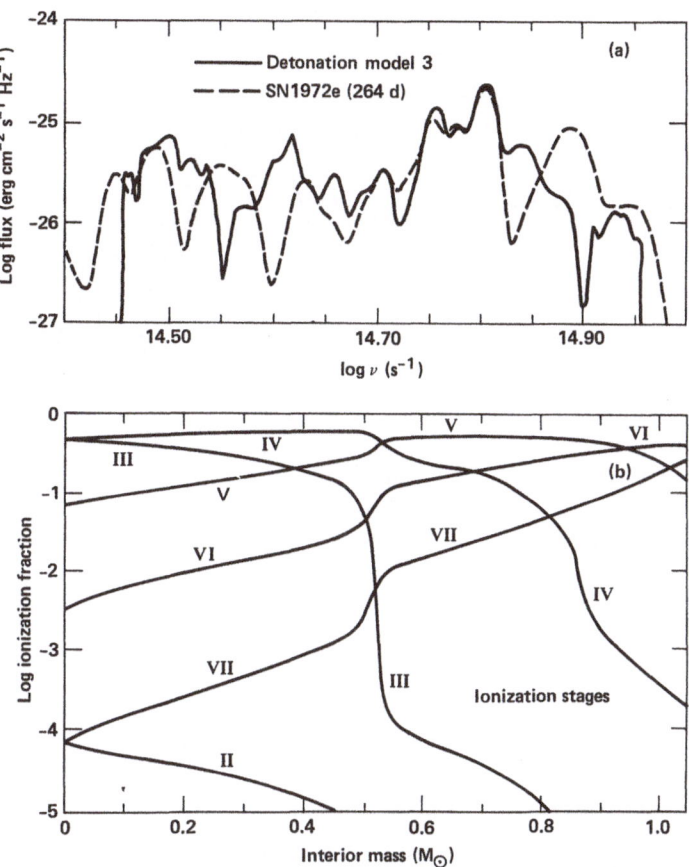

Figure 8. Optical spectrum of detonating Model 3 as compared to that observed (Kirshner *et al* 1973) from supernova 1972e 264 days after its explosion and the calculated ionization distribution of iron group species as a function of interior mass. Rapid expansion of the model results in a diminished abundance of free electrons and less efficient cooling. Thus highly ionized states of iron are abundant in the model that are not seen in the supernova. Line widths in the model are also typically broader than observed. This is *not* a good fit (see also WAW and WTW).

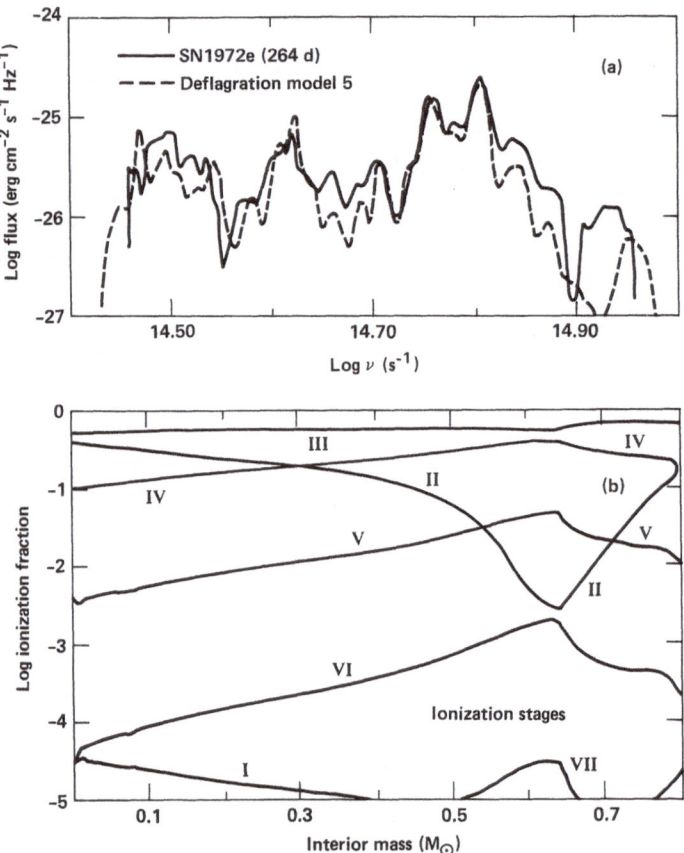

Figure 9. Optical spectrum of deflagrating Model 5 compared to supernova 1972e. The slower expansion gives rise to narrower lines and less highly ionized states of iron and cobalt. This is a much better fit than Figure 8. An absorption feature due to calcium is apparent in the observations at $\log \nu \approx 14.90$ which was not included in the model calculation.

the calcium abundance is observed at what velocity. So far that deconvolution of the observations has not been carried out, but the (simple one-dimensional) deflagrating models may be in some, hopefully slight, trouble here.

3.4 Nucleosynthesis

The chief nucleosynthetic product of these explosions is iron, principally the nucleus ^{56}Fe synthesized, in part, although not exclusively in the deflagrating models, as its radioactive progenitor ^{56}Ni. Owing to their rapid expansion and comparatively low ignition density, electron capture during the explosion of the *detonating* models is of no great consequence except for determining the abundances of trace elements such as ^{57}Fe, ^{59}Co, and ^{58,61,62}Ni. Important synthesis of ^{44}Ca, ^{60}Ni, and ^{64}Zn also occurs but these species are produced as radioactive progenitors (^{44}Ti, ^{60}Zn, and ^{64}Ge) in amounts that are relatively insensitive to the neutron excess. Models 1, 2, and 3 eject 0.24, 0.50, and 1.06 $M_\odot$ of iron group products respectively. Normalizing to ^{56}Fe, the isotopes ^{57}Fe, ^{59}Co, ^{60,61}Ni, and ^{64}Zn are produced in all three models within a factor of 30% of their solar abundances. The species ^{44}Ca is underproduced by a factor ranging from 3 (Model 3) to 16 (Model 1), ^{58}Ni is underproduced by a factor of 2 to 6, and ^{62}Ni, by a factor of 3 to 20. It is especially interesting that these models synthesize so little ^{54}Fe ($< 1\%$ solar). This is partly a consequence of the $\alpha-$*rich freeze out* (Arnett, Truran, and Woosley 1971; Woosley, Arnett, and Clayton 1973) in which all ^{54}Fe is converted into nickel isotopes, and partly a result of the small neutron excess of these models. It is particularly interesting because no other models we have ever studied, including Type II supernova models, appear capable of making ^{56}Fe without ^{54}Fe. A nucleosynthetic component of this sort may be required in order to dilute the ^{54}Fe-rich ejecta of other types of supernovae, especially the deflagrations we shall now discuss.

The nucleosynthesis from (deflagrating) Models 4, 5, and 6 was given in Figures 5 and 6. The final electron mole number (Y_e), is given for the inner core of all three models in Figure 10. The quantity Y_e is related to the electron number density, n_e, by $n_e = \rho N_A Y_e$ and the neutron excess, η, by $Y_e = 1 - 2\eta$. For nuclei having equal numbers of neutrons and protons, $Y_e = 1/2$. Owing to the high ignition density and comparatively slow expansion rate of these models, substantial electron capture occurs during the explosion leading to a decrease in Y_e in the central core, so much so in fact that the dominant iron group nucleus at the center of the star shifts from first ^{56}Ni to ^{54}Fe to ^{56}Fe and, in the extreme case of Model 6, to ^{58}Fe. Electron capture was calculated here using the older version of KEPLER that employs weak interaction rates from Mazurek (1973) and Hansen (1966) and the quasi-equilibrium network described by Weaver, Zimmerman, and Woosley (1978). If anything, these weak rates should *underestimate* the actual electron capture flows. Compared to more realistic rates by Fuller, Fowler, and Newman (1982ab), key reactions, *eg.*, ^{56}Ni(e^-,ν)^{56}Co, are generally slower by a factor of order 5. Thus in reality one might expect the curves of Figure 10 to be shifted to slightly *lower* values of Y_e. (There is some compensation since the formation of a more neutron-rich daughter leads to a substantial reduction in the rate of neutronization. Also capture on free protons, a process that *was* accurately modeled, is a major contributor to the decrease in Y_e, especially for values near 0.5).

The principal nucleosynthetic creation of the deflagrating models is still iron group species but, owing to the neutronization, there is considerable variation of the isotopic composition and none of it is very good news. Model 4 ejects 0.53 $M_\odot$ of ^{4}He, 0.12 $M_\odot$ of ^{16}O, 0.30 $M_\odot$ of ^{20}Ne, 0.16 $M_\odot$ of ^{24}Mg, 0.08 $M_\odot$ of ^{28}Si, 0.017 $M_\odot$ of ^{32}S, 0.0032 $M_\odot$ of ^{36}Ar, 0.0039 $M_\odot$ of ^{40}Ca, 0.16 $M_\odot$ of ^{54}Fe, and 0.61 $M_\odot$ of ^{56}Ni which decays to ^{56}Fe. The ^{54}Fe to ^{56}Fe ratio is an obvious problem. In the sun this ratio is 0.061. In Model 4 it is 0.26 which means that supernovae like Model 4 could have been responsible for synthesizing no more than $\sim$25% of the iron in

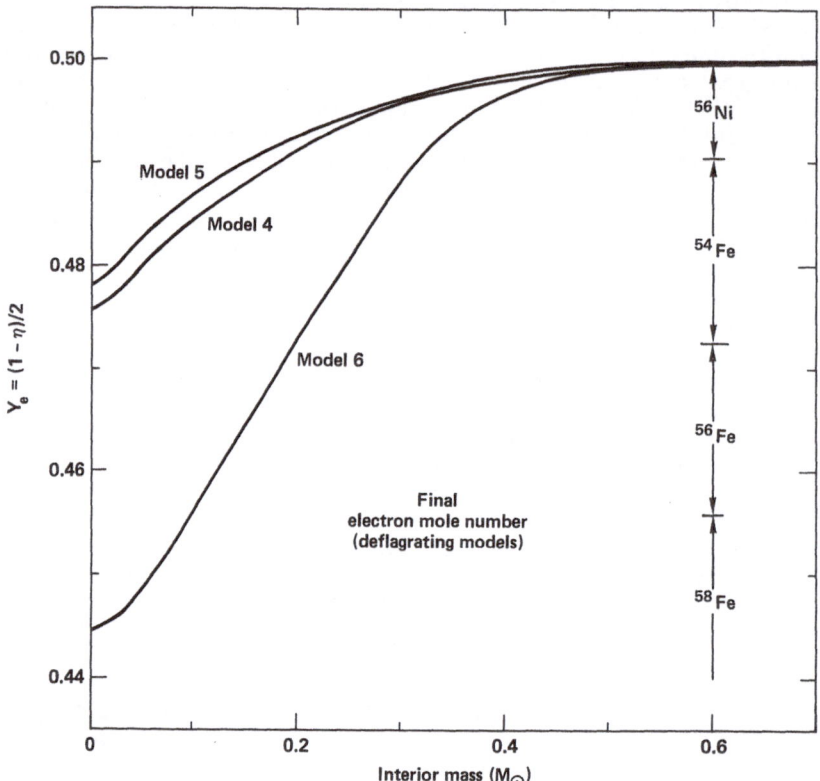

Figure 10. Distribution of the final electron mole number with interior mass in the 3 deflagrating models. The quantity Y_e is initially 0.5 throughout the star (initial composition is of nuclei having $Z = N$) but is decreased by electron capture during the explosion. The dominant iron group isotope in nuclear statistical equilibrium at a given Y_e is indicated by bands. Large overproductions of ^{54}Fe and ^{58}Fe are a troubling aspect of deflagrating models.

You received this card in one of our publications. When you return it for more information, we would highly appreciate if you at the same time indicate below how you heard of the book or books now in your possession.

This information would greatly assist us in serving you better in the future. Thank you for your cooperation.

☐ Advertisements
☐ Bookseller's recommendation
☐ Personal recommendation
☐ Displays in bookshops
☐ Exhibition at:
☐ Reviews in: _____
☐ References in book/journals:
_____
☐ Prospectus received from publisher
☐ Prospectus received from bookseller
☐ Publisher's catalogue
☐ Listing in a subject catalogue of bookseller
☐ Library
☐ Other: _____

stamp

D. Reidel Publishing Company

P.O. Box 17

3300 AA DORDRECHT

THE NETHERLANDS

☐ Philosophy of Science	☐ Social Sciences	☐ Physics
☐ Philosophy of Medicine	☐ Decision Theory	☐ Astronomy
☐ Logic	☐ Operations Research	☐ Earth Sciences
☐ Phenomenology	☐ Business/Consumer Studies	☐ Chemistry
☐ Legal/Political/ Social Philosophy	☐ Statistics	☐ Energy
☐ Linguistics/ Philosophy of Language	☐ Information and Computer Science	☐ Environmental Sciences
☐ Asian Studies	☐ Mathematics	☐ History of Science and Technology
	☐ Mathematical Physics	☐ Special Interests:

Name: _____ (please print)

Institute or Company: _____

Address (home/institute*): _____

_____

_____

*please cross out what does not apply   3/83

our Galaxy. Even so, the remainder of iron would have to be synthesized in objects that make no ^{54}Fe themselves, *eg.*, the detonating models. Nor do Models 5 and 6 fare any better. Model 5 ejects 0.34 and 0.36 $M_\odot$ of unburned ^{12}C and ^{16}O (the near equality given by initial assumptions), 0.0037 $M_\odot$ of ^{20}Ne, 0.0052 $M_\odot$ of ^{24}Mg, 0.027 $M_\odot$ of ^{28}Si, 0.017 $M_\odot$ of ^{32}S, 0.0035 $M_\odot$ of ^{36}Ar, 0.0033 $M_\odot$ of ^{40}Ca, 0.17 $M_\odot$ of ^{54}Fe, and 0.49 $M_\odot$ of ^{56}Ni. This gives roughly the same ^{54}Fe constraint as Model 4. Model 6, because of its larger ignition density, faces even graver difficulties. Amounts of elements lighter than iron are ejected that are comparable to Model 5 but, within the iron group, 0.15 $M_\odot$ of ^{54}Fe, 0.44 $M_\odot$ of ^{56}Fe (with 0.31 $M_\odot$ made as ^{56}Ni and 0.13 $M_\odot$ made as ^{56}Fe itself), and $\sim$*0.07 $M_\odot$ of ^{58}Fe* are produced. This gives a ^{54}Fe/^{56}Fe ratio roughly 5 times that of the sun and a ^{58}Fe/^{56}Fe ratio roughly 40 times that of the sun! We conclude that deflagrating white dwarfs igniting at central densities $\gtrsim 2.5 \times 10^9$ g cm^{-3}, i.e., the central density of Model 5, must have been very rare in nature.

3.5 Type I Summary

We seem to have painted ourselves into a bit of a corner! All models studied here, with the possible exception of Model 1, apparently give reasonable light curves but the class of off-center detonations on carbon-oxygen cores produces unacceptable spectra (although quite reasonable nucleosynthesis) while the carbon-oxygen and neon deflagrations give unacceptable nucleosynthesis (although quite reasonable spectra). Something must be wrong. At this point we do not know the resolution of this dilemma. Obvious, but to us, presently unacceptable, solutions could be i) SNI are not a major source of iron in nature; ii) our nuclear reaction rates and/or atomic data are in much greater error than we believe, or iii) the spectrum of SN1972e is not that of a typical SNI. More likely hypotheses we would suggest are i) the existence of another class of SNI models (helium dwarf deflagrations?) having faster expansion rates and lower central ignition densities than our Models 4-6 but slower velocities than Models 1-3; ii) a low density "atmosphere" surrounding a detonating model which would "tamp" the high expansion velocities of the outer layers; iii) important 2 and 3 dimensional modifications to our current models, especially those involving off-center ignition, and/or iv) a preponderance of deflagrating models with central densities close to that of Model 5 but with more rapid velocities for the deflagration wave than we employed. Obviously the search for the "correct" SNI model is still on.

One of the authors (SEW) gratefully acknowledges the hospitality and stimulating environment of the Erice conference on nucleosynthesis, especially conversations with D. Bond, D. Branch, C. Chiosi, J. Gallagher, and I. Iben. He also is grateful to the Lawrence Livermore National Laboratory for support and computer time over the last 14 months while the work described here was carried out. This work has been supported by the National Science Foundation (AST-81-08509) and the Department of Energy (W-7405-ENG-48).

DISCUSSION

Edmunds: Can you predict what will happen to nucleosynthesis in the new 25 $M_\odot$ star?

Woosley: Qualitatively the results should resemble what was calculated for the old 25 $M_\odot$ star. In detail, the new model may give better production of the silicon to calcium group because the core of the new star is surrounded by material with a shallower density gradient. It is our hope that material which once was a part of the old iron core will now experience explosive oxygen burning with an enhanced production of intermediate mass elements. This remains to be seen. Also you may note (Fig. 2) that the helium core is a little smaller (by about 1 $M_\odot$) and carbon abundance somewhat larger (by about a factor of 2) than reported by Woosley and Weaver 1982b.

Thielemann: The peak in the abundance curve you and Weaver published (Woosley and Weaver 1982b) contains carbon burning products while the minimum contains oxygen burning products. The new $^{12}C(\alpha,\gamma)^{16}O$ rate (Kettner et al 1982), which is enhanced by a factor of 3 to 5 , will leave less carbon and more oxygen after helium burning. Since this changes the fuels for subsequent burning stages will your yield curve be flattened (i.e., improved) by using the new rate?

Woosley: As Willy Fowler has long reminded us, we must get the reaction rates right before we can trust the nucleosynthesis calculations. However, I do not expect the yields of *intermediate* mass nuclei to be grossly affected (I differ with you and Arnett here). These nuclei are produced by *explosive* oxygen burning and, as Arnett, Clayton and I discussed many years ago (1973), it is the peak *temperature* in the explosion which most senitively determines its yield. That is, a carbon composition exposed to the same peak temperature (above $T_9 \approx 3$) as an oxygen composition will give the same nucleosynthesis. Memory of the initial composition is lost. The explosion temperature is set by the pre-explosive density gradient and by the explosion energy of the supernova, both of which are relatively insensitive to the rate for $^{12}C(\alpha,\gamma)^{16}O$. The ratio of *carbon to oxygen* in the ejecta may, of course, change greatly.

Trautvetter: Since you mentioned ^{26}Al as an important candidate for γ-line astronomy, I would like to comment that our group has recently measured the cross section for $^{26}Al(p,\gamma)^{27}Si$ which could eventually destroy freshly produced ^{26}Al. We found, in the energy range $0.27 \lesssim E_p \lesssim 1.8$ MeV, five resonances which are all very weak indicating that ^{26}Al can survive H-burning.

Woosley: That is very interesting, especially for ^{26}Al production in novae. Having done the difficult task of preparing an ^{26}Al target you should now measure the (n,p) cross section since that is the principal destruction mechanism in explosive neon burning which dominates ^{26}Al production in supernovae (Woosley and Weaver 1980).

Renzini: I have 3 short questions: 1) What was the *surface* N/O ratio in the 25 $M_\odot$ star at the time of its explosion? 2) Where is the ^{17}O produced? and 3) Has neon production in the new model changed substantially? The neon production you obtained in earlier models was quite different from Arnett (1978).

Woosley: The surface N/O ratio is close to unity (actually 4:5; see Fig. 2). The species ^{17}O is produced by hydrogen shell burning by the CNO tri-cycle. Neon production has not changed substantially. It is still large.

Renzini: Iben and Truran (1978) have shown that massive AGB stars (if they exist) may make large amounts of ^{22}Ne which, in this case, is mostly primary. Anomalies of ^{22}Ne could then be incorporated into grains forming around such AGB stars.

Woosley: Perhaps, but it might be difficult to incorporate ^{22}Ne, a noble gas, into stellar grains. There exist in the solar system components of neon in meteorites (Ne E) that are almost pure ^{22}Ne. A ^{22}Na carrier which decays after chemical incorporation seems more likely.

Bond: Could you comment on your work with Bodenheimer concerning rotation and oxygen burning driven ejection in massive stars? This is related to the question of how to get Cygnus X-1 black holes with explosions.

Woosley: That work is now published (Bodenheimer and Woosley 1983). The major uncertainty is our treatment of the inner boundry condition which must act, at least partly, to inhibit the accretion flow. We are encouraged that supernova remnants have been observed that do exhibit toroidal symmetry.

Nomoto: What is the propagation velocity of the deflagration wave in your Type I models?

Woosley: It varies, but typically $\sim$ 2000 km s^{-1}, or about 1/3 the speed of sound.

Nomoto: I have also calculated the model of a single detonation leaving a white dwarf remnant behind. Such events occur in a relatively small region of parameter space (i.e., accretion rate *vs* white dwarf mass) and might therefore be expected to be relatively rare.

Woosley: Your calculations may have underestimated the probability of leaving a white dwarf behind because they do not account for the transfer, by hydrodynamical means, of energy in the central portions of the white dwarf to the kinetic energy of ejecta. The total energy of the white dwarf core stripped of its helium (now iron) envelope may be zero, or even positive, and still leave a remnant because excess energy goes into accelerating a portion of the outer layers to high velocity. This happened, in fact, in Model 2 which I believe according to your phase diagram would not have been expected to leave a remnant. Still the parameter space may be restricted. I agree.

Braun: If it is indeed the case that the SNI light curve shape is relatively insensitive to the details of the model, we may have to wait for a more complete sample of SNI spectra before demanding that there be a unique class of progenitors.

Woosley: I certainly agree we need more spectral data, particularly at late times, but Dave Branch tells me that many spectra have been taken near peak light and they all fit one another like gloves. Perhaps it is only the late time spectrum (which is all that Axelrod is presently capable of modeling) that is so discriminatory.

Rayet: What would be the physical reason why Nomoto and Hillebrandt claim the possibility of an r-process site in their 9.6 M$_\odot$ explosion while you do not mention it for your 10 M$_\odot$ model? Is it related to the steep entropy increase during explosion quoted by Nomoto?

Woosley: Evolutionary endpoints vary rapidly in the mass range 9 to 10 M$_\odot$. Our 10 M$_\odot$ star underwent an ordinary core collapse following silicon core depletion.

Nomoto's 9.6 M$_\odot$ model, if my memory serves me correctly, collapsed on electron capture prior to oxygen depletion. The smaller core mass and energy input from nuclear burning during the collapse combined to produce an energetic explosion. I am not surprised that they get production of large amounts of neutron-rich isotopes in such a powerful explosion. In fact, they may find it far too much of a good thing. That is, *overproduction* of neutron-rich isotopes may be a problem.

REFERENCES

Arnett, W. D. 1979, *Ap. J. Lettr.*, **230**, L37.

———. 1980, *Ann. N. Y. Acad. Sci.*, **336**, 366.

Arnett, W. D., Truran, J. W., and Woosley, S. E. 1971, *Ap. J.*, **165**, 87.

Appenzeller, I., and Fricke, K. J. 1972, *Astron. and Ap.*, **21**, 285.

Appenzeller, I., and Tscharnuter, W. 1973, *Astron. and Ap.*, **25**, 125.

Axelrod, T. S. 1980a, PhD thesis, Univ. California at Santa Cruz and Lawrence Livermore Laboratory Report UCRL-52994.

———. 1980b, in *Type I Supernovae*, ed. J. C. Wheeler (Austin: University of Texas), p. 80.

———. 1981, Lawrence Livermore Laboratory Report UCRL-86850 and in preparation for *Ap. J.*.

Barkat, Z., Rakavy, G., and Sack, N. 1967, *Phys. Rev. Lettr.*, **18**, 379.

Bessel, M. S., and Norris, J. 1981, *IAU Colloq. No. 86: Astrophysical Parameters for Globular Clusters*, ed. A. G. D. Philip and D. S. Hayes, (L. Davis Press: Schenectady), 137.

Bodenheimer, P. B., and Woosley, S. E. 1983, *Ap. J.*, **269**, 281.

Borst, L. B. 1950, *Phys. Rev.*, **78**, 807.

Bond, H. E. 1981 *Ap. J.*, **248**, 606.

Bond, J. R., Arnett, W. D., and Carr, B. J. 1982, in *Supernovae: A Survey of Current Research*, ed. M. J. Rees and R. J. Stoneham, (D. Reidel: Dordrecht).

———. 1983, *Ap. J.*, in press.

Bowers, R. L.. and Wilson, J. R. 1982a, *Ap. J. Suppl.*, **50**, 115.

———. 1982b, *Ap. J.*, **263**, 366.

Branch, D. 1980a, in *Supernovae Spectra*, ed. R. Meyerott and G. H. Gillespie, (Am. Inst. Phys.: New York), Proc. No. 63, p. 39.

———. 1980b, in *Type I Supernovae*, ed. J. C. Wheeler (Austin: University of Texas), p 66.

———. 1982, *Proc. XI Texas Symp. on Relativistic Ap.*, to be published by *Ann. N. Y. Acad. Sci.*

Branch, D., Buta, R., Falk, S. W., McCall, M. L., Sutherland, P. G., Uomoto, A., Wheeler, J. C., and Wills, B. J. 1982, *Ap. J. Lettr.*, **252**, L61.

Brown, G. E., Bethe, H. A., and Baym, G. 1982, *Nuclear Physics*, **A375**, 481.

Burbidge, E. M., Burbidge, G. R., Fowler, W. A., and Hoyle, F. 1957, *Rev. Mod. Phys.*, **29**, 547.

Butler, D., Dickens, R. J., and Epps, E. 1978, *Ap. J.*, **225**, 148.

Cameron, A. G. W. 1973, *Spac. Sci. Rev.*, **15**, 121.

Chevalier, R. A. 1981, *Ap. J.*, **246**, 267

Colgate, S. A., and McKee, C. 1969, *Ap. J.*, **157**, 623.

Colgate, S. A., and Petschek, A. 1979, *Ap. J. Lettr.*, **236**, L115.

Colgate, S. A., Petschek, A. G., and Kriese, J. T. 1980, *Ap. J. Lettr.*, 237, L61.

Cox, A. N., and Stewart, J. N. 1970a, *Ap. J. Suppl.*, **19**, 243.

———. 1970b, *Ap. J. Suppl.*, **19**, 261.

Dicke, R. H., and Peebles, P. J. E. 1968, *Ap. J.*, **154**, 891.

Epstein, R. 1979, *MNRAS*, **188**, 305.

Ezer, D., and Cameron, A. G. W. 1971, *Ap. and Spac. Sci.*, **14**, 399.

Falk, S. W., and Arnett, W. D. 1973, *Ap. J. Lettr.*, **180**, L65.

Fowler, W. A. 1966, in *High Energy Astrophysics*, ed. L. Gratton, (Academic Press: New York), 313.

Fricke, K. J. 1973, *Ap. J.*, **183**, 941.

———. 1974, *Ap. J.*, **189**, 535.

Fugimoto, M. Y., and Taam, R. E. 1982, *Ap. J.*, **260**, 249.

Fuller, G. M., Fowler, W. A., and Newman, M. J. 1982a, *Ap. J. Suppl.*, **48**, 279.

———. 1982b, *Ap. J.*, **252**, 715.

Fuller, G. M., Woosley, S. E., and Weaver, T. A. 1982, *Bull. Am. Astron. Soc.*, **14**, 957.

———. 1983, in preparation for *Ap. J.*

Hansen, C. J. 1966, PhD Thesis, Yale University.

Hillebrandt, W. 1982a, *Astron. and Ap.*, **110**, L3.

———. 1982b, in *Supernovae: A Survey of Current Research*, ed. M. J. Ress and R. J. Stoneham, (D. Reidel: Dordrecht).

———. 1983, as presented at XI Texas Symposium on Relativistic Astrophysics, to be published by *Ann. N. Y. Acad. Sci*, preprint.

Iben, I. 1963, *Ap. J.*, **138**, 1090.

———. 1978a, *Ap. J.*, **219**, 213.

———. 1978b, *Ap. J.*, **226**, 996.

———. 1982, *Ap. J.*, **253**, 248.

Iben, I., and Tutukov, S. 1983, preprint, Univ. of Illinois.

Kirshner, R. P., and Oke, J. B. 1975, *Ap. J.*, **200**, 574.

Kirshner, R. P., Oke, J. B., Penston, M. V., and Searle, L. 1973, *Ap. J.*, **185**, 303.

Maza, J., and van den Bergh, S. 1976, *Ap. J.*, **204**, 519.

Mazurek, T. J. 1973, *Ap. and Spac. Sci.*, **23**, 365.

McDonald, J. 1983, *Ap. J.*, **267**, 732.

Meyerott, R. E. 1978, *Ap. J.*, **221**, 975.

———. 1980a, in *Supernovae Spectra*, ed. R. Meyerott and G. H. Gillespie, (Am. Inst. Phys.: New York), Proc. No. 63, p. 49.

———. 1980b, in *Type I Supernovae*, ed. J. C. Wheeler, (University of Texas), p. 72.

Miyaji, S., Nomoto, K., Yokoi, K., and Sugimoto, D. 1980, *Pub. Astron. Soc. Japan*, **32**, 303.

Muller E., and Arnett, W. D. 1982, *Ap. J. Lettr.*, **261**, L109.

Nomoto, K. 1980a, in *Type I Supernovae*, ed. J. C. Wheeler (Austin: University of Texas), p. 164.

———. 1980b, *Spac. Sci. Rev.*, **27**, 563.

———. 1981, in *IAU Symposium 93, Fundamental Problems in the Theory of Stellar Evolution*, ed. D. Sugimoto, D. Q. Lamb, and D. N. Schramm (Dordrecht: Reidel), p. 295.

———. 1982a, *Ap. J.*, **253**, 798.

———. 1982b, *Ap. J.*, **257**, 780.

———. 1983, *Max-Planck Inst. fur Physik und Astrophysik Preprint MPA72*.

Nomoto, K., and Sugimoto, D. 1977, *Publ. Astron. Soc. Japan*, **29**, 765.

Norgaard, H., and Fricke, K. J. 1976, *Astron. and Ap.*, **49**, 337.

Ober, W. 1981, Ph D Thesis, Universitat zu Gottingen.

Ober, W., El Eid, W., and Fricke, K. J. 1983, *Astron. and Ap.*, **119**, 61.

Palla, F., Salpeter, E. E., and Stahler, S. W. 1983, *Ap. J.*, **271**, 632.

Pankey, T. 1962, PhD Thesis, Howard University. *Diss. Abstr.*, **23**, 4.

Papaloizou, J. C. B. 1973, *Mon. Not. R.A.S.*, **162**, 169.

Sanders, R. H. 1970, *Ap. J.*, **162**, 791.

Silk, J. 1977, *Ap. J.*,**211**, 638.

Smarr, L., Wilson, J. R., Barton, R., and Bowers, R. 1981, *Ap. J.*, **246**, 515.

Sneden, C., Lambert, D. L., and Whitaker, R. W. 1979, *Ap. J.*, **234**, 964.

Starrfield, S. G., Sparks, W. M., and Truran, J. W. 1974, *Ap. J. Suppl.*, **28**, 247.

Starrfield S. G., Truran, J. W., and Sparks W. M. 1981, *Ap. J. Lettr.*, **243**, L27.

Stothers, R., and Simon, 1970, *Ap. J.*, **160**, 1019.

Sugimoto, D., and Nomoto, K. 1980, *Spac. Sci. Rev.*, **25**, 155.

Taam, R. E. 1980a, *Ap. J.*, **237**, 142.

———. 1980b, *Ap. J.*, **242**, 749.

Talbot, R. J. 1971, *Ap. J.*, **165**, 121.

Trimble, V. 1982, *Rev. Mod. Phys.*, **54**, 1183.

Wallace, R. K., and Woosley, S. E. 1981, *Ap. J. Suppl.*, **45**, 389.

Wallace, R. K., Woosley, S. E., and Weaver, T. A. 1982, *Ap. J.*, **258**, 696.

Weaver, T. A., Zimmerman, G. B., and Woosley, S. E. 1978, *Ap J.*, **225**, 1021.

Weaver, T. A., and Woosley, S. E. 1979, *Bull. Am. Astron. Soc.*, **11**, 727.

———. 1980a, *Ann. N. Y. Acad. Sci.*, **336**, 335.

———. 1980b, in *Supernovae Spectra*, ed. R. Meyerott and G. H. Gillespie, (Amer. Inst. Phys.: New York), Proceedings No. 63, p. 15.

Weaver, T. A., Axelrod, T. S., and Woosley, S. E. 1980, in *Type I Supernovae*, ed. J. C. Wheeler (Austin: University of Texas), p 113 (WAW).

Weaver, T. A., Woosley, S. E., and Fuller, G. M. 1982, *Bull. Am. Astron. Soc.*, **14**, 957.

———. 1983, in *Numerical Astrophysics: In Honor of J. R. Wilson*, ed. R. Bowers *et al*, (Science Books Intern'l: Portola CA), in press (WWF).

Wheeler, J. C. 1981, *Rept. Prog. Phys.*,**44**, 85.

Wheeler, J. C., and Hansen, C. 1971, *Ap. and Spac. Sci.*, **11**, 373.

Whelan, J., and Iben, I. 1973, *Ap. J.*, **186**, 1007.

Wilson, J. R. 1983, in *Numerical Astrophysics: In Honor of J. R. Wilson*, ed. R. Bowers *et al*, (Science Books Intern'l: Portola CA), in press.

Woosley, S. E. and Weaver, T. A. 1980, *Ap. J.*, **238**, 1017. ——. 1981, *Ann. N. Y. Acad. Sci.*, **375**, 357.

——. 1982a, in *Supernovae: A Survey of Current Research*, ed. M. J. Rees and R. J. Stoneham, (D. Reidel: Dordrecht), p. 79.

——. 1982b, in *Essays on Nuclear Astrophysics*, ed. C. A. Barnes, D. D. Clayton, and D. N. Schramm, (Cambridge Univ. Press), p. 377.

Woosley, S. E., Weaver, T. A., and Taam, R. E. 1980, in *Type I Supernovae*, ed. J. C. Wheeler (Austin: University of Texas), p. 96.

Woosley, S. E., Arnett, W. D., and Clayton, D. D. 1973, *Ap. J. Suppl.*, **231**, 26.

Woosley, S. E., Taam, R. E., and Weaver, T. A. 1983, in preparation for *Ap. J.* (WTW).

SESSION V: PREGALACTIC NUCLEOSYNTHESIS AND CHEMICAL EVOLUTION OF GALAXIES

MASSIVE OBJECTS AND THE FIRST STARS

J. Richard Bond
Institute of Astronomy, Cambridge University, UK.
Institute of Theoretical Physics, Stanford University, USA.

The current status of the theory of VMOs, SMOs and Pop III stars is reviewed. The following subjects are included: observational and theoretical issues associated with Massive Object formation; the semi-analytic treatment of VMO evolution developed in collaboration with Arnett and Carr; the cosmological setting for and possible consequences of Population III stars, with emphasis on the role Massive Objects and their black hole remnants may have played.

1. INTRODUCTION

Supermassive Objects (SMOs) are defined to be 'stars' which are dynamically unstable during core hydrogen burning to general relativistic effects which raise the critical adiabatic index to $4/3+0(GM/Rc^2)$. For Population III abundances – meaning here zero initial metallicity – this implies their mass is $M \gtrsim 10^5 \, M_\odot$. Very Massive Objects (VMOs) are dynamically stable – though pulsationally unstable – through hydrogen burning, but are unstable during their core oxygen phase to the pair instability. The associated mass range is $10^2 \, M_\odot \lesssim M \lesssim 10^5 \, M_\odot$.

Such exotic stars require exotic sites, which two classic papers proposed: Hoyle and Fowler (1963) introduced the concept of SMO and identified them with QSOs; Doroshkevich et al. (1967) suggested that SMOs with mass equal to the Jeans mass at recombination, $\sim 10^6 \, M_\odot$, would be the first objects to form in the universe. These are still the most likely sites, though there are a few candidate VMOs in normal environments. Arguments for and against the existence of Massive Objects are reviewed in §2. The evolution of VMOs during the core hydrogen and helium burning phases, and the fate of these stars, which is determined in the core oxygen phase, is discussed in §3 and §4 respectively. This work was done in collaboration with Dave Arnett and Bernard Carr (Bond, Arnett and Carr 1983, [BAC]). Such cosmological aspects of Population III stars as when and where they form, and what they can do are addressed in §5. Again this work was part of the ABC collaboration (Carr, Bond and Arnett 1983 [CBA], Bond, Carr and Arnett 1983 [BCA],Bond & Carr 1983).

C. Chiosi and A. Renzini (eds.), Stellar Nucleosynthesis, 297–317.
© *1984 by D. Reidel Publishing Company.*

2. THE PROS AND CONS OF MASSIVE OBJECTS

2.1 Massive Objects and Dense Star Clusters

Supermassive black holes are often invoked to power active galactic nuclei such as Seyferts ($M_{bh} \sim 10^6$ $M_\odot$) and QSOs ($M_{bh} \sim 10^8$ $M_\odot$), which suggests a VMO or SMO precursor (Begelman and Rees 1978 [BR]). BR show that such a hole may arise through the following symbolic reaction pathways:
(1) $*+*+...+* \to$ VMO/SMO$\to$bh;   (2) $(*+* \to$gas$)^n \to$cool gas $+ \gamma \to$ VMO/SMO$\to$bh;
(3) $*+$bh (M) $\to$ gas$+$bh(M)$\to$bh(M+dM);   (4) $(*+*)+$bh(M)$\to$gas$+$bh(M)$\to$bh(M+dM).
(1) and (2) are responsible for the initial hole formation, (3) and (4) for its growth by accretion. (1), (2) and (4) rely on star-star collisions, which have not been well studied numerically. However, the gross features are understood. If the collision velocity, i.e. the random velocity of the stars in the cluster, v_c, is below about 0.3 v_e, where v_e is the escape velocity from the surface of a typical star, then physical collisions are not important over the evolution time of the cluster. If $0.3\ v_e < v_c < v_e$, the colliding stars coalesce, with little mass loss. If $v_c > v_e$, the stars disrupt into hot gas.

As a cluster evolves, its central velocity dispersion rises. Since coalescence is important only over a narrow v_c-range, BR suggested (1) is not as important as (2) for black hole production. The gas generated by collisions cools, falls into the cluster center, and makes stars. To get an SMO would require little or no fragmentation; they suppose this can arise if the gas, continually energized by gravitational infall, is maintained at full ionization, so it can be optically thick to electron scattering. One could have a single SMO or VMO, or many VMOs or even many SMOs forming, leaving one or many black hole remnants.

Duncan & Shapiro (1983) have numerically simulated the evolution of a dense cluster of 10^8 solar mass stars, taking as their initial black hole one of $\sim 10^3$ $M_\odot$. The hole grows by accretion as the cluster evolves. They find (4) is more important than tidal disruption of stars by the hole (3). It takes $\sim 10^8 - 10^9$ years for the hole to grow to 10^8 $M_\odot$ assuming it has no difficulties in swallowing, which small mass holes can suffer from. For typical parameters, Bondi accretion results in a rather long build-up time for a 10 $M_\odot$ black hole, suggesting that it may be difficult to begin with a normal massive star - if indeed these do form holes; however, a VMO precursor is all that is required: SMOs are not necessary.

For globular clusters, $v_c \ll v_e$, hence VMO formation could not be through stellar collisions, but would have to be through normal star formation processes. The X-ray emitters are unlikely to be black holes; the preferred candidates are neutron stars which form tight binaries via capture in the cluster cores (Lightman and Grindlay 1982). A 10^3 $M_\odot$ hole could still exist in the centers of globulars though, since it may not be emitting X-rays, due, for example, to the inhibition of close binary formation by tidal disruption or to its ability to swallow gas without radiating.

We may think of 30 Doradus – the giant HII region which so dominates photographs of the LMC – as a young globular cluster. Terlevich (1982) estimates its mass to be $\sim 10^7$ $M_\odot$. The central source, R136a, is a candidate for a VMO with mass ~ 2000 $M_\odot$. If so, it would evolve to a black hole. On the other hand, the cluster center may represent a young globular cluster core in which many O stars and Wolf Rayet stars reside. O stars tend to form in high density regions and are also subject to mass segregation by dynamical friction which centrally condenses them even more.

The pro-VMO story begins with Feitzinger et al. (1980), whose optical observations identified R136a as a compact central component, and led them to suggest it was a VMO. Cassinelli, Mathis and Savage (1981) added UV data and estimated a luminosity consistent with a 2000 $M_\odot$ star. Their analysis has been independently confirmed by Massey and Hutchings (1983), who find 6 giant HII regions in M33 also have R136a-type objects, suggesting the phenomenon may be wide-spread. Panagia et al. (1983) have added IR observations which supports the VMO interpretation with a luminosity $\sim 6 \times 10^7$ $L_\odot$. The most powerful observations use speckle inter-ferometry. Meaburn et al. (1982) infer that R136a has a radius < 500 AU, with the nearest source 4×10^4 AU away. On the other hand, Wiegelt (1981) infers R136a is a double star, with the companions separated by $\sim 2 \times 10^4$ AU.

The con story interprets R136a as just one star in a cluster of central sources some of which the speckle method would not have picked up. Thus the ionizing radiation for 30 Dor would not arise from this single small-scale source. Arguments supporting this view have been presented by Melnick (1983) and Moffat and Seggewiss (1983). Chu and Wolfire (1983) partially resolve R136a into a binary, but do not support the multiple component picture of Moffat and Seggewiss. Melnick suggests that the R136a region could be as large as 10^5 AU in extent and consist of only ~ 6 O stars and ~ 2 Wolf Rayets.

A timing argument may be applied to the viability of the multiple component hypothesis. The dynamical evolution time for a cluster is $\sim 2 \times 10^5$ y $(N_*/100)$ $(R/10^3$ AU$)^{3/2}(M/10 M_\odot)^{-\frac{1}{2}}/\log(N_*/100)$. If we could use the 500 AU number, this is uncomfortably short. For a normal stellar IMF with Melnick's 6 O stars and his 10^5 AU, there is no real difficulty. If we could suppose the regions consisted of O stars only, the system is Trapezium-like. Poveda (1977) gives lifetimes of $\sim 10^6$ y for the Orion Trapezium, with diameter $\sim 10^4$ AU and total mass ~ 170 $M_\odot$. These lifetimes scale with length and inversely with the square root of mass. For 10^3 AU, the time is only $\sim 10^5$ y; for 10^5 AU, it is $\sim 10^7$ y. These should be compared with VMO lifetimes, $\sim 2 \times 10^6$ y. Plausibility can then not be used as an argument until the speckle results can address the Melnick critic-isms. Properties of a 2000 $M_\odot$ Pop I star are given in Table 1. The escape velocity is 3.2×10^3 km s^{-1}, which agrees rather well with the Cassinelli et al. (1981) estimate of wind speed.

2.2 Type V Supernovae, η Carina and IR Sources

    Branch discusses Zwicky's Type V class of supernovae in this volume.
There are only two examples: η Carina in our galaxy and SN1961 in
NGC 1058. η Carina outburst in the optical in 1843. Infra-red emission,
due to dust re-radiation in its mass loss atmosphere, now accounts for
95% of its total luminosity, which is nearly that at ourburst, and
corresponds to the Eddington luminosity of a 140 $M_\odot$ star (Sutton et al.
1974). Andriesse et al. (1978) infer the enormous mass loss rate of
0.08 $M_\odot$/y since 1843. Davidson et al. (1982) find it is nitrogen-rich,
suggestive of CNO processing: η Car should not be a protostar.

    Branch and Greenstein (1971) analyzed the spectrum of the other
Type V supernova, SN1961. The ejected material is helium rich, with
metal abundances consistent with solar. The presupernova star was estimated to have a luminosity corresponding to the Eddington luminosity of
about a 500 $M_\odot$ object (Chevalier 1981), at least from 1937 until 1961.
The width of its lines is only $\sim$2000 km s^{-1}. Its present IR output has
not been investigated. In §3.6, we suggest a mechanism which would yield
these abundance characteristics.

    If we are to accept that η Car may be a VMO which has completed
core hydrogen burning, then we should expect to find main sequence
galactic VMOs. Such stars may have difficulty breaking out of their
protostellar environments in their short nuclear lifetimes; also, large
Ṁ's could make dusty mass loss atmospheres as in η Car. Thus VMOs may be
very bright infra-red sources, with luminosity a considerable fraction
of $L_{ED} = 4\times10^6$ (M/100$M_\odot$) $L_\odot$. Wynn-Williams (1982) has reviewed the
observations of galactic IR sources; the most luminous corresponds to an
Eddington luminosity appropriate to only 7 $M_\odot$.

2.3 Upper Stellar Mass Limit

    Larson and Starrfield (1971) argue that an upper limit to the mass
of stars arises from processes which occur during formation. For example,
when the Kelvin-Helmholtz time for the protostellar core is smaller than
the time for protostellar envelope accretion, the core can ignite hydrogen, generate an HII region, and thus impede further infall. They
estimated 6 $M_\odot$ for this critical mass if the accretion occurs in a dusty
environment where the gas temperature is maintained at $\sim$20K. Numerical
computations of star formation for protostellar masses of 60 $M_\odot$
(Appenzellar and Tscharnuter 1974) and 50 and 150 $M_\odot$ (Yorke and Krügel
1977) do show envelope ejection, with about 70% of the mass being lost;
radiation pressure on grains is the driving mechanism, not HII region
formation.

    In Pop III environments, grain cooling does not occur; the critical
mass rises into the VMO regime for accretion from regions with typical
molecular hydrogen temperatures, and to $\sim$50000$M_\odot$ for 10^{4}K regions. Though
radiation pressure on grains is absent, Doroshkevich and Kolesnik (1976)
claim that Lyman alpha absorption in the neutral regions surrounding the

the HII region will have a similar effect. This ejection mechanism seems easier to avoid, since the neutral region is farther out, is Rayleigh-Taylor unstable, and off-resonance may quickly arise due to the narrowness of the line. Breaking the spherical symmetry via clumpy infall of gas, coagulation of protostellar regions (O stars often form in dense associations), and pressure or shock driven collapse can all conspire to raise the upper mass limit.

There is same evidence that more massive stars do form in lower metallicity environments, and from larger clouds (Larson 1982). Terlevich (1982) claims that giant HII regions of globular cluster scale and larger - so there are none in our galaxy - have an IMF slope which flattens at high mass as the metallicity decreases: d ln $N_*(M)$/d ln M = -4-logZ. With such an IMF, Terlevich and Lynden-Bell (1983) claim the disk and halo G-dwarf problems can be solved, as Schmidt (1963) first suggested. We might expect some of these massive stars to lie in the lower range of VMOs, even if such exotic objects as 2000 $M_\odot$ stars do not form.

2.4 Lower Stellar Mass Limit for Population III Stars

It has often been suggested that there is a lower limit to the stellar masses that can form in metal-free environments. Yoneyama (1972) obtained 60 $M_\odot$ and Hutchins (1976) got 250 $M_\odot$. Though these authors included molecular hydrogen cooling, Palla et al. (1983) have pointed out that, at high density, 3-body processes enhance H_2-formation; if collapsing clouds enter such a regime, the minimum fragment mass could be as low as 0.007 $M_\odot$. Tohline (1980) emphasized the importance of initial conditions, suggesting that typical perturbations within a cloud could not grow fast enough, yielding a minimum mass of 1500 $M_\odot$. Silk (1983) pointed out that clouds which enter the 3-body regime can generate significant perturbations via the thermal instability, and result in masses significantly lower than the Tohline value. Kashlinsky and Rees (1983) have included the typical angular momentum expected from the tidal-torqueing of clouds, and suggest that clouds will collapse to disks before fragmentation occurs. If Pop III stars form earlier than redshift $\sim$100, the temperature of the background radiation will be high enough to delay H_2-formation; this will likely shift the IMF to higher mass. If star formation is initiated in a region which is thick to Thompson scattering, fragmentation becomes difficult. Such conditions may occur around the time of recombination (Hogan 1978).

We conclude that the minimum fragment mass for Pop III stars could be low, and that the maximum mass is higher than with grain cooling and likely lies in the VMO range. The IMF will depend upon cloud mass, epoch of collapse, spectrum and amplitude of subclumps, angular momentum distribution and etc. It is certainly plausible that some VMOs and perhaps some SMOs may form, and not implausible that the IMF will mostly consist of Massive Objects.

3. HYDROGEN AND HELIUM BURNING IN VMOs AND SMOs

3.1 Pulsational Instability

Stars are pulsationally unstable in their fundamental radial mode during core hydrogen burning if $M > M_p$, where $M_p = 91\ M_\odot$ for Pop I, 84 $M_\odot$ for Pop II (Stothers and Simon 1970); Boury (1963) has estimated $M_p \sim 260$ $M_\odot$ for pure hydrogen stars; and pure hydrogen-helium stars will lie somewhere in between. Nonlinear calculations show stabilization of the amplitude of oscillation occurs by shock formation in the envelope (references in BAC). Unfortunately, one cannot determine reliable mass loss rates from such calculations. However, there is no indication that $\dot{M}$ would be so large as to drive VMOs below M_p.

3.2 Eddington Standard Model

The simplest model of VMOs is that of an n=3 polytrope for which

$$\sigma \equiv \beta^{-1} - 1 \equiv s_\gamma/(4Y_T) = 0.19\ (T/keV)^3 (\rho/g\ cm^{-3})^{-1} (4Y_T)^{-1} \quad (3.1)$$

is constant. Here, β is the ratio of gas to total pressure, s_γ is the photon entropy, Y_T (=1.69 for typical Pop III abundances) is the total number of electrons and ions per baryon, and the temperature T is expressed in energy units ($k_B = 1$). The polytropic mass is $M \approx 1.1\ s_\gamma^2\ M_\odot$ at high mass, so the photon entropy ranges from 6 for 100 $M_\odot$ up to ~ 100 for $10^4\ M_\odot$ VMOs, and β is small. Though this model does do fairly well, accurate predictions of VMO parameters require refinements.

3.3 Point Source Model

VMOs have small radiative envelopes attached to large convective cores over which the total entropy is constant. The success of the Eddington model relies on the nuclear entropy, $s_N = Y_T\ \ln(T^{3/2}/\rho) + const$, varying little over the core. The Point Source Model explicitly includes s_N-variations. The central value of photon entropy completely characterizes such models. In BAC, we obtain the following simple relations which characterize the ZAMS structure in terms of the parameter σ evaluated at the stellar center:

$$\sigma\ (M) \approx 0.24\ Y_T^{-1} (M/M_\odot)^{\frac{1}{2}} - 0.61\ ,\quad M > 100\ M_\odot \quad (3.2)$$

$$T_c(\sigma) \approx \begin{cases} 3.8\ (\sigma^2/(1+\sigma))^{0.054} (X_{CN}/10^{-2})^{-0.054} keV,\ \text{Pop I and II} & (3.3a) \\ 11.3\ (\sigma^2/(1+\sigma))^{0.087} (X_{CN}/10^{-9})^{-0.079} keV,\ \text{Pop III} & (3.3b) \end{cases}$$

$$R\ (\sigma) \approx 30\ (\sigma(1+\sigma))^{\frac{1}{2}} (T_c(\sigma)/keV)^{-1}_{max}\ \{1,0.64+0.06\sigma\}\ R_\odot \quad (3.4)$$

$$T_e(\sigma) \cong \begin{cases} 6 \times 10^4\ \sigma^{0.03}\ K & \text{Pop I} & (3.5a) \\ 10^5\ \sigma^{0.04}\ K & \text{Pop III} & (3.5b) \end{cases}$$

$$L \cong L_{ED}(1+\sigma^{-1})^{-1}\ ,\quad L_{ED} = 1.2 \times 10^{38}\ Y_e^{-1}\ M/M_\odot\ erg\ s^{-1} \quad (3.6)$$

Given M, Y_T and the CNO abundance X_{CN}, σ is computed from 3.2, which is
then used to find the central temperature T_c, the central density using
3.1, the radius R, the surface temperature T_e and the luminosity L in
terms of the Eddington luminosity L_{ED}. Here, Y_e is the number of
electrons per baryon. Central and surface temperatures are insensitive
to mass, and the luminosity is nearly Eddington at large mass. Pop III
stars are considerably hotter and smaller than their Pop I and II
counterparts. In Table 1, comparisons with a few numerical models show
good agreement.

Model	$M/M_\odot$	T_c (keV)	L/L_{ED}	$R/R_\odot$	T_e (10^4K)	q_{cci}	M_α/M_i	τ_H(my)
Talbot Pop I	100	3.82	0.32	12.1	5.48	-	-	-
		3.83	0.32	12.6	5.37	0.84	0.49	3.0
KS Pure H	1000	15.0	0.81	14.7	10.7	0.92	0.5	2.8
X_{CN}=8(-11)		15.4	0.76	14.1	10.7	0.98	0.49	3.3
EFO Pop III	100	11.1	0.36	4.10	9.62	0.84	0.40	3.1
X_{CN}=5(-10)		11.1	0.31	4.01	9.38	0.84	0.48	3.3
R136a Pop I	2000	4.15	0.85	76.5	5.86	0.99	0.57	2.0

Table 1: The second line gives the Point Source Model predictions for
the models of Talbot (1971), Kovetz and Shaviv (1977), and El Eid, Fricke
and Ober (1983); their computed values are given on the first line. The
KS star is the least successful fit of all those given in BAC. We are
generally more successful at higher masses. The 2000 $M_\odot$ model is a Point
Source prediction.

3.4 Why $X_{CN} \sim 10^{-9}$ in Population III Stars

The simplest way to see this is to envisage the trajectory of the
star in T_c - X_{CN} space as it evolves from Kelvin-Helmholtz contraction
onto the main sequence. During contraction, the 3α reaction will generate
a very small abundance of carbon, progressively more as evolution proceeds
and the central temperature rises, but never enough for the 3α energy to
compete with the PdV liberated. A monotonically increasing curve is
followed in our T_c - X_{CN} plane. The CNO nuclei catalyze hydrogen burning.
The star can be supported by this luminosity source alone provided its
temperature obeys Eq. (3.3), which defines a monotonically decreasing
trajectory. The true path will pass smoothly from the Kelvin-Helmholtz
curve to the hydrogen burning curve. A lower bound for the CN abundance
on the ZAMS is obtained by the intersection point of these two curves:

$$X_{CNi} \sim 2 \times 10^{-10} \sigma^{0.7} (1+\sigma)^{-0.5} . \tag{3.7}$$

Once the thermal equilibrium trajectory is attained, the 3α reactions
continue to add to X_{CN}, implying further evolution along this path
throughout core hydrogen burning, approximately following

$$X_{CN}(t) \sim X_{CNi}(1+(t/10^4 y)\sigma^{2.9}(1+\sigma)^{-2.5}(X_{CNi}/10^{-9})^{-3.2})^{0.31} \qquad (3.8)$$

These estimates agree reasonably well with the computed models of El Eid et al. (1983).

3.5 Evolution of the Convective Core

The luminosity at the boundary of the convective core can be estimated by requiring that the entropy gradient vanish there. If this is equated with the surface value, an expression is obtained for the convective core mass in terms of the instantaneous value of Y_e in the core:

$$M_{cc}(Y_e)/M(Y_e) = Y_e/Y_{es} \, q_{cci} \qquad (3.9)$$

We have assumed that the luminosity remains approximately constant. Though this method shows $q_{cci} \to 1$ at large mass, it does not predict its value, which must be obtained from point source results; comparisons with models are given in Table 1. The instantaneous value of the star mass is M, which can decrease if there is mass loss; its value is labelled by the central value of Y_e, which is like a time variable in that it mono-tonically decreases as hydrogen is burned. The surface value, Y_{es}, will equal its ZAMS value, Y_{ei}, until the mass loss has eaten down to the mass of the ZAMS convective core, at which point processed nitrogen and helium will become unbared. The final convective core, i.e. the helium core, has mass

$$M_{\alpha}/M_i = 0.5 \, (q_{cci}/Y_{ei}). \qquad (3.10)$$

Eq. (3.9) agrees very well with computed evolutionary models with no mass loss, for example with the BAC 500 $M_{\odot}$ star described below. It also agrees with models with mass loss provided the rate is not in excess of the critical value for which the convective core shrinks at exactly the same rate as the surface:

$$-\dot{M}_{cr} = L/(0.007c^2) \, (2Y_{es})^{-1} = 1.5 \times 10^{-4} (M/10^3 M_{\odot})(L/L_{ED})Y_{es}^{-2} M_{\odot}/y \qquad (3.11)$$

Larger mass loss rates result in essentially completely convective stars; for these, 3.9 still holds with $Y_{es}=Y_e$. If the rate is sometimes below and sometimes above critical, M_{cc} will still follow 3.9, but M_{α} will not be given by 3.10. This is the reason for the slight discrepancy between our prediction and the El Eid et al. result in Table 1.

For Population III stars, the yield of helium is of interest. (See §5). The excess helium – above its original amount Y_i – liberated in winds, per unit mass of wind, is just

$$\langle \Delta Y \rangle = \int_{M_f}^{M_i} 2(Y_{ei}-Y_{es}(t))dM(t)/(M_i-M_f) = (1-Y_i/2)(1-M_f/M_i). \qquad (3.12)$$

The second term is valid only if $-\dot{M} < -\dot{M}_{cr}$, and is a maximum if $M_f = M_\alpha$: $\langle\Delta Y\rangle_M = (1+Y_i)/2$. If $-\dot{M} > -\dot{M}_{cr}$, the yield is less, and the detailed mass loss history is required for computation.

3.6 Envelope Ejection via Super-Eddington Shell Burning

In BAC, we report a calculation of the evolution of a 500 $M_\odot$ Pop I star which used an implicit hydro code with time-dependent convection included in a diffusion approximation based upon convective blob transport times. Mass loss was not included. The evolution up to core H exhaustion follows closely the description given by the analytics of the previous sections; even Eq. (3.9) holds in spite of the implicit inclusion of overshoot in our convection.

The ignition of 2.5 $M_\odot$ in a shell generated a luminosity jump of 29%, though only 10% could be tolerated for stability to be maintained locally. Convection would transport the excess to the surface, but the time-dependent criterion restricted the zone's growth to mass fraction 0.71. A total of 3.5×10^{52} ergs was generated in the shell. The gravitational binding energy of the envelope was only 0.5×10^{52} ergs. Thus 3×10^{52} ergs = 3×10^{-5} Mc² was the net liberated energy. This efficiency is comparable to that of a Type II SN. The entire region above the helium core was ejected. By the time envelope recombination occurred, the expansion was supersonic. However, implicit hydro is subject to numerical damping so our velocity estimates are not reliable. We expect Pop II VMOs to experience a similar phenomenon.

In Pop III stars, CNO abundances are so low that helium burning occurs first, with strong shell burning only following once oxygen has been dredged-up out of the core. The subsequent dynamical evolution would then be similar to our Pop I model. Woosley and Weaver (1982) [WW] have evolved a 500 $M_\odot$ Pop III star with a code similar to that used by BAC. They do find large dredge-up and strong shell burning, which drives their envelope to very low density but does not lead to ejection. Explicit inclusion of overshoot drastically modifies this result (Woosley, private communication). The BAC burning and dynamics occurred over about one Kelvin-Helmholtz time ($\sim10^4$y), theirs took about six, long enough for complete dredge-up of the helium-carbon core to deposit $\sim10\%$ metallicity by mass in the envelope.

The treatment of time-dependent convection, which is so crucial for this mechanism, is a major uncertainty in these calculations, and accounts in part for the differences between the WW and BAC results. The following questions need to be answered: Over what mass and metallicity range does this phenomenon occur? How sensitive is the outcome to changes in convective criteria? Will a large fraction of metals be lost at the same time? The latter issue relates to primordial helium production, since one cannot generate large amounts unless the metal-to-helium ratio ejected is very small. The 10^4y timescale is short enough to ensure that the wind will ultimately generate a very energetic Sedov-like blast, which could have a large impact upon the pregalactic medium (§5).

The light curve of these events has not been computed, and in any case will depend upon the amount of mass loss in the main sequence phase – as will the velocity of the ejecta. We do expect helium-rich ejecta which have metal abundances similar to solar for Pop I models. This phenomenon is thus a candidate for the SN1961 Type V event.

3.7 Helium Core Phase

If helium cores are laid bare by this mechanism, they will be pulsationally unstable (Stothers and Simon 1970). On the other hand, it does not require much of an envelope to stabilize against oscillation (Simon and Stothers 1969). Helium cores can also be described by n=3 polytropes for which $Y_T = 0.75$, $M_\alpha \sim 9.9 \sigma^2$ $M_\odot$ for high mass, and the central temperature is approximately $T_{\alpha c} = 18 \sigma^{0.08}$ keV.

3.8 VMO/SMO Boundary

General relativistic polytropes – solution to the TOV equations of hydrostatic structure with $p = K \rho^{\Gamma_1}$, with ρ the total mass-energy density and Γ_1 assumed constant – can be applied to the determination of the maximum mass which can be stably supported by hydrogen burning. Bludman (1973) shows that instability arises if $\Gamma_1 - 4/3 < 1.73 \, p_c / \rho_c \, c^2$ as long as the latter term is small. Using Eq. (3.2), a convenient expression for the critical mass above which a star is GR unstable in terms of the central temperature follows:

$$M_{cr} (T_c) = 1.6 \times 10^6 \, Y_T \, 1\text{keV}/T_c \, M_\odot . \tag{3.13}$$

With the aid of the T_c relations, the boundary mass becomes

$$M_{cr} \begin{cases} \sim 4 \times 10^5 \, (X_{CN}/10^{-2})^{0.05} \, M_\odot & \text{Pop I and II,} \\ \sim 1 \times 10^5 \, (X_{CN}/10^{-9})^{0.07} \, M_\odot & \text{Pop III.} \end{cases} \tag{3.14}$$

With $X_{CN} \sim 10^{-11}$, we estimate $\sim 8 \times 10^4$ $M_\odot$ in agreement with the value given by Ober for this case in this proceedings. Our results also agree with the calculations of Fricke (1973). Applying the same method to helium cores yields $\sim 5 \times 10^4$ $M_\odot$.

3.9 SMO Fate

If $M > M_{cr}$, spherically symmetric stars will never be stable. Indeed, it depends upon the protostellar conditions whether a quasistable Kelvin-Helmholtz phase develops at all. Unless the gas is thick to Thomson scattering, the core would tend to run away from the envelope, yielding either an accreting VMO or, if the flow is coherent enough, an SMO whose kinetic energy gained on infall can drive it well beyond the stable thermal equilibrium point for $M < M_{cr}$. The fate in the latter case must be determined by dynamical calculations. Until the recent numerical results of Fuller and Woosley reported by Woosley in this volume, only a few detailed calculations were available (Appenzeller and

Fricke 1972a,b, Appenzeller and Tscharnuter 1973, Schmidt 1973, Shapiro
and Teukolsky 1979), which only sparsely populated the M-Z plane. Fricke
(1973) solved the Post-Newtonian virial and energy equations assuming
an n=3 polytrope density structure to map out SMO fates in this plane,
and the new work confirms the qualitative features of his phase diagram.
Below the mass given approximately by 3.14, there is equilibrium. A
triple point exists at some critical metallicity along this line, Z_{cr}.
Fricke estimates its value to be 0.015; Fuller and Woosley get a somewhat smaller value. In any case, it is in the Pop I regime. A line
following $M \sim Z$ extends from this triple point upward, separating regions
in which complete collapse to a black hole occurs from regions in which
thermonuclear disruption occurs.

Thus, nonrotating SMOs of Pop III and II collapse completely. In
a limited mass range, Pop I can explode. Rotation dramatically increases
the range of exploders, up to $\sim 10^8$ $M_\odot$! With slow rotation, Fricke (1974)
still finds Pop III will form black holes. Implicit 2-D codes will
greatly aid our understanding of the fates of SMOs. Rotation can not
only lead to the effective braking of infall, but in some circumstances
can lead to disk formation, with a core more VMO-like than SMO-like.
The addition of magnetic fields further complicates the evolution, leading to the concept of spinar which Ozernoi and Usov (1971) and Bisnovatyi-
Kogan and Blinnikov (1973) have identified with quasars themselves.

Thus the SMO problem couples the star formation issue, with its
many complications and unknowns, directly to the question of fate.
Nørgaard and Fricke (1976) have pointed out that exploding SMOs make
large amounts of ^{7}Li: ^{7}Li/^{4}He $\gtrsim 10^3 (^7$Li/^{4}He$)_\odot$. If rotation can make Pop
III SMOs explode, we can thus get a mechanism for cosmological ^{7}Li
production, but the same stars cannot also give significant amounts of
helium.

4. VMO FATE: PAIR SUPERNOVA OR BLACK HOLE

4.1 Entropology of Oxygen Cores

The same type of argument used to get convective core evolution in
hydrogen burning can be applied to helium burning; in this case, Y_e is
constant, so $M_0 \approx M_\alpha$. The 3α reactions generate mostly oxygen, though
there is some carbon and neon. It is a relatively good approximation to
treat the oxygen core as dynamically distinct from the overlying layers,
which makes the fate dependent primarily upon M_0; and such issues as the
amount of mass loss in prior phases, whether slow as in a normal wind or
catastrophic as in §3.6, just determine the relation of M_0 to the precursor star.

We can obtain a crude estimate of when an O core would go GR unstable before it goes pair unstable by inserting $T_c \sim 100$ keV into Eq.
(3.14): $M_0 \sim 8000$ $M_\odot$. The 100 keV is a compromise between two opposing
tendencies: the pairs are produced by photons on the tail of their

distribution, so $T_c \ll 2 m_e$; enough of the star must produce pairs for the global Γ_1 to fall below 4/3.

The generation and transformation of entropy provides a useful picture of core collapse. The total entropy has three components: $s = s_\gamma + s_e + s_N$; $s_\gamma \sim T^3/\rho$ is a measure of the photon pressure contribution; s_e includes the contribution of ionization electrons, and of e^+e^- pairs; s_N has only a logarithmic dependence on T. Only in the latest phases of collapse towards a black hole do $\nu\bar{\nu}$ losses become important; prior to this, s changes only due to nuclear burning. Fig. 1, taken from BAC, gives the entropy evolution of a typical element in a core which will collapse to a black hole. The regime of pair production occurs at constant entropy of course, but pressure is robbed since s_γ is transferred to s_e. Indeed, around T=100 keV, it costs 16 units of entropy for each pair that is made. The entropy jump at oxygen burning and the smaller one at silicon burning add pressure to resist collapse. The burning rates are so temperature sensitive that the jump is practically discontinuous at the 'ignition' temperature for these reactions. The nuclear entropy goes down, reflecting the fact that freezing the nucleons into the Si configuration gives fewer translational degrees of freedom than if they are locked into 0. The phase of 'alpha-quenching', when iron peak nuclei photodissociate into alphas is in fact isentropic since it occurs in nuclear statistical equilibrium, but s_N must go up for the nuclear state is less ordered; this robs s_γ and pair entropy - the pairs are by this time extremely relativistic, and so their entropy reflects their pressure contribution. If we continued this s-ρ diagram to higher density, s would begin to drop dramatically as $\nu\bar{\nu}$ losses become important; this is especially reflected in s_γ and s_e. At a few MeV, the $\alpha \to$ n,p transition is met. Again this is an isentropic transmutation, but the $s_N \sim 3$ for alphas must jump up to $\sim$10 for free nucleons, accompanied by a further plunge in s_γ and the pressure. Urca neutrino losses add to this plunge - though this can be alleviated somewhat by the mass element becoming thick to neutrino flow. In any case, so much pressure has been

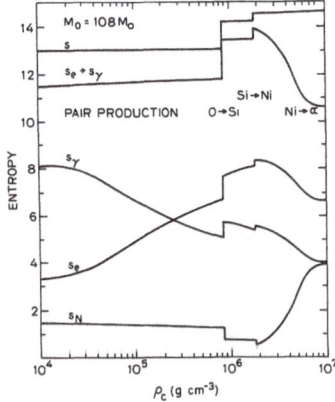

Fig. 1 Entropy evolution of a collapsing mass element.

lost that collapse is essentially in free-fall.

If a core explodes, the path of an element in s-ρ space is similar to the above description until at some stage its inflow is halted, with the maximum density reached being a function of position in the star. For those elements that attain high enough temperatures to burn, their outward trajectory is along a higher adiabat. Not only do they gain extra pressure associated with this extra entropy as the pairs recombine, but they also gain an extra amount since their nuclear entropy is lower than it was in the oxygen state. This large difference between outgoing pressure and its infall value results in a clean separation between cores which explode and those that collapse completely.

4.2 Core Binding Energy

To estimate the global dynamics of a given core, we need to piece together s(ρ) trajectories for all elements. For accuracy, a full evolutionary model is needed. However, a much simpler model works very well due to the clean-cut separation. Details of this model are given in Bond et al. (1982) and BAC. Since convection generates isentropic regions, s is assumed constant. A ρ(r) structure is needed: this is provided by an n=3 polytrope. The core mass is related to the initial entropy at the point of marginal core instability. A global binding energy is calculated as a function of central density. Fig. 2 shows such curves for values of the initial entropy and core mass near the critical value which separates exploding from collapsing cores. For the 108 $M_\odot$ collapse, the kinetic energy gained from sliding down the first hill easily takes the core over the hump created by nuclear burning, so it falls down toward the black hole pit; the side of the hill continually steepens as free-fall is approached. The kinetic energy of the 88 $M_\odot$ core is predicted to drop to zero as a result of burning: the inflow then turns to outflow, and the outgoing hill is much steeper than the incoming one due to the entropy gain.

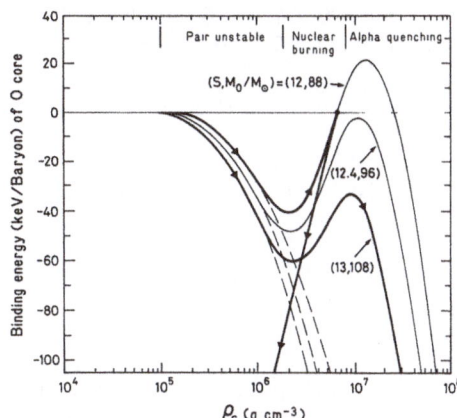

Fig. 2 Binding energy curves separate explosion from collapse.

We obtain $M_{Oc} \approx 100 M_\odot$. Ober et al. (1983) find their 90 $M_\odot$ O core explodes (their stellar mass at this point is 112 $M_\odot$ after mass loss), but the next one in their sequence, $\sim$110 $M_\odot$, does not. WW have a 90 $M_\odot$ core which explodes, and a 105 $M_\odot$ core which just explodes. Identification of the oxygen core mass in the latter case is difficult, since helium burning during the dynamical phase generated some extra carbon above what would normally be identified with the O core. Arnett (1973)'s $M_O = 93 M_\odot$ ($M_\alpha = 100 M_\odot$) explodes. These very different calculations, one semi-analytic, one using mass loss, one using only a helium core, all give about the same answer, so this boundary may be taken as reasonably well established. Rotation drives M_{Oc} up. The exact amount remains to be calculated, though simple arguments in BAC indicate that reasonable specific angular momenta appropriate to O stars do not drive it up enormously. However, it is possible that the outer regions could explode whilst inner ones collapse, unlike the either/or nonrotating case.

4.3 Exploding VMOs

Arnett's 1973 $M_O = 58 M_\odot$, $M_\alpha = 64 M_\odot$ core left a very loosely bound 2.2 $M_\odot$ Si remnant. All other calculations between $M_O = 40 M_\odot$ and 100 $M_\odot$ have shown complete disruption. In Table 2, the yields - expressed as $(X_j/X_{j\odot})/(X_O/X_{O\odot})$ - of WW, Ober et al. (1983) and Arnett (1973) are compared. The CN differences can be attributed to differing convection criteria. Abundances of the products of oxygen burning and incomplete silicon burning are in good agreement. A major feature is that not very much iron is generated so the O/Fe ratio is large. Also, this is one of the few places in stellar theory where Si burning does not go to completion, so the calcium group is abundantly produced. WW emphasized that this alone is a sufficient reason for taking the role of such stars in the nucleosynthetic history seriously.

	M_O	^{12}C	^{14}N	^{16}O	^{20}Ne	^{24}Mg	^{28}Si	^{32}S	^{36}Ar	^{40}Ca	>Si
WW	90	0.24	0.052	1	0.28	0.96	4.1	2.6	1.3	1.5	9.5
Ober	90	0.11	5(-6)	1	0.094	0.64	4.6	3.1	1.0	0.21	8.9
Arnett	93	0.096	-	1	0.041	0.76	-	-	-	-	9.6

Table 2: Abundances by mass relative to solar normalized to oxygen.

BAC also give an analytic estimate of the explosive kinetic energy:
$E_{exp} \sim 6 \times 10^{52} ((M_O/100 M_\odot)^{2.8} - (M_O/100 M_\odot)^{1.2}/4)$ ergs.

5. POPULATION III STARS

5.1 Theories of Galaxy Formation

About 90% of the mass in the universe is made of dark matter. It is unlikely to be intergalactic gas. The two main hypotheses are that it consists of massive collisionless relics of the Big Bang, or of the remnants of a first generation of stars. Stellar remnants would have been baryonic at the epoch of primordial nucleosynthesis. The standard model for the evolution of structure in the universe is that small amplitude waves grow to nonlinearity by gravitational instability. These waves may be sound waves, in which case the density fluctuations are adiabatic, or entropy waves, in which case the fluctuations are isothermal. Adiabatic perturbations damp on small scales, and the first structures to form are the scale of large clusters, pancakes, which then break up into small gas clouds from which the first stars arise at redshift $\sim$5. In pure baryonic universes, this theory can be ruled out by observations of the fluctuation level in the 3K background radiation. Isothermal perturbations do not damp, so small scales are the first to collapse, with larger-scale structure arising from the hierarchical clustering of smaller structures. The first generation of stars would form at $z\sim$100 in such a theory.

Massive collisionless relics of the Big Bang can be of three main types: hot (neutrinos), warm (gravitinos), or cold (primordial black holes, monopoles, axions, gravitinos). The classification is based upon their velocity dispersion (Bond and Szalay 1983). If the universe is dominated by such particles, then the adiabatic theory becomes viable again. Indeed, it becomes the preferred theory since most ideas for fluctuation generation predict sound waves will have much higher amplitudes than entropy waves (Bond, Kolb and Silk 1982). In the hot picture, the first structures are of supercluster scale, and the first stars form from their break-up at $z\sim$5. In the warm picture, cluster scales collapse first, and the first stars could form at $z\sim$20. In the cold picture, small scales are the first to collapse, with most Pop III stars forming at $z\sim$30, and larger structures forming by hierarchical clustering. These redshift limits are imposed by microwave background fluctuation constraints.

Though the first stars would form at very different epochs in these pictures, it is remarkable that the clouds from which they arise are predicted to be of about the same mass, $\sim 10^6$ $M_\odot$. This is the Jeans mass just after recombination, the characteristic mass for clouds of 10^4 K which form by thermal instability in a galactic collapse, the minimum mass required for the molecular hydrogen cooling rate to exceed freefall for cloud collapse in the cold picture, and is near the Sunyaev-Zeldovich mass of fragments in pancake pictures. The mass spectrum of stars which arise from cloud collapse will depend on parameters other than the mass as was discussed in §2.4, and these can differ in the various scenarios. Norman and Silk (1981) show how many galactic properties can be understood if there existed a phase in which $\sim 10^6$ $M_\odot$ gaseous subunits were colliding. In particular, the existence of abundance gradients in galaxies points to substantial star formation occurring as the galaxy was collapsing. This does not rule out a pregalactic phase of star formation, however.

5.2 Consequences of Pop III

The following points are addressed more fully in CBA, and complete references are given there.

(1) *Population II metallicity:* Truran and Cameron (1971) suggested that the G-dwarf problem is a consequence of Pop II abundances having been generated by a pregalactic generation of Pop III. There are other ways of solving this problem however. On the other hand, if we agree that at most Pop II metallicity, $\sim 10^{-4}$ by mass, can be produced by a pregalactic generation, then a severe constraint is imposed upon the number of stars which can supernova. This includes massive stars in the range ~ 15 to ~ 50 to 100 $M_\odot$. Tornambe (1983) has pointed out that the intermediate $4-8$ $M_\odot$ stars of Population III can carbon detonate even if their Pop I and II counterparts lose enough mass in the 2-shell phase to avoid it — since Pop III have too low a CNO abundance to drive significant H shell burning. Any IMF with substantial power in either of these exploding mass ranges then can have only a small integrated mass density. Normal IMFs fall into this category.

(2) *Abundance anomalies:* The oxygen-to-iron ratio (Sneden et al. 1979) and sulphur-to-oxygen ratio (French 1980) are both apparently higher in lower metallicity environments, which argues for more massive stars in the past. This does not necessarily require that they be Pop III or VMOs however.

(3) *Microwave Distortions:* Woody and Richards (1979) reported a 2σ distortion from black body exists near the peak of the 3K background. The only viable theories proposed to generate such a feature utilize the nuclear energy from Pop III created at $z\sim 100$ which is thermalized by grains (Rowan-Robinson et al. 1979). A large stellar mass density relative to the closure density is required by such schemes, $\Omega_*\sim 0.1$. A normal IMF could not do this, as it would run up against the metallicity constraint. Accretion onto black holes can alleviate this difficulty somewhat, but the generation of holes also requires an exotic IMF.

(4) *Dark Matter:* To avoid the metallicity constraint while making $\Omega\sim 0.1$ from remnants imposes either the Jupiter solution with the IMF skewed to very low masses, or the black hole solution, with the IMF skewed to $M > M_c$. Here, M_c is that initial stellar mass whose oxygen core mass is above $M_{0c}\sim 100$ $M_\odot$. It is at least 200 $M_\odot$. The constraint resulting from stellar light is not as strong as the metallicity one, though limits on M dwarfs in our galaxy's halo do require a Jupiter IMF to fall off very steeply below a solar mass. At the other extreme, all of the black hole precursors could not have been SMOs above $\sim 10^6$ $M_\odot$, since their continued passage through the disk of our galaxy would transfer energy to the stars, puffing up the disk more than is observed. A hole larger than $\sim 10^7$ $M_\odot$ would have spiralled into the galactic center by dynamical friction, and its influence would have been detected. This primarily leaves the VMO range to account for most of the holes. If a significant metal release arises in the evolution of those VMOs which collapse due either to the super-Eddington phenomenon of 3.6 or to rotation effects, then pregalactic black holes will not account for a large fraction of the dark matter. An advantage of black holes is that their ability to accrete gas allows the efficiency with which the medium turns gas into stars to be lower than required by the Jupiter solution.

(5) *Energy injection into the pregalactic medium:* Non-gravitational forces will give rise to some of the structure in the universe. How dominant such effects will be depends upon the form, efficiency and epoch of energy release and upon its transformation into other forms. Dorosh-kevich et al. (1967) suggested that nuclear energy from Massive Objects could be a significant source of density fluctuations. The radiation generated by VMOs and by SMOs (for $\sim 10^4$y) and by accretion onto their black hole remnants create HII regions which overlap, leaving the universe fully ionized (6). Gas transport via Thomson scattering can then generate large scale perturbations. Hogan (1983) and Hogan and Kaiser (1983) have studied this effect; they require $\Omega_* \sim 0.1$ at $z \sim 100$ to create significant structure this way. Ostriker and Cowie (1981) and Ikeuchi (1981) have studied how kinetic energy from explosions generates shocks in the expanding IGM; shells are swept-up, fragment, form new explosions which creates shocks on larger scales. The result is the amplification of small scale structure to large via hydrodynamical energy injection. Winds can have similar effects. For both HII regions and supernova explosions, the metallicity constraint is a limitation. For example, SN explosions give at most 30 $(Z/10^{-4})$ eV per baryon, where Z is the metallicity we allow to come from this explosive phase. The Type V SN discussed in 3.6 can swallow all the metals in the black hole while liberating ~ 30 keV per baryon that has gone into the star. If SMOs of Pop III can explode with sufficient rotation, their efficiency can be very high. (Woosley reports $\sim 10^{-3}$). Black hole energy injection via jets and winds provide another source which is unconstrained by metal ejection.

(6) *Reionization of the Universe:* The Gunn-Peterson test shows that the universe is very highly ionized at least out to $z \sim 3$. The nuclear energy from stars provides a less stringent constraint than gravitational energy released in galaxy formation: $\Omega_* \sim 10^{-5}$ in ionizing stars is required, though this quantity depends upon epoch and the clumpiness of the gas. If ionization occurs late ($z < 16$), recombination of the IGM will not occur. HII regions created by such stars could generate a significant redshifted Lyα feature — see CBA for details.

(7) *Helium generation:* To get large amounts, a scheme involving VMO winds with metals locked up in holes is required. BCA discuss the conditions under which 25% helium could be generated. This requires about 80 to 90% of the original gas to go into a first generation of VMOs, which is large but not implausible for hierarchical theories. One gets a large popula-tion of black holes at the same time, the precise density depending upon the subsequent accretion history. The universe would need a large lepton number per baryon to avoid helium production in the first three minutes (Fowler 1970); its photon entropy per baryon could still be the $\sim 10^9$ required by the 3K density. On the other hand, if conventional primordial nucleosynthesis holds, one could invoke VMOs to generate fluctuations at low metallicity if the apparent observational scatter in the Y-Z plane turns out to be real.

(8) *Deuterium generation:* Audouze and Silk (1984) consider cosmic rays from stars such as VMOs that have already produced helium incident upon pure hydrogen, generating deuterium by spallation. In this way, they avoid Li over-production by $\alpha + \alpha$ reactions, which is a major difficulty for the subject of deuteronomy (Epstein 1977). Rees (1984) considers a neutron cloud which forms by high energy collisions in the neighborhood of black holes produced by VMOs or SMOs: $n + p \rightarrow d + \gamma$. Bond, Carr and Hogan (1983) consider the photodissociation of helium to create ^{3}He as well as deuterium by $\sim$30 MeV gamma rays generated by black holes, extending earlier work of Ozernoi and Chernomordick (1975). ^{7}Li can be generated by SMOs (3.9) and perhaps by processes associated with black holes (Rees 1984).

(9) *Gravitational Waves:* A stochastic background of redshifted waves can come from black hole formation events, such as VMO and SMO collapse. Bond and Carr (1983) suggest that, by analogy with O stars, VMOs and hence their holes will often be in close binaries, which can generate a much larger gravitational wave signal due to coalescences long after formation. Such waves would be detectable by laser interferometry and Doppler tracking.

5.3 Politics of Population III

The role of the first stars in the history of the universe depends upon the unknown IMF. The *conservative* supposes the IMF remains the same, that the first stars form with galaxies, and that Pop III is a short phase which generates a little Z but no real abundance anomalies. The *liberal* allows his IMF to skew somewhat to high mass, entertains the notion of a pregalactic generation, but above all respects the all powerful metallicity constraint. He can get a little bit of Z, a higher O/Fe ratio and the calcium group from massive stars, and may accept that the odd VMO forms; and SMOs may arise in AGNs. Both conservatives and liberals can ionize the universe and keep it ionized if the light is generated at z < 15. The *radical* allows absolute freedom for his IMF. He can get dark matter from black holes or Jupiters, noting that the inferred dynamical value of Ω, $\sim$0.1, is not inconsistent with the value of the baryon density parameter inferred from primordial nucleosynthetic arguments; and, the observed baryon density now - including stars and the X-ray emitting gas in clusters of galaxies - is only about one-tenth of this. The radical can point to spectral distortions in the microwave background as justification for his position. The *anarchist* chooses the radical IMF, and attempts to generate ^{4}He from VMOs, and D, ^{3}He and ^{7}Li from other post-recombination events. Ober, Falk and Audouze, this volume, choose a liberal IMF, and find too many metals, and marginally too much light to get all of the ^{4}He this way. BCA choose a radical IMF, and are freed from both constraints. The anarchist may require rather too many coincidences for his view to be compelling. Big Bang Nucleosynthesis forms a *totalitarian* regime which places strict controls on the allowed variation of the baryon density parameter to get the abundance set {^{4}He, D, ^{3}He, ^{7}Li} to be compatible with observations. If this parameter turns out to be much larger than that inferred now, we will be

confronted with a baryon gap, as Yang et al. (1983) have emphasized.
The radical may become the neo-orthodox. At present the *democrat* would
vote for a liberal-conservative coalition, getting a little bit of Z,
some abundance anomalies, and no dark matter.

I take pleasure in acknowledging my continuing conversations on
these subjects with Craig Hogan, Wolfgang Ober, Martin Rees, Joe Silk,
Roberto Terlevich and Stan Woosley. In particular, Martin Rees has often
made the thick transparent. This review derives mostly from work labelled
by permutations of {A,B,C}, and I thank the other elements of this set
for an enjoyable interaction.

REFERENCES

Andriesse, C.D., Donn, B.D. and Viotti, R. 1978, M.N.R.A.S. 185, 771.
Appenzeller, I. and Fricke, K. 1972a, Astron. Ap. 18, 10.
                                 1972b, Astron. Ap. 21, 285.
Appenzeller, I and Tscharnuter, W. 1973, Astron. Ap. 25, 125.
                                   1974, Astron. Ap. 30, 423.
Arnett, W.D. 1973, in *Explosive Nucleosynthesis*, ed. Arnett, W.D. and
        Schramm, D.N. (Austin: University of Texas Press).
Audouze, J. and Silk, J. 1984, in Proc. 3rd Moriond Conference,
        J. Audouze and J. Tran Thanh Van (eds.), Dordrecht: Reidel.
Begelman, M.C. and Rees, M.J. 1978, M.N.R.A.S. 185, 847.
Bisnovatyi-Kogan, G.S. and Blinnikov, S.I. 1973, Sov. Astron. 17, 304.
Bludman, S. 1973, Ap. J. 183, 637.
Bond, J.R., Arnett, W.D. and Carr, B.J. 1982, in *Supernovae*, ed. Rees,
        M. and Stoneham, R. (Dordrecht: Reidel).
Bond, J.R., Kolb, E.W. and Silk, J. 1982, Ap. J. 255, 341.
Bond, J.R., Arnett, W.D. and Carr, B.J. 1983, Ap. J., in press.
Bond, J.R., Carr, B.J. and Arnett, W.D. 1983, Nature, in press.
Bond, J.R. and Carr, B.J. 1983, preprint.
Bond, J.R., Carr, B.J. and Hogan, C. 1983, preprint.
Bond, J.R. and Szalay, A.S. 1983, Ap. J., in press.
Boury, A. 1963, Annales d'astrophysique 26, 354.
Branch, D. and Greenstein, J.L. 1971, Ap. J. 167, 89.
Carr, B.J., Bond, J.R. and Arnett, W.D. 1983, Ap.J., in press.
Cassinelli, J.P., Mathis, J.S. and Savage, B.D. 1981, Science 212, 1497.
Chevalier, R. 1981, Fund. Cosmic Physics, 7, 1.
Chu, Y.H. and Wolfire, M. 1983, Bull A.A.S. 15, 644.
Davidson, K., Walborn, N.R. and Gull, T.R. 1982, Ap. J. 254, L47.
Doroshkevich, A.G., Zeldovich, Ya.B. and Novikov, I.N. 1967, Sov. Astron.
        11, 233.
Doroshkevich, A.G. and Kolesnik, I.G. 1976, Sov. Astron. 20, 4.
Duncan, M.J. and Shapiro, S.L. 1983, Ap. J. 268, 565.
El Eid, M.F., Fricke, K.J. and Ober, W.W. 1983, Astron. Ap. 119, 54.
Epstein, R.I. 1977, Ap. J. 212, 595.
Feitzinger, J.V., Schlosser, W., Schmidt-Kaler, Th. and Wrinkler, C.
        1980, Astron. Ap. 84, 50.
Fowler, W.A. 1970, Comments Ap. and Space Phys. 2, 134.
French, H.B. 1980, Ap. J. 240, 41.

Fricke, K.J. 1973, Ap. J. 183, 941.
         1974, Ap. J. 189, 535.
Hogan, C. 1978, M.N.R.A.S. 185, 889.
Hogan, C. 1983, M.N.R.A.S. 202, 1101.
Hogan, C. and Kaiser, N. 1983, Ap. J., in press.
Hoyle, F. and Fowler, W.A. 1963, M.N.R.A.S. 125, 169.
Hutchins, J.B. 1976, Ap. J. 205, 103.
Ikeuchi, S. 1981, Publ. Astr. Soc. Japan 33, 211.
Kashlinsky, A. and Rees, M.J. 1983, preprint.
Kovetz, A. and Shaviv, G. 1971, Ap. Space, Sci. 14, 378.
Larson, R.B. and Starrfield, S. 1971, Astron. Ap. 13, 190.
Larson, R.B. 1982, M.N.R.A.S. 200, 159.
Lightman, A.P. and Grindlay, J.E. 1982, Ap. J. 262, 145.
Massey, P. and Hutchings, J.B. 1983, DAO preprint.
Meaburn, J., Hebden J.C., Morgan, B.L. and Vine, H. 1982, M.N.R.A.S. 200
      1P.
Melnick, J. 1983, ESO Messenger 32, 11.
Moffat, A.F.J. and Seggewiss, W. 1983, preprint.
Nørgaard, H. and Fricke, K.J. 1976, Astron. Ap. 49, 337.
Norman, C. and Silk, J. 1981, Ap. J. 247, 59.
Ober, W.W., El Eid, M.F. and Fricke, K.J. 1983, Astron. Ap. 119, 61.
Ostriker, J.P. and Cowie, L.L. 1981, Ap. J. Lett. 243, L127.
Ozernoi, L.M. and Chernomordick, V.V. 1975, Sov. Astron. 19, 693.
Ozernoi, L.M. and Usov, V.V. 1971, Ap. Space Sci. 12, 267.
Palla, F., Salpeter, E.E. and Stahler, S.W. 1983, Cornell preprint
      CRSR 804.
Panagia, N., Tanzi, E.G. and Tarenghi, M. 1983, ESO preprint.
Poveda, A. 1977, Rev. Mex. de Astron. y Astrophys. 3, 189.
Rees, M.J. 1984, in Proc. 3rd Moriond Conference, J. Audouze and
      J. Tran Thanh Van (eds.), Dordrecht: Reidel.
Rowan-Robinson, M., Negroponte, J. and Silk, J. 1979, Nature 281, 635.
Schmidt, M. 1963, Ap. J. 137, 758.
Schmidt, J. 1973, Astron. Ap. 27, 351.
Shapiro, S.L. and Teukolsky, S.A. 1979, Ap. J. 234, L177.
Silk, J. 1983, preprint.
Simon, N.R. and Stothers, R. 1969, Ap. J. 155, 247.
Sneden, C., Lambert, D. and Whitaker, R.W. 1979, Ap. J. 234, 964.
Stothers, R. and Simon, N.R. 1970, Ap. J. 160, 1019.
Sutton, E. Becklin, E. and Neugebauer, G. 1974, Ap. J. 190, L69.
Talbot, R.J. 1971, Ap. J. 165, 121.
Terlevich, R. 1982, Ph.D. thesis, Cambridge University.
Terlevich, R. and Lynden-Bell, D. 1983, preprint.
Tohline, J.E. 1980, Ap. J. 239, 417.
Tornambe, A. 1983, preprint.
Truran, J.W. and Cameron, A.G.W. 1971, Ap. Space Sci. 14, 179.
Wiegelt, G. 1983, quoted in Melnick (1983).
Woody, D. and Richards, P.L. 1979, Phys. Rev. Lett. 42, 925.
Woosley, S.E. and Weaver, T.A. 1982, in *Supernovae*, ed. Rees, M.J. and
      Stoneham, R.J. (Dordrecht: Reidel).
Wynn-Williams, C.G. 1982, Ann. Rev. Astron. Ap. 20, 587.
Yang, J., Turner, M.S., Steigman, G., Schramm, D.N. and Olive, K.A. 1983,

        preprint.
Yoneyama, T. 1972, Publ. Astron. Soc. Japan, $\underline{24}$, 87.
Yorke, H.W. and Krügel, E. 1977, Astron. Ap. $\underline{54}$, 183.

Question.

Ober: How much matter of the Universe has to go into stars in your
        model to produce 25% He and not to overproduce the heavy element
        abundance for Pop II stars?

Bond: We, of course, must avoid in any assumed IMF any stars which
        explode. Thus, the masses must be well above the critical mass,
        $M_c > 200\ M_\odot$, below which stars undergo pair supernovae, since
        these can turn as much as 50% of their initial mass into metals.
        This is devastating, since we cannot tolerate more than Pop II
        abundance production. Choosing 10^{-4} for Pop II, say, this would
        impose the requirement that $< 2 \times 10^{-4}$ of the initial gas can go
        into such stars. At the same time, we require that $\sim 80\%$ of the
        initial gas must pass through stars with $M > M_c$ which undergo en-
        velope ejection. We also cannot tolerate more than $\sim 10^{-4}$ metal
        dredge-up in those that produce black holes. Rotation during
        final collapse to the hole could, perhaps, lead to metal production
        imposing a further constraint. The "ifs" in all this do stretch
        the imagination very much, lending support to Big Bang helium
        synthesis, but there is, as yet, no internal inconsistency in
        this way of getting 25%.

STELLAR He PRODUCTION IN A COLD UNIVERSE?

W. W. Ober and H. J. Falk
Max-Planck-Institut für Astrophysik
8046 Garching b. München
Fed. Rep. Germany

I. INTRODUCTION

A hot Big-Bang model of the Universe is quite successful in explaining the observed Helium abundance together with the 2.7 K microwave background radiation. But one might think of other scenarios as well, for example the mass-energy conversion by nucleosynthesis in stars formed in a cold (no photons or particle pairs) Universe out of density fluctuations (Carr & Rees, 1977; Rees, 1978; White & Rees, 1978). Carr (1977) has discussed Helium production in such a Universe by cosmological nucleosynthesis and showed that Helium decreases as the lepton to baryon ratio (L/N) increases - above L/N > 1.5 no Helium is cosmologically produced at all. He concluded that the observed Helium abundance could only be produced if the ratio was in the range 1.20 < L/N < 1.23, but then an overproduction of heavier elements would be expected. Helium production in massive mass losing stars has been considered by Talbot and Arnett (1971) whose semi-analytic treatment gave a Helium nucleosynthetic yield as great as 0.5. However, Carr (1977) has questioned the certainty of their yields.

We have investigated the evolution of pure hydrogen stars (X = 1. , Y = 0. , Z = 0.) in the mass range of 10 to 500 solar masses including mass loss by stellar winds. The lower mass limit is justified by the fact that the Jeans Mass in a cold Universe never rises above the Chandrasekhar Mass which is 5.6 $M_\odot$ if all nucleons are in proton form (Carr, 1977; Silk, 1977). The enrichment of the pregalactic medium with Helium and heavier elements has been calculated using a nuclear reaction network to follow in detail the chemical evolution.

II. RESULTS

The lifetimes of pure hydrogen stars are roughly twice as large as those of POP I composition objects (t_{ms} = X*(2.+X) (Falk & Mitalas, 1981)). During the main sequence phase the luminosities are in the range 3×10^3 to 1×10^7 $L_\odot$ while the effective temperatures are between 30000 K and

C. Chiosi and A. Renzini (eds.), Stellar Nucleosynthesis, 319–324.
© *1984 by D. Reidel Publishing Company.*

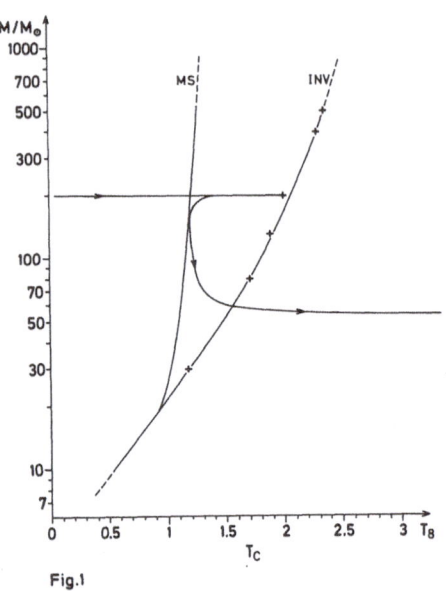

Fig.1

Fig. 1: The central temperature T_c
versus the initial mass M_i

80000 K and the central temperatures
exceed 10^8 K for stars more massive
than 20 $M_\odot$ (fig. 1). The curve INV
refers to the maximum temperature
achieved before CNO-burning dominates. The curve MS is the locus of
minimum main sequence temperatures.
The evolutionary track of a star of
initially 200 $M_\odot$ up to the end of
He-burning is shown.

Stars up to about 20 $M_\odot$ burn
their hydrogen via the PP-chain
(Ezer & Cameron, 1971). More massive
objects contract further until He-
burning generates at a mass fraction
of $X_{He} = 10^{-4}$ a tiny amount of
X_{CNO} (10^{-13}) after which the CNO-
cycle takes over from the PP-chain
and He-burning (mainly the triple-
alpha reaction). The contraction is
stopped and the stars expand on a
Kelvin-Helmholtz timescale towards
their main sequence. The hydrogen
is then consumed at $X_{CNO} = 10^{-11}$
by the normal CNO-cycle operating
at temperatures greater than 10^8 K
(fig. 1).

The mass loss rates computed by the algorithm of Chiosi (1981) are
in the range of 10^{-4} to 10^{-5} $M_\odot$/yr during this evolutionary phase so that
the stars lose considerable amounts of their initial masses ($\Delta M/M < 0.5$).
Due to the large convective cores of massive stars the outblown material
is enriched in Helium. In fig. 2 the mass fraction of Helium (Y), ejected
via mass loss and assumed completely mixed, is plotted versus the initial
mass M_i at the end of He-burning. The following burning stages proceed so
fast that mass loss can be neglected. At high masses the ejected Helium
fraction levels off at 0.3. This can be explained by the fact that the
relative mass lost during the main sequence evolution decreases with increasing mass.

To obtain the enrichment due to a stellar mass distribution, Y is
folded with the Initial Mass Function (IMF):

$N(m)$ dm $= M^{-\alpha}$ dm,

where α had the values $\alpha = 2.35, 2.00, 1.80$ (see Ober et al., 1983).
Integration over the mass range considered gives the Helium yield

$\langle Y \rangle = \int Y * \psi (m)$ dm,

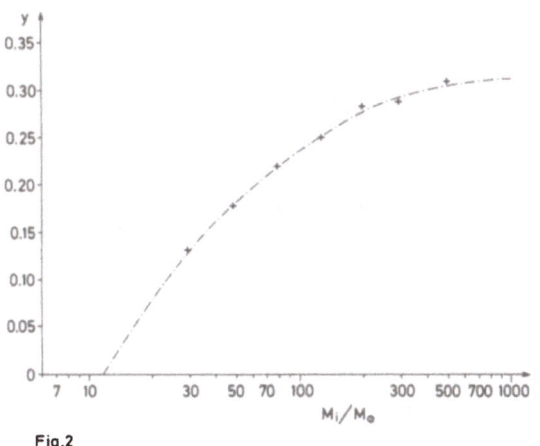

Fig. 2: The Helium
mass fraction of the
ejected and assumed
completely mixed matter
as a function of the
initial mass M_i in
solar units.

Fig.2

where $\psi(m) = M^{(1-\alpha)}$.

The integrand $Y * \psi(m)$ plotted in fig. 3 shows that the more the mass
distribution is biased towards the more massive stars the greater is the
He yield.

       The final mass at the end of He-burning (M_α) is plotted in fig. 4 as
a function of the initial mass M_i. Also indicated is the known (expected)
fate of the objects in the mass range considered. Stars having masses
greater than 150 $M_\odot$ at the end of He-burning collapse towards black holes.
From 150 $M_\odot$ down to 50 $M_\odot$ is the domain of the Pair Creation Supernovae
(PCSN). The fate of stars in the mass range $10 < M_\alpha/M_\odot < 50$ is not known
very well, but they will probably form black holes. Below $M_\alpha = 10\ M_\odot$ the

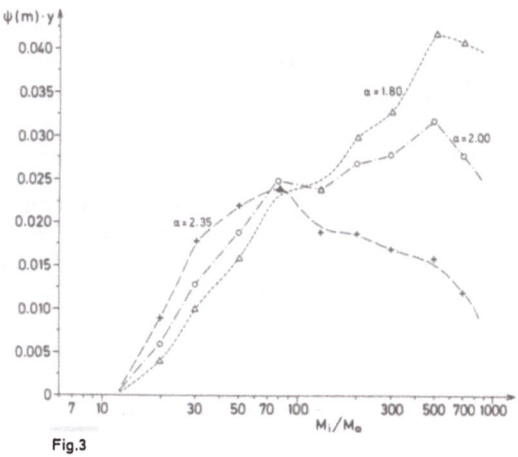

Fig. 3: The Helium
enrichment by mass
loss for different
slopes of the IMF
labelled with the
spectral index α.

Fig.3

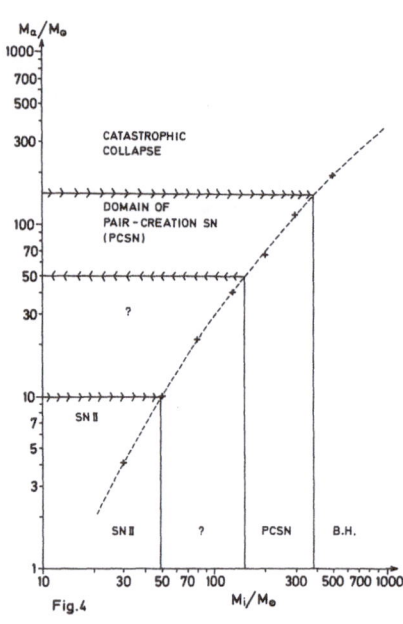

$M_\alpha/M_\odot$

CATASTROPHIC
COLLAPSE

DOMAIN OF
PAIR-CREATION SN
(PCSN)

?

SN II

SN II   ?   PCSN   B.H.

$M_i/M_\odot$

Fig.4

stars end their lives in "normal" SN I or SN II explosions.

Taking abundance calculations for SN II by Weaver & Woosley (1980) and Ober et al. (1983) for the PCSN we get together with the IMF the enrichment of the pregalactic medium with heavy elements (fig. 5). As shown in fig. 4 there are two distinct mass ranges which contribute to the heavy element enrichment: the normal supernovae and the PCSN. The enrichment with Helium and heavy elements for different slopes of the IMF are listed in Table 1.

TABLE 1

α	2.35	2.00	1.80
$\langle Y \rangle$	0.16	0.20	0.22
$\langle Z \rangle$	0.17	0.15	0.14

Fig. 4: M_i vs M_α mass relation for the adopted mass loss algorithm.

Helium is well in the range of observational data for $\alpha \lesssim 2.0$. This requires that all the matter of the Universe has to be processed in pure hydrogen stars to get the enrichment factors listed, i.e. the efficiency must be of order 100%. This seems unlikely but not impossible. On the other hand, the heavy element abundances are in extreme contradiction with the observations of even the most metal rich stars.

III. DISCUSSION

The energy density of the observed 2.7 K microwave background radiation is 0.25 eV/cm^3. This requires an energy input at the epoch z of $3.8 \times 10^{-2} * (1+z)$ MeV/particle if the Universe is just closed ($H_O = 75$ km/(s Mpc)). Because stellar nucleosynthesis yields about 7 MeV/particle, about $5.4 \times 10^{-3} * (1+z)$ mass fractions (at critical density) of the Universe could generate the entire microwave background radiation at redshift (1+z). With the required efficiency of 100% to account for 0.2 He mass fraction only 20% of the original protons have to be processed. On energetic grounds this leads to a formation redshift of z = 36 which is in good agreement with values derived by Layzer and Hively (1973) and by Alfvén and Mendis (1977). If black hole formation were to have taken place one easily could have much higher redshift values z (Carr, 1977) due to the much higher efficiency of mass to energy conversion by black holes compared to nucleosynthesis.

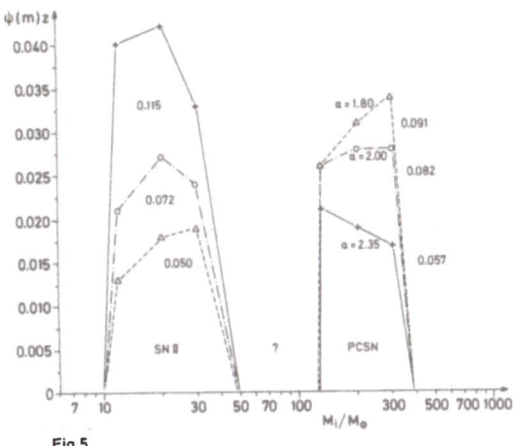

Fig.5

Fig. 5: The enrichment with heavy elements. The distinct contributions reflect the possible supernovae types, SN I, II and PCSN, respectively.

In order to produce the observed Helium by stellar nucleosynthesis special conditions in a cold Universe must be fulfilled. The L/N ratio has to be greater than 1.5. Though Helium enrichment is possible by steady mass loss in stellar winds, the enrichment remains low, even with the very high mass loss rates adopted. To end up with the required Helium abundance nearly all of the matter in the Universe has to be processed in such pure hydrogen stars. But in that case the enrichment with heavy elements by supernovae explosions is at least one order of magnitude too large compared to the most metal rich stars. For the low metal halo stars the discrepancy increases by three orders of magnitude. If large heavy element enrichment by supernovae explosions could be avoided, for instance by a very special primordial IMF then the problem disappears. A small number of supernovae is still required to produce some heavy elements needed to form grains which are necessary to thermalize the radiation energy. It is, however, questionable whether complete thermalization could have occured since that time (Rees, 1978).

Though it might be possible to generate the microwave background radiation there is still the shortcoming of a reasonable ratio of the newly synthesized Helium to heavy elements. Therefore it is very likely that the main contribution of the observed Helium abundance arises from primordial cosmological nucleosynthesis and that probably only a few tenths of a percent of the observed Helium mass fraction is a result of stellar nucleosynthesis.

REFERENCES

Alfvén, H. and Mendis, A.: 1977, Nature 266, 698
Carr, B.J.: 1977, Astron. Astrophys. 60, 13

Carr, B.J. and Rees, M.J.: 1977, Astron. Astrophys. 61, 705
Chiosi, C.: 1981, Astron. Astrophys. 93, 163
Ezer, D. and Cameron, A.G.W.: 1971, Astrophys. Space Sci. 14, 399
Falk, H.J. and Mitalas, R.: 1981, Monthly Not. Roy. Astron. Soc. 196, 225
Layzer, D. and Hively, R.: 1973, Astrophys. J. 179, 361
Ober, W.W., El Eid, M.F., Fricke, K.J.: 1983, Astron. Astrophys. 119, 61
Rees, M.J.: 1978, Nature 275, 35
Silk, J.I.: 1977, Astrophys. J. 211, 638
Talbot, R.J. and Arnett, W.D.: 1971, Nature Phys. Sci. 229, 150
White, S.D. and Rees, M.J.: 1978, Monthly Not. Roy. Astron. Soc. 183, 342
Weaver, T.A. and Woosley, S.E.: 1980, Ann. N.Y. Acad. Sci. 336, 335

CHEMICAL ABUNDANCES AND GALAXY EVOLUTION

John S. Gallagher, III
Department of Astronomy
University of Illinois - Urbana

ABSTRACT

   The chemical abundance characteristics of two comparatively
structurally simple classes of galaxies, the ellipticals and irregulars, are briefly reviewed. Neither of these types of galaxies have
properties which are fully consistent with the quantitative predictions of basic chemical evolution models. Possible sources for discrepancies are discussed in terms of fundamental evolutionary processes and limitations inherent in current observational data.

I. INTRODUCTION

   In attempting to develop a consistent picture for the formation
and evolution of galaxies, one quickly becomes aware that most galaxies contain an almost biological intricacy of observable features and
interrelated physical processes. Chemical abundances of heavy elements or "metals" can provide an important means for understanding
evolutionary histories of galaxies despite their complexity, since
metal abundances in principle depend in a comparatively straightforward way on integrals of the star formation rate over time (Talbot and
Arnett 1973). In fact simple analytical models have met with considerable success in describing abundance patterns in real galaxies
(e.g. see reviews by Audouze and Tinsley 1976; Tinsley 1980; Pagel
1981).

   As a result of our special location, by far the best and most extensive data on abundances are available for the Milky Way. Unfortunately the Milky Way, like other spirals, has diverse structural properties; e.g. both disk and spheroidal components are present and each
has its own characteristic stellar populations and abundances. Given
the current rather primitive state of theories for galaxy evolution,
it is then also of interest to examine properties of structurally simple galaxies, even though less detailed information is available. The

C. Chiosi and A. Renzini (eds.), Stellar Nucleosynthesis, 325–340.
© *1984 by D. Reidel Publishing Company.*

objective of this paper is to present a brief and admittedly somewhat
biased overview of chemical abundance patterns in two comparatively
simple types of galaxies, the ellipticals and normal irregulars.

II. ABUNDANCES IN ELLIPTICAL GALAXIES

a) Variations with Global Parameters

Elliptical galaxies (Es) can be ordered by a variety of observable parameters, the two main historical choices being the optical luminosity (which is essentially a measure of the number of $\sim 1 M_{\odot}$ stars
in each galaxy) or the projected stellar velocity dispersion (a function of the mass and binding energy). It has been known for some time
that a statistical relationship exits between optical luminosities and
colors of Es, in which fainter Es are bluer (e.g. Sandage and Visvanathan 1978; Wirth and Gallagher 1983, hereafter WG, and references
therein). The precise form of the color-luminosity correlation, however, is uncertain, both because it is intrinsically noisy and because
there are problems in defining samples of inherently faint Es.

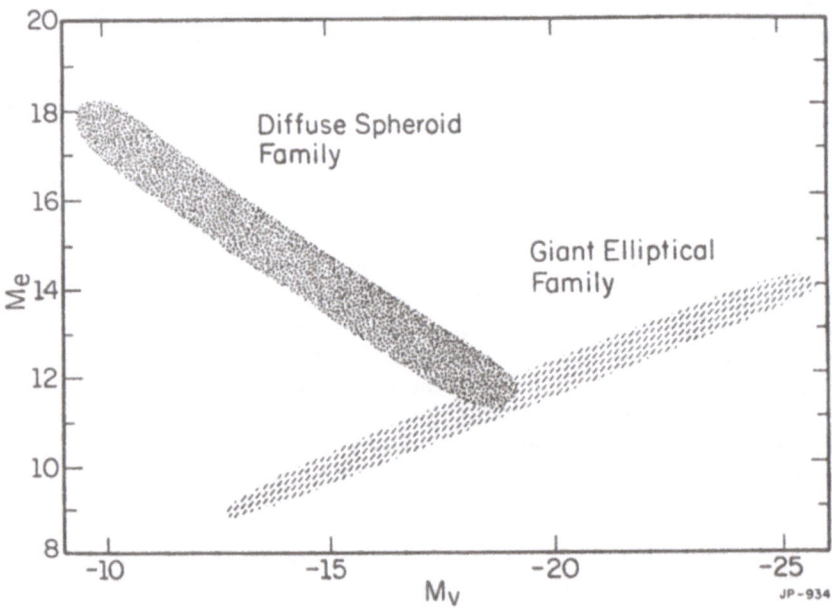

FIG. 1 – Schematic drawing of absolute visual magnitudes of E
galaxies versus surface brightness in mag arcmin^{-2} showing
the two E galaxy family branches (from Wirth 1982).

The latter point is illustrated in Figure 1 where galaxies classified as Es are plotted schematically in terms of mean surface
brightness M_e as a function of luminosity. This plot defines two

structural families of Es: the classical giant Es which includes all
luminous E galaxies and extends to near dwarf luminosities where M32-
like compact Es are found (WG), and the diffuse dwarf Es. In classical Es mean surface brightness declines with increasing luminosity,
and the projected light distributions are close to those predicted for
slightly truncated isothermal spheres (i.e. the profiles are well fit
by high central concentration King models or the de Vaucouleurs $r^{1/4}$
law). Classical Es usually have red optical colors in the sense that
they are all redder than the majority of Galactic globular clusters
(cf. Frogel, Persson, and Cohen 1980). Diffuse Es follow Baade's
(1963) sequence of parallel decreases in surface brightness and luminosity, have quasi-exponential (or equivalently low central concentration King model) brightness profiles, and have blue optical colors
which overlap with those of globular clusters (Caldwell 1982, 1983;
Wirth and Shaw 1983). A more detailed description of these two families is given by WG.

While diffuse Es are easily recognized as dwarfs due to their
distinctive brightness profiles, it is not obvious that they have followed the same evolutionary paths as the classical Es. A problem may
therefore arise if Es from these two family branches are mixed, as has
been the case in most studies of the dependence of global parameters
of Es on luminosity. As a result most of the conclusions in the literature are to some degree biased. Since diffuse Es are systematically bluer than their classical counterparts at a given absolute magnitude, the inclusion of mixed E families, for example, increases the
slope of the luminosity-color correlation and thus of the implied relationship between luminosity and metallicity.

Empirical estimates of metallicity levels in Es are difficult to
obtain, since one must (at present) rely primarily on properties of
the integrated light (absorption feature strengths, colors). These in
general will depend on both specific characteristics of the stellar
population (such as age mixture) and on the mean metallicity of the
stars, $\bar{Z}_*$ (see O'Connell 1980). Based on our experience with old star
clusters (Aaronson et al. 1978; Searle, Wilkinson, and Bagnuolo 1980)
and studies of galaxian absorption line strengths (Faber 1973, 1977;
Strom et al. 1978) the U-V color seems to be a reasonable metallicity
indicator, at least in systems with homogeneous old stellar populations. Caution is still in order since, for example, the form of the
horizontal branch or elapsed time since the last epoch of major star
formation may complicate matters.

For purposes of illustration we will adopt the Aaronson et al.
(1978) calibration of U-V versus metallicity. $\bar{Z}_*$ can then be seen to
range from ~ 1/2 solar for dwarf classical Es to (with considerable
uncertainty) somewhat above solar in luminous classical Es, and in
general the scatter in color at a given absolute magnitude is almost
as large as the total trend between M_B = -15 and -23. Interestingly,
the correlation between optical color and velocity dispersions for
classical Es is very much tighter than the color-luminosity rela-

tionship (Tonry and Davis 1981; Terlevich et al. 1981). Both groups present the reasonable interpretation that metallicity levels are strongly influenced by gas retention and thus by gravitational binding energy (see also Larson 1974; Vigroux et al. 1981) but are less dependent on the total number of stars in the system.

Taken at face value, the mean metal levels as estimated from $\bar{Z}_*$ in classical Es present some challenges for simple chemical evolution models. First, since $\bar{Z}_* \gtrsim Z_\odot$ in luminous Es, the energy requirements for metal production are prodigous, $E_{nuc} \sim 10^{-2}\bar{Z}_* M_* c^2 \sim 10^{62}$ erg! This energy must be lost from the forming galaxy without disrupting the protogalactic system, leading to excessive loss of metals, or (perhaps) producing too much photon luminosity. These requirements are more easily met by galaxies which form comparatively slowly, i.e. on time scales which are longer than dynamical collapse times (Tinsley and Larson 1979; Wirth 1982), although additional fine tuning is needed to provide the tight color-velocity dispersion correlation. Second as M. Edmunds has emphasized during these proceedings, chemical evolution models perdict $\bar{Z}_* \leq y$, where y is the metal yield (e.g. Audouze and Tinsley 1976; Tinsley 1980; Struck-Marcell 1981). In the Milky Way we evidently have y $\sim Z_\odot$, and the possible existence of galaxies with $\bar{Z}_* > Z_\odot$ therefore should be cause for concern. For example, this could be symptomatic of massive star-enriched IMFs during the formation of Es. But in any case the high global $\bar{Z}_*$ suggests that Es retained, mixed, and successfully astrated much of their metals. The moderate metallicity levels of hot intracluster gas on the other hand indicates that significant metals may have been lost from Es (De Young 1978), which would further constrain us to models with high initial metal yields.

The diffuse dwarf Es are equally as difficult to deal with as their classical cousins. From our experience in the Local Group, these systems are likely to contain trace young stellar populations that affect in unpleasant ways the integrated spectral indexes and colors from which abundances are derived (Burstein 1977; Wirth and Shaw 1983). However, if we ascribe the colors found by Caldwell (1982, 1983) as being primarily a reflection of old star metallicity levels, then the diffuse dwarf Es overlap the Galactic globular cluster colors, which implies $\bar{Z}_* < Z_\odot$. Thus star formation probably did not go to completion in these systems, but was halted by removal of gas into some form other than stars with near normal IMFs. It is therefore quite possible, as a number of researchers have hypothesized, that the diffuse Es are generically related to some other class of galaxy (such as dwarf irregulars) with differences between current structures being solely a result of premature gas removal via external (stripping) or internal (starbursts plus galactic winds) processes. If this view is correct, then the diffuse Es are in essence dwarf analogs to S0s (i.e. derived from disk-like galaxies) which should not be considered in the same evolutionary category as the classical Es (true spheroidals).

b) Metallicity Distributions within Classical Es

Although modest radial bluings in optical colors are well known features of many classical Es, these galaxies do not in general become as blue as metal-poor globular clusters, even at large radii (e.g. Miller and Prendergast 1963; Strom and Strom 1978; Gallagher, Faber, and Burstein 1979; Wirth 1982). Unlike disk-dominated spirals, classical Es evidently do not have a metal-poor halo fringe, which raises the question as to where and how many metal poor stars do exist in these galaxies. In other words, it would appear that classical Es may suffer from severe cases of "G-dwarf" syndrome due to a possible large deficiency of metal-poor stars.

The metal-poor stars, however, could prove quite challenging to find in Es if they are spatially well mixed, and there are growing indications that this is indeed the true situation. First the Stroms, Forte, and their collaborators (Strom, et al. 1981; Forte et al. 1981) have argued that globular clusters in Es are bluer and therefore evidently more metal-poor than typical field stars at the same projected radius. Thus metal-rich Es seem to contain globular clusters of at least intermediate metallicities. Second, ultraviolet measurements of classical Es obtained with the OAO, ANS and IUE satellite observatories show the presence of a comparatively flat mid-UV continuum which is depressed by a factor of ~ 5 in flux relative to the optical near UV spectrum (see Bertola, Capaccioli, and Oke 1982). This has been interpreted as arising from either an old, intermediate or low metallicity stellar population or from residual star formation that leads to the presence of hot stars (Code and Welch 1979; Gunn, Stryker, and Tinsley 1981).

It is not at present easy to distinguish between these two options. The M31 bulge, however, shares many of the spectral characteristics of classical Es, and yet a good case can be made that young stars are lacking (Deharveng et al. 1982). The spectral energy distribution of the M31 bulge in the optical and UV is, however, consistent with a modest admixture of metal-poor and hot, highly evolved low mass stars (Wu et al. 1980). We can therefore still admit the possiblity that metal rich classical Es do contain metal-poor stars, even though firm proof of this point and quantitative estimates for the stellar metallicity distributions are not yet available. If, on the other hand, Gunn, Stryker, and Tinsley are correct and the mid-UV luminosity is mainly due to young stars, then we may be faced with explaining the existence of galaxies which are exceedingly deficient in metal-poor stars. These would certainly require major departures from simple instantaneous recycling, closed system chemical abundance models. As in other such circumstances, we will call on the Space telescope to resolve this issue, which allows us to leave the problem for now!

III.   ABUNDANCES IN IRREGULAR GALAXIES

a)   General Characteristics

Magellanic type irregulars (Irrs) in an evolutionary sense lie at an opposite extreme from the Es. Irrs are rich in gas and often have striking young stellar populations which give rise to a characteristic chaotic optical appearance. They are, like the Es, relatively chemically homogeneous in that the gas metallicities, as deduced from emission line strengths in HII regions, are spatially quite constant within a given Irr galaxy (Pagel et al. 1978; Talent 1980). As a result the Irrs present perhaps the best available cases for tests of chemical evolution models for abundances in gas, and have been prime targets for a number of groups working in this area (e.g. Lequeux et al. 1979; French 1980; Pagel, Edmunds, and Smith 1980; Kinman and Davidson 1981; Hunter, Gallagher, and Rautenkranz 1981, hereafter HGR). Furthermore, since the Irr-related intergalactic-HII regions have the lowest known gas metallicities (which are still considerably higher than the most metal-poor Milky Way stars), the Irrs have served as important estimators of the primordial He abundance and the He enrichment rate $\Delta Y/\Delta Z$ (see Pagel 1982).

In the standard scenario the Irrs are represented by closed systems, one zone models. The gas metallicity Z_g thus depends only on the ratio of stellar to gas masses and on the yield from each generation of stars, or in the simplest case $Z_g = y \ln(f^{-1})$ where f is the gas mass fraction. It is proper to first emphasize that this approach has been quite successful in describing many of the ensemble features of Irr galaxies. Now that more data are becoming available, it is equally proper to ask where the problems may lie, and this is the aspect of Irr galaxy chemical properties which I will emphasize here. This reexamination is in part motivated by the recognition that many Irr galaxies have higher gas oxygen abundances than predicted by the standard model (Hunter 1982b; Matteucci and Chiosi 1983).

b)   Gas Abundances and Evolutionary Processes in Irrs

The one zone, closed system model requires the observers to obtain seemingly straightforward information about galaxies: a gas metallicity and the mass fraction of gas. M. Edmunds has already reviewed some of the difficulties surrounding the process of extracting metal abundances from measured emission line intensities in HII regions. Usually the most reliable data are obtained for oxygen, and even in favorable cases the uncertainties are $\gtrsim 10\%$ while more commonly the errors may be several times larger (e.g. when the faint [OIII] $\lambda 4363$ line is not accurately measured). Perhaps the best extragalactic HII region spectra have been obtained in connection with efforts to set the primordial He abundance (e.g. Rayo, Peimbert, and Torres-Peimbert 1982). Here, aside from possible sources of error in modelling the HII regions, the intrinsic accuracy of the current relative flux calibrations proably limits the precision of line ratio measurements and thus He/H to something like 5%.

Gas mass fractions are more difficult to determine. Normally the gas content is taken to be equal to measured 21 cm HI masses (in some cases without even correcting for the inevitable mass contribution from He). Since molecular gas is likely also to be present (e.g. the recent CO detection in the Irr NGC 1569 by Young, Gallagher and Hunter 1984), this approach probably systematically underestimates the total gas mass. As many Irrs are already metal-rich relative to their gas contents, increasing the gas mass will only make matters worse.

A more subtle problem is posed by how one fits a single zone mo- del to a real galaxy. Normally HII regions are confined to well with- in the optical radius of the galaxy, while the HI may extend to many Holmberg radii (Huchtmeier, Seiradakis, and Materne 1981; Hunter 1982a,b). As emphasized in HGR, it is then not clear what part of the gas distribution actually plays an active role in the chemical evolu- tion of the galaxy; i.e. are metals uniformly mixed into all of the gas, or is outer gas merely a passive reservoir. In general the ac- tive gas fraction is probably somewhat less than the total gas mass, which will tend to correct the observations back toward the model pre- dictions.

Similarly, the mass fraction of gas used in comparing with the models should refer to the ratio of gas to gas plus stars that have taken part in the chemical enrichment of a galaxy. It is not obvious that the usual assumption of setting the mass of stars plus gas equal to the dynamical mass M_T within the optical diameter is well founded. For example, Tinsley (1981) and Lin and Faber (1983) hypothesize that even small galaxies, such as the Irrs, contain significant amounts of dark matter which may not be in the form of stars or even nucleons. This view, however, is controversial; Gallagher, Hunter and Tutukov (1983) find satisfactory fits to Irr masses with constant star forma- tion rate, Salpeter IMF stellar population models. At issue here are facts of order 2, and thus we can adopt the dynamical mass as an upper bound on the amount of chemically "active" matter in Irrs, but also recognize that as much as 1/2 of the dynamical mass easily could be in the form of inactive dark matter.

Given this impressive list of uncertainties, why then is any cor- relation found between Z_g and gas mass fraction? The answer may in part lie in a fortunate selection of galaxies for pioneering studies and in part reflect a statistical trend for metal-poor galaxies to be systematically gas-rich. This latter point qualitatively supports the application of standard chemical evolution models. An examination of individual cases unfortunately reveals some disturbing results. For example, I Zw 18 is at the lower limit of known gas metallicities, but is not extraordinarily rich in HI (Lequeux and Viallefond 1980; Matte- ucci and Chiosi 1983). In fact the standard closed system, instanta- neous recycling model predicts an oxygen abundance of $\gtrsim$ 10 times that actually observed. Clearly simple evolutionary models, which depend

only on a fixed (but IMF-dependent) yield and gas mass fraction to describe the state of a system, do not <u>quantitatively</u> reproduce the gas
chemical properties of even uncomplicated Irr galaxies.

Once the standard model is abandoned, the range of possibilies
dramatically expands and key physical processes are much more difficult to identify. Another regularity among Irr galaxies, the correlation between M_T and metallicity, has been noted by several investigators, but has received less emphasis than the Z_g-gas fraction relationship. This is puzzling since for galaxies with $M_T \lesssim 10^{10}$ M_Θ, M_T
and Z_g are quite closely tied, especially as compared with the wide
scatter in gas fraction at a given Z_g. M_T certainly is a major, if
not dominant, factor in controlling metallicities in Irr galaxies
(HGR).

Total mass may influence metallicity in a variety of ways, the
most obvious of which is through the binding energy. There is ample
empirical evidence that the formation of high mass stars in nearly
coeval, spatially localized groups can produce major disruptions in
the cool interstellar medium in the disks of even massive galaxies.
These are observed as "holes" in the cool gas surrounding large OB associations (Meaburn 1980; Brinks 1981). Once such a hole has been
formed, gas from supernova may preferentially flow through the chimney
created by the hole and into the halo where, in galaxies of sufficiently low mass, gas will be lost from the system. D. DeYoung and I
are currently modelling this process to see if selective loss of enriched matter could reduce the effective yields and thus explain the
decrease in gas phase abundances in lower mass galaxies. In galaxies
of intermediate mass, galactic fountains of enriched matter from OB
associations also may act to spread metals throughout the systems,
thereby leading to chemical homogeneity. This picture contrasts with
the model developed by Terlevich and Melnick (1981) in which giant HII
regions are considered as gravitationally bound subunits that may
evolve with considerable independence, but there are good reasons to
believe the open view presented here is more typical of conditions in
Irrs (Gallagher and Hunter 1983).

Since mass and galaxy structural parameters, such as density distribution, may be linked, we expect evolutionary patterns also will
vary with galaxy mass. If star formation efficiency were to decline
with M_T, then a trend for decreasing metallicity could also occur. We
know very little about factors which influence global star formation
processes, and thus any statements in this area are very preliminary.
There is, however, evidence that the common assumption of star formation rates varying as global gas density squared is not correct, but
rather low mass density galaxies appear to make new stars at nearly
constant rates (Kennicutt 1983; Gallagher et al. 1983).

The empirical basis for this hypothesis can be crudely illustrated by comparing the stellar population mixes in high and low mass density galaxies. For simple closed system models, the denser galaxies

should evolve more rapidly, and thus be more metal rich and contain an older stellar population. Spectrophotometric studies, such as the work reported by Hunter (1982b; and see Figure 2) in her Ph.D. thesis,

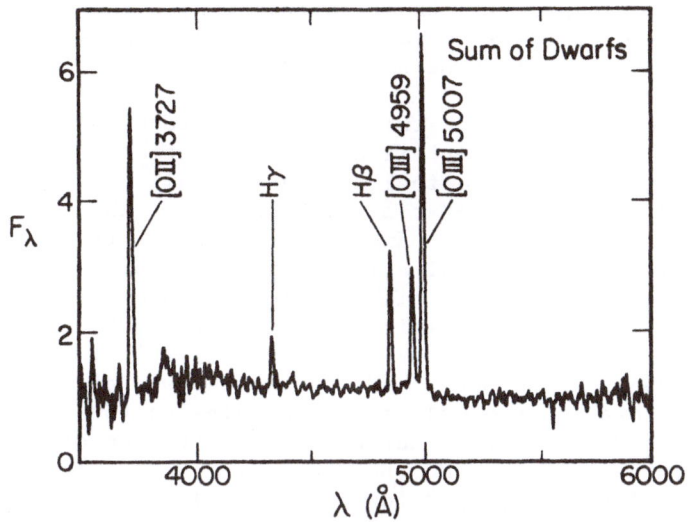

FIG. 2 - Mean relative spectral energy distribution for dwarf Irr galaxies based on spectrophotometry of 7 galaxies obtained by Hunter (1982b) with the KPNO 0.9-m telescope. Both the emission lines and stellar continuum with CaII absorption and Balmer jump are similar to features found in giant Irrs. The dwarfs are difficult to observe since their central surface brightnesses rarely exceed the night sky brightness.

and the extensive UBV photometric measurements by de Vaucouleurs, de Vaucouleurs, and Buta (1981) show that giant and dwarf Irrs have remarkably similar stellar content. It therefore appears that the star formation histories of most normal ($M_T \lesssim 10^{10}$ M_θ) Irr galaxies are well represented by near constant mean astration rates over ~ 10^{10} yr time scales in the central regions. The _absolute_ level of star formation activity, on the other hand, is likely to be dependent on M_T.

We do know that lower M_T systems also have lower metallicities. This then could in part reflect a variation in the _global_ star formation efficiency. A speculative scenario which is consistent with the above discussion is to consider star formation as a critical phenomenon. Below some threshold gas density, astration is very inefficient; phenomenologically this results in the high HI to stellar mass ratios common in the outer regions of galaxies and in the central concentration of star forming activity in Irrs (Einasto 1972; Hunter 1982a,b). Above the critical density, the number of young stars may be set by the stars themselves via feedback processes which modify the state of the interstellar medium, but are only weakly dependent on the large

scale galaxian environment (e.g. Seiden and Gerola 1982; Suchkov and
Shchekinov 1979; Shore 1981; Cox 1983). Note that in this picture it
is not necessary for star formation to propagate in any organized manner. Smaller, less massive galaxies will tend to have more of their
gas supplies at low densities; thus the fraction of the available active mass capable of being involved in star formation is reduced, and
the metallicity level will be depressed by dilution.

Finally we briefly mention two other traditional factors in galaxy chemical evolution: gas infall and IMF variations. It is quite
possible that Irr galaxies are fueled by gas infall from cool, gravitationally bound halos (Gallagher et al. 1983), but these will be pol-
luted by metals ejected from the disk. Bound infall therefore mainly
increases the amount of gas available to dilute metals and the retention of metals ejected from the disk. Halos of lower mass galaxies
will collapse more slowly, and this might lead to an M_T-Z_g relationship. Infall from unbound gas is not very likely for the low mass
systems we are considering here. IMF variations are often considered
in attempting to explain the colors of Irr galaxies (Searle, Sargent,
and Bagnulo 1973; Huchra 1977), but at present studies of the luminosity function in the LMC and SMC (Butcher 1977; Dennefeld and Tammann
1980) as well as integral properties of normal Irrs (Lequeux et al.
1981; Gallagher et al. 1983) suggest that a near Salpeter IMF applies
in most circumstances. We also note that in cases where we are in
trouble with production of metals in Irrs, it is because systems with
large numbers of <u>observed</u> OB stars have low Z_g and thus IMF variations
do not resolve the problem (e.g. II Zw 70: O'Connell, Thuan, and Goldstein 1978; NGC 1569: HGR). For the present it is therefore reasonable to continue to presume that <u>large</u> fluctuations in the average IMF
are not common in low mass density, actively star forming systems during the present epoch. We are therefore almost certainly not facing
the most complex possible situation.

IV. CONCLUSIONS

Currently most interpretations of metal abundances in galaxies
are based on theoretical models which do not explicitly include galaxian evolution; e.g. the standard instantaneous recycling prescriptions
depend only on integrals of the star formation rate. While the standard models have provided an entry into understanding abundances, they
do not yield good quantitative fits to the observations, even for
structurally simple galaxies of the types discussed here. It may
therefore now be appropriate to expand our domain to include not only
open models (i.e. those which allow for <u>both</u> gas inflows and outflows), but also models which integrate abundances into specific evolutionary scenarios. Such models can, for example, allow for proper
accounting of time dependent mass flows between the stellar and gaseous phases (see Truran and Cameron 1971) as well as incorporating other reasonable complexities, including the impact of energy supplied
from massive stars on galactic gas inflow or outflow rates (cf. Loose,

Krügel, and Tutukov 1982). A calculation with steeply declining star formation rates would, for example, be quite useful in exploring the IMF problem in the metal-rich classical Es (cf. DeYoung 1978).

In the Irrs we have seen the empirical data hint at variations in global star formation efficiencies in galaxies with similar star formation histories. Here relaxation of the single zone and closed system approximations seem most appropriate (Matteucci and Chiosi 1983), especially if coupled with further efforts to describe dependences of the star formation process on global galaxian properties. Fortunately we can expect good progress in observational investigations of Irrs, which should help bring our understanding of these primitive galaxies to a level which is near that of the spirals.

Finally we should continue to be concerned with the IMF. A good case can be made that a standard IMF does describe mean properties of many galaxies which are currently supporting active star formation. These systems, however, are all rather similar in that astration is occurring in low density disks and appears to have achieved some type of equilibrium state which allows it to continue for long times at nearly constant rates (Kennicutt 1983; Gallagher et al. 1983). In particular, the possibility that star formation may change its character under non-equilibrium or unusual circumstances needs to be kept in mind (especially with regard to young galaxies), and may even prove accessible to observational test in disturbed galaxies which may be out of equilibrium (such as interacting galaxies or peculiar cluster galaxies like NGC 1275; Sarazin and O'Connell 1983; Wirth, Kenyon, and Hunter 1983), or in the extreme environments of galaxy nuclei (Rieke and Lebofsky 1979; Tutukov and Krügel 1981).

It is a pleasure to thank my former students, Deidre Hunter and Allan Wirth, for their collaboration in research on many of the ideas discussed here. C. Chiosi, A. Renzini, P. Seiden, J. Truran and S. Tutukov have also contributed to the development of this paper through many interesting discussions on problems relating to the evolution of galaxies. Research on galaxy evolution at the University of Illinois has been made possible by the unflagging support of the Department of Astronomy, and by the U.S. National Science Foundation through grants AST 81-15325 and AST 82-14127.

REFERENCES

Aaronson, M., Cohen, J., Mould, J., and Malkan, M. 1978, Astrophys. J. 223, 824.
Audouze, J. and Tinsley, B. M. 1976, Ann. Rev. Astron. Astrophys. 14, 43.
Baade, W. 1963, "Evolution of Stars and Galaxies." Ed. C. Payne-Gaposchkin, Harvard University Press.

Bertola, F., Capaccioli, M., and Oke, J. B. 1982, Astrophys. J. <u>254</u>, 494.

Brinks, E. 1981, Astron. Astrophys. <u>95</u>, L1.

Butcher, H. 1977, Astrophys. J. <u>216</u>, 372.

Burstein, D. 1977, In "The Evolution of Galaxies and Stellar Populations." Ed. B. M. Tinsley and R. B. Larson, Yale Observatory, p. 191.

Caldwell, N. 1982, Ph. D. Thesis, Yale University.

Caldwell, N. 1983, Astron. J. <u>88</u>, 804.

Code, A. D. and Welch, G. A. 1979, Astrophys. J. <u>228</u>, 95.

Cox, D. 1983, Astrophys. J. Letters <u>265</u>, L61.

Deharveng, J. M., Joubert, M., Monnet, G., and Donas, J. 1982, Astron. Astrophys. <u>106</u>, 16.

Dennefeld, M. and Tammann, G. A. 1980, Astron. Astrophys. <u>83</u>, 275.

de Vaucouleurs, G., de Vaucouleurs, A., and Buta, R. 1981, Astron. J. <u>86</u>, 1429.

DeYoung, D. S. 1978, Astrophys. J. <u>223</u>, 47.

Einasto, J. 1972, Astrophys. Letters <u>11</u>, 195.

Faber, S. M. 1973, Astrophys. J. <u>179</u>, 731.

Faber, S. M. 1977, In "The Evolution of Galaxies and Stellar Populations." Ed. B. M. Tinsley and R. B. Larson, Yale Observatory, p. 157.

Forte, J. C., Strom, S. E., and Strom, K. M. 1981, Astrophys. J. Letters <u>245</u>, L9.

French, H. 1980, Astrophys. J. <u>240</u>, 41.

Frogel, J. A., Persson, S. E., and Cohen, J. G. 1980, Astrophys. J. <u>240</u>, 785.

Gallagher, J. S., Faber, S. M., and Burstein, D. 1980, Astrophys. J. <u>235</u>, 743.

Gallagher, J. S. and Hunter, D. A. 1983, Astrophys. J., in press.

Gallagher, J. S., Hunter, D. A., and Tutukov, A. 1983, preprint.

Gunn, J. E., Stryker, L. L., and Tinsley, B. M. 1981, Astrophys. J. <u>244</u>, 48.

Huchra, J. 1977, Astrophys. J. <u>217</u>, 928.

Huchtmeier, W., Sieradakis, J., and Materne, J. 1981, Astron. Astrophys. <u>102</u>, 134.

Hunter, D. A. 1982a, Astrophys. J. <u>260</u>, 81.

Hunter, D. A. 1982b, Ph.D. Thesis, University of Illinois - Urbana.

Hunter, D. A., Gallagher, J. S., and Rautenkranz, D. 1982, Astrophys. J. Suppl. <u>49</u>, 53.

Kennicutt, R. 1983, Astrophys. J., in press.

Kinman, T. D. and Davidson, K. 1981, Astrophys. J. <u>243</u>, 127.

Larson, R. B., 1974, Monthly Notices Roy. Astron. Soc. <u>169</u>, 229.

Lequeux, J., Peimbert, M., Rayo, J. F., Serrano, A. and Torres-Peimbert, S. 1979, Astron. Astrophys. <u>80</u>, 155.

Lequeux, J. and Viallefond, F. 1980, Astron. Astrophys. 91, 269.

Lequeux, J., Maucherat-Joubert, M., Deharveng, J. M., and Kunth, D. 1981, Astron. Astrophys. <u>103</u>, 305.

Lin, D. N. C. and Faber, S. M. 1983, Astrophys. J. Letters <u>266</u>, L21.

Loose, H. H., Krügel, E., and Tutukov, A. 1982, Astron. Astrophys. <u>105</u>, 342.

Matteucci, F. and Chiosi, C. 1983, Astron. Astrophys. 123, 121.
Meaburn, J. 1980, Monthly Notices Roy. Astron. Soc. 192, 365.
Miller, R. H. and Prendergast, K. H. 1963, Astrophys. J. 136, 713.
O'Connell, R. W. 1980, Astrophys. J. 236, 430.
O'Connell, R. W., Thuan, T. X., and Goldstein, S. J. 1978, Astrophys. J. Letters 226, L11.
Pagel, B. E. J. 1981, In "Structure and Evolution of Normal Galaxies." Ed. M. Fall and D. Lynden-Bell, Cambridge Univ. Press, p. 211.
Pagel, B. E. J. 1982, Phil. Trans. Roy. Soc. Lond. A 307, 19.
Pagel, B. E. J., Edmunds, M. G., Fosbury, A. E. and Webster, B. L. 1978, Monthly Notices Roy. Astron. Soc. 184, 569.
Pagel, B. E. J., Edmunds, M. G., and Smith, G. 1980, Monthly Notices Roy. Astron. Soc. 193, 219.
Rayo, J. F., Peimbert, M. and Torres-Peimbert, S. 1982, Astrophys. J. 255, 1.
Rieke, G. and Lebofsky, M. 1979, Ann. Rev. Astron. Astrophys. 17, 477.
Sandage, A. R. and Visvanathan, W. 1978, Astrophys. J. 223, 707.
Sarazin, C. L. and O'Connell, R. W. 1983, Astrophys. J. 268, 552.
Seiden, P. E. and Gerola, H. 1982, Fund. Cosmic Phys. 7, 241.
Searle, L., Sargent, W. L. W., and Bagnuolo, W. 1973, Astrophys. J. 179, 427.
Searle, L., Wilkinson, A., and Bagnuolo, W. G. 1980, Astrophys. J. 239, 803.
Shore, S. 1981, Astrophys. J. 249, 93.
Strom, K. M. and Strom, S. M. 1978, Astron. J. 83, 73.
Strom, K. M., Strom, S. E., Wells, D. C. and Romanishin, W. 1978, Astrophys. J. 224, 782.
Strom, S. E., Forte, J. C., Harris, W. E., Strom, K. M., Wells, D. C., and Smith, M. G. 1981, Astrophys. J. 245, 416.
Struck-Marcell, C. 1981, Monthly Notices Roy. Astron. Soc. 197, 487.
Suchkov, A. A. and Shchekinov, Yu. A. 1979, Sov. AJ 23, 665.
Talbot, R. J., Jr. and Arnett, W. D. 1973, Astrophys. J. 186, 51.
Talent, D. 1980, Ph.D. Thesis, Rice University.
Terlevich, R. T. and Melnick, J. 1981, Monthly Notices Roy. Astron. Soc. 195, 389.
Terlevich, R. T., Davies, R. L., Faber, S. M., and Burstein, D. 1981, Monthly Notices Roy. Astron. Soc. 196, 381.
Tinsley, B. M. 1980, Fund. Cosmic Phys. 5, 287.
Tinsley, B. M. 1981, Monthly Notices Roy. Astron. Soc. 194, 63.
Tinsley, B. M. and Larson, R. B. 1979, Monthly Notices Roy. Astron. Soc. 186, 503.
Tonry, J. L. and Davis, M. 1981, Astrophys. J. 246, 680.
Truran, J. W. and Cameron, A. G. W. 1971, Astrophys. Space Sci. 14, 179.
Tutukov, A. and Krügel, E. 1981, Sov. A. J. 24, 539.
Vigroux, L., Chieze, J. P. and Lazareff, B. 1981, Astron. Astrophys. 98, 119.
Wirth, A. 1981, Astron. J. 86, 981.
Wirth, A. 1982, Ph.D. Thesis, University of Illinois - Urbana.
Wirth, A. and Shaw, R. 1983, Astron. J. 88, 171.

Wirth, A., Kenyon, S. J., and Hunter, D. A. 1983, Astrophys. J. <u>269</u>, 102.
Wirth, A. and Gallagher, J. S. 1983, preprint.
Wu, C.-C., Faber, S. M., Gallagher, J. S., Peck, M., and Tinsley, B. M. 1980, Astrophys. J. <u>237</u>, 290.

DISCUSSION

EDMUNDS: How does the "lack" of gradients discovered by Wirth in Es compare with the observations of others?

GALLAGHER: Wirth's results on Es are in good statistical agreement with other studies, such as those by the Stroms. Note that here we have been quoting color differences between a central position and a point with a standard estimated stellar space density, which is generally smaller than the maximum color change. However, some type of normalization is necessry in intercomparing galaxies.

DANZIGER: Is the story for M32 proposed by Faber (1973) of a much more luminous galaxy tidally stripped by M31 still the same, without complications?

GALLAGHER: Yes, M32 is definitely tidally stripped. This process, however, probably has had a large effect only on the luminosity with both metallicity an central velocity dispersion remaining largely unchanged. The main difference from Faber's studies lies in our interpretation of M32-type galaxies, rather than the NGC 205-like diffuse dwarfs, as true representatives of low luminosity classical elliptical galaxies. In support of this view, Wirth and Gallagher have found some examples of dwarf classical ellipticals which are not tidally stripped.

RENZINI: Concerning the UV light in old populations: I think that the shortward of $\sim$ 2000 Å the most likely stellar contributors are PN nuclei (with or without a nebula around) and accreting WD's, part of which may be the SNI precursors. At 2000 $\lesssim \lambda \lesssim$ 3500 Å light may come from HB stars in a (trace) population which is metal poor, or from metal rich HB stars which undergo mass transfer in a binary system, or even from metal rich HB stars if the HB becomes blue again when metallicity increases above solar (but we don't know if this is the case).

GALLAGHER: We agree that PN nuclei are main contributors to the $\lambda \lesssim$ 1500 Å light in old stellar systems. Your point about the possible role of binaries is interesting and certainly deserves careful examination. Perhaps we will eventually be able to learn about the metal-rich BHB stars from studies of the Galactic and M31 bulge stellar content.

MATHEWS: I would like to ask if you could clarify a point. In your talk you mentioned that the Schmidt model with n = 2 was widely accepted. From what you have said it would appear that there is no reason not to take n = 1, or perhaps the best value would be n = 0.

GALLAGHER: For a global application of the Schmidt relationship with n = 2 (which is not what Schmidt was modelling in his paper!), one would expect the SFR to decline steeply as gas is exhausted in a galaxy. Thus recent observational indications of near constant SFRs in disk galaxies might be interpreted as evidence for lower values of n.

EDMUNDS: (1) How reliable are gas fraction estimates? (2) Does the gas fraction versus total mass correlation influence the significance of the metallicity-total mass correlation? (3) The decrease in SFR, as estimated from SFR scaling as gas density squared, may not be a severe constraint if the system is not closed (e.g. inflow)?

GALLAGHER: (1) Not very. Molecules, gas fraction accessible to star formation, and numbers of low mass stars are major sources of uncertainty. (2) Yes, I believe this is a problem but its severity is at present uncertain since our sample is still small. (3) In the case of Irrs this is true, but the comparatively flat radial distributions of OB star densities may still present a problem.

TORRES-PEIMBERT: You mentioned earlier that the He abundance determinations are uncertain by 5% due to calibration uncertainties and yet today you mention that your $H\beta$ fluxes have errors of 5%. I find that overly optimisstic on your part since you also have the problem of very faint objects (fainter than sky level) and of subtracting the Balmer line contributions from a non-negligible continuum.

GALLAGHER: That is an older viewgraph. The typical precision probably is at best 10% in the high surface brightness Irrs. You are certainly correct that the errors are larger in low surface brightness dwarf Irr spectra.

WOOSLEY: About how much energy would it take to blow a "hole" in the galaxy as you described? Would it take one or many supernovae? $\sim 10^{52}$ erg? How do these "holes" relate to the "tunnels" that were discussed some years ago linking supernova remnants?

GALLAGHER: Off hand, I would guess that 10^{52-53} ergs are required, although the conditions for the hot gas to break out from the

disk and into the halo are uncertain. Thus collective interactions between stars (winds, multiple sypernovae) appear to be necessary. This should be a related aspect of
tunnels, and in fact is only a slight modification of galactic coronae and fountains which were theoretically described some time ago by Spitzer and by Field and Shapiro.

S. TORRES-PEIMBERT: (1) You are saying that it is not possible to
obtain any line flux to better than 5% due to the uncertainties in the calibration stars, and therefore that observers are underestimating their error bars. Certainly
this possible source would not affect relative He/H abundances that have been derived in the same observing run.
(2) Many authors have not found any He vs Z dependence
from different objects. However if you compare their mean
values it is possible to see that those that have deter-
mined high $\langle Z \rangle$ have also determined high $\langle Y \rangle$. There is an
increasing trend in the mean values. Kunth and Sargent
(1983) obtain high Y values for low Z objects but find no
slope. Perhaps they are overcorrecting for their neutral
helium and since their objects cover a very small range in
Z, it is not expected that they could find a Y vs Z relation.

GALLAGHER: (1) I am especially worried about flux calibrations over
small wavelength intervals, which can be seen to be a problem from intercomparisons of flux scales based on different
standard stars. Perhaps 5% is too pessimistic, but this
will be a factor at the level of a few percent. Differential comparisons of line ratio measurements should indeed
be less affected by flux scale variations, and so (aside
from interpretive and modelling difficulties) should be
more accurately measurable than the absolute abundances.
(2) It would also seem to me that it is best to deter-
mine $\Delta Y / \Delta Z$ by spanning the widest possible range in metallicity using similar equipment so as to minimize systematic
errors. Certainly there is compelling theoretical and observational evidence that $\Delta Y / \Delta Z$ is non-zero, although I
still feel the value is very poorly known. With regard to
your second point, the Kunth and Sargent program appears to
have been targeted at absolute He abundance measurements
and thus is not especially well suited to studies of
$\Delta Y / \Delta Z$. The differences between the various estimates of Y
at low Z (~ 0.1 Z_0) in my opinion illustrate the difficulties associated with measurements of Y to levels of 10%
or better.

THREE PROBLEMS IN THE CHEMICAL EVOLUTION OF GALAXIES

M. G. Edmunds and B. E. J. Pagel
University College, Cardiff and Royal Greenwich Observatory

In these talks we shall discuss three topics of current interest; (i) the abundance of nitrogen in galaxies, and its interpretation in terms of stellar nucleosynthesis; (ii) the abundances of elements in the central regions of galaxies; (iii) the effects of gas flows and yield on the chemical evolution of galaxies.

Since it will be a useful framework in all three topics, we first briefly review the so-called "simple" model of chemical evolution. For more details see e.g. Pagel and Patchett 1975, Tinsley 1980, Pagel 1981. The model has a single, closed zone with no accretion or loss of material. Instantaneous recycling is assumed i.e. production of metals and mass loss from stars occur on a timescale short compared to the system's overall evolution. A primary element is one whose synthesis does not depend on the initial composition of the star in which synthesis takes place. If z is the mass fraction of such an element, if the gas mass in the system is μ (initially unity) and the mass of material in stars is αS, where α is the fraction of material formed into stars which is not subsequently returned to the interstellar medium, then for a primary element

$$\frac{d(z\mu)}{ds} = p'(1 - z) - \alpha z \tag{1}$$

Here p' is the mass of the primary element produced per mass of material formed into stars. It is convenient to define the "yield" $p = p'/\alpha$ for an element. For a closed system

$$\mu = 1 - \alpha S \tag{2}$$

and (1) and (2) solve, with initial condition z = o to give

$$z = p \ln\left(\frac{1}{\mu}\right) = p \ln\left(\frac{1}{f}\right) \tag{3}$$

where f is the gas fraction, the ratio of gas mass to total mass in the system (in this case $f = \mu/1$)

C. Chiosi and A. Renzini (eds.), Stellar Nucleosynthesis, 341–358.

A secondary element requires the pre-existence of a seed element for its synthesis. For example, the conversion of ^{16}O to ^{14}N during CNO cycling is the secondary production of nitrogen from primary seed i.e. carbon and/or oxygen. If P'_N is a suitably defined production coefficient we now write (instead of (1)), with Z_N, Z_{ox} mass fraction of nitrogen and oxygen respectively

$$\frac{d(Z_N\mu)}{ds} = P'_N \cdot Z_{ox} - \alpha Z_N \qquad (4)$$

Solving with the previous equations implies that the relative amounts of nitrogen and oxygen in closed systems of varying f should behave as

$$\frac{Z_N}{Z_{ox}} = \frac{P_N \cdot Z_{ox}}{2P_{ox}} \qquad (5)$$

If both elements were primary, two relations of the form (3) would hold and

$$\frac{Z_N}{Z_{ox}} = \frac{P_N}{P_{ox}} \qquad (6)$$

So for primary production of nitrogen we expect Z_N/Z_{ox} = constant in different systems, for nitrogen to be secondary we expect Z_N/Z_{ox} proportional to Z_{ox}.

OBSERVED NITROGEN ABUNDANCES

We do not know a priori whether secondary or primary production of nitrogen will dominate, although we must expect some secondary production if the CNO cycle does operate in stars. There are cases like the planetary nebula in NGC 6822 (Dufour and Talent 1980) which shows more than twice as much nitrogen as could have been made by processing all its initial oxygen or carbon - implying primary production. The three major sources of spectral data are (i) stars - although we would not want to trust red giants for galactic trends, since the giants may be self-enriched and their observed composition may not accurately represent the interstellar medium out of which they formed. (ii) HII regions - but not planetary nebulae. We do not favour deriving galactic N/O gradients from type 1 planetaries, since again self-enrichment (perhaps due to their binary nature) may occur. (iii) the cold interstellar medium. Radio observation of molecular lines can give isotope ratios, although there can be serious saturation effects. Tosi (1982), concludes that $^{12}C/^{13}C$ ratios indicate secondary production of ^{13}C, but the $^{14}N/^{15}N$ ratio and its interpretation remains uncertain.

STELLAR NITROGEN ABUNDANCES

The observational situation is very confused. A problem has existed ever since Harmer and Pagel (1970) found two old halo stars of similar overall metal abundance [Fe/H] ~ - 1.2, but the subgiant ν Indi showed [N/Fe] ~ - 0.8, while the subdwarf HD 25329 has [N/Fe] ~ + 0.5, a factor of twenty different. Tomkin and Bell (1973) and Clegg (1977) have found subdwarf stars with nitrogen abundances like ν Indi, while Snedden (1974) found dwarf stars with [N/Fe] ~ O and giants with [N/Fe] > O, sometimes too large to be explained by secondary processing alone on the basis of standard abundance ratios in the stars. Recently Bessell and Norris (1982) located two metal-poor dwarfs [Fe/H] ~ - 2.0 with [N/Fe] ~ + 1.7 to 2.0. Since [C/H] follows [Fe/H] (Peterson and Snedden 1978) the limit for secondary production $Z_N \leqslant Z$ (initial C + N + O) implies that in these stars initially [O/Fe] > 0.7 to 1.0, probably greater than the systematic trend of [O/Fe] $\leqslant$ 0.6 observed in old stars (Snedden, Lambert and Whitaker 1979). Since anomalies in nitrogen do not occur in all old dwarf stars, it is tempting to seek an explanation in initial in homogeneities of C and O (subsequently processed to N in the stars) in the gas out of which the stars formed. Such an explanation has been put forward to explain the very wide variation of N (a factor of 10) on the giant branch of the globular clusters M92 (Carbon et al 1982), M15 (Trefzger et al 1983) and perhaps to a lesser extent (the variation is not so great) in old field halo giants (Kraft et al 1982). There is some evidence for anticorrelation of N abundance with C abundance - as would be expected in secondary processing, (NGC 6752, 47 Tuc, M71 Norris et al 1981, Smith and Norris 1982), but both internal processing and primordial variation will probably be required for a full explanation. For field giant anomalies one might try to invoke binary evolution effects, but the current view is that binaries are very rare in globular clusters.

Primary production of ^{14}N in stars is believed to be possible in supernovae and also by CNO processing ^{12}C which has been created in the star by helium burning (Truran and Cameron 1971) - this still fits our definition of primary because the amount of nitrogen produced does not depend (except through any metallicity effect on stellar evolution) on the initial metallicity of the star. But such primary production in the star itself cannot explain peculiar N/O ratios in dwarf stars unless they have been through a giant phase, and then mixed back down onto the main sequence.

HII REGION ABUNDANCES

Pioneering work by Searle (eg 1971) suggested radial abundance gradients in spiral galaxies which were steeper in N than in O, but he was not able to take full account of variations in T_{eff}. Later measurements obtaining T_e from λ[OIII] 4363 and empirical methods (see Pagel and Edmunds 1981) found that N/O certainly varied less than O/H in any given system. The ionization correction N/O = N+/O+ used is supported by recent detailed HII region models. The lack of a marked N/O gradient in our own Galaxy has recently been nicely confirmed by Shaver et al (1983). The gas abundances in galactic nuclei will be discussed below, but there is really no conslusive evidence for N/O ever exceeding perhaps

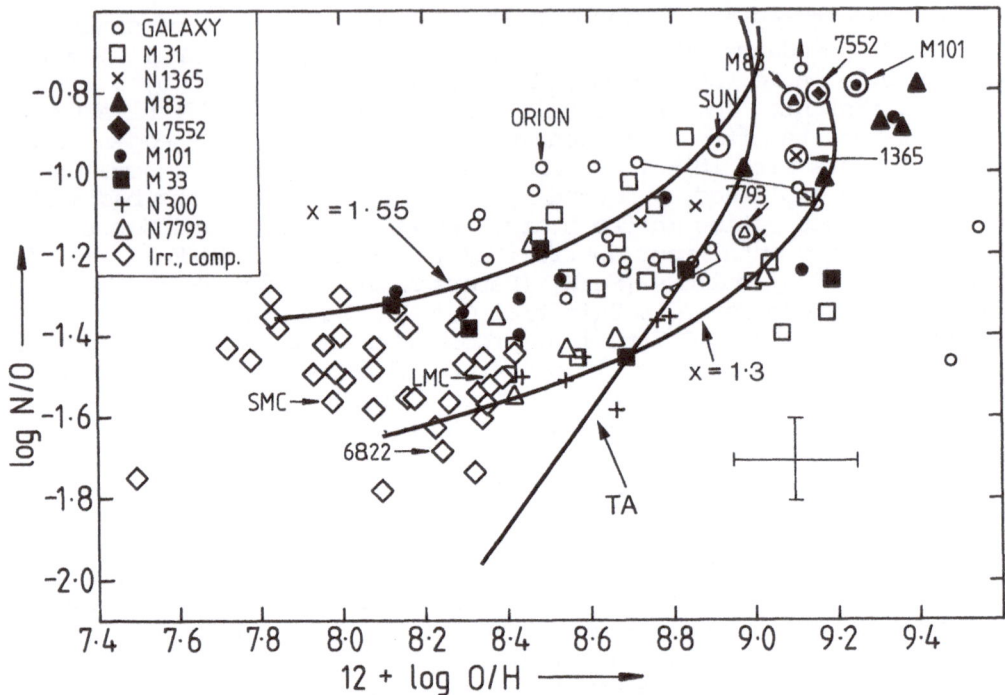

Fig. 1 Nitrogen and oxygen abundances determined by HII region analysis
(a revised version of Fig.3 in Pagel and Edmunds 1981). The line marked
TA is a simple model with the instantaneous recycling approximation
relaxed (Talbot and Arnett 1974), assuming an exponentially decreasing
star formation rate. The lines marked x = 1.3, 1.55 are models from
Alloin et al (1979) with IMF slopes characterised by x.

1.5 to 2 times its solar value. The general impression one gets from
the data (see Fig. 1) is that there is a large spread in N abundances at
a given O abundance and that the N/O ratio across a galaxy does not vary
as much as expected from secondary processing.

INTERPRETATIONS

(1) Secondary Processing

      We could first ask whether a purely secondary process is sufficiently
effective to produce the observed abundance of nitrogen. For   solar
neighbourhood observed values of $Z_{ox} \simeq 0.01$ (i.e. almost half of all
metals), $f = 0.1$, the simple model from equation (3) implies $P_{ox} \simeq 0.004$.
Then since $Z_N/Z_{ox} \sim 0.1$, equation (5) implies $P_N = 0.09$. Thus nearly 9%
of oxygen going into stars must be processed into N and ejected. Since
$\alpha = 80\%$ of the oxygen going into stars must remain as long lived stars
or remnants, this implies that the process would have to be remarkably
efficient. The simple model with the instantaneous recycling approximation

relaxed has been calculated by Talbot and Arnett (1974), and their prediction is shown as a curve in Fig. 1. It certainly does not fit all
the data, but it is interesting that it gives a reasonable lower limit
for part of the diagram.

(2) Variation of Initial Mass Fraction

Representing the initial mass fraction for star formation by

$$N(m)\,dm \propto m^{-(1 + x)}\,dm \tag{7}$$

Alloin et al (1979) followed by Lequeux et al (1979) calculated models
with secondary production plus some "prompt" (i.e. from relatively short
lived stars) primary production. Their predictions are shown in Fig. 1,
and modest variation of x - which alters the relative numbers of O and
N producing stars - allows all the data to be fitted. But it is hard
to understand why there should be just the right variation in x across
a given system to give an approximately constant N/O.

(3) Time-Delayed Primary Nitrogen

We suggested (Edmunds and Pagel 1978) that equation (6) should have
the form

$$\frac{Z_N}{Z_{OX}} = \frac{P_N}{P_{OX}} \cdot F(t) \tag{8}$$

where $F(t)$ takes account of the time delay before primary nitrogen is
released to the interstellar medium because of the evolution time of
the synthesising stars. Provided $F(t)$ is reasonably the same across a
given galaxy, then Z_N/Z_{OX} would be constant, but different values of
Z_N/Z_{OX} between systems would imply different ages, in the sense of the
time elapsed since the bulk of star formation occured. But the problem
here is that the timescales are awkward. Fig. 2 shows the evolution
of a system with mass fraction as in (7), assuming $F(\text{now}) = 1$ for the
Galaxy, and $F(0) = 0.2$ (this latter could be a bit less on the basis
of NGC 55). It is assumed that, for stars less than a particluar mass,
a constant fraction (which need never exceed 2 to 3×10^{-3}) of the star's
mass is released as primary nitrogen. It will be noted that the implied
"ages" for systems like the LMC, SMC, NGC 6822 are very small unless
the upper mass limit at which nitrogen production occurs is low. Truran
and Cameron's primary production model was intended to apply to intermediate mass (3 - 9 M☉) stars, and it is not obvious that the range could
be pushed down to 1 - 2 M☉.

(4) Other Models

Recently two rather more complicated models have been suggested.
White and Andouze (1983) invoke secondary production with inflow of
fresh gas into a galaxy and stochastic mixing, while Serrano and Peimbert
(1983) use inflow plus secondary processing plus relaxing instantaneous

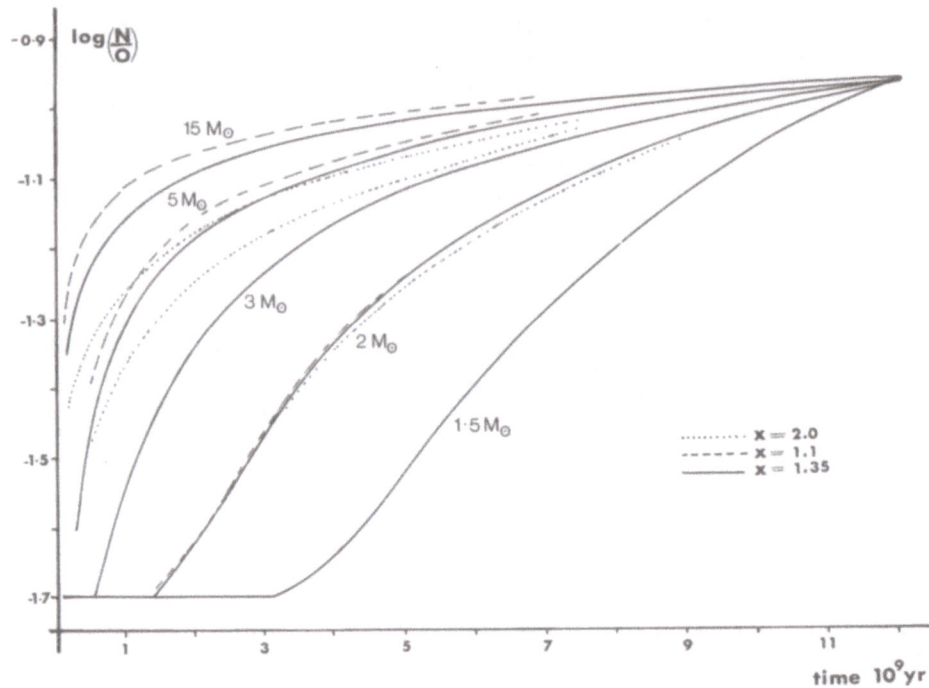

<u>Fig. 2</u> Predicted nitrogen to oxygen ratios as a fraction of time for
synthesis of nitrogen in stars whose upper mass limit is the labelled
value on the curves. The curves are shown for different values of the
IMF slope x. It is assumed in these calculations that all the stars
were formed at t = 0, with almost immediate oxygen synthesis.

recycling plus a metallicity dependent oxygen yield (i.e. oxygen is no
longer truly primary). Such models need more study, but we shall
argue later that inflow may only worsen the secondary production problem,
and the metallicity dependence of overall yield is not certain (Edmunds
and Pagel 1983), particularly as the required form may aggravate the solar
neighbourhood "G-dwarf" problem.

CONCLUSIONS ABOUT NITROGEN SYNTHESIS

     No one scheme seems really satisfactory. Significant amounts of
primary nitrogen are required in varying amounts. There may be some
secondary contribution since N/O gradients are not completely flat
(although this could be age or IMF effect), and since CNO processing
does surely occur, as evidenced by isotope ratios in evolved giant
atmospheres (e.g. Tomkin, Luck and Lambert 1976).

ABUNDANCES IN THE CENTRAL REGIONS OF GALAXIES

The abundances or the gas in the central regions of external galaxies can be estimated from the emission line spectrum (if present), and the abundances in the stars (or at least, some mean value) from photometry or absorption line spectra. It is only for our own Galaxy that individual stars towards the central regions have been analysed. Of the two components, gas and stars, analysis of the former may at present be the more reliable.

EMISSION LINE ANALYSIS

The emission line spectra shown by spiral nuclei can broadly be divided into four types, several of which may exist simultaneously and be spatially resolved in nearby galaxies. (i) Seyfert 1 with broad Balmer lines and strong lines of high excitation potential (ii) Seyfert 2 where Balmer lines show similar widths to the forbidden lines, and lines of high excitation are again present (iii) LINERS, (defined by Heckman 1980) similar to Seyferts, but with lines of characteristically low ionization and found in perhaps 30% of large galaxies. (iv) conventional HII regions, photoionized by stars.

Analysis of the HII regions is reasonably secure, although due to the high abundances in galactic nuclei the temperature sensitive emission lines are weak and generally far too faint to observe against the inevitable underlying stellar absorption spectrum. But empirical methods using strong line ratios (e.g. Pagel, Edmunds and Smith, 1980) can be applied, although the very high abundance end of this calibration has been revised slightly downwards (Edmunds and Pagel 1983). HII regions can be found as close as 10" to a nucleus in several galaxies, and the abundances follow the trend of the abundance gradients observed further out in the disks.

For Seyferts and Liners, good photoionization models are only just becoming available. We will concentrate here on liners, where the resemblence to supernova remnant spectra originally led to the suggestion of shock excitation. Indeed, the presence of strong [NII] lines in nuclei had been known since the work of Burbidge and Burbidge. Interpretation of these nuclei as shock excited, however, leads to anomalous abundances, and it is becoming clear from the work of Ferland and Netzer (1983), Halpern and Steiner (1983), Keel (1983) and our own observations (with A. Diaz and M.M. Phillips) that these nuclei are in fact photoionized by a hard spectrum - i.e. x-ray heating of the neutral and singly ionized regions. We show in Fig. 3 a diagnostic diagram of the form proposed by Baldwin, Phillips and Terlevich (1981), together with recent observations in, or near, galactic nuclei by ourselves and others. It can be seen that in several cases we have observations of HII regions near the nucleus, and that the sudden increase in [OIII] and [NII] strength closer to and in the nucleus can be satisfactorally accounted for as the influence of hard photoionization without the need for abundances much

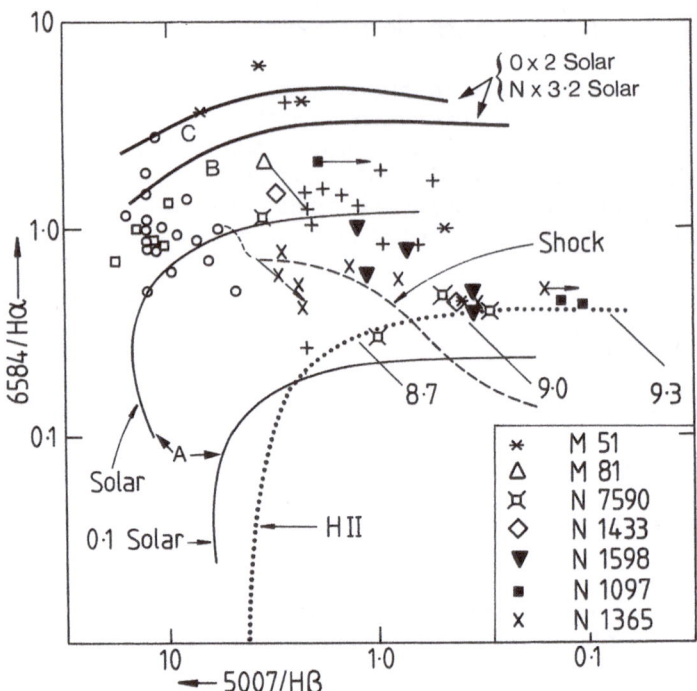

Fig. 3 Diagnostic diagram of [NII]/Hα line strength against [OIII]/Hβ for emission line gas in or near galactic nuclei. The solid lines represent non-thermal photoionizing spectrum predictions by Ferland and Netzer (1983) or Netzer (private communication). The dotted line is for gas ionized by hot stars, with approximate position for three $\log(O/H) + 12$ values marked. Circles represent classic Seyferts, and the + crosses liners. The non-thermal spectra are marked (A) $\nu^{-1.5}$ (B) $\nu^{-1.5} \exp(-\nu/300)$ (C) $\nu^{-0.8} \exp(-\nu/50)$, with ν in Ryd.

above those implied in the HII regions. It is even possible (with hard enough radiation) to reach the extreme [NII] line strengths in the nucleus of M51 without unreasonably high nitrogen abundance. We summarize by suggesting that the gas abundances of N and O in the nuclei of spirals range from solar to perhaps twice, or at most three times, solar.

STELLAR ABUNDANCES IN GALACTIC NUCLEI

As background to the idea of stellar populations we mention the reviews of King (1971), Gratton (1979), Faber (1977), Pagel and Edmunds (1981) and Mould (1982), and on nuclei in particular the recent succinct article by O'Connell (1982a).

The absorption spectra of an intermediate nucleus (e.g. M31) looks more like that of the sun than that of a globular cluster, and Morgan and Mayal (1957) noticed that the integrated spectrum of a nucleus could

be assigned a rough spectral class that correlated with the dominance of
the spheroid - e.g. K stars in giant ellipticals and SO's, A to F stars
in Sc spirals. A classic problem has been to disentangle the integrated
colours and spectrum of a population of stars in order to infer its age
and metallicity. The intrinsic brightness of short-lived massive blue
stars causes them to dominate the colour of young populations, but a
similar blueness can result from metal deficiency in an old population.
The light output from old populations is dominated by giant stars (provided the initial mass function is not too steep, i.e. $x \leqslant 3$ in relation
(7)), and is reddened with increasing metal abun ance in the stars by
three effects. (i) The close spaced   atomic absorption lines in the
blue and violet spectrum "block" more blue light as they increase in
strength with more metals, but the radiation must escape somewhere and
it comes through stronger in the red. (ii) The position of the giant
branch in the HR diagram is determined by the opacity of stellar material,
and moves redward with increased metals. (iii) The blue extent of the
horizontal branch evolution of stars depends on their metal content, in
a metal-poor population there are more hot, blue horizontal branch stars.
Attempts at a calibration of these effects has been made by, for example,
Aaronson et al (1978), Frogel et al (1980). So far these calibrations
have been based on populations of a single metalicity - this may be a
serious shortcoming if the models are used to analyse populations which
almost inevitably contain a mix of stars of different metalicity (what
we shall call the metalicity structure), and the relative luminosity
contribution by stars of different metalicity will depend on both the
stars' metalicities and the metalicity structure. If the population
analysed is not all of the same age, then youth will contribute its own
blueness, and indeed there is evidence (e.g. Burstein 1979, Fig. 2) that
even some ellipticals (particularly dwarfs) show evidence of fairly
recent star formation. In general, however, absorption line strength
observations support the idea that colours are an indicator of metalicity
for elliptical and spheroidal systems.

The best absorption line index seems to be Mg_2, a narrow band which
includes the Mg "b" triplet and the band-head of MgH. A theoretical
calibration was made by Mould (1978) using population models from globular clusters, but there are real difficulties since there have been
uncertainties about the absolute abundance level in globular clusters
(the scale used by Mould may well be about right, see Cohen 1983), and
also since most galactic nuclei are considerably more metal rich than
any globular clusters, so calibration must therefore be an extrapolation.
Both line strength and colour indices suggest that the nuclear abundances
of ellipticals reach up as high as 2 or 3 times solar, and that there is
some correlation of the abundance level with absolute magnitude of the
system.

THE GALAXY

An important analysis of individual K giant stars in the nuclear
bulge of our own Galaxy, as viewed through "Baade's Window" has been

started by Whitford and Rich (1983). Using infrared colours as a temperature indicator for the stars, and essentially the Mg_2 index as a
metalicity indicator they find a wide spread of abundances, from perhaps
1/10 solar to several times solar. The observed metal-rich stars will
be very useful components for future popualtion models, and we look
forward eagerly to a suitably unbiased sample of stars which could be
used to determine the metalicity structure of the nuclear bulge.

OTHER SPIRAL NUCLEI

The problem of non-uniqueness of population determination when there
is limited observational data has been discussed for example by Peck
(1980) for M31, and although additional information at all wavelengths
helps (such as UV data from OAO, ANS, IUE) in such an analysis, the
considerable UV flux in many galaxies could be interpreted as due to
either old, hot very evolved stars (e.g. Welch 1982 for M31) or evidence
of recent star formation. A difficulty was posed by McClure, Cowley
and Crampton's (1980, but see also, 1982) suggestion that the bulges
of spirals really followed the metalicity absolute-magnitude dependence
of ellipticals and contained only (or predominantly) very metal poor
stars like globular clusters. This has been contested on colour and
spectral grounds for M33 by O'Connell (1982a, 83) and for NGC 7793 by
our own work (Diaz et al 1982) where the profiles of the hydrogen and
Ca+ lines show the spectrum is dominated by early F, and not globular
cluster-like, stars. The implication is that high resolution observations
of line profiles are valuable in sorting out population mixes. O'Connell's
work suggests a moderate metal deficiency [Fe/H] ~ - 0.5 for M33 bulge
stars, of about the level expected from elliptical galaxies a similar
M_v to its bulge (observational points in Fig. 4 of Pagel and Edmunds
1981).

SUMMARY OF NUCLEAR ABUNDANCES

(1) Nuclei contain metal rich gas (up to 2 or 3 times solar),
often photoionized by a non-thermal source.

(2) Young stars are often present in spiral nuclei.

(3) Elliptical and spiral galaxies have stellar populations
in the nuclei which contain both metal poor and metal rich
stars, often up to 2 or 3 times solar abundance. Those stars
which form part of the spheroidal or "bulge" component (i.e. as
distinct from disc stars, or stars intimately associated with
the nuclear region) may have a "mean" metalicity which correlated
with the absolute magnitude of the spheroidal component, and
also its velocity dispersion(Terlevich et al 1981)

THE YIELD PROBLEM

The existence of a stellar population of high "mean" abundance in galactic nuclei poses a problem. It is not difficult to show from the simple model that the following limit applies on mean abundance in stars:-

$$\bar{Z} = \frac{\int Z ds}{\int ds} \leqslant p$$

Such a mean may be naïve, and not what we really observe, since the abundance indicators and relative contributions of stars may be a complicated function of metalicity. For the solar neighbourhood, if the simple model held we would expect $Z_{ys} = p\ln(1/f)$, and with relevant values say $f \sim 0.1$, abundance in young stars $Z_{ys} \simeq 0.03$ at most $(Z_{\odot} \simeq 0.02)$, so $p = 0.013$. Thus we expect $\bar{Z} \leqslant 0.6 Z_{\odot}$. If $\bar{Z}$ is several times solar in nuclei it would imply $p \sim 0.04$ to 0.06 there, and the observed gas abundances and gas fractions in irregular galaxies imply $p \simeq 0.004$. Is this yield variation real? How could it be caused? We now discuss the influence of gas flows on chemical evolution, a mechanism favoured by one of us (MGE) to explain some of this yield variation, and then finally comment that both of us acknowledge that some actual variation of the yield must be occuring, perhaps through variation of the IMF.

A SIMPLE INFLOW/OUTFLOW MODEL

An enlightening single zone model can be constructed by combining models of Hartwick (1976) and Twarog (1980), which illustrates the features found in more sophisticated two-zone models which allow for the inflow of enriched gas (Edmunds and Morgan 1983). We assume that star formation occurs at a rate proportional to the rate of inflow of unprocessed gas into the zone (stars, after all, need gas to form out of), and that outflow of material occurs at a rate proportional to the star formation rate (say due to the effect of stellar and supernova winds). The geometry could be imagined as flow from a spheroid into a disk, or flow from the outer parts of a disk into the inner parts, etc. Taking star formation rate as $\frac{ds}{dt}$, then inflow $\gamma\frac{ds}{dt}$, outflow $\lambda\frac{ds}{dt}$.

Equations (1) and (2) become, suitably modified

$$\frac{d(Z\mu)}{ds} = p'(1 - Z) - \alpha Z - \lambda Z \tag{10}$$

$$\mu = 1 - \alpha s + \gamma s - \lambda s \tag{11}$$

The solution is (assuming $Z \ll 1$)

$$Z = \frac{p\alpha}{\gamma} (1 - \mu^{\frac{\gamma}{\alpha + \lambda - \gamma}}) \tag{12}$$

$$\mu = \frac{f}{1 + (1 - f)(\frac{\lambda}{\alpha} - \frac{\gamma}{\alpha})} \tag{13}$$

and the total mass of the system m, on reaching gas fraction f gives

$$\mu = mf \tag{14}$$

It is convenient to introduce the idea of an "effective" yield P_{eff} as the yield implied by the system had we analysed it as a simple closed model, i.e.

$$P_{eff} = \frac{Z}{\ln (1/f)} \tag{15}$$

where Z, f are given by (12) and (13). We plot in Fig. 4 P_{eff} over p, the "true" yield , as a function of m and R, the ratio of outflow to inflow, $R = \lambda/\gamma$. From this we note the interesting result that the simple model $\gamma = \lambda = 0$ gives the highest metalicity in the gas for a given gas fraction, and that any combination of inflow of unenriched gas or outflow gives a decreased effective yield. It can be shown in this inflow/outflow model that the mean stellar abundance defined as in equation (9) obeys

$$\bar{z} \lesssim \frac{p}{1 + \lambda/\alpha} \tag{16}$$

So if the centres of galaxies indicate the true value of p (i.e. with $\lambda = 0$), then $P_{eff} \simeq 0.04$ or 0.06, and a solar neighbourhood "effective yield" $P_{eff} \simeq 0.01$ would require that both inflow and outflow would have to be invoked. Also if this high value of p were also assumed for irregulars, considerable outflow as well as inflow would be needed. It is possible to formally explain the low effective yield for irregulars, and the low metalicities of dwarf ellipticals (from equation (16)) by just having lots of outflow. But one worries that large outflow of gas would gravitationally unbind the system (Hills 1980).

Sample metalicity structures for the model are shown in Fig. 5, from which it is clear that a model with inflow, or suitable inflow and outflow, is a much better match to the metalicity structure in the solar neighbourhood than the simple model (as recognized by Larson (1972) and Lynden-Bell (1975)), and pure outflow always makes the match worse. It may also be shown (in the instantaneous recycling approximation) that the dependence of a secondary to primary element abundance ratio (e.g. O/N as function of O/H) is always as steep as, or steeper than, that predicted by equation (5) for the simple model.

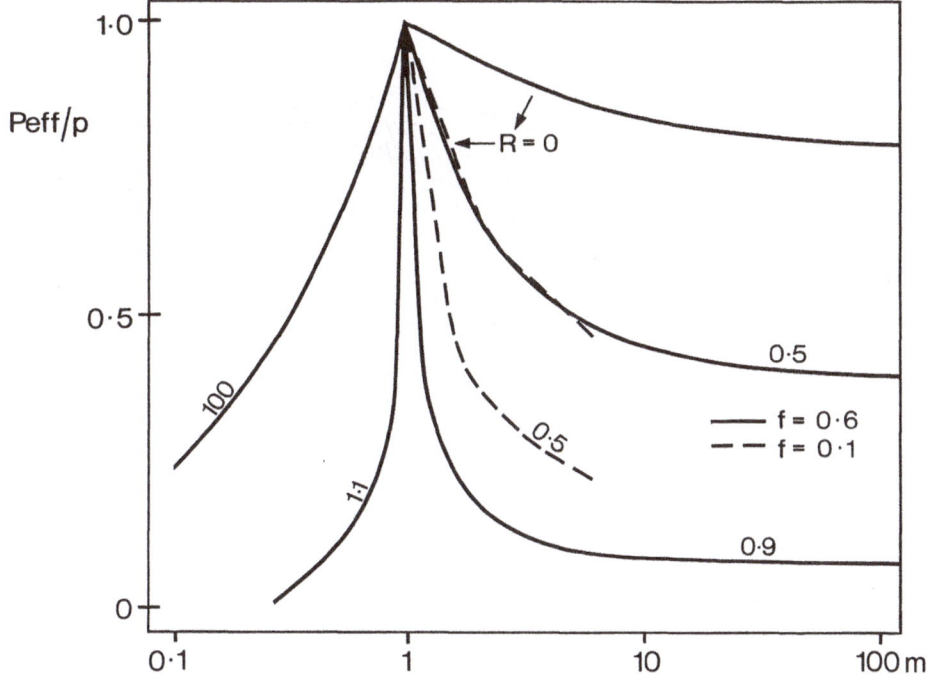

Fig. 4 The simple inflow/outflow model. Predicted effective yield as a function of final mass m on reaching gas fraction f (initial mass unity) and curves are marked with their R, ratio of outflow to inflow.

REMAINING PROBLEMS

        Does "true" yield vary? It may well do. Webster and Smith (1983) have noticed that HII region abundances seem to correlate with surface density. There is some evidence from spiral and irregular galaxy gas abundances and gas fractions that the yield may depend upon surface mass density (Edmunds and Pagel 1983), perhaps due to the effect of the local gravitational potential on the initial mass function for star formation. The implications of such a variation for e.g. secondary element formation and correlation of metalicity with other parameters should be investigated. The yield could be metalicity dependent if the evolution of stars (e.g. through mass loss rates) is significantly influenced by metal abundance.
        Knowledge of the metalicity structure of other systems will be very valuable in constraining inflow and outflow models. The Galactic bulge and the Magellanic Clouds are good places to start.
        If mean stellar abundances higher than the true yield are required, it seems that the only way to do it is to enrich the gas before allowing stars to form. This is essentially the way in which Larson (1974) generated gradients in his collapse models, but it is not clear that star formation was held up long enough, and the inflowing gas made rich enough, to give the observed range of abundances in our own Galactic bulge, or the high mean abundances observed in other nuclei.

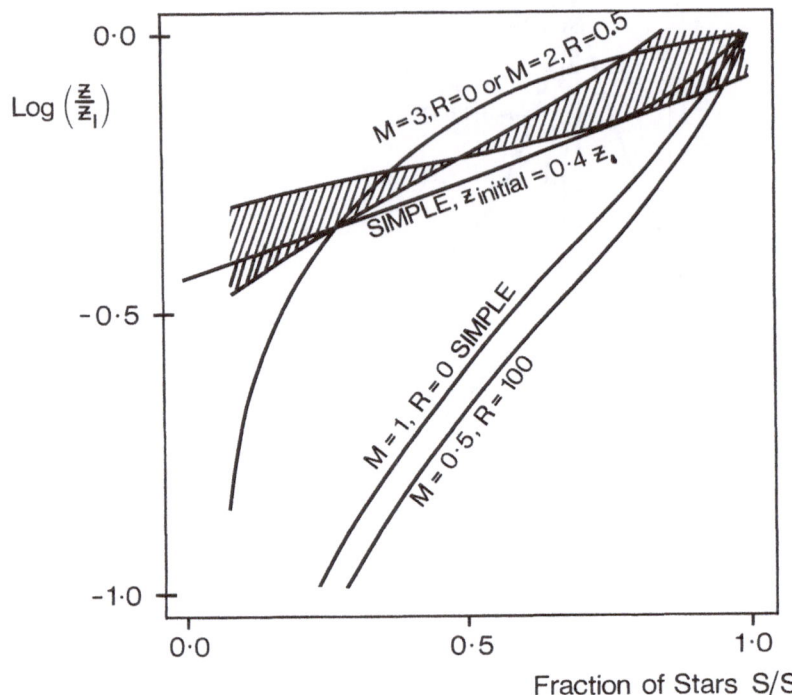

Fig. 5 Metalicity structure given by simple and selected inflow/outflow models. The hatched region represents approximate observational results of Pagel and Patchett (1975) for the solar neighbourhood.

REFERENCES

Aaronson M. et al. (1978) Ap. J. 223, 824
Alloin D. et al. (1979) Astr. Ap. 78, 200
Balwin J. et al. (1981) P.A.S.P. 93, 5
Bessel M. and Norris J. (1982) Ap. J. 263, L29
Burbidge E.M. and Burbidge G.R. (1962) Ap. J. 135, 694
Burstein D. (1979) Ap. J. 232, 74
Carbon D.F. et al. (1982) Ap. J. Suppl. 49, 207
Cohen J. G. (1983) Ap. J. 270, 654
Clegg R.E.S. (1977) M.N.R.A.S. 181, 1
Diaz A. et al. (1982) M.N.R.A.S. 201,49p
Dufour R.J. and Talent D.L. (1980) Ap. J. 235, 22
Edmunds M.G. and Morgan D.J. (1983) in preparation
Edmunds M.G. and Pagel B.E.J. (1978) M.N.R.A.S. 185, 77p
Edmunds M.G. and Pagel B.E.J. (1983) in preparation
Faber S.M. (1977) in"Evolution of Galaxies and Stellar Populations",
        ed. R.B. Larson and B.M. Tinsley, Yale Univ. Press

Ferland G. and Netzr H. (1983) Ap. J. 264, 105
Frogel H. et al. (1980) Ap. J. 240, 785
Gratton L. (1974)in "Structure and Evolution of Galaxies", ed. G. Setti
        and L. Woltjer, D. Reidel P. C. Dordrecht, Holland
Halpern J.P. and Steiner J.E. (1983) Ap. J. 269, L37
Harmer D. and Pagel B.E.J. (1970) Nature 225, 349
Hartwick F.D.A. (1976) Ap. J. 209, 418
Heckman T.M. (1980) Astr. Ap. 87, 152
Hillis J.G. (1980) Ap. J. 235, 986
Keel W.C. (1983) Ap. J. 269, 466
King I.R. (1971) P.A.S.P. 83, 377
Kraft R.P. et al. (1982) P.A.S.P. 94, 55
Larson R.B. (1972) Nature 236, 7
Larson R.B. (1974) M.N.R.A.S. 166, 585
Lequeux J. et al. (1979) Astr. Ap. 80, 155
Lynden-Bell D. (1975) Vistas in Astronomy 19, 299
McClure R.D. et al. (1980) Ap. J. 236, 112
McClure R.D. et al. (1982) Ap. J. 263, 325
Morgan W.W. and Mayal N.U. (1957) P.A.S.P. 69, 291
Mould J. (1978) Ap. J. 220, 434
Mould J. (1982) Ann. Rev. Astron. Ap. 20
Norris J. et al. (1981) Ap. J. 244, 205
O'Connel R.W. (1982a) Ap. J. 257, 89
O'Connel R.W. (1982b) preprint
O'Connel R.W. (1983) Ap. J. 267, 80
Pagel B.E.J. and Edmunds M.G. (1981) Ann. Rev. Astron. Ap. 19, 77
Pagel B.E.J. et al. (1980) M.N.R.A.S. 193, 219
Pagel B.E.J. and Patchett B.E. (1975) M.N.R.A.S. 172, 13
Pagel B.E.J. (1981) in "Structure and Evolution of Normal Galaxies",
        ed. M. Fall and D. Lynden-Bell, Cambridge Univ. Press
Peck M.L. (1980) Ap. J. 238, 79
Peterson R.C. and Snedden C. (1978) Ap. J. 225, 913
Searle L. (1971) Ap. J. 168, 327
Serrano A. and Peimbert M. (1983) preprint
Shaver P.A. et al. (1983) M.N.R.A.S. 204, 53
Smith G.K. and Norris J. (1982) Ap. J. 254, 149 and 594
Sneden C. et al. (1979) Ap. J. 234, 964
Talbot R.J. and Arnett W.D. (1974) Ap. J. 190, 605
Terlevich R. et al. (1981) M.N.R.A.S. 196, 381
Tinsley B.M. (1980) Fundaments of Cosmic Physics 5, 287
Tomkin J. et al. (1976) Ap. J. 210, 694
Tosi M. (1982) Ap. J. 254, 699
Trefzger C.F. et al. (1983) Ap. J. 266, 144
Truran J.W. and Cameron A.G.W. (1971) Ap. Sp. Sci. 14, 179
Twarog B. (1980) Ap. J. 242, 242
Webster B.L. and Smith M.G. (1983) M.N.R.A.S. in press
Welch G.A. (1982) Ap. J. 259, 77
White S.D. and Audouze J. (1983) M.N.R.A.S. 203, 603
Whitford A. E. and Rich R.M. (1983) preprint

DISCUSSION

Gallagher: If nitrogen production is characterized by long time scales, then will radial mixing in disk galaxies tend to level the ratio of $|N/O|$? Is it then proper to consider the radial variations in $|N/O|$ within spirals as being indicative only of the nuclear production processes, or should one perhaps adopt about integral $|N/O|$ values for spirals as well as irregulars.

Edmunds: The radial mixing cannot be too important a factor because strong $|O/H|$ gradients are observed in some disk galaxies, and these would tend to have been flattened out if the mixing were important.

Iben: You have been talking about O to N relations, why don't you also consider C to N?

Edmunds: Because it is observationally much more difficult. Most of the arguments I have used are based on HII region abundances, and carbon abundances are only accessible in the ultra-violet. Any carbon abundances derived are then uncertain because of the unknown amounts locked up in graphite or, other, interstellar grains.

Iben: Could the destruction of O^{16} into N^{14} somehow lead to a late steady state with $|N/O|$ constant?

Edmunds: I don't think so. It is possible to rewrite the equations for the simple model with a destruction term for the oxygen which is processed to nitrogen. The equations then become

$$Z_{Ox} = p_{Ox} (1 - f^{p_N}) / p_N$$

$$Z_N/Z_{Ox} = (p_N \ln (1/f)) / (1 - f^{p_N}) - 1$$

Fitting the solar neighbourhood values to determine p_{Ox} and p_N, and then looking at the behaviour of Z_N/Z_{Ox} with Z_{Ox} shows negligibly different behaviour from the simple secondary model. It might be interesting to try again with an inflow/outflow model, but the real point is that the nitrogen production only uses up a small part of the oxygen produced.

Renzini: Have you got any information on $|Fe/H|$ in these galaxies? The point is that $|Fe/H|$ may affect stellar evolution in the sense of reducing the production of primary N for increasing $|Fe/H|$.

Edmunds: We have no detailed information on the $|Fe/H|$ for the disk galaxies (other than for our own, near the sun), and really only for the Magellanic Clouds for irregulars. The $|Fe/H|$ is only accessible through stellar spectra, while the $|N/O|$ estimates come from nebulae.

Mathews: I would like to ask whether the N/O ratio vs. O/H could be explained with secondary nitrogen and an increasing birth rate function. Such a birth rate function is consistent with continuity of the IMF. The increase in N/O due to older stars could then be compensate by increased oxygen production from massive stars.

Edmunds: There are two problems i) is the time scale if 3 - 8 $M_\odot$ say stars give the secondary nitrogen then the return time to the ISM is still comparatively short compared to galactic evolution timescales. So unless the birthrate had increased dramatically recently, the N and O would still look like a secondary correlation. ii) It would be surpri-

sing that the birthrate had increased in just the right way across galaxies to provide a fairly constant N/O despite various O/H.
Gallagher: Isn't true (that is to state this a a question) that Peck's
point about non-uniqueness in population synthesis models in part concerns
the redundant nature of the information contained in a spectrum? Also,
he finds that if a population synthesis model does not fit a spectrum,
then there is no unique way to determine what features of the model are
incorrect. Thus, in a sense, aren't these general limits on what one can
learn about a stellar population from its spectrum, even though matters
can certainly be improved by going for example to higher spectral resolution?
Edmunds: Yes, perhaps I am being unduly pessimistic about population synthesis. I can place useful constraints, particularly if (as you suggest)
high resolution is available.
Danziger: You seem to be saying at the conclusion of your talk that there
was not a great future for population synthesis, and yet earlier you
quoted appraingly results of published population synthesis work. So
what is it that you would like to have that you think will be impossible
to obtain?
Edmunds: Population synthesis can give a rough indication of overall metallicity, but it cannot (at least at its present level of sophistica-
tion) give details of the metallicity structure of the population.
Renzini: The M giants found by Frogel and Whitford in the galactic bulge
are rather bright AGB stars (M_b up to -5) for which one can infer an initial mass of 1.5 $M_\odot$. This corresponds to an age of about 2 billion years.
Then using these stars as ingredients in synthetizing populations in
ellipticals would imply that the presence of young (intermediate age)
stars is assumed in such galaxies.
Edmunds: The age you suggest is indeed rather worringly young for bulge
stars. But further investigation of these stars is obviously essential,
since they are the best templates to use for the metal-rich component
in population synthesis.
Gallagher: 1) Even though zero metal gas inflow is a standard feature of
chemical evolution models, it should be note that gas resevoirs with even
1/10 $Z_\odot$ are extraordinarily uncommon at the present epoch. This certainly
shows that both inflows and outflows must be considered, and that one
might best invoke a time dependent enrichment of the gas resevoirs.
2) You might also expect surface brightness-metallicity correlations if,
gravitational binding energies are playing a role in determining gas
metallicities.
Edmunds: To 1): An inflow of 1/10 $Z_\odot$ gas into the disk at the present
time would be effectively equivalent to an inflow of unenriched gas, although the difference from zero metallicity would of course be important
at early times when the disk metallicity were much smaller. To 2): Yes,
but beware of unbinding the whole system?
Woosley: I would like to go back to the problem of abundances. Since galactic formation transpired in $> 10^8$ yr (and one has a similar time scale for rotational mixing) during which a generation of massive stars can
occur roughly every 3 million years, one might expect considerable inho-
mogeneity in the spatial distribution of composition at early times when
Pop II was forming. Stars $\geq 100\ M_\odot$ for example (Pop III) are expected to

produce copious O^{16} but very little N^{14} and Fe^{56} . Later, following
CNO processing in Pop II, these inhomogeneities in O^{16} can translate
to inhomogeneities in N^{14} which may produce a large spread of $|N/Fe|$,
for example. How does this affect your arguments for primary nitrogen?
Edmunds: The effect you suggest could have been important early on at
low metallicities, but the "bulk" of O and N have been synthetized since
that time - and the arguments are based on galaxies where the metallicity is well above Pop III.
Renzini: How the galactic bulge could have managed to produce stars with
$\overline{Z \cong 3 Z_\odot}$?
Edmunds: A few stars of $3 Z_\odot$ would not be a problem, trouble arises if
the mean metallicity of the stars approaches 2 or 3 $Z_\odot$. In this case 1)
the actual yield must be higher than in the solar vicinity or 2) the
actual yield must be higher everywhere and the combined effects of inflow
and outflow must have lowered the apparent yield in the solar neighbourhood or 3) star formation only occured in the galactic bulge after very
enriched gas had collected there.
Mathews: I would like to ask how realistic is to suppose that the star
formation rate is proportional to the infall rate, when we observe in the
solar vicinity star formation mechanisms which do not appear to directly
relate to infall for example spiral density waves, expanding HII regions,
or supernova shocks.
Edmunds: Probably not very realistic at all! But the model has mathema-
tical convenience, and although we are investigating star formation histories, I suspect the qualitative behaviour will not be greatly different.

PROBLEMS IN THE CHEMICAL EVOLUTION OF GALAXIES

C. Chiosi [1,2] and F. Matteucci [3]

1) Institute of Astronomy, University of Padova, Italy
2) International School for Advanced Studies, Trieste, Italy
3) Istituto di Astrofisica Spaziale C.N.R., Frascati, Italy

Introduction

The chemical evolution of a galaxy is governed by the evolution of individual stars as well as collective processes such as star formation rate,
stellar mass function and dynamics of the gas-stars system. Such a problem
has been the subject of a great deal of work as reported in many up to date reviews, among which we recall Tinsley (1980), Pagel and Edmunds (1981)
and Chiosi and Jones (1983), and it goes beyond the scope of this note to
explore its many intricacies in detail. Therefore, instead of rewiewing
the recent developments achieved in this subject, we shall discuss here
to some extents only a few points that in our opinion are of current interest, namely:
i) the determination of chemical yields per stellar generation for various
major elements;
ii) the chemical history of the galactic disk;
iii) the chemical evolution of magellanic and dwarf irregular galaxies.
Prior to this we shall briefly review the so called simple model of galactic evolution as a useful framework to the three topics.

1. The Simple Model

Considering our ignorance of the basic physical processes intervening in
the galactic evolution, the simple model is an effective tool which goes
a long way to emphasizing some key issues of galactic chemical evolution.
Let M be the total mass of the system which may increase as a result of
gaseous accretion at rate f(t) or even decrease as a result of gas ejection at a rate w(t). The rates f(t) and w(t) have to be modelled on the
basis of some idea for building up galaxies and for ejecting gas from
these, respectively. The time variation of M is

$$\dot{M} = f(t) - w(t) \quad . \tag{1}$$

Gas, having total mass M_g is consumed by forming stars at a rate $\Psi(t)$, by
ejecting gas out of the system at a rate w(t), but it is also replenished
at a rate E(t) by the ejecta from evolving stars, therefore

C. Chiosi and A. Renzini (eds.), Stellar Nucleosynthesis, 359–381.
© *1984 by D. Reidel Publishing Company.*

$$\dot{M}_g = - \Psi(t) + E(t) + f(t) - w(t) \quad . \tag{2}$$

The abundance by mass of any elemental species X_i is governed by the equation

$$(X_i M_g)^\cdot = X_i \Psi(t) + E_i(t) + X_{if} f(t) - X_i w(t) , \tag{3}$$

where $E_i(t)$ represents the rate of increase in abundance by gas ejected from evolving stars, whereas X_{if} is there to account for the fact that the accreted gas could be enriched. The quantities $E(t)$ and $E_j(t)$ are somewhat complicated since they depend on the number of stars dying at time t. These stars are born at a time τ_m prior to t, where τ_m is the lifetime of a star of mass m. The number of such stars formed was proportional to the stellar mass function $\Phi(m)$ and star formation rate Ψ at time $t-\tau_m$ and these stars when they die leave a remnant of mass m_r. The material thrown out, part of which can be even ejected before the very final stage, can be regarded as consisting of a fraction $(1 - p_{im})$ of the star's mass which is ejected but which has not been processed and therefore has the same metallicity as the gas from which the star formed, and a fraction p_{im} that is both ejected and turned into new formed heavy elements. Accordingly $E(t)$ and $E_i(t)$ are

$$E(t) = \int_{m_t}^{\infty} (m-m_r) \, \Psi(t-\tau_m) \, \Phi(m) \, dm , \tag{4}$$

$$E_i(t) = \int_{m_t}^{\infty} [(m-m_r-m \, p_{im}) \, Z(t-\tau_m) + m \, p_{im}] \, \Psi(t-\tau_m) \, \Phi(m) \, dm . \tag{5}$$

In the approximation of instantaneous recycling ($\tau_m \ll t$, where m₁ is about 1 $m_\odot$) they become

$$E(t) = \Psi(t) \int_{m_1}^{\infty} (m-m_r) \, \Phi(m) \, dm = \Psi(t) \, R \tag{6}$$

where R is called the "return fraction", and

$$E_i(t) = R \, X_i \, \Psi + y_i \, (1-R) \, (1-X_i) \, \Psi \tag{7}$$

where

$$y_i = \int_{m_1}^{\infty} m \, p_{im} \, \Phi(m) \, dm \, / \, (1-R) \tag{8}$$

is called the "net yield" of the element i. With simple algebric manipulations eqs. (2) and (3) can be written

$$\dot{M}_g = - (1-R) \, \Psi(t) + f(t) - w(t) \tag{9}$$

$$M_g \dot{X}_i = y_i \, (1-R) \, (1-X_i) \, \Psi(t) + (X_{if} - X_i) \, f(t) \quad . \tag{10}$$

With the aid of eq. (10), the ratio between the enrichment rate of any two elements can be derived. In the general case of the infall rate being proportional to the star formation rate ($f = \lambda \Psi$) we get

$$\frac{dX_i}{dX_j} = \frac{y_i \, (1-R) \, (1-X_i) + (X_{if} - X_i) \, \lambda}{y_j \, (1-R) \, (1-X_j) + (X_{jf} - X_j) \, \lambda} \quad . \tag{11}$$

The correspondent expressions for no infall ($\lambda = 0$) and for an infall rate just balancing the star formation rate ($f = (1-R)\Psi$) can be easily derived from eq. (11). These relationships are particularly useful to study the helium to heavy element enrichment ratio $\delta Y/\delta Z$. These equations have been discussed by numerous authors; among these we sholud mention the pioneering papers by Talbot and Arnett (1971, 1973, 1975), Pagel and Patchett (1975), Lynden-Bell (1975), the comprehensive review by Tinsley (1980) and the recent papers by Chiosi (1980), Serrano and Peimbert (1981) and Yokoi et al (1983). Among the various quantities intervening in the construction of eqs. (1) to (11), the net yields y_i are of particular interest here as they rest on the stellar nucleosynthesis make up.

2. The Net Yields Of Elements

Each generation of stars contributes to the chemical enrichment of the galaxy by processing new material in the stellar interiors and returning to the interstellar gas a fraction of the mass containing both processed and un processed matter during the various stages of mass ejection (stellar wind, planetary nebula, supernova explosion). What we need is a complete specification as a function of the stellar mass of the fractionary amount of each elemental species returned to the interstellar medium as well as the fraction of mass going into a remnant, thus subtracted for ever to further nuclear processing. The return fraction R and the net yields y_i's can be written according to the formalism of Talbot and Arnett (1973), which is ideally suited for studies of chemical evolution. The advantage with this formalism compared to other simple formulations is that many important conclusions regarding chemical abundances can be reached without having to consider detailed chemical models. This formalism describes the ejected nucleosynthesis products from stars in terms of a production matrix involving all elemental species. The production matrix element $Q_{ij}(m)$ specifies which fraction of the star initially present in form of species j is ejected as species i. Thus if X_j denotes the initial abundance by mass of species j, the quantity

$$r_i(m) = \Sigma_j \, Q_{ij}(m) \, X_j \qquad\qquad (12)$$

fixes the fractional mass ejected into the interstellar medium by a star of mass m in form of species i. The production matrix element per stellar generation is

$$q_{ij} = \int_{m_1}^{\infty} Q_{ij}\,(m) \, \phi(m) \, dm \qquad . \qquad (13)$$

The total fractionary mass ejected by a generation of stars is

$$R = \Sigma_i \int_{m_1}^{\infty} r_i(m) \cdot \phi(m) \, dm \qquad , \qquad (14)$$

whereas the net yield is

$$y_i = \Sigma_{i \,\neq j} \int_{m_1}^{\infty} Q_{ij}(m) \, X_j \, \phi(m) \, dm \, / \, (1-R) \qquad (15)$$

where $\phi(m) = m \, \Phi(m)$. The confrontation of relations (14) and (15) with
(6) and (8) allows us to link the present formalism to that of the basic
equations with instantaneous recycling. Since stars with mass $m < m_1$ are
expected not to contribute to the nucleosynthetic enrichment (their Q_{ij}'s
being zero), the integration of eqs. (14) and (15) can be performed over
the whole range of mass encompassed by $\phi(m)$. m_1 in fact stands for the
lower mass limit of born stars. The stellar mass function obeys the normalization condition

$$\int_1^\infty \phi(m) \, dm = \zeta \qquad\qquad (16)$$

where ζ is the fraction of mass of $\phi(m)$ in stars more massive than $1 \, m_\odot$.
The problem is now reduced to assign prescriptions for $\phi(m)$ and for Q_{ij}'s.

a) The Stellar Mass Function

The stellar mass function $\Phi(m) \, dm$, which gives the number of stars in the
mass interval m to $m+dm$, is often represented by a power law of the mass

$$\Phi(m) \propto m^{-(1+x)} \qquad\qquad (17)$$

together with the assumption that it does not depend on time, metallicity
or anything else. This approximation is not necessarily the best. The work
of Burki (1977, 1978) and Boisse et al. (1980) may indeed be interpreted
as evidence for variations in $\Phi(m)$ with the ambient conditions. Various
$\Phi(m)$'s have been proposed over the years, among which we recall Salpeter
(1955), Serrano (1978), Lequeux (1979), Miller and Scalo (1979), Tinsley
(1980) and Garmany et al. (1982). This last one only for stars more massive than about 25 $m_\odot$. In the following, results will be presented for a
few illustrative cases.

b) The Production Matrix $Q_{ij}(m)$

Since the work of Talbot and Arnett (1973) new developments in stellar
evolution have occured allowing for a more up to date estimate of the matrix. It is beyond the scope of this note to review the progress made in
understanding the late stages of evolution of stars all over the range of
mass for which we refer to the exhaustive reviews of Iben and Renzini
(1983), Nomoto (1984), Wheeler (1982), Woosley and Weaver (1983), Woosley
et al. (1984) and Maeder (1983). It suffices to recall here the major no-
velties brought into the field. In the range of low to intermediate mass
stars ($1 < m < 8 \, m_\odot$) the study of Iben and Truran (1978) of the thermal
pulsing regime till the stage of planetary nebula and white dwarf formation; the study of Renzini and Voli (1981) on the evolution of surface
abundances of several important elements from the main sequence till the
stage of envelope ejection in occurrence of dredge up episodes, envelope
burning process and mass loss by stellar wind during the red giant and
asymptotic giant branch evolution. The intricacies of the evolution of
stars in the range of mass 8 to 12 $m_\odot$ are summarized by Nomoto (1984).
These stars in fact cannot evolve straightforwardly to form an iron core
as in more massive stars, but follow a much more complicated scheme due to
the effect of the relatively stronger degeneracy in the core. On the side
of massive stars ($12 < m < 100 \, m_\odot$), we recall the bare core calculations
of Arnett (1978) in which pure He stars have been followed till the pre-

supernova stage, and the work of Woosley and Weaver (1983) and Woosley et al. (1984 and reference therein) on the nucleosynthetic history of complete models of massive stars evolved at constant mass till the stage of supernova explosion. In the meantime, a great renewval in our knowledge of the evolution (and nucleosynthesis) of massive stars has occured due to the inclusion of the effects of mass loss by stellar winds during the core H and He-burning phases (see Chiosi (1981) and Maeder (1981a) for up to date reviews of the subject). Calculations of the $Q_{ij}(m)$ elements incorporating one or more of the above improvements are by Chiosi and Caimmi (1979), Mallik (1980, 1981) and Chiosi and Matteucci (1982, 1983). Derivations of stellar and net yields of helium and heavier elements lumped together, which do not necessarily rest on the matrix formalism, are by Serrano and Peimbert(1981) and Maeder(1981b,1983,1984). Rather than going into a lengthy discussion of the procedures used to derive the matrix elements as a function of the star's mass, it is worth recalling here only a few points of major interest and uncertainty as well. Firstly, only stellar models of single stars have been considered, while on the contrary the effects of binary mass transfer on the element production may play a significant yet not wholly assessed role. This problem starts now being considered. We recall the study of Vanbeveren and Olson (1980) for the case of massive stars and of Greggio and Renzini (1983) for the chemical enrichment by binary models of type I supernovae. Secondly, it is assumed that $Q_{ij}(m)$'s do not vary with the chemical composition of the star. This approximation is not necessarily true. The effect of a low metal content has been recently evaluated by Chiosi and Matteucci (1982, 1983) and Maeder (1983) for massive stars suffering mass loss by stellar wind. The difference with respect to the standard case are not very large. These results are however still uncertain. No estimates of the yields for low and intermediate mass stars with low metal content have been made so far. Thirdly, all of these stellar models rest on classical assumptions concerning the basic input physics. Uncertainties and modifications due to mixing processes (overshoot, turbulent diffusion, etc.) start now being considered. Lastly, theoretical yields are customarily derived only for several important elements such as He^4, C^{12}, C^{13}, O^{16}, Mg and Si or iron group nuclei taken as a single element. Further simplifications are made when comparing theoretical results with observational data by introducing the concept of average nucleosynthetic event or typical nucleosynthetic star's mass. No detailed comparison of the whole pattern of abundances of heavy elements in the solar mixture (Cameron, 1973) has been made so far with theoretical predictions properly weighting the contribution from stars of different mass. In the construction of the matrix elements intervene several important parameters which deserve some general remarks. These are: the remnant mass m_r, the mass of the He core m_α, the mass of the carbon-oxygen core m_{co}, the composition of the ejecta.

i) The mass of the remnant

The relation between the initial mass m and the mass as well as the nature of the remnant is still far from being clearly established. The following assumption is customarily made: a neutron star of 1.4 $m_\odot$ is assumed to be left by all stars more massive than 8 $m_\odot$. Thus the possibility of black hole formation is neglected. Below 8 $m_\odot$, stars are assumed to leave a white dwarf remnant if their initial mass is smaller than $m_{up} \leqslant 8\ m_\odot$. The value for m_{up} is not firmly assessed at the present time. This range of mass, m_{up} and the mass of the white dwarf depend on the amount of mass lost by stellar wind during the red giant and asymptotic giant branch phases. See Iben and Renzini (1983) for an exhaustive discussion of this topic. The situation in the range of mass $m_{up} < m < 8\ m_\odot$ is still highly uncertain. It goes from complete disruption to the production of a quasi static iron core or to the creation of a stable iron white dwarf. See Wheeler (1982, and references therein) for more details on this point.

ii) The mass of the He core

The mass of the He core corresponds to that portion of the star inside the H-burning shell in which H is turned into He. m_α for a given mass depends to various extents upon many factors, such as the treatment of convection and overshoot, dredge up episodes, penetration of the external convection, outward migration of the H-burning shell and mass loss during the various evolutionary phases. In the mass range $m < 8\ m_\odot$, m_α derived from Renzini and Voli (1981) takes into account the effects of mass loss by stellar wind during red giant and asymptotic giant branch stages, envelope convection penetration and envelope burning. In the range 8 to 10 $m_\odot$ m_α can be estimated from the models of Sugimoto and Nomoto (1980), whereas for more massive stars it can be derived from the models of Maeder (1981a) and Bressan et al. (1981) which take mass loss and convective overshoot into account. The relation between the initial mass and m_α for stars in this latter range of mass and in presence of mass loss has been for long a matter of debate (Maeder, 1981b,1983,1984). The controversial point is whether or not m_α for mass losing models is smaller than for constant mass ones. Chiosi et al. (1978) found that mass loss during the core H-burning phase reduces the size of the convective core and in consequence the mass of the He core at the beginning of the He-burning phase. In principle, due to the outward migration of the H-burning shell in presence of a smooth profile of chemical composition in the star, m_α may considerably increase during the core He-burning phase. Chiosi's et al. (1978) models had on the contrary a sizable intermediate convective zone overtopping the H-burning shell and thus preventing the H-burning shell from migrating outwards. The mass m_α did not increase during the core He-burning phase. This is why they felt reasonably safe to adopt the new $m(m_\alpha)$ relationship when reading off Arnett's (1978) data to transfer the results of bare cores to those of standard stars. Maeder (1981a) using different mass loss rates and slightly different input physics found different results, in that models mimiced the behaviour of constant mass ones. In a subsequent paper Bressan et al. (1981) computed models with overshoot and mass loss thus establishing a new $m(m_\alpha)$ relationship, not very different from that obtained for constant mass models and from that of Maeder (1981a) without overshoot. Owing to the many uncertainties still affecting both the theory of overshoot and the mass loss rates, we believe that all the above results are equally legitimate.

iii) The mass of the carbon-oxygen core

The carbon-oxygen core is defined as the portion of the star in which He-burning and possibly later ones took place. m_{co} in the range of low and

intermediate mass stars can be easily derived from current models. It is
in fact coincident with the remnant mass for $m < m_{up}$ and 1.4 $m_\odot$ up to
about 10 $m_\odot$. In the range of more massive stars it can be derived from the
models of Maeder (1981a). In this mass range m_{co} may suffer from the same
type of uncertainty that affected m (m_α) as mass loss during the core
He-burning phase may appreciably change the mass of the carbon-oxygen core
left behind at the end of the core He-burning phase.
Fig. 1 displays m_r, m_α , m_{co} as a function of the initial mass which result
from the previous discussion and for the particular case of η_R = 0.6 (mass
loss parameter) and α_R = 1.5 (mixing length parameter of the envelope convection) of Renzini and Voli (1981), and mass loss from massive stars according to case B of Maeder (1981a). We like to stress the qualitative

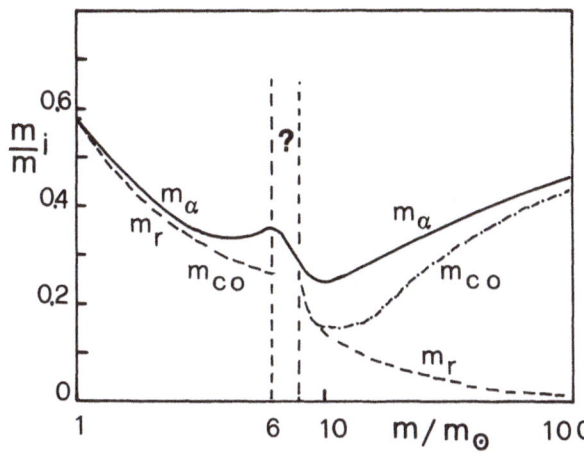

Fig. 1 Run of the fractio-
        nary mass of the He
core, C-O core and remnant
as a function of the star
mass

nature of this diagram which may change substancially if any of the assumptions discussed above will turn out not to hold.

iv) The composition of the ejecta

The composition of the ejected material is still rather uncertain, this
reflecting indeed the uncertainty in the final fate of stars over the various ranges of mass. In Fig. 2 we plot for purposes of illustration the
relative abundances ($\Delta M_i^{ej}/m$) of several important elements. These are
derived from Renzini and Voli (1981) for stars in the range 1 < m < m_{up}.
Only their case η_R = 0.33 and α_R = 1.5 is shown for simplicity. In the
mass range 10 to 100 $m_\odot$ the abundances of C, O, Ne, Mg and Si are derived
from Arnett (1978) implemented with the data of Woosley and Weaver (1983)
and adapted to the $m(m_\alpha$) relation of Fig. 1. The He abundance is taken
from Maeder (1981b) to account for the variations in He caused by the red
supergiant convection and mass loss by stellar wind. This latter may even
change the abundance of some heavier elements (C and O) in presence of
substancial mass loss during the core He-burning phase. This point will be
discussed to some extent later on. The abundances for the mass range m_{up}
to 10 $m_\odot$ are intentionally not displayed as they are still more uncertain

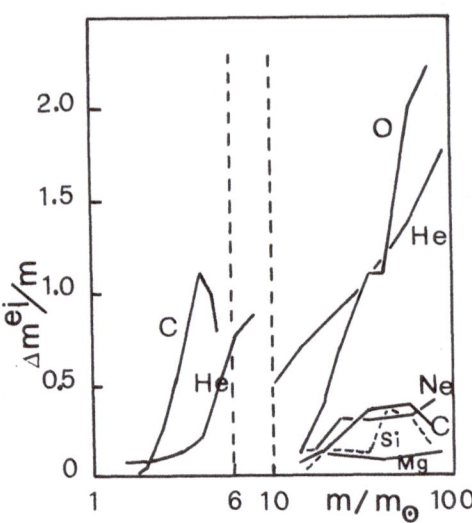

*Fig. 2 Elemental abundances of
the ejecta for several
important elements*

than the real fate of these stars.
in fact, besides contributing to the
interstellar enrichment through the
material lost in the wind (Renzini
and Voli, 1981), these stars may be
also significant producers of heavy
elements (C, O, Ne,..) in amounts and
proportions that depend on the evolution through the final stages. See
Nomoto (1980), Wheeler (1982) and
Yokoi et al. (1983) for illustrative
examples. The minimal assumption is
customarily made of taking into account
only the contribution from the wind
in the lack of more precise information on the real fate of these stars.
Therefore, in the examples illustrated
below the data of Renzini and Voli
(1981) are used for the stars in the
mass range m_{up} to 8 $m_\odot$, whereas some
plausible extrapolation is made for
those in the range 8 $m_\odot$ to 10 $m_\odot$.
Recently, particular attention has
been payed to the effect of mass loss
by stellar wind on the chemical yields for massive stars (Meader,1983,1984
and references therein). Maeder's estimates of $(\Delta M_i^{ej}/m)$'s are shown in
Fig. 3, where the contribution to the various elements by stellar wind (w)
is indicated separated from that by supernova explosion (s). The contribution by the wind to heavy elements is rather small and therefore not
displayed. Finally, the repartition among C, O and heavier elements shown
in Fig. 3 has been obtained with the aid of Arnett's (1978) data starting
from the global estimates of Maeder (1983). It is evident that the stellar yields of metals are reduced by mass loss as much He is ejected in
the wind and thus subtracted to further nuclear processing. Conversely,
the yield of He increases. It is worth recalling here that the He yield
properly accounts for the effect of convective dredge up during the red
supergiant phase. Moreover, since Maeder's yields are based on a detailed
sheet balance in the course of evolution, they do not rest on the adoption
of a particular $m(m_\alpha)$ relationship. They however depend on the rate and
modality of mass loss.

c) Elemental Yields per Stellar Generation

The products of the stellar yields discussed in so far by a stellar mass
function determine the net yield per stellar generation for any given element. These are presented in Tables 1 and 2 for several combinations of
stellar mass function and stellar input (mass loss rates, overshoot, nucleosynthesis data, etc.). Table 1 contains the characterizing parameters for
each set of yields, whereas Table 2 lists the "net yields" of He^4, C^{12},
C^{13}, N^{14}, O^{16}, Ne^{20}+ Mg^{24} and the Si-Fe group of elements. In addition to
this, the ratio y_{He}/y_Z, y_Z being the cumulative yield of metals, is shown.
The data of Tables 1 and 2 are taken from Chiosi and Matteucci (1983).

Table 1
(Basic Parameters of Yield Calculations)

Case	1	2	3	4	5	6	7	8	9	10	11

Stellar Mass Function

	1	2	3	4	5	6	7	8	9	10	11
x	0.6 a	0.6	0.6	1.35c	1.35	1.35	1.35	1.35	1.35	0.6	1.35
	2.0 b	2.0	2.0							2.0	
ζ	0.44	0.44	0.44	0.50	0.50	0.50	0.25	0.25	0.25	0.44	0.50
m_l	0.007	0.007	0.007	0.16	0.16	0.16	0.02	0.02	0.02	0.007	0.16
m_u	100	100	100	100	100	100	100	100	100	100	100

Stellar Models and Nucleosynthesis Sources for Massive Stars

	1	2	3	4	5	6	7	8	9	10	11
$\dot{M}$	O	H	I	O	H	I	O	H	I	M	M
Ov	No	No	Yes	No	No	Yes	No	No	Yes	No	No
Nu	A,W	A,W	A,W	A,W	A,W	A,W	A,W	A,W	A,W	M,A	M,A

Stellar Models and Nucleosynthesis Sources for Low-Interm. Mass Stars
$n_R = 0.33$ $\alpha_R = 1.5$

a) for $m < 1.8\ m_O$; b) for $m > 1.8\ m_O$; c) over the whole range of mass
H) High mass loss rates ; I) Intermediate mass loss rates ; Ov) Overshoot
as in Bressan et al. (1981) ; M) Mass loss rate and ejecta (wind) from
Maeder (1983, 1984); Nu) Nucleosynthesis from Arnett (1978) (A) and
Woosley and Weaver (1983) (W).

Table 2
(Net Yields for Several Major Elements)

Case	1	2	3	4	5	6	7	8	9	10	11
$He^4(-2)$	1.1	1.0	1.1	4.1	3.4	4.4	1.5	1.3	1.6	1.1	4.5
$C^{12}(-3)$	1.7	1.1	1.9	8.6	4.3	9.9	3.2	1.6	3.7	1.3	6.6
$O^{16}(-3)$	3.4	0.9	4.5	31.	10.	39.	12.	3.9	15.	21.	21.
$N^{14}(-4)$	9.8	9.9	9.8	28.	29.	28.	11.	11.	11.	9.9	30.
$C^{13}(-5)$	11.	9.6	11.	49.	40.	53.	19.	15.	20.	11.	55.
Ne+Mg (-3)	1.3	0.4	1.6	9.8	3.7	11.	3.7	14.	4.3	0.8	6.8
Si-Fe (-4)	5.1	15.	6.6	37.	15.	47.	14.	5.8	18.	3.1	26.
Y_Z (-3)	8.0	3.5	9.7	57.	23.	69.	21.	8.7	26.	5.6	40.
Y_{He}/Y_Z	1.3	2.8	1.1	0.7	1.5	0.6	0.7	1.5	0.6	1.9	1.1

The data of Table 1 and Table 2 are from Chiosi and Matteucci (1983)

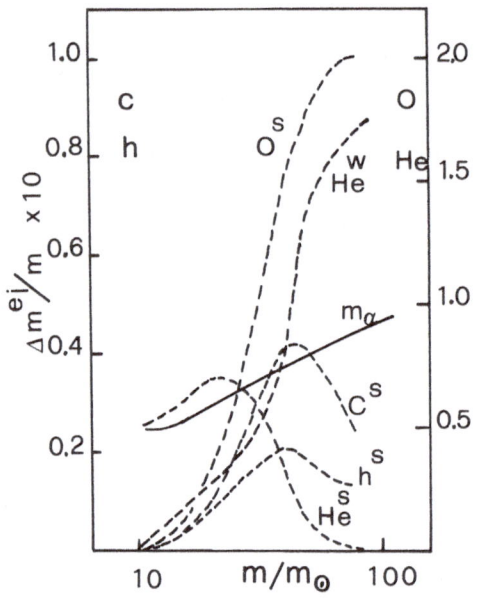

Fig. 3   Elemental abundances in
          the ejecta from massive
stars in occurrence of significant
mass loss by stellar wind

d)   The δY/ δZ Problem

It has long been suggested that the helium abundance Y is correlated to
the heavy element abundance Z. From observational determinations of chemical abundances in HII regions of the galaxy and Magellanic Clouds,
Peimbert and Torres-Peimbert (1976, 1977) and Torres-Peimbert and Peimbert
(1977) estimated δY/δZ = 3. This ratio, which has been often used to put
constraints on chemical models of galaxies and to determine the primordial
He abundance, is still a matter of contention. On one side, recent observational determinations for a group of dwarf irregular galaxies having the
lowest metallicity, seem to indicate a large scatter in Y which somewhat
weakens the above correlation (Kunth and Sargent, 1983). On the other
side, the high value found by the mexican group has so far eluded simple
explanations. In th context of the simple closed model of galactic evolution, we write

$$\delta Y/\delta Z \ = dY/dZ = Y_{He} \ (1-R) \ / \ y_Z \ .$$ (18)

Therefore , the ratio can be used to test theoretical predictions for y_{He}
to y_Z ratio, and in turn to test the validity of assumptions relative
to the stellar mass function and stellar nucleosynthetic input. Various
estimates of the ratio aimed at testing the effect of diverse hypotheses
have been made over the years. They range from 0.7 to 3.3. Arnett's (
1978) data and Salpeter's (1955) mass function lead to 0.7 (Mallik, 1980).
Heavy mass loss by stellar wind and /or a much steeper mass function in
the domain of massive stars give 1.8 (Chiosi and Caimmi, 1979) or 2 to
3 (Chiosi, 1979), (Mallik, 1981), (Serrano and Peibert, 1981). Maeder
(1981b) improving upon the effect of mass loss from massive stars lowered

the ratio to about 1. Chiosi and Matteucci (1982) incorporating the effect of low and intermediate mass stars as well as mass loss in massive stars raised the ratio to about 2 and tried to constrain the stellar mass function. More recently, Maeder (1983a,b) revised the whole problem and estimated the value of 1.3. He used the mass function of Miller and Scalo (1979) below 25 $m_\odot$ and of Garmany et al. (1982) above, and adopted the same nucleosynthesis scenario as in Chiosi and Matteucci (1982). If we rely on Maeder's estimate, it seems that the observational determination of 3 can be hardly obtained with current stellar models. An obvious way out of the difficulty is to lower the amount of heavy elements ejected by massive stars by playing with the mass limit above which core collapse produces black holes. This limit in fact is still uncertain. Following Maeder (1984) we can assume as a first approximation that only the stellar winds contribute to the nucleosynthetic enrichment while all remaining material before core collapse is trapped for ever into a black hole. In this case y_{He}/y_Z increases from 1.3 when the limit is 150 $m_\odot$ to 3 when the limit is lowered to 20 $m_\odot$. Although this kind of exercise is highly speculative, it seems worth pursuing along this line of thought as according to current scenarios for the evolution of massive stars in presence of mass loss enormous amounts of matter are lost in the wind before the completion of the core He-burning phase (Chiosi, 1981; Maeder, 1981a,1983,1984). The question arises whether these remnants composed of almost bare C-O cores would evolve toward the final stage (collapse and explosion) in the same way they would if still buried in the original star. Simply we do not know it. Furthermore, we must recall that when comparing $\delta Y/\delta Z$ to the prediction of the simple closed model as in eq. (18) we drastically approximate the complexity of the problem of galactic evolution. In reality, $\delta Y/\delta Z$ is a complicated function of $\Psi(t)$, $f(t)$, and to some extents of the $E_i(t)$'s (the true enrichment rates). This problem was firstly addressed by Hacyan et al. (1976) and Serrano and Peimbert (1981). These latter investigated

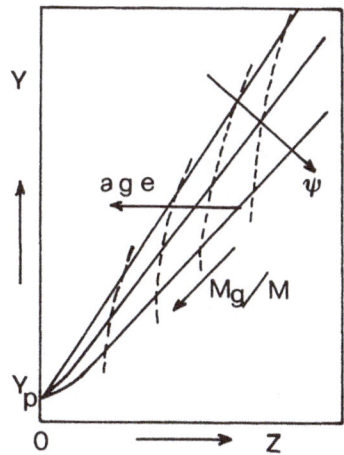

the effect of the rate of star formation on Y(Z). Adopting a closed model of chemical evolution and keeping constant all other parameters, they found that Y(Z) varies with increasing star formation rate as schematically shown in Fig. 4. The effect of infall of primordial gas at various rates starts now being taken into account. To illustrate the point, we study the dependence of (dY/dZ) given by eq. (11) on λ , the ratio of infall to star formation rate . This is performed assuming the yields of He and metals of Maeder (1983,1984), y_{He} = 0.0189 and y_Z= 0.0143 and two typical abundances of He and metals. The results are displayed in Fig. 5 for the case of an evolved system (Y = 0.28 and Z = 0.02) and an unevolved one (Y = 0.25 and Z = 0.005). The abundances for this latter case are estimated averaging the observational determinations of Kunth and Sargent (1983). The dependency of (dY/dZ) on Y_p, the primordial helium

Fig. 4   Schematic dependence of Y(Z) on age, rate of star formation and fractionary gas content

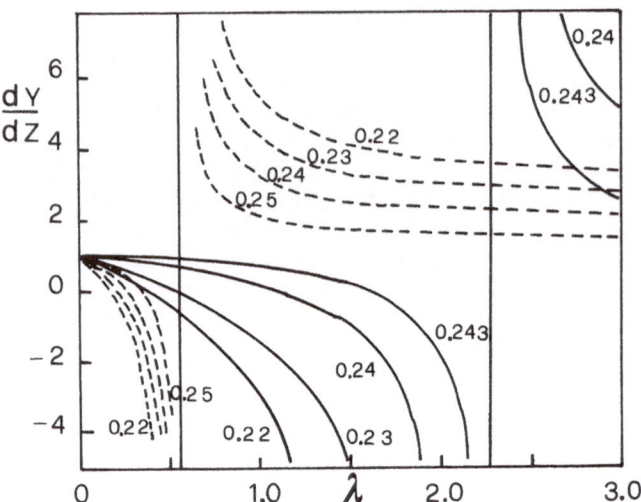

Fig. 5   *The dependence of dY/dZ on the ratio of infall to star formation rates. Full lines show the case Y = 0.25 and Z = 0.005, whereas dashed lines indicate the case Y = 0.28 and Z = 0.02. The value for the abundance of primordial helium is annotated along each curve*

abundance is also shown. It is soon evident that dY/dZ strongly depends upon λ and Y_p. Since the relative importance of infall with respect to star formation rate may vary from galaxy to galaxy and within any given galaxy it may change with time, we expect dY/dZ not to be constant. In addition to this, for any given λ , the response of the system depends on its chemical state (compare the case with Z = 0.005 to that with Z = 0.02). Several aspects of the problem are clarified by this preliminary analysis. Firstly, the observational $\delta Y/\delta Z$ should in any case be compared with the locus of formal solutions of eq. (11) in the Y(Z) plane. $\delta Y/\delta Z$ can be safely related to dY/dZ only in the case of $\lambda = 0$ and y_{He} and y_Z constant in the course of evolution. In all other cases, this straightforward procedure might be dangerous especially if aimed at deriving the value of primordial helium, as dY/dZ may vary with time and from galaxy to galaxy. Secondly, the large scatter in Y(Z) plane found by Kunth and Sargent (1983) for dwarf irregular galaxies may not be entirely of observational nature, but it may also reflect a real scatter in dY/dZ among these glaxies. These objects in fact are likely to suffer from intense episodes of stellar activity and perhaps unusual rates of infall (Gordon and Gottesman, 1981). If so, in addition to possible variation in the yields, λ may span a wide range of values in this type of galaxies. In the light of this, it seems that th use of $\delta Y/\delta Z$ to derive Y_p and to infer constraints on the chemical yields, and in turn on the stellar mass function and stellar evolution input, is still premature if not hopelessly inconclusive.

3. The Chemical History Of The Galactic Disk

In recent years, evidence has been accumulating that the nucleosynthesis production of several important elements, such as carbon, oxygen,
nitrogen and iron has varied during the life of the galaxy. Sneden et al.
(1979) have shown that $|O/Fe|$ in the sun is different from that in the
metal poor halo stars, indicating that the first stars in the disk formed
out of gas already O-rich relative to Fe. Furthermore, Clegg et al.
(1981), analysing a sample of unevolved F and G type disk stars found that
while the abundances of C and N closely follow that of Fe, the abundance
of O varies much more slowly than Fe. The theoretical interpretation of
these observational trends has been recently studied by Twarog and Wheeler (1982) in the context of the standard nucleosynthetesis scenario,
which sees all heavy elements as produced by massive stars, and of the
simple model of galactic evolution with constant rates of star formation
and mass accretion. They found that
i) Clegg's et al. (1981) data were consistent with the notion of yield
ratios constant with time;
ii) the products of Arnett's (1978) stellar yields into Miller's and Scalo's (1979) mass function leads to overproduction of O and C relative to
Fe;
iii) the high $|O/Fe|$ ratio for the most metal poor stars implies that O
was truly overproduced during the early stages of galactic evolution.
However these conclusions may be subjected to some criticism at least for
the following reasons: a) more up to date yields are now available; b)
Clegg's et al. (1981) data suggest indeed yields and yield ratios variable with time; c) the simple model of galactic evolution with constant
rates of star formation and mass accretion is perhaps a too crude approximation of reality. Prior to any other consideration, several aspects of
problem must be clarified. A potential source of ambiguity in the comparison of theoretical results with observational data are the absolute
abundances of various elements within the sun. We adopt Cameron's (1973)
compilation to set the abundances of C, O and Fe in the sun, and we assume
that the solar fraction of hydrogen (by mass) is 0.73. Thus the abundances relative to hydrogen of the three elements of interest here are

$$|C/H|_\odot = -2.22 \qquad |O/H|_\odot = -1.83 \qquad |Fe/H|_\odot = -2.57$$

in the usual notation. According to this, the data of Clegg et al. (19
81) are normalized to the above solar abundances and the following least
squares fits of the data are derived

$$|C/H| \; = 0.122 + 0.795 \, |Fe/H| \qquad\qquad (19)$$

$$|O/H| \; = 0.046 + 0.450 \, |Fe/H| \qquad\qquad (20)$$

for $|Fe/H| > -1$. Accordingly, $|O/Fe| = 0.046 - 0.65 \, |Fe/H|$. Further
constraint on theoretical models is provided by the $|Fe/H|$ versus
age relationship of Twarog (1980). This , however, has been adapted to

the assumption that the disk is 13×10^9 yr old and that the age of the
sun is 4.5×10^9 yr. To follow up on the idea that the yields of C, O and
Fe might have varied during galactic evolution due to variations in the
stellar mass function and/or stellar nucleosynthesis, we first calculate
yields appropriate for a population of stars formed out of primordial
gas. This is not an easy task as little is known of the properties of
such stars and a considerable amount of extrapolation is needed. There
are reasons for believing that very massive stars ($m > 100\ m_{\odot}$) could form
in the early evolution of galaxies (Carr et al., 1983) with a mass spectrum skewed in favour of these. Nevertheless, low mass stars could also
have formed as indicated by the study of Palla et al. (1983). In the light
of this, we do not commit ourselves with any particular mass range but
allow for the formation of stars all over the possible values of mass.
Since the mass spectrum is not known, we simply assume a power law of the
mass and test various choices for the slope and mass boundaries. As the
evolution of stars with little or no metals turns out not to be greatly
different from normal evolution, the following scheme has been adopted:
i) low and intermediate mass stars are supposed to evolve like those of
normal chemical composition, whereas their chemical properties are inferred from the case $n_R = 0.33$ and $\alpha_R = 0$ of Renzini and Voli (1981).
ii) massive stars are supposed to evolve at constant mass during the core
H and He burning phases, whereas Arnett(1981) data are used to derive the
composition of the ejecta at final stage.
iii) very massive stars ($100\ m_{\odot} < m < 300\ m_{\odot}$) are likely to evolve during
core H and He-burning phases undergoing severe mass loss by stellar wind,
and are known to suffer electron pair instability in later phases which
leads to supernova explosion (El Eid et al., 1983; Ober et al., 1983;
Carr et al., 1983). The nucleosynthesis data are taken from El Eid et al.
(1983) and Ober et al. (1983). For more details on the evolution of very
massive stars see Bond (1983), whereas for a more exhaustive justification
of the above assumptions see Chiosi (1983). With the aid of the nucleosynthesis results recalled above, Chiosi and Matteucci (1983) have computed a large grid of yields for various stellar mass functions. Two cases
are reported in Table 4 for purposes of illustration.

Table 4

(Net yields for stars with primordial composition)

$x(\phi)$	ζ	m_1	m_u	He(-2)	C(-3)	O(-2)	Si-Fe(-3)	α_R
1.35	0.20	0.01	200	1.40	2.14	1.27	1.06	0.
1.35	0.44	0.10	200	4.13	6.30	3.74	3.12	0.

Equations (1) to (3) complemented by assumptions for the law of star formation, infall and gas loss are used to study the chemical history of the
galactic disk. Following Chiosi (1980), the rate of mass accretion is

$$f(t) = (M - M_o)\ \tau^{-1}(\ 1 - \exp(-t_g/\tau))^{-1}\exp(-t/\tau)\ ,\qquad (21)$$

where τ is the time scale of mass accretion, t_g the age of the disk (13×10^9 yr), M and M_o are the present and initial mass respectively. This formulation is particularly suited as it can be related to the dynamical nature of the process of disk formation. The rate of star formation is assumed to be proportional to the gas mass

$$\Psi = \eta \ M_g \tag{22}$$

where η is a free parameter. Finally no galactic wind is supposed to occur $(w(t) = 0)$. In addition to this, the following relation is added to eqs. (1) to (3) and (21) and (22)

$$y_i = A_i \ X_i + B_i \tag{23}$$

which constitutes the major difference of the present formulation with respect to standard ones. Given M and M_o the parameters τ and η follow from imposing that the model reproduces the present values of M_g/M, f, Ψ and ratio of the mean past to present star formation rate. We assume $M = 100 \ m_o/pc^2$ (Innanen, 1973), $M_g = 10 \ m_o/pc^2$ (Pagel and Patchett, 1975), $< \Psi >/\Psi = 2.5$ (Tinsley, 1976), $M_o = 10 \ m_o/pc^2$ (this parameter does not affect very much the final results), $f = 2 \ m_o/pc^2/10^9$ yr (Oort, 1970) and $\Psi = 5 \ m_o/pc^2/10^9$yr (Miller and Scalo, 1979). With the aid of these values, we derive $\eta = 0.4$ and $\tau = 4 \times 10^9$ yr. To determine the coefficients A_i and B_i of relation (22) we proceed as follows. We start imposing that at the present time the yields and abundances are those typical of young stars. For these we assume $X_C = 4.5(-3)$, $X_O = 1.1(-2)$ and $X_{Fe} = 2.0(-3)$ (Pagel and Edmunds, 1981). The correspondent yields are those of case 9 of Table 2. The initial abundances are obtained from the least squares fits of the data for $|Fe/H| = -1$, these are $|C/H| = -0.67$ and $|O/H| = -0.40$. The correspondent yields are considered as free parameters of the problem. When a particular choice is made, the coefficients A_i and B_i are fixed and the model is calculated. This procedure is repeated by changing the initial yields till when a guess model is derived which reasonably matches all major observational constraints. This guess model provides us with $y_i(X_i)$ relationships which are then adjusted till when the calculated abundances are in satisfactory accord with relations (19) and (20). The theoretical X_i/H are derived with the aid of the relation $X_H = 0.77 - 0.005 \ t$ of Twarog and Wheeler (1982), where t is in billion of years. We find that the combination of case 9 of Table 2 with case 2 of Table 4 gives very satisfactory results. These are displayed in Figs. 6 and 7 together with the observational data. The variation of the yields over the disk life is shown in Fig. 8. In particular, the model predicts a present metal yield $y_Z = 0.007$, a ratio $< \Psi >/\Psi = 1.5$ and obeys the relation $|O/Fe| = -0.56 \ |Fe/H| + 0.07$. The observational data seem to require a more abundant production of C, which can be perhaps obtained using the data of Renzini and Voli (1981) for $\alpha_R = 0$ all over the galaxy life (dotted line in Fig. 6). Finally, we like to stress that had we assumed constant yields, we would have obtained the same results as in Twarog and Wheeler (1982). The most important result of this analysis lies in the kind of dependence upon age that we have found for the yields in order to make theoretical models compatible with the observational data. Whether

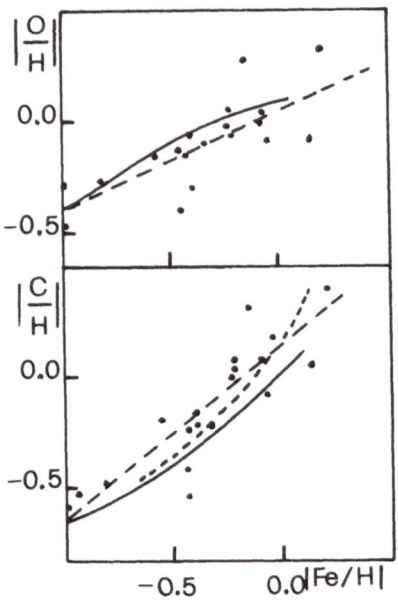

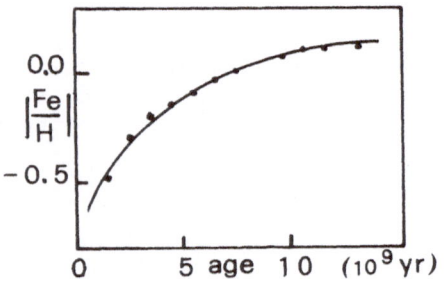

Fig. 7 Fe vs age relation for the
        model with variable yields.
The dots are the data of Twarog (1980)

Fig. 6 O vs Fe (top) and C vs Fe
        (bottom) relations for the
model with variable yields.   The
dotted line is the case α_R = 0.
The dots are the observational data of Clegg et al. (1981)

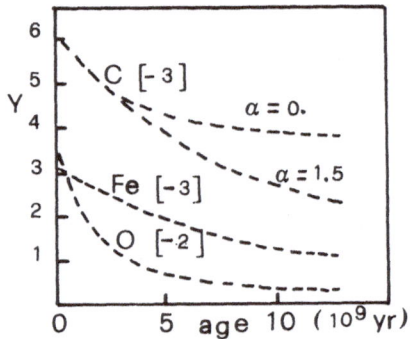

Fig. 8 Yields of C, O and Fe as a
        function of the age for the
model with variable yields. Two values of α_R are indicated

this is due to variations in the stellar mass function or in the stellar
nucleosynthesis data or both cannot be easily disentangled. According to
the results of Fig. 8, even for the present stellar population the yields
of C, O and Fe are expected to be a factor of 1.6, 4.1 and 1.6 lower than
those of Table 2, in the order. It is an easy matter to recognize that
this would require a more substancial variation than a simple adjustement
of the mass function in the domain of massive stars. In fact the results
do not change very much when adopting the mass function of Garmany et al.
(1982). Rather, our analysis seems to indicate that C, O and Fe should be
produced in proportions somewhat different from those predicted by the
nucleosynthesis scenario that we have adopted. Looking at the yields of
the three elements in more detail, C is ejected in almost equal amounts by
low to intermediate mass stars and massive stars, due to the weighting of
mass function toward lower masses, whereas O and Fe are exclusively ejected by massive stars. To lower the yield of O by a factor of 3 to 4 one

could argue that only massive stars in the mass range $10 < m < m^*$ explode, whereas more massive objects collapse into black holes. This is reasonably justified by the uncertainty with which both the shock strength and location of the mass cut (which determines the remnant mass) are known (Woosley et al., 1984). We can perform this excercise by making the extreme assumption that all material which is not lost in the wind and remains in the star at time of explosion collapse into a black hole. We find that m^* determined by the request of a lower oxygen production ranges from 40 $m_\odot$ to 50 $m_\odot$. The correspondent m^*'s required by the decrease of C and Fe production are relatively higher but still compatible with the previous estimate. If so, explosive nucleosynthesis leaving a neutron star of 1.4 $m_\odot$ should be limited to stars in the mass range 10 to 40 -50 $m_\odot$. Amazingly and puzzling enough, this mass limit coincides with the mass above which mass loss by stellar wind during core H and He burning phases becomes a dominant process in a massive star's life. Any further speculation about the possibility that huge mass loss by stellar wind may eventually affect the kind of final explosion is premature. As for the dependence of y_i's upon age shown in Fig. 8, this could be understood as due to the combined effects of variations in the stellar mass function and nucleosynthesis input. In fact, a result of our model is that the fraction ζ of massive stars decreases from 0.44 to 0.25, whereas the lower and upper limits, m_l and m_u, decrease from 0.1 to 0.02 $m_\odot$ and from 200 to 100 $m_\odot$, respectively. This means that the stellar mass function found to be consistent with the observations was in the past moderately skewed toward more massive stars. Systematic variations in the nucleosynthetic make up with the star's chemical composition and/or galaxy age are difficult to quantify due to the lack of sufficient evolutionary models, even if they are likely to occur.

4.   The Chemical Evolution Of Dwarf Irregular Galaxies

The interpretation of chemical properties of irregular galaxies (dwarf and magellanic) has so far eluded simple schematizations. These galaxies, among which blue compacts play a particular role, are characterized by active or very active star formation at the present time, blue colours, high gas contents and low metallicities. However, exceptions like IZw18 are known to exist which disobey the general rule. In fact, they seem to posses too low a metallicity for their estimated gas to total mass ratios. As already suggested long time ago by Searle et al. (1973) dwarf irregular galaxies are likely to suffer from sporadic though very intense episodes of stellar activity. The nature of the process causing bursts of star formation is not yet clear, even if current modelling indicates both dynamical stimulation among interacting systems (Larson and Tinsley, 1978) and /or stochastic self propagating star formation mechanism of Gerola et al. (1981) as plausible candidates. This latter in particular appears to be very appealing as not all galaxies having bursts of star formation are interacting systems. In fact, Gerola's et al. (1981) mechanism predicts that the bursting mode of star formation is a natural consequence of the stochastic star formation process when occuring in systems of small size. In the light of this , it has been suggested that star formation in bursts can account for the large scatter in several fundamental properties (colours, luminosities, metallicities, gas contents) existing among irregular galaxies. However, it is an easy matter to show that with the exception of colours and luminosities, the scatter in metal content and fractionary gas content cannot be explained by the bursting mode of star formation alone, but more complex models have to be devised. This problem has been studied to some extents by Matteucci and Chiosi (1983) who collected from various sources in the literature the basic data for some 45 irregular galaxies and examined them in the light of Gerola's et al. (1981) theory. Amongst others, the Z versus Log M_g/M relation is customarily used to test models of chemical evolution. This is shown in Fig. 9 (full dots). This diagram is constructed assuming the value of 100 Km/sec/Mpc for the Hubble constant, the relation Log Z = 1.42 + Log N(O)/N(H) of Lequeux et al. (1979) to transfer the observational O abundance into metallicity, and $M_g = 1.3\ M_{HI}$, where M_{HI} is the atomic hydrogen mass from 21 cm line observations whereas the factor 1.3 is there to take helium into account. A detailed discussion of the uncertainty affecting all these quantities can be found in Matteucci and Chiosi (1983). Finally, it is worth recalling that the large HI haloes surrounding blue compact and irregular galaxies (Gordon and Gottesman,1981)

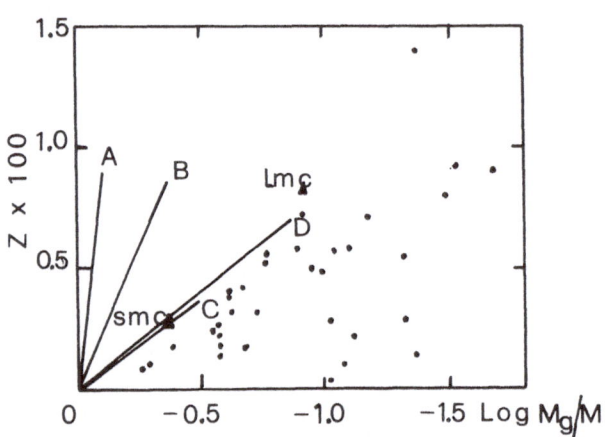

Fig. 9 Z vs Log M_g/M relation for irregular galaxies. The full dots and triangles are the observational data. The solid lines visualize the models discussed in the text

can be perhaps taken as indicating evidence of infall of primordial gas. Theoretical models based on eqs. (1) to (3), relation (21) for the rate of infall and the following prescription for the rate of star formation

$$\Psi(t) = \nu'\ s(t)\ M_g(t) \qquad\qquad (24)$$

have been calculated for various stellar mass functions. No instantaneous recycling approximation was used. s(t) of eq. (24) represents the fraction of the system undergoing star formation in bursts as derived from the numerical experiments of Gerola et al. (1981), whereas ν' is an efficiency parameter to be fixed by the comparison with the observational data. In the present formulation, ν' is equivalent to $\nu\ M(t_g)$ of Matteucci and Chiosi (1983). The results of these numerical models are shown in Fig. 9. As expected, the slope of Z vs Log M_g/M relation depends mostly upon the mass function through the parameters x and ζ , whereas a secondary role is played by ν' or ν and τ . Star formation in bursts does not change this rule as models are expected to move in jumps along a given line determined by the mass function and stellar input, that is the metallicity yield. The cases shown in Fig. 9 are calculated assuming x = 1.35 and ζ = 0.50

(A), x = 1.35 and ζ = 0.25 (B), x of Serrano-Lequeux with ζ = 0.44 and
ν = 4 (C), the same but with ν = 10 (D). From a detailed comparison of the
model results with the properties of SMC and LMC, Matteucci and Chiosi
(1983) fixed ν = 10, τ = 4 to 6 billion years and the yield of metal of
case C and/or D as the best parameters of the problem. It is well evident
that in no case the scatter shown by the observational data can be reproduced by the bursting mode of star formation alone but other physical
processes must be at work. These likely are: i) systematic variations of
the yield of metals due to both different mass function and stellar evolution properties from galaxy to galaxy; ii) variations in the relative
importance of the star formation rate versus infall rate; iii) variations
in the relative importance of the star formation rate versus the galactic
wind rate; or any combination of the three . These effects possibly super-
posed to the bursting mode of star formation. The effect of each of the

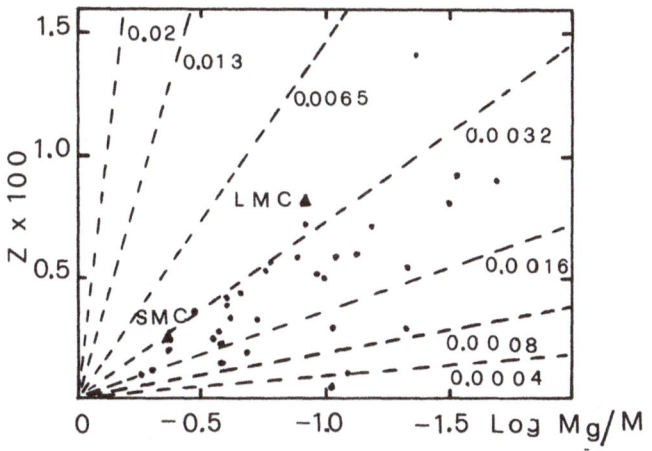

Fig. 10 Theoretical Z vs Log M_g/M relationships for models with various
         yields of metals and bursting mode of star formation. These mo-
dels are calculated with f(t) = 0 and w(t) = 0. y_z is annotated along
each curve. The dots and triangles are the observational data

above alternatives on the Z versus Log M_g/ M relationships has been analyzed separately on models with star formation occuring in bursts as in
the stochastic star formation mechanism of Gerola et al. (1981). The results are shown in Figs. 10, 11, and 12 in the order. The effect of different infall rates and galactic wind rates are evaluated on models with
the same yield of metals (y_z = 0.0032) for simplicity. Which of the three
alternatives has to be preferred cannot be assessed at the present time.
While there are not compelling reasons for believing that the yields of
metals can vary from galaxy to galaxy by the factor of ten which is needed
to encompass the whole range of metallicities, inflow and galactic wind
are likely to occur at rates which may vary from galaxy to galaxy as in

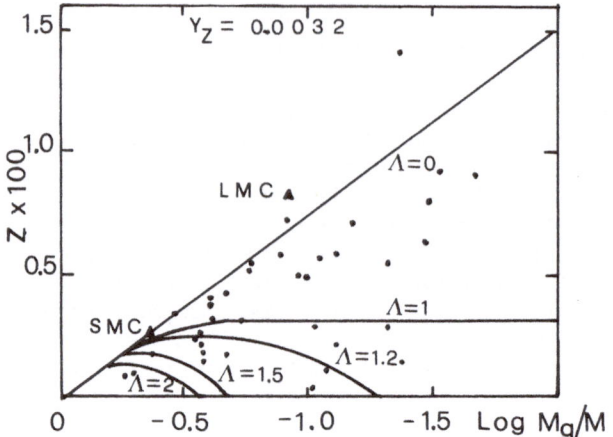

Fig. 11   Theoretical Z vs Log M_g/M relationships for models with infall
          and bursting mode of star formation. The rate of star formation
is assumed to be proportional to the infall rate, $\Psi(t) = f(t)/(1-R)/\Lambda$.
Only the case with $y_Z = 0.0032$ is shown. Dots and triangles are the
observational data

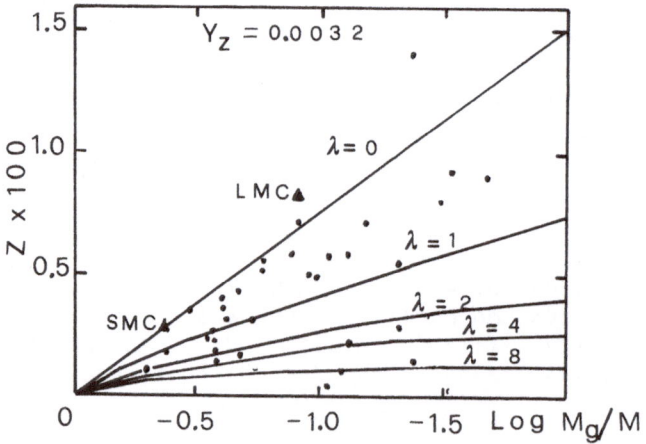

Fig. 12   Theoretical Z vs Log M_g/M relationships for models with galactic
          wind and bursting mode of star formation. The rate of gas loss
is assumed to be proportional to the rate of star formation, $w(t) = (1-R)$
$\lambda \Psi(t)$. Only the case with $y_Z = 0.0032$ is shown. Dots and triangles are
the observational data

our numerical experiments.

This work has been financially supported by the National Council of Research of Italy (C. N. R. - G. N. A.)

References

Arnett, W.D., 1978, Astrophys. J. 219, 1008
Boisse, P., Gispert, R., Coron, N., Wynbergen, J.J., Serra, G., Ryter, G., Puget, J.L., 1981, Astron. Astrophys. 94, 265
Bond, J.R., 1984, in "Stellar Nucleosynthesis", Erice Workshop, ed. C. Chiosi and A. Renzini, D. Reidel Publ. Comp., Dordrecht, Holland, this volume.
Bressan, A., Bertelli, G., Chiosi, C., 1981, Astron. Astrophys. 102, 25
Burki, G., 1977, Astron. Astrophys. 57, 135
Burki, G., 1978, Astron. Astrophys. 62, 159
Cameron, A.G.W., 1973, Space Sci. Rev. 15, 121
Carr, B.J., Bond, J.R., Arnett, W.D., 1983, Astrophys. J. in press
Chiosi, C., 1979, Astron. Astrophys. 80, 252
Chiosi, C., 1980, Astron. Astrophys. 83, 206
Chiosi, C., 1981, in "The Most Massive Stars", ESO Workshop, ed. S. D'Odorico, D. Baade and K. Kjar, p. 27
Chiosi, C., 1983, in "The First Stellar Generations",Frascati Workshop, ed. V. Caloi and F. D'Antona, Mem. Soc. Astr. It. v. 54,p. 251
Chiosi, C., Caimmi, R., 1979, Astron. Astrophys. 80, 234
Chiosi, C., Jones, B., 1983, in "The Origin and Evolution of Galaxies", ed. B.J.T. Jones and J.E. Jones, D. Reidel Publ. Comp., Dordrecht, Holland, p. 197
Chiosi, C., Matteucci, F., 1982, Astron. Astrophys. 105, 140
Chiosi, C., Matteucci, F., 1983, in preparation
Chiosi, C., Nasi, E., Sreenivasan, S.R., 1978, Astron. Astrophys. 63, 103
Clegg, R.E.S., Lambert, D.L., Tomkin, J., 1981, Astrophys. J. 250, 262
El Eid, M.F., Fricke, K.J., Ober, W.W., 1983, Astron. Astrophys. 119, 54
Garmany, C.D., Conti, P.S., Chiosi, C., 1982, Astrophys. J. 263, 777
Gerola, H., Seiden, P.E., Schulmann, L.S., 1980, Astrophys. J. 242, 517
Greggio, L., Renzini, A., 1983, Astron. Astrophys. 118, 217
Gordon, D., Gottesman, S.T., 1981, Astron. J. 86, 2
Hacyan, S., Dultzin-Hacyan, D., Torres-Peimbert, S., Peimbert, M., 1976, Rev. Mex. Astron. Astrofis. V. 1, 355
Iben, I. Jr., Renzini, A., 1983, Ann. Rev. Astron. Astrophys. in press
Iben, I.Jr., Truran, J.W., Astrophys. J. 220, 980
Innanen, K.A., 1973, Astrophys. Space Sci. 22, 393
Kunth, D., Sargent, W.L.W., 1983, preprint
Larson, R.B., Tinsley, B.M., 1978, Astrophys. J. 219, 46
Lequeux, J., 1979, Astron. Astrophys. 80, 35
Lequeux, J., Peimbert, M., Rayo, J.F., Serrano, A., Torres-Peimbert, S., 1979, Astron. Astrophys. 80, 155
Lynden-Bell, D., 1975, Vistas in Astronomy 19, 299
Maeder, A., 1981a, in "The Most Massive Stars", ESO Workshop, ed. S. D'Odorico, D. Baade and K. Kjar, p. 173
Maeder, A., 1981b, Astron. Astrophys. 101, 385
Maeder, A., 1984, in "Stellar Nucleosynthesis", Erice Workshop, ed. C.

Chiosi and A. Renzini, D. Reidel Publ. Comp., Dordrecht, Holland, this volume.

Maeder, A., 1983, in "Primordial Helium", ESO workshop, ed. P.A. Shaver, D. Kunth and K. Kjar, p. 89

Mallik, D.C.W., 1980, Astrophys. Space Sci. 69, 133

Mallik, D.C.W., 1981, J. Astrophys. Astron. 2, 171

Matteucci, F., Chiosi, C., 1983, Astron. Astrophys. 123, 121

Miller, G.E., Scalo, J.M., 1979, Astrophys. J. Suppl. 41, 513

Nomoto, K., 1980, in "Proceedings of the Texas Workshop on Type I Super-novae", ed. J.C. Wheeler, Univ. of Texas Press, Austin, p. 164

Nomoto, K., 1984, in "Stellar Nucleosynthesis", Erice Workshop, ed. C. Chiosi and A. Renzini, D. Reidel Publ. Comp., Dordrecht, Holland, this volume.

Ober, W.W., El Eid, M.F., Fricke, K.J., 1983, Astron. Astrophys. 119, 61

Oort , J.H., 1970, Astron. Astrophys. 7, 381

Palla, F., Salpeter, E.E., Stahler, S.W.,1983, preprint

Pagel, B.E.J., Edmunds, M.G., 1981, Ann. Rev. Astron. Astrophys. 19, 77

Pagel, B.E.J., Patchett, B.E., 1975, Monthly Notices Roy. Astron. Soc. 172, 13

Peimbert, M., Torres-Peimbert, S., 1976, Astrophys. J. 203, 581

Peimbert, M., Torres-Peimbert, S., 1977, Monthly Notices Roy. Astron. Soc. 179, 217

Renzini, A., Voli, M., 1981, Astron. Astrophys. 94, 175

Salpeter, E.E., Astrophys. J. 121, 161

Searle, L., Sargent, W.L.W., Bagnuolo, W.G., 1973, Astrophys. J. 179, 427

Sneden, C., Lambert, D.L., Whitaker, R.W., 1979, Astrophys. J. 234, 964

Serrano, A., 1978, PH. D. Thesis

Serrano, A., Peimbert, M., 1981, Rev. Mex. Astron. Astrofis. 5, 109

Sugimoto, D., Nomoto, K., 1980, Space Sci. Rev. 25, 155

Talbot, R.J., Arnett, W.D., 1971, Astrophys. J. 170, 409

Talbot, R.J., Arnett, W.D., 1973, Astrophys. J. 197, 551

Talbot, R.J., Arnett, W.D., 1975, Astrophys. J. 233, 888

Tinsley, B.M., 1976, Astrophys. J. 208, 797

Tinsley, B.M., 1980, in "Fundamentals of Cosmic Physics", v. 5, p. 287

Torres-Peimbert, S., Peimbert, M., 1977, Rev. Mex. Astron. Astrofis. v. 5, 181

Twarog, B.A., 1980, Astrophys. J. 242, 242

Twarog, B.A., Wheeler, J.C., 1982, Astrophys. J. 261, 638

Vanbeveren, D., Olson, G.L., 1980, Astron. Astrophys. 81, 228

Wheeler, C.J., 1982, in "Reports on Progress in Physics", in press

Woosley, S.E., Weaver, T.A., 1983, in "Supernovae: A Survey of Current Research", ed. M.J. Rees and R.J. Stoneham, D. Reidel Publ. Comp., Dordrecht, Holland, p. 79

Woosley, S.E., Axelrood, T.S., Weaver, T.A., 1984, in "Stellar Nucleo-synthesis", Erice Workshop, ed. C. Chiosi and A. Renzini, D. Reidel Publ. Comp., Dordrecht, Holland, this volume.

Yokoi, K., Takahashi, K., Arnould, M., 1983, Astron. Astrophys. 117, 65

DISCUSSION

WOOSLEY: How do you expect your results to change if only a fraction of iron comes from massive stars and the bulk comes from Type I Supernovae produced by white dwarfs provoked into explosion by accretion?

CHIOSI: There is not an easy answer to this question as the yield of iron would depend on the mass function holding in that range of mass, on the frequency of binary systems having appropriate parameters for producing Type I Supernovae, and finally on the amount of iron ejected in this case. Too many uncertainties for a simple prediction.

OBER: If mass loss reduces significantly the size of the C-O core, how large would be the effect over the mass range for possible supernova explosions of massive stars?

CHIOSI: Mass loss by stellar wind in massive stars may significantly affect the mass size of the C-O core say above initially 50 $m_\odot$ under the rates currently estimated. Since this would actually occur during the so called WR phase (late or whole core He-burning phase) for which both the mass loss rates and duration are not very well known, it is rather difficult to derive quantitative estimates of this effect.

EDMUNDS: (To S. Woosley) If the bulk of iron group comes from low mass stars, what sets the neutron excess before NSE? i.e. are there likely to be overmetallicity versus odd/even effects in the iron group?

WOOSLEY: Electron capture occuring during nuclear statistical equilibrium tends to erase memory of the previous metallicity. Thus the odd/even effects in the iron group are not very sensitive to the original metallicity of the star.

GALLAGHER: I just wanted to comment that on your graph of a standard stochastic model the star formation rate seems to vary rapidly by large factors ($\gtrsim 4$). From studies of stellar populations it appears likely that most galaxies at typical luminosities ($\sim$ LMC) in the the present DIG samples have considerably lower short term fluctuations in star formation rates, perhaps a factor of 2 or less. This might also reduce the ability of stochastic models to explain the scatter in metallicity.

THE CHEMICAL EVOLUTION OF SPIRAL GALAXIES: THE GALAXY, M31,
M33, M83, and M101

A.I. DIAZ[1,2]   and   M. TOSI[3]

[1] Royal Greenwich Observatory, Herstmonceux, U.K.
[2] Astronomy Centre, University of Sussex, U.K.
[3] Osservatorio Astronomico Universitario, Bologna, Italy

One of the possible applications of stellar nucleosynthesis models
is the study of the behaviour of chemical elements during the evolution
of galaxies. The combination of correct models of stellar nucleosynthesis and of galactic evolution should allow to reproduce the observed
properties of each element, in particular the absolute value and the
radial distribution of the element abundance.

In this study, we have computed chemical evolution models of the
Galaxy and four nearby spirals (M31, M33, M83 and M101) in order to
reproduce the oxygen abundances derived from the observations of HII
regions in the disks of these galaxies.

We have assumed for all these spirals the same age (13 Gyr) and the
same initial mass function, IMF, derived for the Galaxy (Tinsley, 1980).

The law of star formation (e.g. constant or exponential in time,
proportional to gas density, etc.) and the infall rate of gas from
outside the disk are the only free parameters of our models (for a
detailed description of the models, see Tosi, 1982, and Diaz and Tosi,
1983, hereafter Paper I and II respectively). Furthermore, the infall
rate f is allowed to range only between 0 (closed models) and an upper
limit derived for each galaxy from its present mass M_{now} through the
expression

$$M_0 = M_{now} - \int_0^{now} f(t)\, dt$$

where M_0 is the initial mass of the galaxy, which obviously cannot be
negative.

383

C. Chiosi and A. Renzini (eds.), Stellar Nucleosynthesis, 383–388.
© *1984 by D. Reidel Publishing Company.*

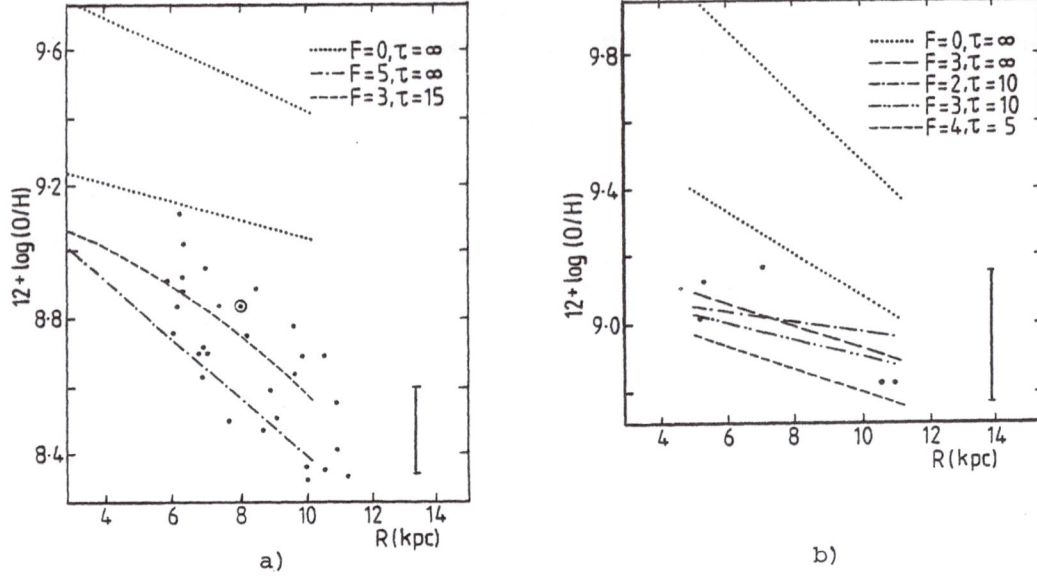

Fig.1. a) For the Galaxy: radial variation of the oxygen abundances as derived from the indicated models. The infall rate is expressed in M☉ kpc^{-2} yr^{-1} and the star formation e-folding time τ in Gyr. The dots represent the observed HII region abundances, whose average error bar is shown in the lower right corner. The solar oxygen abundance is indicated by the usual sun symbol.
       b) Same as a), but for M31.

       Once a law of star formation is adopted, its values $\psi(t)$ for each region of a galaxy are univocally derived from the gas and total mass distributions presently observed in that galaxy, through eq.(4) of Paper II.

       For the stellar contribution to the elements in the interstellar medium during and at the end of the star lives, we have adopted the results of stellar nucleosynthesis models by:
a) Renzini and Voli (1981), for low and intermediate mass stars;
b) Arnett (1978), for massive stars with constant mass evolution or, alternatively, Chiosi and Caimmi (1979), for massive stars evolving with mass loss.

       Figures 1 to 4 show the oxygen abundances resulting from the combination of these models for our five galaxies. Also shown are the oxygen abundances derived from the [OII + OIII] /H$_\beta$ line intensity ratios observed in the HII regions of each galaxy.

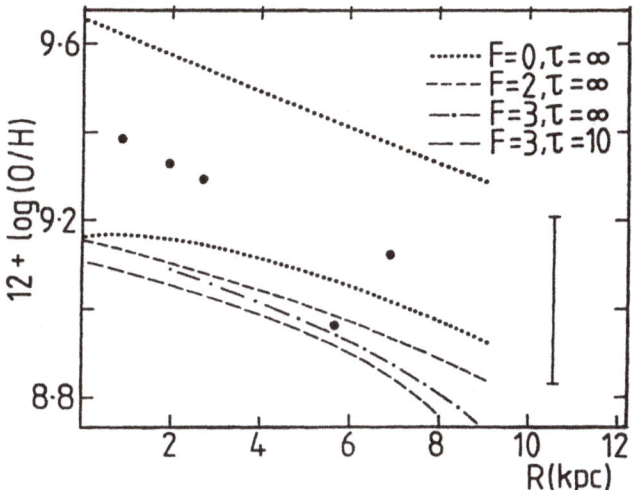

Fig.2. Same as Fig.1, but for M83.

        The upper dotted line of each figure shows the result of combining
Arnett's (1978) yields and a model with constant star formation and no
infall. All the other lines refer to models with Chiosi and Caimmi's
(1979) yields and different choices of the infall rate per unit area F
and the e-folding time τ of an exponentially decreasing star formation
rate. For sake of simplicity, we do not show in the figures the curves
resulting by combining these choices of F and τ with Arnett's models;
however, their location in the plot can be visualized by shifting
upwards as a solid pattern the "Chiosi and Caimmi" set of curves and
letting coincide the two dotted lines.

        It is apparent from the figures that for no galaxy of our sample
the oxygen abundances derived from stellar evolution models without mass
loss are compatible with the observed abundances.

        On the other hand, in the case of the Galaxy (Fig.1a) and M31
(Fig.1b) the stellar yields computed by Chiosi and Caimmi taking mass
loss into account perfectly fit both the observed abundances and
gradient, when inserted in a galactic evolution model with slowly
decreasing star formation rate (τ = 15 Gyr for the Galaxy and 10 Gyr
$\leqslant \tau \leqslant \infty$ for Andromeda) and a constant infall $F = 3.10^{-3}$ $M\odot kpc^{-2} yr^{-1}$ (cfr.
Paper II for details). This model of the Milky Way also reproduces the
age-metallicity relation derived by Twarog (1980) for the solar
neighbourhood and the overall metallicity gradients observed for young
stars (e.g. Panagia and Tosi, 1981) and for old stars (Mayor, 1976).

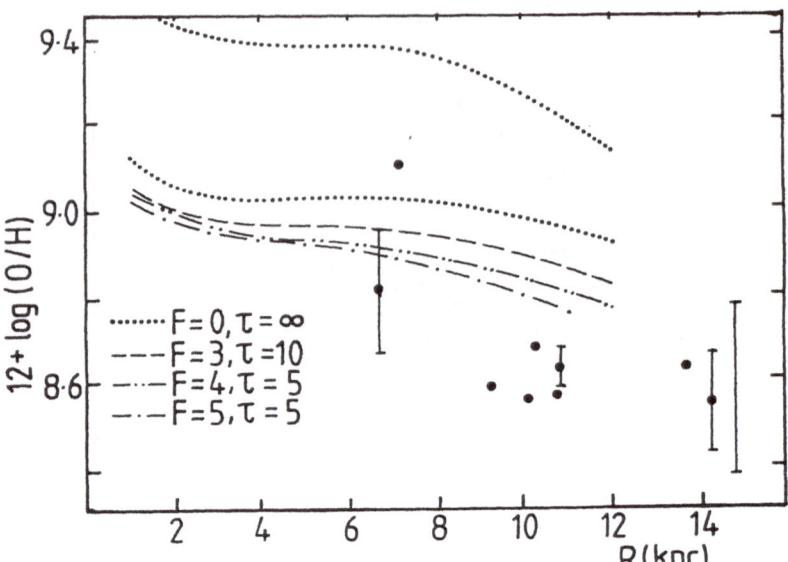

Fig.3. Same as Fig.1, but for M101.

In the case of M83 (Fig.2) and M101 (Fig.3), the model results are consistent with the observations, although the agreement is not as good as in the previous two spirals. For M83, the indication is toward a constant star formation and a low infall ($F \leqslant 2.10^{-3}$ M$\odot$kpc^{-2}yr^{-1}). An improvement of the agreement between models and observations can be obtained by relaxing the assumption that in all our spirals the initial mass function be equal to that derived in the solar neighbourhood and adopting for M83 an IMF with more massive stars and/or with a higher upper mass limit.

For M101, instead, an IMF with less massive stars would give a better agreement with the observational data, since even with the highest possible infall ($F=4.10^{-3}$ M$\odot$kpc^{-2}yr^{-1}) and a rapidly decreasing star formation rate ($\tau =5$ Gyr) the model abundances are higher than (although consistent with) those observed in HII regions.

An IMF with less massive stars should be invoked for M33 (Fig.4) as well, to overcome the discrepancy between models and HII region abundances. The fact that M33 has a very low surface brightness somehow supports a sensible reduction of the massive stars in the IMF. On the other hand, it is also possible that the present disagreement be due to the very high uncertainty (cfr. Paper II) on the gas and total mass distributions available for this galaxy.

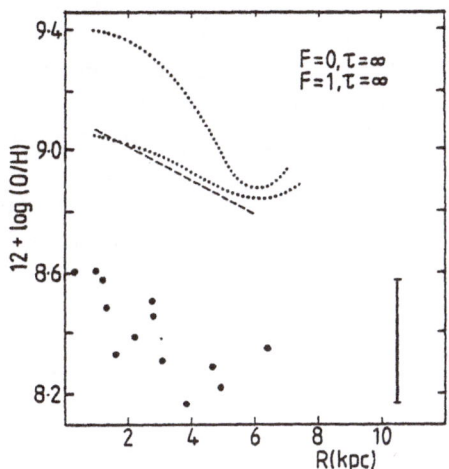

Fig.4. Same as Fig.1, but for M33.

REFERENCES

Arnett, W.A., 1978, Astrophys.J. **219**, 1008
Chiosi, C., Caimmi, R., 1979, Astron.Astrophys. **80**, 234
Diaz, A.I., Tosi, M., 1983, submitted to M.N.R.A.S. (Paper II)
Mayor, M., 1976, Astron.Astrophys. **48**, 301
Panagia, N., Tosi, M., 1981, Astron.Astrophys. **96**, 306
Renzini, A., Voli, M., 1981, Astron.Astrophys. **94**, 175
Tinsley, B.M., 1980, Fund.Cosmic Phys. **5**, 287
Tosi, M., 1982, Astrophys.J. **254**, 699 (Paper I)
Twarog, B.A., 1980, Astrophys.J. **242**, 242

DISCUSSION

     Woosley: How much has the average galactic metallicity changed since the sun was formed?

     Tosi: Twarog's age-metallicity relation gives an increase of the stellar metallicity by a factor of about 1.5 in the last 5 billion years.

     Henry: How unique are your models? How confident are you that similar results cannot be obtained by sliding parameters around?

Tosi: Since we have two free parameters, the uniqueness of the solution is guaranteed when the observational constraints are more than two. For the Galaxy, the observed independent quantities presently reproduced by one model are four: oxygen abundances, oxygen gradient, metallicity gradient and age-metallicity relation, and we can therefore feel quite safe.

Bond: Have you included zone-to-zone transport in any of your calculation? In particular, I am thinking of fountain effects in which one shoots metals above the disk and lets them rain down over much of the disk. What effects would this have on your results?

Tosi: We haven't included either radial exchange of gas between rings or fountain effects, because the present knowledge of these phenomena is still too poor; in any case, the results would be significantly affected only if the rate of these flows is comparable to the SFR.

Edmunds: Given that the gas fraction is fixed by observations, the effect of radial mixing (if both in and out) will be to flatten the gradient. If only one way, it is not so easy to tell.

Torres-Peimbert: I would like to point out that there is an electron temperature gradient in the galaxies that is explained partly by the metallicity effect in HII regions; but there is also an excitation gradient that can perhaps be attributed to different IMFs.

Tosi: The problem with a radially varying IMF is that assuming more (less) massive stars toward the galactic center, the resulting oxygen gradient would be steeper (flatter) then actually observed in HII regions. Viceversa, with our uniform IMF the observed gradient is perfectly reproduced.

Braun: Have you explored the possibility of allowing the assumed galactic age to vary to obtain an improved fit?

Tosi: Yes, but any variation of the age implies through eq.(4) of Paper II an opposite variation of the SFR, and the corresponding effects in the computation of the present oxygen abundances tend to compensate.

ACTINIDE R-PROCESS CHRONOMETERS AND β-DELAYED FISSION

F.-K. Thielemann
Max-Planck-Institut für Physik und Astrophysik
Institut für Astrophysik
8046 Garching b. München
Fed. Rep. Germany

When including new predictions for β-delayed fission and neutron emission rates in an r-process calculation, the production ratios for $^{232}Th/^{238}U$ and $^{244}Pu/^{238}U$ are drastically lowered. Employing the chronometric pairs $^{232}Th/^{238}U$, $^{235}U/^{238}U$, $^{244}Pu/^{238}U$ and $^{129}I/^{127}I$ within a simple exponential model for chemical evolution of the Galaxy results in an age ranging roughly from 14 to 20×10^9. This is much larger than previous age determinations from actinide chronometers. Within the same model this value is consistant with predictions from the $^{187}Re/^{187}Os$ chronometric pair and comparable with recent determinations of globular cluster ages and a Hubble constant of $H_o < 70$ km s^{-1} Mpc^{-1}. In more complex galactic evolution models the chronometers seem to lead to similar results, however, problems with the observed stellar age-metallicity relation are indicated.

1. R-PROCESS PRODUCTION RATIOS OF $^{232}Th/^{238}U$, $^{235}U/^{238}U$, AND $^{244}Pu/^{238}U$

Heavy nuclei above Pb and Bi are only produced in the r(apid neutron capture) process which is supposed to take place in explosive scenarios where for very short periods a high neutron density becomes available. In such an environment rapid neutron capture ($\tau_{n,\gamma} < \tau_\beta$) enables the population of nuclei in the r-process path far from stability. After decay back to the stability line the characteristic r-abundance peaks for stable nuclei are formed, due to the build-up of nuclei at neutron magic numbers, where the longest half-lives are encountered. While there are many suggestions for the site of the r-process [(a)matter from type II supernova explosions originating from the innermost ejected layers near the highly neutronized neutron star remnant (Hillebrandt, Takahashi, Kodama, 1976; see for a review Hillebrandt 1978 and references therein), (b)release of a large amount of neutrons via the reaction $^{22}Ne(\alpha,n)^{25}Mg$ during the passage of the SN shock front through the He-burning shell (Truran et al. 1978; Thielemann et al. 1979; Blake et al., 1981), (c)release of neutrons from the reaction $^{13}C(\alpha,n)^{16}O$ in the helium zone (Cowan et al.,1983)], all of these scenarios have associated problems [(a)neutronization gradient $\overline{N}(r)/\overline{Z}(r)$ and position of the mass cut between the neutron star and ejected material, (b)existence of an equilibrium between neutron capture and photodesinte-

389

gration down to low neutron densities, (c)existence of the required
amount of ^{13}C in the He-zone].

The application performed here, however, requires only an abundance
distribution in the r-process path which reproduces solar r-abundances.
Such a distribution was obtained by performing an r-process calculation
in explosive He-burning as in Hillebrandt et al. (1981) but including also
the effect of β-delayed fission. This effect of fissioning after β-decay
becomes important for heavy nuclei above Z=80. Part of the results obtained by Thielemann et al. (1983a), by using β-strength function calculations
from Klapdor et al. (1981, 1982), fission barrier heights from Howard and
Möller (1980) and the neutron optical potential by Jeukenne et al. (1977)
are displayed in Fig. 1.

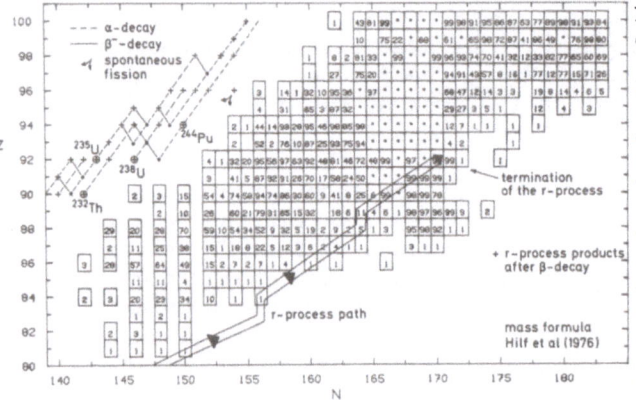

Fig.1: The termination
of the r-process by
β-delayed fission at
Z≈92, N≈170. Crosses
indicate nuclei populat-
ed by β-decay of nuclei
in the r-process path.
The long-lived Th, U,
and Pu isotopes are built
up by α- and β-decay
chains. The numbers in
boxes give the percentage
of fission subsequent to
β-decay in the fissioning
daughter nucleus. Dots
denote 100% fission.

A consequence of the predicted sink of 100% β-delayed fission around
Z=94, N=168 (where the second minimum in the fission barrier is lower than
the first minimum and consequently becomes the ground state of the nucleus)
is that the r-process would terminate around Z≈92.

Detailed results of the r-process calculation and abundances after
β-decay can be found in Thielemann et al. (1983b), here we want to concentrate on the r-process chronometers. Further α-decay chains after β-decay
to the stability line will populate the long lived isotopes ^{232}Th ($t_{1/2}$ =
1.405 10^9y), ^{238}U ($t_{1/2}$ = 4.46 10^9y) , ^{235}U ($t_{1/2}$ = 7.038 10^8y)*. The
production ratios obtained with the data from Fig. 1 are displayed in
Table 1 in comparison to earlier publications. As can be seen from Table 1,
^{244}Pu is the most sensitive case for the influence of nuclear fission
which was not taken into account by most of the previous investigations.
Wene and Johansson (1976) and Krumlinde et al. (1981) already included
β-delayed fission but only assuming a single hump barrier and (or) different β-decay strength functions. The ^{232}Th/^{238}U ratio being most important
for the galactic age prediction because of the long ^{232}Th half-life also
reduces drastically. ^{247}Cm/^{238}U not given in Table 1 also has a rather
low production ratio (≈ 10^{-2}), which is consistent with Chen and

*, and ^{244}Pu ($t_{1/2}$ = 8.26 10^7y)

TABLE 1

PRODUCTION RATIOS OF THE ACTINIDE CHRONOMETERS

$^{232}Th/^{238}U$	$^{235}U/^{238}U$	$^{244}Pu/^{238}U$	Work
1.58	1.64	(0.96)	Burbridge et al. (1957)
1.65	1.65	–	Fowler and Hoyle (1969)
1.90	1.89	0.96	Seeger and Schramm (1970)
1.65	1.42	0.90	Fowler (1972)
1.70	0.89	0.53	Wene and Johansson (1976)
1.80	1.42	0.90	Fowler (1978)
$1.9^{+0.2}_{-0.4}$	$1.5^{+0.5}_{-0.6}$	$0.9^{+0.1}_{-0.2}$	adopted standard Symbalisty and Schramm (1981)
1.50	1.10	0.40	Krumlinde et al. (1981)
1.39	1.24	0.12	Thielemann et al. (1983)

Wasserburg (1980) who did not detect ^{247}Cm and its fission products
($t_{1/2} = 1.56 \times 10^7 y$) in meteorites. Typical uncertainties in the production
ratios might be of the order 0.20. Using a different level density dis-
cription at low excitation energies [merging a back-shifted Fermi gas
formula into a constant temperature form (Lynn, 1980)] leads to ratios
of $^{232}Th/^{238}U = 1.30$ and $^{244}Pu/^{238}U = 0.16$.

2. AGE PREDICTION FROM THE SIMPLE EXPONENTIAL MODEL OF GALACTIC NUCLEOSYN-
   THESIS

   When knowing the production ratios of chronometric pairs and their
existing abundance ratios at the time of the solidification in the solar
system, information can be deduced about the duration of (r-process)
nucleosynthesis in the Galaxy and also the total age. Those interpretations
are however dependent on assumptions for the time dependence of the
nucleosynthesis production function during the evolution of the Galaxy.
This also holds true for the "model independent" approach (Schramm and
Wasserburg, 1970; see also Symbalisty and Schramm,1981) when mean ages
for different chronological pairs shall be converted to total galactic
ages (Tinsley, 1977, 1980). Thus we try here the interpretation in a
simple model ansatz and discuss the uncertainties and the comparison to
more complex models in the final section.

   A simple model is the "exponential model" introduced by Fowler (1972),
which assumes continuous nucleosynthesis over a duration Δ with an exponential decrease corresponding to t/T_R. Local and time inhomogeneities are
thought to average out due to mixing, except for the passage of the last
spiral arm, represented by a final spike which contributes the fraction S
to the nucleosynthesis of a stable nucleus (see also Reeves, 1978, 1979).
A period of free decay δ follows until the formation of the solar system
4.55 $10^9 y$ ago (at the time $t = \Delta+\delta$). The abundance ratio of a chronometric
pair at the formation of the solar system can be expressed by the (r-process) production ratio and the four "unknowns" Δ, T_R, S, and δ (see
Fowler, 1972; Thielemann et al., 1983b)

$$N_A(\Delta + \delta)/N_B(\Delta + \delta) = P_A/P_B \; f(\Delta, T_R, S, \delta, \lambda_A)/f(\Delta, T_R, S, \delta, \lambda_B) \qquad (1)$$

λ_A being the decay rate of the radioactive nucleus A.
With the knowledge of $N_A(\Delta + \delta)/N_B(\Delta + \delta)$ from meteoritic studies the model
parameters Δ, T_R, S and δ can be determined by four chronometric pairs;
usually $^{129}I/^{127}I$ with $P_A/P_B = 1.26$ (Käppeler et al., 1982) and
$t_{1/2}(^{129}I)=1.57 \; 10^7 y$ is added as fourth pair to $^{232}Th/^{238}U$, $^{235}U/^{238}U$ and
$^{244}Pu/^{238}U$. The ratios from meteoritic studies are $^{232}Th/^{238}U=2.32$ (Anders
and Ebihara, 1982), 2.22 (Cameron, 1982), $^{235}U/^{238}U=0.317$ (Anders and
Ebihara, 1982), 0.315 (Cameron, 1982), $^{244}Pu/^{238}U=0.005\pm0.001$ (Hudson et
al., 1982), $^{129}I/^{127}I=(0.8-2.3) \; 10^{-4}$ (Jordan et al., 1980). While the
$^{235}U/^{238}U$ ratio is quite accurately known, $^{129}I/^{127}I$ has a high uncertainty (due to inhomogeneities in the early solar nebula or condensation times
comparable with the ^{129}I half-life; see also Clayton, 1983), previous
$^{244}Pu/^{238}U$ values were up to 3-5 times larger (probably because of
chemical fractionation) and the $^{232}Th/^{238}U$ is reduced from 2.50 (see
Symbalisty and Schramm, 1981). Table 2 therefore displays solutions to the
system of equations (1) with varying $^{232}Th/^{238}U$, $^{244}Pu/^{238}U$ and $^{129}I/^{127}I$
meteoritic ratios.

TABLE 2

$N_A(\Delta + \delta)/N_B(\Delta + \delta)$ $^{232}Th/^{238}U$	$^{244}Pu/^{238}U$	$^{129}I/^{127}I$	Δ^+	S	δ^+	T_R^+
(a)	2.50	0.005	*	16.10	0.062	0.162	>
		0.004	*	13.24	0.068	0.193	>
(b)	2.32	0.005	*	13.20	0.062	0.158	>
		0.006	*	13.18	0.058	0.129	>
		0.008	1.0 10^{-4}	7.87	0.124	0.167	1.88
			1.8	8.96	0.105	0.150	3.57
(c)	2.22	0.005	*	11.54	0.060	0.147	>
(d)	2.32	0.005	*	16.30	0.065	0.211	>
(e)	2.32	0.005	1.0 10^{-4}	10.24	0.067	0.153	44.90
			1.8	10.69	0.060	0.144	>

$^+$in units of $10^9 y$, > absolute value > $10^{20} y$, (d) and (e) with changed
production ratios (see text), * variations within $(1.0-1.8)\times10^{-4}$ do not
change the results in the given digits.

For the most probable meteoritic values in case (b) with $^{244}Pu/^{238}U=$
0.005±0.001 very close results are given fœr Δ, S and T_R ($e^{-t/T_R}\approx$const,
$t < 10^{11} y$). This constancy of the effective nucleosynthesis rate is much
more preferable over a strong decrease found in earlier publications (see
e.g. Fowler, 1972, 1978) as pointed out by Hainebach and Schramm (1977).
Variations in the Pu/U ratio to higher values, as previously used, lead
to smaller ages. Cases (a)-(c) show the large dependence on the $^{232}Th/$
^{238}U meteoritic ratio, from the previous "standard value" of 2.50 (see
e.g. Symbalisty and Schramm, 1981) to 2.22 (Cameron, 1982). The ratio of
2.50 led to the results given in an earlier publication (Thielemann et al.
1983b). Cases (d) and (e) display the uncertainties in production ratios:
(d) a different level density treatment is used (see sect. 1) leading to
$^{232}Th/^{238}U=1.30$ and $^{244}Pu/^{238}U=0.16$, (e) $^{232}Th/^{238}U=1.50$ is employed.
Table 2 suggests a value of $\Delta = (13\pm4) \; 10^9 y$ within this model ansatz,

being almost twice as large as previous predictions (Fowler, 1972, 1978) which neglected β-delayed fission. An initial burst of star formation and nucleosynthesis would reduce this value. The beginning of r-process nucleosynthesis in our Galaxy is then given by

$$t_G = \Delta + \delta + 4.55 \ 10^9 y \approx (17.6 \pm 4) \ 10^9 y \tag{2}$$

which is a lower limit on the age of the Galaxy. The largest ages for globular clusters which contain the oldest stars in our Galaxy are another constraint. Reported values are $(17 \pm 2) \ 10^9 y$ (Sandage, 1982), $(15-18) \ 10^9 y$ (Van den Berg, 1983), and $(14-19) \ 10^9 y$ (Janes and Demarque, 1983). The present result is also in accordance with the Hubble-time of the Universe $t_U = (19.5 \pm 3) \ 10^9 y$ given by Sandage and Tammann (1982).

3. COMPARISON WITH OTHER CHRONOMETERS AND MORE COMPLEX CHEMICAL EVOLUTION MODELS

The independent $^{187}Re/^{187}Os$ chronometer has been analyzed in the same model by several authors giving $\Delta = 10.4 \ 10^9 y$ (Winters et al., 1980), $10.8 \ 10^9 y$ (Browne and Berman, 1982), $(8.9 \pm 2.0) \ 10^9$ (Winters and Macklin, 1982). The analysis of the above given values was done, assuming $T_R/\Delta = 0.43$ as results from Fowler (1978) for the actinide chronometers. Using the constant nucleosynthesis production rate ($T_R \approx \infty$) from this work would shift those values to $\Delta \approx 15 \ 10^9 y$, being consistent with the result given here [$\Delta = (13 \pm 4) \ 10^9 y$]. There are however several uncertainties associated with the Re/Os pair: (a) neutron capture of the thermally populated first excited state of ^{187}Os at 9.8 keV determines the cross section ratio $\sigma*(186)/\sigma*(187) = F_\sigma \ \sigma_{lab}(186)/\sigma_{lab}(187)$ at stellar temperatures; (b) the ^{187}Re half-life is ionization and therefore temperature dependent and reduced in stellar environments (see e.g. Yokoi et al., 1983). Recent inelastic neutron scattering data (Hershberger et al., 1983) indicate a value of $F_\sigma \approx 0.82$ as already theoretically predicted by Woosley and Fowler (1979). Yokoi et al. (1983) undertook an extensive study of the Re/Os chronometer. The analysis was done in a detailed galactic evolution model contrary to the simple "exponential model" applied here. This adds a large degree of complexity and uncertainty with nucleosynthesis yields from mass loss by stellar winds, from type I and type II supernovae with the associated uncertainties in stellar evolution calculations, initial mass functions and the time dependence of the SFR. The introduction of a temperature dependent ^{187}Re half-life also demands knowledge about the location of the ^{187}Re nuclei during the galactic evolution. Nevertheless, this is the only way to proceed, when the final aim of really understanding chemical evolution of the Galaxy shall be reached.

Yokoi et al. (1983) also included $^{232}Th/^{238}U$ and $^{235}U/^{238}U$ with different production ratios in their chemical evolution model. Their Fig. 6a shows that a $^{235}U/^{238}U$ production ratio of 1.24 (see Sect. 1) fits well into the allowed region, while Fig. 6b explains that a galactic age of $t_G > 15 \ 10^9 y$ is demanded with a $^{232}Th/^{238}U$ production ratio of ≈ 1.40.

The Re/Os ages can be deduced from their Fig. 9 via f_s which includes the value of F_σ. With the change from $1.0 \le F \le 1.15$ to ≈ 0.82 the Re/Os chronometer also demands $t_G > 15 \; 10^9 y$.

Yokoi et al. (1983) do not take into account ages above $15 \; 10^9 y$ because their chemical evolution model does not reproduce the age-metalicity relation for the galactic disk in those cases. It is, however, evident that the oldest age bins in the age-metalicity relation by Twarog (1980) have the largest age uncertainties, which are, contrary to the metalicities, theoretically predicted. The partially existing inconsistencies might be removed when each of the three entering fields are known with higher precision: (a) nuclear physics envolved in the r-process (b) stellar evolution (c) galactic evolution.

The actinide chronometer production ratios are affected by the nuclear mass law, fission barrier heights and the β-strength function predictions. In this respect the effect of upcoming β-strength function calculations (Krumlinde and Möller, 1983; Mathews et al., 1983) will be of great interest. Improved and extended calculations for the late burning stages of stellar evolution and type I and type II supernova explosions will improve the precision of nucleosynthesis yields. Galactic evolution calculations might have to drop the instantaneous recycling approximation.

I want to thank M. Arnould and W.A. Fowler for helpful discussions. M.T. McEllistrem provided experimental results before publication. Part of the presented β-delayed fission values come from a previous collaboration with J. Metzinger and H.V. Klapdor.

REFERENCES

Anders, E., Ebihara, M.: 1982, Geochim. Cosmochim. Acta 46, 2363
Blake, J.B., Woosley, S.E., Weaver, T.A., Schramm, D.N.: 1981, Astrophys.
    J. 248, 315
Browne, J.C., Berman, B.L.: 1981, Phys. Rev. C23, 1434
Burbidge, E.M., Burbidge, G.R., Fowler, W.A., Hoyle, F.: 1957, Rev. Mod.
    Phys. 29, 547
Cameron, A.G.W.: 1982, in "Essays in Nuclear Astrophysics" (Cambridge
    University Press), p. 23
Chen, J.H., Wasserburg, G.J.: 1980, Lunar Planet Sci. XI, 131
Clayton, D.D.: 1983, Astrophys. J. 268, 381
Cowan, J.J., Cameron, A.G.W., Truran, J.W.: 1983, Astrophys. J. 265, 429
Fowler, W.A.: 1972, in "Cosmology, Fusion and Other Matters" ed. F. Reines,
    (Boulder: Colorado Assoc. Univ. Press) p. 67
Fowler, W.A.: 1978, in Proc. Robert A. Welch Foundation Conference on
    Chemical Research XXI Cosmochemistry, Houston, p. 61
Hainebach, K.L., Schramm, D.N.: 1977, Astrophys. J. 212, 347
Hershberger, R.L., Macklin, R.L., Balakishnan, M., Hill, N.W., McEllistrem,
    M.T.: 1983, to be published
Hilf, E.R., von Groote, H., Takahashi, K.: 1976, CERN 76-13, 142

Hillebrandt, W., Klapdor, H.V., Oda, T., Thielemann, F.-K.: 1981, Astron. Astrophys. 99, 195
Hillebrandt, W.: 1978, Space Sci. Rev. 21, 639
Hillebrandt, W., Takahashi, K., Kodama, T.: 1976, Astron. Astrophys. 52, 63
Howard, W.M., Möller, P.: 1980, At. Data Nucl. Data Tables 25, 219
Hudson, B., Hohenberg, C.M., Kennedy, B.M., Podosek, F.A.: 1982, Lunar Planetary Sci. XIII, 346
Janes, K., Demarque, P.: 1983, Astrophys. J. 264, 206
Jeukenne, J.P., Lejeune, A., Mahaux, C.: 1977, Phys. Rev. C16, 80
Jordan, J., Kirsten, T., Richter, H.: 1980, Z. Naturforsch. 35a, 145
Käppeler, F., Beer, H., Wisshak, K., Clayton, D.D., Macklin, R.L., Ward, R.A.: 1982, Astrophys. J., 257, 821
Klapdor, H.V., Metzinger, J., Oda, T.: 1982, Z. Phys. A. 309, 91
Klapdor, H.V., Oda, T., Metzinger, J., Hillebrandt, W., Thielemann, F.-K.: 1981, Z. Phys. A229, 213
Krumlinde, J., Möller, P., Wene, C.O., Howard, W.M.: 1981, CERN 81-09, 260
Krumlinde, J., Möller, P.: 1983, submitted to Nucl. Phys. A
Lynn, J.E.: 1980, in "Nuclear Physics for Applications", IAEA-SMR-43 Vienna, p. 353
Mathews, G.J., Bloom, S.D., Hausman, R.F.: 1983, submitted to Phys. Rev. C
Myers, W.D., Swiatecki, W.J.: 1967, Ark. Fys. 36, 343
Reeves, H.: 1978, in "Protostars and Planets", ed. T. Gehrels, (Tucson: Univ. Arizona Press) p. 389
Reeves, H.: 1979, Astrophys. J. 237, 229
Sandage, A.: 1982, Astrophys. J. 252, 553
Sandage, A., Tammann, G.A.: 1982, Astrophys. J. 256, 339
Schramm, D.N., Wasserburg, G.J.: 1970, Astrophys. J. 162, 57
Seeger, P.A., Schramm, D.N.: 1970, Astrophys. J. 160, L157
Symbalisty, E.M.D., Schramm, D.N.: 1981, Rep. Prog. Phys. 44, 293
Thielemann, F.-K., Arnould, M., Hillebrandt, W.: 1979, Astron. Astrophys. 74, 175
Thielemann, F.-K., Metzinger, J., Klapdor, H.V.: 1983a, Z. Phys. A309, 301
Thielemann, F.-K., Metzinger, J., Klapdor, H.V.: 1983b, Astron. Astrophys. 123, 162
Tinsley, B.M.: 1977, Astrophys. J. 216, 548
Tinsley, B.M.: 1980, Fund. Cosmic Phys. 5, 287
Truran, J.W., Cowan, J.J., Cameron, A.C.W.: 1978, Astrophys. J. (Letters) 222, L63
Twarog, B.A.: 1980, Astrophys. J. 242, 242
Yokoi, K., Takahashi, K., Arnould, M.: 1983, Astron. Astrophys. 117, 65
Van den Berg, D.A.: 1983, Astrophys. J. Suppl. 51, 29
Wene, C.O., Johannson, S.A.E.: 1976, CERN 76-13, 584
Winters, R.R., Macklin, R.L., Halperin, J.: 1980, Phys. Rev. C21, 563
Winters, R.R., Macklin, R.L.: 1982, Phys. Rev. C25, 208
Woosley, S.E., Fowler, W.A.: 1979, Astrophys. J. 233, 411

DISCUSSION

Iben: Your model has four parameters and you use four
"observed" isotopic ratios to constrain them. By choosing another,
perhaps equally "reasonable" model, with a different set of parameter,
would it be possible to alter your conclusion to, say, to
$\sim (15\pm4)\times10^9$yr?*

Thielemann: The present study has as the main result, that when considering β-delayed fission, much larger ages are obtained than previously. In
addition, the values obtained from the actinide chronometers and the
^{187}Re/^{187}Os pair come quite close within the same model. But, besides
nuclear and meteoritic uncertainties, the model ansatz, of course, influences the result. For example, an initial burst of nucleosynthesis
would lead to lower ages. It is, however, encouraging that the most
recent determinations of globular cluster ages give similar results.

Renzini: Could cosmochronology methods safely exclude an age as small
as 10×10^9 years?

Thielemann: This would demand for a large change, but as mentioned before,
besides nuclear and meteoritic uncertainties, the model ansatz can influence the result. Really safe conclusions might only be drawn, when a
complex galactic evolution model can explain the abundance ratios of
chronological pairs at the time of meteorite formation in the solar
system as well as the chemical evolution of the Galaxy (age-metalicity
relation etc...).

Mathews: I would like to ask about the self-consistency of the nuclear
physics, employed in this calculation. The deformation energies affect
both the beta-decay rates and the fission barriers. Ideally, both the
ground state deformations and the fission barriers ought to be computed
in the same model. It is not clear to me from what you have said whether
that is the case in your calculation.

Thielemann: This is not the case, the available β-strength function
calculations from Klapdor et al. (1982) do not employ the ground state
deformations by Howard and Möller (1980) which produce their masses and
fission barrier heights. In addition Howard and Möller (1980) seem to
predict the neutron-drip line too far out from the stability line (compared to more recent work), therefore the ground state masses are taken
from Hilf et al. (1976). Consistent sets of data are not available yet;
I tried to simulate systematic errors by varying the fission barrier
heights within 1 MeV. This led to changes of the order of 0.10 in the
most important ^{232}Th/^{238}U production ratio.

*In the talk a result of $(16.8-22.8)\times10^9$y was given corresponding to a
meteoritic ratio of ^{232}Th/^{238}U = 2.50, which has been used in previous
studies.

SUBJECT (ELEMENT) INDEX

The major elements and their isotopes are listed in order of increasing atomic number. The index gives the number of the first page of those contributions in which the various elements are indicated relative to either observational determinations or model results.